Physics Beyond The Standard Model

" Theories to Explain Deficiencies in The Standard Model "

Edited by Paul F. Kisak

Contents

0.1 Physics beyond the Standard Model

Physics beyond the Standard Model (BSM) refers to the theoretical developments needed to explain the deficiencies of the Standard Model, such as the origin of mass, the strong CP problem, neutrino oscillations, matter–antimatter asymmetry, and the nature of dark matter and dark energy.[1] Another problem lies within the mathematical framework of the Standard Model itself—the Standard Model is inconsistent with that of general relativity, to the point that one or both theories break down under certain conditions (for example within known spacetime singularities like the Big Bang and black hole event horizons).

Theories that lie beyond the Standard Model include various extensions of the standard model through supersymmetry, such as the Minimal Supersymmetric Standard Model (MSSM) and Next-to-Minimal Supersymmetric Standard Model (NMSSM), or entirely novel explanations, such as string theory, M-theory, and extra dimensions. As these theories tend to reproduce the entirety of current phenomena, the question of which theory is the right one, or at least the "best step" towards a Theory of Everything, can only be settled via experiments, and is one of the most active areas of research in both theoretical and experimental physics.

0.1.1 Problems with the Standard Model

Despite being the most successful theory of particle physics to date, the Standard Model is not perfect.[2] A large share of the published output of theoretical physicists consists of proposals for various forms of "Beyond the Standard Model" new physics proposals that would modify the Standard Model in ways subtle enough to be consistent with existing data, yet address its imperfections materially enough to predict non-Standard Model outcomes of new experiments that can be proposed.

Phenomena not explained

The Standard Model is inherently an incomplete theory. There are fundamental physical phenomena in nature that the Standard Model does not adequately explain:

- *Gravity.* The standard model does not explain gravity. The approach of simply adding a "graviton" (whose properties are the subject of considerable consensus among physicists if it exists) to the Standard Model does not recreate what is observed experimentally without other modifications, as yet undiscovered, to

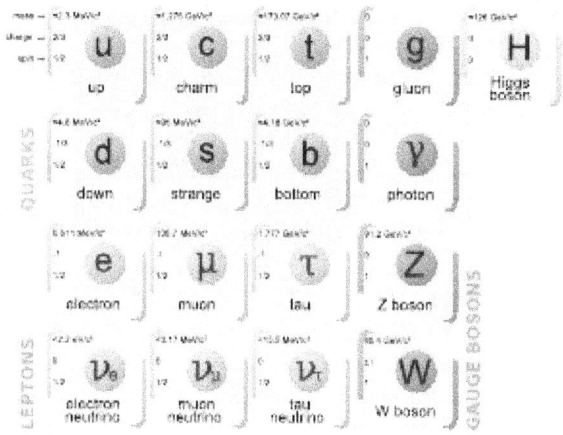

The Standard Model of elementary particles

the Standard Model. Moreover, instead, the Standard Model is widely considered to be incompatible with the most successful theory of gravity to date, general relativity.[3]

- *Dark matter and dark energy.* Cosmological observations tell us the standard model explains about 5% of the energy present in the universe. About 26% should be dark matter, which would behave just like other matter, but which only interacts weakly (if at all) with the Standard Model fields. Yet, the Standard Model does not supply any fundamental particles that are good dark matter candidates. The rest (69%) should be dark energy, a constant energy density for the vacuum. Attempts to explain dark energy in terms of vacuum energy of the standard model lead to a mismatch of 120 orders of magnitude.[4]

- *Neutrino masses.* According to the standard model, neutrinos are massless particles. However, neutrino oscillation experiments have shown that neutrinos do have mass. Mass terms for the neutrinos can be added to the standard model by hand, but these lead to new theoretical problems. For example, the mass terms need to be extraordinarily small and it is not clear if the neutrino masses would arise in the same way that the masses of other fundamental particles do in the Standard Model.

- *Matter–antimatter asymmetry.* The universe is made out of mostly matter. However, the standard model predicts that matter and antimatter should have been created in (almost) equal amounts if the initial conditions of the universe did not involve disproportionate matter relative to antimatter. Yet, no mechanism sufficient to explain this asymmetry exists in the Standard Model.

Experimental results not explained

No experimental result is widely accepted as definitively contradicting the Standard Model at the "five sigma" level, widely considered to be the threshold of a "discovery" in particle physics. But because every experiment contains some degree of statistical and systemic uncertainty, and the theoretical predictions themselves are also almost never calculated exactly and are subject to uncertainties in measurements of the fundamental constants of the Standard Model (some of which are tiny and others of which are substantial), it is mathematically expected that some of the hundreds of experimental tests of the Standard Model will deviate to some extent from it, even if there were no "new physics" to be discovered.

At any given time there are a number of experimental results that are significantly different from the Standard Model expectation, although many of these have been found to be statistical flukes or experimental errors as more data has been collected. On the other hand, any "beyond the Standard Model" physics would necessarily first manifest experimentally as a statistically significant difference between an experiment and the theoretical prediction.

In each case, physicists seek to determine if a result is a mere statistical fluke or experimental error on the one hand, or a sign of new physics on the other. More statistically significant results cannot be mere statistical flukes but can still result from experimental error or inaccurate estimates of experimental precision. Frequently, experiments are tailored to be more sensitive to experimental results that would distinguish the Standard Model from theoretical alternatives.

Some of the most notable examples include the following:

- *Muonic hydrogen* – the Standard Model makes precise theoretical predictions regarding the atomic radius size of ordinary hydrogen (a proton-electron system) and that of muonic hydrogen (a proton-muon system in which a muon is a "heavy" variant of an electron). However, the measured atomic radius of muonic hydrogen differs significantly from that of the radius predicted by the Standard Model using existing physical constant measurements by what appears to be as many as seven standard deviations.[5] Doubts about the accuracy of the error estimates in earlier experiments, which are still within 4% of each other in measuring a truly tiny distance, and a lack of a well motivated theory that could explain the discrepancy, have caused physicists to be hesitant to describe these results as contradicting the Standard Model despite the apparent statistical significance of the result and a lack of any clearly identified possible source of experimental error in the results.

- *Anomalous magnetic dipole moment of muon* – the experimentally measured value of muon's anomalous magnetic dipole moment ("muon g-2") is significantly different from the Standard Model prediction.[6]

- *BaBar data suggests possible flaws in the Standard Model* – results from a BaBar experiment may suggest a surplus over Standard Model predictions of a type of particle decay ($B \rightarrow D^{(*)}\tau^-\tau\nu$). In this, an electron and positron collide, resulting in a B meson and an antimatter B meson, which then decays into a D meson and a tau lepton as well as a tau antineutrino. While the level of certainty of the excess (3.4 sigma in statistical language) is not enough to claim a break from the Standard Model, the results are a potential sign of something amiss and are likely to affect existing theories, including those attempting to deduce the properties of Higgs bosons.[7] In 2015, LHCb reported observing a 2.1 sigma excess in the same ratio of branching fractions.[8]

- Proton radius – the proton's charge radius measured using electron probes is different than when measured using muons.[9]

Theoretical predictions not observed

Observation at particle colliders of all of the fundamental particles predicted by the Standard Model has been confirmed. The Higgs boson is predicted by the Standard Model's explanation of the Higgs mechanism, which describes how the weak SU(2) gauge symmetry is broken and how fundamental particles obtain mass; it was the last particle predicted by the Standard Model to be observed. On July 4, 2012, CERN scientists using the Large Hadron Collider announced the discovery of a particle consistent with the Higgs boson, with a mass of about 126 GeV/c^2. A Higgs boson was confirmed to exist on March 14, 2013, although efforts to confirm that it has all of the properties predicted by the Standard Model are ongoing.[10]

A few hadrons (i.e. composite particles made of quarks) whose existence is predicted by the Standard Model, which can be produced only at very high energies in very low frequencies have not yet been definitively observed, and "glueballs"[11] (i.e. composite particles made of gluons) have also not yet been definitively observed. Some very low frequency particle decays predicted by the Standard Model have also not yet been definitively observed because insufficient data is available to make a statistically significant observation.

Theoretical problems

Some features of the standard model are added in an ad hoc way. These are not problems per se (i.e. the theory works fine with these ad hoc features), but they imply a lack of understanding. These ad hoc features have motivated theorists to look for more fundamental theories with fewer parameters. Some of the ad hoc features are:

- *Hierarchy problem* – the standard model introduces particle masses through a process known as spontaneous symmetry breaking caused by the Higgs field. Within the standard model, the mass of the Higgs gets some very large quantum corrections due to the presence of virtual particles (mostly virtual top quarks). These corrections are much larger than the actual mass of the Higgs. This means that the bare mass parameter of the Higgs in the standard model must be fine tuned in such a way that almost completely cancels the quantum corrections. This level of fine-tuning is deemed unnatural by many theorists.

- *Number of parameters* – the standard model depends on 19 numerical parameters. Their values are known from experiment, but the origin of the values is unknown. Some theorists have tried to find relations between different parameters, for example, between the masses of particles in different generations.

- *Quantum triviality* – suggests that it may not be possible to create a consistent quantum field theory involving elementary scalar Higgs particles.

- *Strong CP problem* – theoretically it can be argued that the standard model should contain a term that breaks CP symmetry—relating matter to antimatter—in the strong interaction sector. Experimentally, however, no such violation has been found, implying that the coefficient of this term is very close to zero. This fine tuning is also considered unnatural.

0.1.2 Grand unified theories

Main article: Grand Unified Theory

The standard model has three gauge symmetries; the colour SU(3), the weak isospin SU(2), and the hypercharge U(1) symmetry, corresponding to the three fundamental forces. Due to renormalization the coupling constants of each of these symmetries vary with the energy at which they are measured. Around 10^{16} GeV these couplings become approximately equal. This has led to speculation that above this energy the three gauge symmetries of the standard model are unified in one single gauge symmetry with a simple group gauge group, and just one coupling constant. Below this energy the symmetry is spontaneously broken to the standard model symmetries.[12] Popular choices for the unifying group are the special unitary group in five dimensions SU(5) and the special orthogonal group in ten dimensions SO(10).[13]

Theories that unify the standard model symmetries in this way are called Grand Unified Theories (or GUTs), and the energy scale at which the unified symmetry is broken is called the GUT scale. Generically, grand unified theories predict the creation of magnetic monopoles in the early universe,[14] and instability of the proton.[15] Neither of these have been observed, and this absence of observation puts limits on the possible GUTs.

0.1.3 Supersymmetry

Main article: Supersymmetry

Supersymmetry extends the Standard Model by adding another class of symmetries to the Lagrangian. These symmetries exchange fermionic particles with bosonic ones. Such a symmetry predicts the existence of *supersymmetric particles*, abbreviated as *sparticles*, which include the sleptons, squarks, neutralinos and charginos. Each particle in the Standard Model would have a superpartner whose spin differs by 1/2 from the ordinary particle. Due to the breaking of supersymmetry, the sparticles are much heavier than their ordinary counterparts; they are so heavy that existing particle colliders may not be powerful enough to produce them.

0.1.4 Neutrinos

In the standard model, neutrinos have exactly zero mass. This is a consequence of the standard model containing only left-handed neutrinos. With no suitable right-handed partner, it is impossible to add a renormalizable mass term to the standard model.[16] Measurements however indicated that neutrinos spontaneously change flavour, which implies that neutrinos have a mass. These measurements only give the relative masses of the different flavours. The best constraint on the absolute mass of the neutrinos comes from precision measurements of tritium decay, providing an upper limit 2 eV, which makes them at least five orders of magnitude lighter than the other particles in the standard model.[17] This necessitates an extension of the standard model, which not only needs to explain how neutrinos get their mass, but also why the mass is so small.[18]

One approach to add masses to the neutrinos, the so-called seesaw mechanism, is to add right-handed neutrinos and have these couple to left-handed neutrinos with a Dirac mass term. The right-handed neutrinos have to be sterile, meaning that they do not participate in any of the standard model interactions. Because they have no charges, the right-handed neutrinos can act as their own anti-particles, and have a Majorana mass term. Like the other Dirac masses in the standard model, the neutrino Dirac mass is expected to be generated through the Higgs mechanism, and is therefore unpredictable. The standard model fermion masses differ by many orders of magnitude; the Dirac neutrino mass has at least the same uncertainty. On the other hand, the Majorana mass for the right-handed neutrinos does not arise from the Higgs mechanism, and is therefore expected to be tied to some energy scale of new physics beyond the standard model, for example the Planck scale.[19] Therefore, any process involving right-handed neutrinos will be suppressed at low energies. The correction due to these suppressed processes effectively gives the left-handed neutrinos a mass that is inversely proportional to the right-handed Majorana mass, a mechanism known as the see-saw.[20] The presence of heavy right-handed neutrinos thereby explains both the small mass of the left-handed neutrinos and the absence of the right-handed neutrinos in observations. However, due to the uncertainty in the Dirac neutrino masses, the right-handed neutrino masses can lie anywhere. For example, they could be as light as keV and be dark matter,[21] they can have a mass in the LHC energy range[22][23] and lead to observable lepton number violation,[24] or they can be near the GUT scale, linking the right-handed neutrinos to the possibility of a grand unified theory.[25][26]

The mass terms mix neutrinos of different generations. This mixing is parameterized by the PMNS matrix, which is the neutrino analogue of the CKM quark mixing matrix. Unlike the quark mixing, which is almost minimal, the mixing of the neutrinos appears to be almost maximal. This has led to various speculations of symmetries between the various generations that could explain the mixing patterns.[27] The mixing matrix could also contain several complex phases that break CP invariance, although there has been no experimental probe of these. These phases could potentially create a surplus of leptons over anti-leptons in the early universe, a process known as leptogenesis. This asymmetry could then at a later stage be converted in an excess of baryons over anti-baryons, and explain the matter-antimatter asymmetry in the universe.[13]

The light neutrinos are disfavored as an explanation for the observation of dark matter, due to considerations of large-scale structure formation in the early universe. Simulations of structure formation show that they are too hot—i.e. their kinetic energy is large compared to their mass—while formation of structures similar to the galaxies in our universe requires cold dark matter. The simulations show that neutrinos can at best explain a few percent of the missing dark matter. However, the heavy sterile right-handed neutrinos *are* a possible candidate for a dark matter WIMP.[28]

0.1.5   Preon Models

Several preon models have been proposed to address the unsolved problem concerning the fact that there are three generations of quarks and leptons. Preon models generally postulate some additional new particles which are further postulated to be able to combine to form the quarks and leptons of the standard model. One of the earliest preon models was the Rishon model.[29][30][31]

To date, no preon model is widely accepted or fully verified.

0.1.6   Theories of everything

Theory of everything

Main article: Theory of everything

Theoretical physics continues to strive toward a theory of everything, a theory that fully explains and links together all known physical phenomena, and predicts the outcome of any experiment that could be carried out in principle. In practical terms the immediate goal in this regard is to develop a theory which would unify the Standard Model with General Relativity in a theory of quantum gravity. Additional features, such as overcoming conceptual flaws in either theory or accurate prediction of particle masses, would be desired. The challenges in putting together such a theory are not just conceptual - they include the experimental aspects of the very high energies needed to probe exotic realms.

Several notable attempts in this direction are supersymmetry, string theory, and loop quantum gravity.

String theory

Main article: String theory

Extensions, revisions, replacements, and reorganizations of the Standard Model exist in attempt to correct for these and other issues. String theory is one such reinvention, and many theoretical physicists think that such theories are the next theoretical step toward a true Theory of Everything. Theories of quantum gravity such as loop quantum gravity and others are thought by some to be promising candidates to the mathematical unification of quantum field theory and general relativity, requiring less drastic changes to existing theories.[32] However recent work places stringent limits on the putative effects of quantum gravity on the speed of light, and disfavours some current models of quantum gravity.[33]

Among the numerous variants of string theory, M-theory, whose mathematical existence was first proposed at a String Conference in 1995, is believed by many to be a proper "ToE" candidate, notably by physicists Brian Greene and Stephen Hawking. Though a full mathematical description is not yet known, solutions to the theory exist for specific cases.[34] Recent works have also proposed alternate string models, some of which lack the various harder-to-test features of M-theory (e.g. the existence of Calabi–Yau manifolds, many extra dimensions, etc.) including works by well-published physicists such as Lisa Randall.[35][36]

0.1.7 See also

- Antimatter tests of Lorentz violation

- Beyond black holes

- Fundamental physical constants in the standard model

- Higgsless model

- Holographic principle

- Little Higgs

- Lorentz-violating neutrino oscillations

- Minimal Supersymmetric Standard Model

- Peccei–Quinn theory

- Preon

- Standard-Model Extension

- Supergravity

- Seesaw mechanism

- Supersymmetry

- Superfluid vacuum theory

- String theory

- Technicolor (physics)

- Theory of everything

- Unsolved problems in physics

- Unparticle physics

0.1.8 References

[1] Womersley, J. (February 2005). "Beyond the Standard Model" (PDF). *Symmetry Magazine.* Retrieved 2010-11-23.

[2] Lykken, J. D. (2010). "Beyond the Standard Model". *CERN Yellow Report.* CERN. pp. 101–109. arXiv:1005.1676. CERN-2010-002.

[3] Sushkov, A. O.; Kim, W. J.; Dalvit, D. A. R.; Lamoreaux, S. K. (2011). "New Experimental Limits on Non-Newtonian Forces in the Micrometer Range". *Physical Review Letters* **107** (17): 171101. arXiv:1108.2547. Bibcode:2011PhRvL.107q1101S. doi:10.1103/PhysRevLett.107.171101. It is remarkable that two of the greatest successes of 20th century physics, general relativity and the standard model, appear to be fundamentally incompatible. But see also Donoghue, John F. (2012). "The effective field theory treatment of quantum gravity". *AIP Conference Proceedings* **1473**: 73. arXiv:1209.3511. doi:10.1063/1.4756964. One can find thousands of statements in the literature to the effect that "general relativity and quantum mechanics are incompatible". These are completely outdated and no longer relevant. Effective field theory shows that general relativity and quantum mechanics work together perfectly normally over a range of scales and curvatures, including those relevant for the world that we see around us. However, effective field theories are only valid over some range of scales. General relativity certainly does have problematic issues at extreme scales. There are important problems which the effective field theory does not solve because they are beyond its range of validity. However, this means that the issue of quantum gravity is not what we thought it to be. Rather than a fundamental incompatibility of quantum mechanics and gravity, we are in the more familiar situation of needing a more complete theory beyond the range of their combined applicability. The usual marriage of general relativity and quantum mechanics is fine at ordinary energies, but we now seek to uncover the modifications that must be present in more extreme conditions. This is the modern view of the problem of quantum gravity, and it represents progress over the outdated view of the past."

[4] Krauss, L. (2009). *A Universe from Nothing.* AAI Conference.

[5] Randolf Pohl; Ronald Gilman; Gerald A. Miller; Krzysztof Pachucki (2013). "Muonic hydrogen and the proton radius puzzle". *Annu. Rev. Nucl. Part. Sci.* **63**. arXiv:1301.0905. Bibcode:2013ARNPS..63..175P. doi:10.1146/annurev-nucl-102212-170627. The recent determination of the proton radius using the measurement of the Lamb shift in the muonic hydrogen atom startled the physics world. The obtained value of 0.84087(39) fm differs by about 4% or 7 standard deviations from the CODATA value of 0.8775(51) fm. The latter is composed from the electronic hydrogenate atom value of 0.8758(77) fm and from a similar value with larger uncertainties determined by electron scattering.

[6] Thomas Blum; Achim Denig; Ivan Logashenko; Eduardo de Rafael; B. Lee Roberts; Thomas Teubner; Graziano Venanzoni. "The Muon (g-2) Theory Value: Present and Future". arXiv:1311.2198.

[7] Lees, J. P.; et al. (BaBar Collaboration) (1970). "Evidence for an excess of B → D$^{(*)}$τ⁻τv decays". *Physical Review Letters* **109** (10). arXiv:1205.5442. Bibcode:2012PhRvL.109j1802L. doi:10.1103/PhysRevLett.109.101802.

[8] . arXiv:1506.08614. Bibcode:2015PhRvL.115k1803A. doi:10.1103/PhysRevLett.115.111803. Missing or empty |title= (help)

[9] . arXiv:1502.05314. Bibcode:2015PrPNP..82...59C. doi:10.1016/j.ppnp.2015.01.002. Missing or empty |title= (help)

[10] O'Luanaigh, C. (14 March 2013). "New results indicate that new particle is a Higgs boson". CERN.

[11] Marco Frasca. "What is a Glueball?" (March 31, 2009) http://marcofrasca.wordpress.com/2009/03/31/what-is-a-glueball-2/

[12] Peskin, M. E.; Schroeder, D. V. (1995). *An introduction to quantum field theory*. Addison-Wesley. pp. 786–791. ISBN 978-0-201-50397-5.

[13] Buchmüller, W. (2002). "Neutrinos, Grand Unification and Leptogenesis". arXiv:hep-ph/0204288 [hep-ph].

[14] Milstead, D.; Weinberg, E.J. (2009). "Magnetic Monopoles" (PDF). Particle Data Group. Retrieved 2010-12-20.

[15] P.. Nath; P. F.. Perez (2006). "Proton stability in grand unified theories, in strings, and in branes". *Physics Reports* **441** (5–6): 191–317. arXiv:hep-ph/0601023. Bibcode:2007PhR...441..191N. doi:10.1016/j.physrep.2007.02.010.

[16] Peskin, M. E.; Schroeder, D. V. (1995). *An introduction to quantum field theory*. Addison-Wesley. pp. 713–715. ISBN 978-0-201-50397-5.

[17] Nakamura, K.; et al. (Particle Data Group) (2010). "Neutrino Properties". Particle Data Group. Retrieved 2010-12-20.

[18] Mohapatra, R. N.; Pal. P. B. (2007). *Massive neutrinos in physics and astrophysics*. Lecture Notes in Physics **72** (3rd ed.). World Scientific. ISBN 978-981-238-071-5.

[19] Senjanovic, G. (2011). "Probing the Origin of Neutrino Mass: from GUT to LHC". arXiv:1107.5322 [hep-ph].

[20] Grossman, Y. (2003). "TASI 2002 lectures on neutrinos". arXiv:hep-ph/0305245v1 [hep-ph].

[21] Dodelson, S.; Widrow, L. M. (1993). "Sterile neutrinos as dark matter". *Physical Review Letters* **72**: 17. arXiv:hep-ph/9303287. Bibcode:1994PhRvL..72...17D. doi:10.1103/PhysRevLett.72.17.

[22] Minkowski, P. (1977). "μ → e γ at a Rate of One Out of 10^9 Muon Decays?". *Physics Letters B* **67** (4): 421. Bibcode:1977PhLB...67..421M. doi:10.1016/0370-2693(77)90435-X.

[23] Mohapatra, R. N.; Senjanovic, G. (1980). "Neutrino mass and spontaneous parity nonconservation". *Physical Review Letters* **44** (14): 912. Bibcode:1980PhRvL..44..912M. doi:10.1103/PhysRevLett.44.912.

[24] Keung, W.-Y.; Senjanovic, G. (1983). "Majorana Neutrinos And The Production Of The Right-handed Charged Gauge Boson". *Physical Review Letters* **50** (19): 1427. Bibcode:1983PhRvL..50.1427K. doi:10.1103/PhysRevLett.50.1427.

[25] Gell-Mann, M.; Ramond, P.; Slansky, R. (1979). P. van Nieuwenhuizen; D. Freedman, eds. *Supergravity*. North Holland.

[26] Glashow, S. L. (1979). M. Levy, ed. *Proceedings of the 1979 Cargèse Summer Institute on Quarks and Leptons*. Plenum Press.

[27] Altarelli, G. (2007). "Lectures on Models of Neutrino Masses and Mixings". arXiv:0711.0161 [hep-ph].

[28] Murayama, H. (2007). "Physics Beyond the Standard Model and Dark Matter". arXiv:0704.2276 [hep-ph].

[29] Harari, H. (1979). "A Schematic Model of Quarks and Leptons". *Physics Letters B* **86** (1): 83–86. Bibcode:1979PhLB...86...83H. doi:10.1016/0370-2693(79)90626-9.

[30] Shupe, M. A. (1979). "A Composite Model of Leptons and Quarks". *Physics Letters B* **86** (1): 87–92. Bibcode:1979PhLB...86...87S. doi:10.1016/0370-2693(79)90627-0.

[31] Zenczykowski, P. (2008). "The Harari-Shupe preon model and nonrelativistic quantum phase space". *Physics Letters B* **660** (5): 567–572. arXiv:0803.0223. Bibcode:2008PhLB..660..567Z. doi:10.1016/j.physletb.2008.01.045.

[32] Smolin, L. (2001). *Three Roads to Quantum Gravity*. Basic Books. ISBN 0-465-07835-4.

[33] Abdo, A. A.; et al. (Fermi GBM/LAT Collaborations) (2009). "A limit on the variation of the speed of light arising from quantum gravity effects". *Nature* **462** (7271): 331–4. arXiv:0908.1832. Bibcode:2009Natur.462..331A. doi:10.1038/nature08574. PMID 19865083.

[34] Maldacena, J.; Strominger, A.; Witten, E. (1997). "Black hole entropy in M-Theory". *Journal of High Energy Physics* **1997** (12): 2. arXiv:hep-th/9711053.

Bibcode:1997JHEP...12..002M.   doi:10.1088/1126-6708/1997/12/002.

[35] Randall, L.; Sundrum, R. (1999). "Large Mass Hierarchy from a Small Extra Dimension". *Physical Review Letters* **83** (17): 3370. arXiv:hep-ph/9905221. Bibcode:1999PhRvL..83.3370R. doi:10.1103/PhysRevLett.83.3370.

[36] Randall, L.; Sundrum, R. (1999). "An Alternative to Compactification". *Physical Review Letters* **83** (23): 4690. arXiv:hep-th/9906064. Bibcode:1999PhRvL..83.4690R. doi:10.1103/PhysRevLett.83.4690.

0.1.9 Further reading

- Lisa Randall (2005). *Warped Passages: Unraveling the Mysteries of the Universe's Hidden Dimensions.* HarperCollins. ISBN 0-06-053108-8.

0.1.10 External resources

- Standard Model Theory @ SLAC

- Scientific American Apr 2006

- LHC. Nature July 2007

- Open Questions

- Working group - schedule

- Les Houches Conference. Summer 2005

Chapter 1

Evidence Beyond The Standard Model

1.1 Mass generation

In theoretical physics, a **mass generation** mechanism is a theory that describes the origin of mass from the most fundamental laws of physics. Physicists have proposed a number of models that advocate different views of the origin of mass. The problem is complicated because mass is strongly connected to gravitational interaction, and no theory of gravitational interaction reconciles with the currently popular Standard Model of particle physics.

There are two types of mass generation models: gravity-free models and models that involve gravity.

1.1.1 Gravity-free models

In these theories, as in the Standard Model itself, the gravitational interaction either is not involved or does not play a crucial role.

- The Higgs mechanism is based on a symmetry-breaking scalar field potential, such as the quartic. The Standard Model uses this mechanism as part of the Glashow–Weinberg–Salam model to unify electromagnetic and weak interactions. This model was one of several that predicted the existence of the scalar Higgs boson.

- Technicolor models break electroweak symmetry through new gauge interactions, which were originally modeled on quantum chromodynamics.[1][2]

- Coleman–Weinberg mechanism (spontaneous symmetry breaking through radiative corrections).

- Models of composite W and Z vector bosons.[3]

- Top quark condensate.

- Asymptotically safe weak interactions [4][5] based on some nonlinear sigma models.[6]

- Symmetry breaking driven by non-equilibrium dynamics of quantum fields above the electroweak scale.[7][8]

- Unparticle physics and the unhiggs[9][10] models posit that the Higgs sector and Higgs boson are scaling invariant, also known as unparticle physics.

- UV-Completion by Classicalization, in which the unitarization of the WW scattering happens by creation of classical configurations.[11]

1.1.2 Models that involve gravity

- Extra-dimensional Higgsless models use the fifth component of the gauge fields in place of the Higgs fields. It is possible to produce electroweak symmetry breaking by imposing certain boundary conditions on the extra dimensional fields, increasing the unitarity breakdown scale up to the energy scale of the extra dimension.[12][13] Through the AdS/QCD correspondence this model can be related to technicolor models and to *UnHiggs* models, in which the Higgs field is of unparticle nature.[14]

- Unitary Weyl gauge. If one adds a suitable gravitational term to the standard model action with gravitational coupling, the theory becomes locally scale-invariant (i.e. Weyl-invariant) in the unitary gauge for the local SU(2). Weyl transformations act multiplicatively on the Higgs field, so one can fix the Weyl gauge by requiring that the Higgs scalar be a constant.[15]

- Preon and models inspired by preons such as the Ribbon model of Standard Model particles by Sundance Bilson-Thompson, based in braid theory and compatible with loop quantum gravity and similar theories.[16] This model not only explains the origin of mass, but also interprets electric charge as a topological quantity (twists carried on the individual ribbons), and colour charge as modes of twisting.

- In the theory of superfluid vacuum, masses of elementary particles arise from interaction with a physical vacuum, similarly to the gap generation mechanism in superfluids.[17] The low-energy limit of this theory suggests an effective potential for the Higgs sector that is different from the Standard Model's, yet it yields the mass generation.[18][19] Under certain conditions, this potential gives rise to an elementary particle with a role and characteristics similar to the Higgs boson.

1.1.3 See also

- Mass

- Higgs mechanism

- Spontaneous symmetry breaking

1.1.4 References

[1] Steven Weinberg (1976), "Implications of dynamical symmetry breaking", *Physical Review* **D13** (4): 974–996, Bibcode:1976PhRvD..13..974W. doi:10.1103/PhysRevD.13.974.
S. Weinberg (1979), "Implications of dynamical symmetry breaking: An addendum", *Physical Review* **D19** (4): 1277–1280, Bibcode:1979PhRvD..19.1277W. doi:10.1103/PhysRevD.19.1277.

[2] Leonard Susskind (1979), "Dynamics of spontaneous symmetry breaking in the Weinberg-Salam theory", *Physical Review* **D20** (10): 2619–2625, Bibcode:1979PhRvD..20.2619S. doi:10.1103/PhysRevD.20.2619.

[3] Abbott, L. F.; Farhi, E. (1981), "Are the Weak Interactions Strong?", *Physics Letters B* **101** (1–2): 69, Bibcode:1981PhLB..101...69A. doi:10.1016/0370-2693(81)90492-5

[4] Calmet, X. (2011), "Asymptotically safe weak interactions", *Mod. Phys. Lett.* **A26**: 1571–1576, arXiv:1012.5529, Bibcode:2011MPLA...26.1571C. doi:10.1142/S0217732311035900

[5] Calmet, X. (2011), "An Alternative view on the electroweak interactions", *Int.J.Mod.Phys.* **A26**: 2855–2864, arXiv:1008.3780, Bibcode:2011IJMPA..26.2855C. doi:10.1142/S0217751X11053699

[6] Codello, A.; Percacci, R. (2008), "Fixed Points of Nonlinear Sigma Models in d>2", *Physics Letters B* **672** (3): 280–283, arXiv:0810.0715, Bibcode:2009PhLB..672..280C. doi:10.1016/j.physletb.2009.01.032

[7] "Bifurcations and pattern formation in particle physics: An introductory study". *EPL (Europhysics Letters)* **82**: 11001. Bibcode:2008EL......8211001G. doi:10.1209/0295-5075/82/11001.

[8] http://www.ejtp.com/articles/ejtpv7i24p219.pdf

[9] http://arxiv.org/PS_cache/arxiv/pdf/0807/0807.3961v2.pdf

[10] http://arxiv.org/PS_cache/arxiv/pdf/0901/0901.3777v2.pdf

[11] Dvali, Gia; Giudice, Gian F.; Gomez, Cesar; Kehagias, Alex (2011). "UV-Completion by Classicalization". arXiv:1010.1415. Bibcode:2011JHEP...08..108D. doi:10.1007/JHEP08(2011)108.

[12] Csaki, C.; Grojean, C.; Pilo, L.; Terning, J. (2004), "Towards a realistic model of Higgsless electroweak symmetry breaking", *Physical Review Letters* **92** (10): 101802, arXiv:hep-ph/0308038, Bibcode:2004PhRvL..92j1802C. doi:10.1103/PhysRevLett.92.101802, PMID 15089195

[13] Csaki, C.; Grojean, C.; Murayama, H.; Pilo, L.; Terning, John (2004). "Gauge theories on an interval: Unitarity without a Higgs", *Physical Review D* **69** (5): 055006, arXiv:hep-ph/0305237, Bibcode:2004PhRvD..69e5006C. doi:10.1103/PhysRevD.69.055006

[14] Calmet, X.; Deshpande, N. G.; He, X. G.; Hsu, S. D. H. (2008), "Invisible Higgs boson, continuous mass fields and unHiggs mechanism". *Physical Review D* **79** (5): 055021, arXiv:0810.2155, Bibcode:2009PhRvD..79e5021C. doi:10.1103/PhysRevD.79.055021

[15] Pawlowski, M.; Raczka, R. (1994). "A Unified Conformal Model for Fundamental Interactions without Dynamical Higgs Field", *Foundations of Physics* **24** (9): 1305–1327, arXiv:hep-th/9407137, Bibcode:1994FoPh...24.1305P. doi:10.1007/BF02148570

[16] Bilson-Thompson, Sundance O.; Markopoulou, Fotini; Smolin, Lee (2007), "Quantum gravity and the standard model", *Class. Quantum Grav.* **24** (16): 3975–3993, arXiv:hep-th/0603022, Bibcode:2007CQGra..24.3975B. doi:10.1088/0264-9381/24/16/002.

[17] A. V. Avdeenkov and K. G. Zloshchastiev, *Quantum Bose liquids with logarithmic nonlinearity: Self-sustainability and emergence of spatial extent*, J. Phys. B: At. Mol. Opt. Phys. **44** (2011) 195303, ArXiv:1108.0847.

[18] K. G. Zloshchastiev, *Spontaneous symmetry breaking and mass generation as built-in phenomena in logarithmic nonlinear quantum theory*, Acta Phys. Polon. B **42** (2011) 261-292 ArXiv:0912.4139.

[19] V. Dzhunushaliev and K.G. Zloshchastiev (2012). *Singularity-free model of electric charge in physical vacuum: Non-zero spatial extent and mass generation*. ArXiv:1204.6380.

1.2 Strong CP problem

In particle physics, the **strong CP problem** (CP standing for **charge parity**) is the puzzling question of why quantum chromodynamics (QCD) does not seem to break the CP-symmetry.

According to quantum chromodynamics there could be a violation of CP symmetry in the strong interactions. However, there is no experimentally known violation of the CP-symmetry in strong interactions. As there is no known reason for it to be conserved in QCD specifically, this is a "fine tuning" problem known as the strong CP problem.

The strong CP problem is sometimes regarded as an unsolved problem in physics.[1]

1.2.1 What is CP violation?

Main article: CP violation

CP-symmetry states that the laws of physics should be the same if a particle were interchanged with its antiparticle (C symmetry), and then left and right were swapped (P symmetry). In particle physics, CP violation (CP standing for **Charge Parity**) is a violation of the postulated **CP-symmetry** (or **Charge conjugation Parity symmetry**): the combination of C-symmetry (charge conjugation symmetry) and P-symmetry (parity symmetry).

1.2.2 How CP can be violated in QCD

QCD does not violate the CP-symmetry as easily as the electroweak theory; unlike the electroweak theory in which the gauge fields couple to chiral currents constructed from the fermionic fields, the gluons couple to vector currents. Experiments do not indicate any CP violation in the QCD sector. For example, a generic CP violation in the strongly interacting sector would create the electric dipole moment of the neutron which would be comparable to 10^{-18} e·m while the experimental upper bound is roughly one trillionth that size.

This is a problem because at the end, there are natural terms in the QCD Lagrangian that are able to break the CP-symmetry.

$$\mathcal{L} = -\frac{1}{4} F_{\mu\nu} F^{\mu\nu} - \frac{n_f g^2 \theta}{32\pi^2} F_{\mu\nu} \tilde{F}^{\mu\nu} + \bar{\psi}(i\gamma^\mu D_\mu - m e^{i\theta'\gamma_5})\psi$$

For a nonzero choice of the θ angle and the chiral quark mass phase θ' one expects the CP-symmetry to be violated.

If the chiral quark mass phase θ' can be converted to a contribution to the total effective θ angle, it will have to be explained why this effective angle is extremely small instead of being of order one; the particular value of the angle that must be very close to zero (in this case) is an example of a fine-tuning problem in physics. If the phase θ' is absorbed in the gamma-matrices, one has to explain why θ is small, but it will not be unnatural to set it equal to zero.

If at least one of the quarks of the standard model were massless, θ would become unobservable; i.e. it would vanish from the theory. However, empirical evidence strongly suggests that none of the quarks are massless and so this solution to the strong CP problem fails.

1.2.3 Proposed solutions

There are several proposed solutions to solve the strong CP problem. The most well-known is Peccei–Quinn theory.[2] involving new scalar particles called axions. A solution that does not involve new particles was presented in 2003.[3]

1.2.4 References

[1] Mannel, Thomas (2–8 July 2006). "Theory and Phenomenology of CP Violation" (PDF). *Nuclear Physics B, vol. 167*. The 7th International Conference on Hyperons, Charm And Beauty Hadrons (BEACH 2006). Lancaster: Elsevier. pp. 170–174. doi:10.1016/j.nuclphysbps.2006.12.083. Retrieved 15 Aug 2015.

[2] Peccei, Roberto D.; Quinn, Helen R. (1977). "CP Conservation in the Presence of Pseudoparticles". *Physical Review Letters* **38** (25): 1440–1443. Bibcode:1977PhRvL..38.1440P. doi:10.1103/PhysRevLett.38.1440.

[3] Banerjee, H; Chatterjee, D; Mitra, P (30 October 2003). "Is there still a strong CP problem?". *Physics Letters B* **573**: 109–114. arXiv:hep-ph/0012284. doi:10.1016/j.physletb.2003.08.058.

1.3 Neutrino oscillation

Neutrino oscillation is a quantum mechanical phenomenon whereby a neutrino created with a specific lepton flavor (electron, muon, or tau) can later be measured to have a different flavor. The probability of measuring a particular flavor for a neutrino varies periodically as it propagates through space.[1]

First predicted by Bruno Pontecorvo in 1957,[2] neutrino oscillation has since been observed by a multitude of experiments in several different contexts. Notably, the existence of neutrino oscillation resolved the long-standing solar neutrino problem.

Neutrino oscillation is of great theoretical and experimental interest, as the precise properties of the process can shed light on several properties of the neutrino. In particular, it implies that the neutrino has a non-zero mass, which requires a modification to the Standard Model of particle physics.[1] The experimental discovery of neutrino oscillation, and thus neutrino mass, by the Super-Kamiokande Observatory and the Sudbury Neutrino Observatory was recognized with the 2015 Nobel Prize for Physics.[3]

1.3.1 Observations

A great deal of evidence for neutrino oscillation has been collected from many sources, over a wide range of neutrino energies and with many different detector technologies.[4] The 2015 Nobel Prize in Physics was shared by Takaaki Kajita and Arthur B. McDonald for their early pioneering observations of these oscillations.

Solar neutrino oscillation

The first experiment that detected the effects of neutrino oscillation was Ray Davis's Homestake Experiment in the late 1960s, in which he observed a deficit in the flux of solar neutrinos with respect to the prediction of the Standard Solar Model, using a chlorine-based detector.[5] This gave rise to the Solar neutrino problem. Many subsequent radiochemical and water Cherenkov detectors confirmed the deficit, but neutrino oscillation was not conclusively identified as the source of the deficit until the Sudbury Neutrino Observatory provided clear evidence of neutrino flavor change in 2001.[6]

Solar neutrinos have energies below 20 MeV and travel approximately 1 A.U. between the source in the Sun and detector on the Earth. At energies above 5 MeV, solar neutrino oscillation actually takes place in the Sun through a resonance known as the MSW effect, a different process from the vacuum oscillation described later in this article.[1]

Atmospheric neutrino oscillation

Large detectors such as IMB, MACRO, and Kamiokande II have observed a deficit in the ratio of the flux of muon to electron flavor atmospheric neutrinos (see *muon decay*). The Super-Kamiokande experiment provided a very precise measurement of neutrino oscillation in an energy range of hundreds of MeV to a few TeV, and with a baseline of the diameter of the Earth; the first experimental evidence for atmospheric neutrino oscillations was announced in 1998.[7]

Reactor neutrino oscillation

Many experiments have searched for oscillation of electron anti-neutrinos produced at nuclear reactors. Such oscillations give the value of the parameter θ_{13}. Neutrinos produced in nuclear reactors have energies similar to solar neutrinos, of around a few MeV. The baselines of these experiments have ranged from tens of meters to over 100 km.

In 2012, the Daya Bay experiment announced a discovery that $\theta_{13} \neq 0$ with a significance of 5.2σ;[8] these results have since been confirmed by RENO and Double Chooz.[9]

Beam neutrino oscillation

Neutrino beams produced at a particle accelerator offer the greatest control over the neutrinos being studied. Many experiments have taken place that study the same oscillations as in atmospheric neutrino oscillation using neutrinos with a few GeV of energy and several-hundred-km baselines. The MINOS, K2K, and Super-K experiments have all independently observed muon neutrino disappearance over such long baselines.[1]

Data from the LSND experiment appear to be in conflict with the oscillation parameters measured in other experiments. Results from the MiniBooNE appeared in Spring 2007 and contradicted the results from LSND, although they could support the existence of a fourth neutrino type, the sterile neutrino.[1]

In 2010, the INFN and CERN announced the observation of a tau particle in a muon neutrino beam in the OPERA detector located at Gran Sasso, 730 km away from the source in Geneva.[10]

T2K, using a neutrino beam directed through 295 km of earth and the Super-Kamiokande detector, measured a non-zero value for the parameter θ_{13} in a neutrino beam.[11] NOvA, using the same beam as MINOS with a baseline of 810 km, is sensitive to the same.

1.3.2 Theory

Neutrino oscillation arises from a mixture between the flavor and mass eigenstates of neutrinos. That is, the three neutrino states that interact with the charged leptons in weak interactions are each a different superposition of the three neutrino states of definite mass. Neutrinos are created in weak processes in their flavor eigenstates[nb 1]. As a neutrino propagates through space, the quantum mechanical phases of the three mass states advance at slightly different rates due to the slight differences in the neutrino masses. This results in a changing mixture of mass states as the neutrino travels, but a different mixture of mass states corresponds to a different mixture of flavor states. So a neutrino born as, say, an electron neutrino will be some mixture of electron, mu, and tau neutrino after traveling some distance. Since the quantum mechanical phase advances in a periodic fashion, after some distance the state will nearly return to the original mixture, and the neutrino will be again mostly electron neutrino. The electron flavor content of the neutrino will then continue to oscillate as long as the quantum mechanical state maintains coherence. Since mass differences between neutrino flavors are small in comparison with long coherence length for neutrino oscillations this microscopic quantum effect becomes observable over macroscopic distances.

Pontecorvo–Maki–Nakagawa–Sakata matrix

Main article: Pontecorvo–Maki–Nakagawa–Sakata matrix

The idea of neutrino oscillation was first put forward in 1957 by Bruno Pontecorvo, who proposed that neutrino-antineutrino transitions may occur in analogy with neutral kaon mixing.[2] Although such matter-antimatter oscillation has not been observed, this idea formed the conceptual foundation for the quantitative theory of neutrino flavor oscillation, which was first developed by Maki, Nakagawa, and Sakata in 1962[13] and further elaborated by Pontecorvo in 1967.[14] One year later the solar neutrino deficit was first observed,[15] and that was followed by the famous paper of Gribov and Pontecorvo published in 1969 titled "Neutrino astronomy and lepton charge".[16]

The concept of neutrino mixing is a natural outcome of gauge theories with massive neutrinos and its structure can be characterized in general.[17] In its simplest form it is expressed as a unitary transformation relating the flavor and mass eigenbasis, and can be written

$$|\nu_\alpha\rangle = \sum_i U_{\alpha i}^* |\nu_i\rangle$$

$$|\nu_i\rangle = \sum_\alpha U_{\alpha i} |\nu_\alpha\rangle .$$

where

- $|\nu_\alpha\rangle$ is a neutrino with definite flavor. α = e (electron), μ (muon) or τ (tauon).

- $|\nu_i\rangle$ is a neutrino with definite mass m_i , $i = 1, 2, 3$.

- The asterisk (*) represents a complex conjugate. For antineutrinos, the complex conjugate should be dropped from the first equation, and added to the second.

$U_{\alpha i}$, represents the *Pontecorvo–Maki–Nakagawa–Sakata matrix* (also called the *PMNS matrix, lepton mixing matrix,* or sometimes simply the *MNS matrix*). It is the analogue of the CKM matrix describing the analogous mixing of quarks. If this matrix were the identity matrix, then the flavor eigenstates would be the same as the mass eigenstates. However, experiment shows that it is not.

When the standard three neutrino theory is considered, the matrix is 3×3. If only two neutrinos are considered, a 2×2 matrix is used. If one or more sterile neutrinos are added (see later) it is 4×4 or larger. In the 3×3 form, it is given by:[18]

$$
\begin{aligned}
U &= \begin{bmatrix} U_{e1} & U_{e2} & U_{e3} \\ U_{\mu 1} & U_{\mu 2} & U_{\mu 3} \\ U_{\tau 1} & U_{\tau 2} & U_{\tau 3} \end{bmatrix} \\
&= \begin{bmatrix} 1 & 0 & 0 \\ 0 & c_{23} & s_{23} \\ 0 & -s_{23} & c_{23} \end{bmatrix} \begin{bmatrix} c_{13} & 0 & s_{13}e^{-i\delta} \\ 0 & 1 & 0 \\ -s_{13}e^{i\delta} & 0 & c_{13} \end{bmatrix} \begin{bmatrix} c_{12} & s_{12} & 0 \\ -s_{12} & c_{12} & 0 \\ 0 & 0 & 1 \end{bmatrix} \begin{bmatrix} e^{i\alpha_1/2} & 0 & 0 \\ 0 & e^{i\alpha_2/2} & 0 \\ 0 & 0 & 1 \end{bmatrix} \\
&= \begin{bmatrix} c_{12}c_{13} & s_{12}c_{13} & s_{13}e^{-i\delta} \\ -s_{12}c_{23}-c_{12}s_{23}s_{13}e^{i\delta} & c_{12}c_{23}-s_{12}s_{23}s_{13}e^{i\delta} & s_{23}c_{13} \\ s_{12}s_{23}-c_{12}c_{23}s_{13}e^{i\delta} & -c_{12}s_{23}-s_{12}c_{23}s_{13}e^{i\delta} & c_{23}c_{12} \end{bmatrix} \begin{bmatrix} e^{i\alpha_1/2} & 0 & 0 \\ 0 & e^{i\alpha_2/2} & 0 \\ 0 & 0 & 1 \end{bmatrix}
\end{aligned}
$$

where $cij = \cos \theta ij$ and $sij = \sin \theta ij$. The phase factors α_1 and α_2 are physically meaningful only if neutrinos are Majorana particles — i.e. if the neutrino is identical to its antineutrino (whether or not they are is unknown) — and do not enter into oscillation phenomena regardless. If neutrinoless double beta decay occurs, these factors influence its rate. The phase factor δ is non-zero only if neutrino oscillation violates CP symmetry; this has not yet been observed experimentally. If experiment shows this 3×3 matrix to be not unitary, a sterile neutrino or some other new physics is required.

Propagation and interference

Since $|\nu_i\rangle$ are mass eigenstates, their propagation can be described by plane wave solutions of the form

$$|\nu_i(t)\rangle = e^{-i(E_i t - \vec{p}_i \cdot \vec{x})} |\nu_i(0)\rangle .$$

where

- quantities are expressed in natural units ($c = 1, \hbar = 1$)

- E_i is the energy of the mass-eigenstate i ,

- t is the time from the start of the propagation,

- $\vec{p}_i$ is the three-dimensional momentum,

- $\vec{x}$ is the current position of the particle relative to its starting position

In the ultrarelativistic limit, $|\vec{p_i}| = p_i \gg m_i$, we can approximate the energy as

$$E_i = \sqrt{p_i^2 + m_i^2} \simeq p_i + \frac{m_i^2}{2p_i} \approx E + \frac{m_i^2}{2E},$$

where E is the total energy of the particle.

This limit applies to all practical (currently observed) neutrinos, since their masses are less than 1 eV and their energies are at least 1 MeV, so the Lorentz factor γ is greater than 10^6 in all cases. Using also $t \approx L$, where L is the distance traveled and also dropping the phase factors, the wavefunction becomes:

$$|\nu_i(L)\rangle = e^{-im_i^2 L/2E}|\nu_i(0)\rangle.$$

Eigenstates with different masses propagate at different speeds. The heavier ones lag behind while the lighter ones pull ahead. Since the mass eigenstates are combinations of flavor eigenstates, this difference in speed causes interference between the corresponding flavor components of each mass eigenstate. Constructive interference causes it to be possible to observe a neutrino created with a given flavor to change its flavor during its propagation. The probability that a neutrino originally of flavor α will later be observed as having flavor β is

$$P_{\alpha \to \beta} = |\langle \nu_\beta(t)|\nu_\alpha \rangle|^2 = \left| \sum_i U_{\alpha i}^* U_{\beta i} e^{-im_i^2 L/2E} \right|^2.$$

This is more conveniently written as

$$P_{\alpha \to \beta} = \delta_{\alpha\beta} - 4 \sum_{i>j} \mathrm{Re}(U_{\alpha i}^* U_{\beta i} U_{\alpha j} U_{\beta j}^*) \sin^2 \left(\frac{\Delta m_{ij}^2 L}{4E} \right)$$
$$+ 2 \sum_{i>j} \mathrm{Im}(U_{\alpha i}^* U_{\beta i} U_{\alpha j} U_{\beta j}^*) \sin \left(\frac{\Delta m_{ij}^2 L}{2E} \right).$$

where $\Delta m_{ij}^2 \equiv m_i^2 - m_j^2$. The phase that is responsible for oscillation is often written as (with c and $\hbar$ restored)

$$\frac{\Delta m^2 c^3 L}{4\hbar E} = \frac{\mathrm{GeV\,fm}}{4\hbar c} \times \frac{\Delta m^2}{\mathrm{eV}^2} \frac{L}{\mathrm{km}} \frac{\mathrm{GeV}}{E}$$
$$\approx 1.27 \times \frac{\Delta m^2}{\mathrm{eV}^2} \frac{L}{\mathrm{km}} \frac{\mathrm{GeV}}{E},$$

where 1.27 is unitless. In this form, it is convenient to plug in the oscillation parameters since:

- The mass differences, Δm^2, are known to be on the order of 1×10^{-4} eV^2
- Oscillation distances, L, in modern experiments are on the order of kilometers
- Neutrino energies, E, in modern experiments are typically on order of MeV or GeV.

If there is no CP-violation (δ is zero), then the second sum is zero. Otherwise, the CP asymmetry can be given as

$$A_{\mathrm{CP}}^{(\alpha\beta)} = P(\nu_\alpha \to \nu_\beta) - P(\bar{\nu}_\alpha \to \bar{\nu}_\beta)$$
$$= 4 \sum_{i>j} \mathrm{Im}(U_{\alpha i}^* U_{\beta i} U_{\alpha j} U_{\beta j}^*) \sin \left(\frac{\Delta m_{ij}^2 L}{2E} \right)$$

In terms of Jarlskog invariant

$$\mathrm{Im}(U_{\alpha i} U_{\beta i}^* U_{\alpha j}^* U_{\beta j}) = J \sum_{\gamma,k} \varepsilon_{\alpha\beta\gamma} \varepsilon_{ijk}$$

the CP asymmetry is expressed as

$$A_{\mathrm{CP}}^{(\alpha\beta)}$$
$$= 16J \sum_\gamma \varepsilon_{\alpha\beta\gamma} \sin \left(\frac{\Delta m_{21}^2 L}{4E} \right) \sin \left(\frac{\Delta m_{31}^2 L}{4E} \right) \sin \left(\frac{\Delta m_{32}^2 L}{4E} \right)$$

Two neutrino case

The above formula is correct for any number of neutrino generations. Writing it explicitly in terms of mixing angles is extremely cumbersome if there are more than two neutrinos that participate in mixing. Fortunately, there are several cases in which only two neutrinos participate significantly. In this case, it is sufficient to consider the mixing matrix

$$U = \begin{pmatrix} \cos\theta & \sin\theta \\ -\sin\theta & \cos\theta \end{pmatrix}.$$

Then the probability of a neutrino changing its flavor is

$$P_{\alpha \to \beta, \alpha \neq \beta} = \sin^2(2\theta) \sin^2 \left(\frac{\Delta m^2 L}{4E} \right) \text{ units) (natural.}$$

Or, using SI units and the convention introduced above

$$P_{\alpha \to \beta, \alpha \neq \beta} = \sin^2(2\theta) \sin^2 \left(1.27 \frac{\Delta m^2 L}{E} \frac{[\mathrm{eV}^2][\mathrm{km}]}{[\mathrm{GeV}]} \right).$$

This formula is often appropriate for discussing the transition $\nu_\mu \leftrightarrow \nu_\tau$ in atmospheric mixing, since the electron neutrino plays almost no role in this case. It is also appropriate for the solar case of $\nu_e \leftrightarrow \nu_x$, where ν_x is a superposition of ν_μ and ν_τ. These approximations are possible because the mixing angle θ_{13} is very small and because two of the mass states are very close in mass compared to the third.

Classical analogue of neutrino oscillation

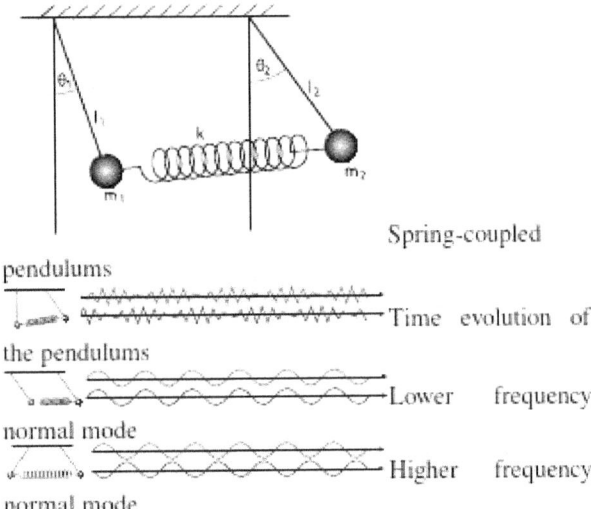

Spring-coupled pendulums

Time evolution of the pendulums

Lower   frequency normal mode

Higher   frequency normal mode

The basic physics behind neutrino oscillation can be found in any system of coupled harmonic oscillators. A simple example is a system of two pendulums connected by a weak spring (a spring with a small spring constant). The first pendulum is set in motion by the experimenter while the second begins at rest. Over time, the second pendulum begins to swing under the influence of the spring, while the first pendulum's amplitude decreases as it loses energy to the second. Eventually all of the system's energy is transferred to the second pendulum and the first is at rest. The process then reverses. The energy oscillates between the two pendulums repeatedly until it is lost to friction.

The behavior of this system can be understood by looking at its normal modes of oscillation. If the two pendulums are identical then one normal mode consists of both pendulums swinging in the same direction with a constant distance between them, while the other consists of the pendulums swinging in opposite (mirror image) directions. These normal modes have (slightly) different frequencies because the second involves the (weak) spring while the first does not. The initial state of the two-pendulum system is a combination of both normal modes. Over time, these normal modes drift out of phase, and this is seen as a transfer of motion from the first pendulum to the second.

The description of the system in terms of the two pendulums is analogous to the flavor basis of neutrinos. These are the parameters that are most easily produced and detected (in the case of neutrinos, by weak interactions involving the W boson). The description in terms of normal modes is analogous to the mass basis of neutrinos. These modes do not interact with each other when the system is free of outside influence.

When the pendulums are not identical the analysis is slightly more complicated. In the small-angle approximation, the potential energy of a single pendulum system is $\frac{1}{2}\frac{mg}{L}x^2$, where g is the standard gravity, L is the length of the pendulum, m is the mass of the pendulum, and x is the horizontal displacement of the pendulum. As an isolated system the pendulum is a harmonic oscillator with a frequency of $\sqrt{g/L}$. The potential energy of a spring is $\frac{1}{2}kx^2$ where k is the spring constant and x is the displacement. With a mass attached it oscillates with a period of $\sqrt{k/m}$. With two pendulums (labeled a and b) of equal mass but possibly unequal lengths and connected by a spring, the total potential energy is

$$V = \frac{m}{2}\left(\frac{g}{L_a}x_a^2 + \frac{g}{L_b}x_b^2 + \frac{k}{m}(x_b - x_a)^2 \right).$$

This is a quadratic form in xa and xb, which can also be written as a matrix product:

$$V = \frac{m}{2}(x_a \quad x_b)\begin{pmatrix} \frac{g}{L_a} + \frac{k}{m} & -\frac{k}{m} \\ -\frac{k}{m} & \frac{g}{L_b} + \frac{k}{m} \end{pmatrix}\begin{pmatrix} x_a \\ x_b \end{pmatrix}.$$

The 2×2 matrix is real symmetric and so (by the spectral theorem) it is "orthogonally diagonalizable". That is, there is an angle θ such that if we define

$$\begin{pmatrix} x_a \\ x_b \end{pmatrix} = \begin{pmatrix} \cos\theta & \sin\theta \\ -\sin\theta & \cos\theta \end{pmatrix}\begin{pmatrix} x_1 \\ x_2 \end{pmatrix}$$

then

$$V = \frac{m}{2}(x_1 \quad x_2)\begin{pmatrix} \lambda_1 & 0 \\ 0 & \lambda_2 \end{pmatrix}\begin{pmatrix} x_1 \\ x_2 \end{pmatrix}$$

where λ_1 and λ_2 are the eigenvalues of the matrix. The variables x_1 and x_2 describe normal modes which oscillate with frequencies of $\sqrt{\lambda_1}$ and $\sqrt{\lambda_2}$. When the two pendulums are identical ($La = Lb$), θ is 45°.

The angle θ is analogous to the Cabibbo angle (though that angle applies to quarks rather than neutrinos).

When the number of oscillators (particles) is increased to three, the orthogonal matrix can no longer be described by a single angle; instead, three are required (Euler angles). Furthermore, in the quantum case, the matrices may be complex. This requires the introduction of complex phases in addition to the rotation angles, which are associated with CP violation but do not influence the observable effects of neutrino oscillation.

1.3.3 Theory, graphically

Two neutrino probabilities in vacuum

In the approximation where only two neutrinos participate in the oscillation, the probability of oscillation follows a simple pattern:

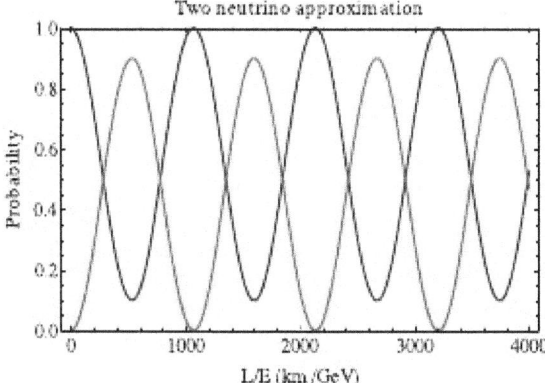

The blue curve shows the probability of the original neutrino retaining its identity. The red curve shows the probability of conversion to the other neutrino. The maximum probability of conversion is equal to $\sin^2 2\theta$. The frequency of the oscillation is controlled by Δm^2.

Three neutrino probabilities

If three neutrinos are considered, the probability for each neutrino to appear is somewhat complex. Here are shown the probabilities for each initial flavor, with one plot showing a long range to display the slow "solar" oscillation and the other zoomed in to display the fast "atmospheric" oscillation. The oscillation parameters used here are consistent with current measurements, but since some parameters are still quite uncertain, these graphs are only qualitatively correct in some aspects. These values were used:

- $\sin^2 2\theta_{13} = 0.10$ (Controls the size of the small wiggles.)

- $\sin^2 2\theta_{23} = 0.97$.

- $\sin^2 2\theta_{12} = 0.861$.

- $\delta = 0$ (If it is actually large, these probabilities will be somewhat distorted and different for neutrinos and antineutrinos.)

- $\Delta m2$
  $12 = 7.59 \times 10^{-5}$ eV^2.

- $\Delta m2$
  $32 \approx \Delta m2$
  $13 = 2.32 \times 10^{-3}$ eV^2.

- Normal mass hierarchy.

[19]

1.3.4 Observed values of oscillation parameters

- $\sin^2 (2\theta_{13}) = 0.093 \pm 0.008$.[20] PDG combination of Daya Bay, RENO, and Chooz results.

- $\sin^2 (2\theta_{12}) = 0.846 + 0.021$
  -0.021.[20] This corresponds to θ_{sol} (solar), obtained from KamLand, solar, reactor and accelator data.

- $\sin^2 (2\theta_{23}) > 0.92$ at 90% confidence level, corresponding to $\theta_{23} \equiv \theta_{atm} = 45 \pm 7.1°$ (atmospheric)[21]

- $\Delta m2$
  $21 \equiv \Delta m2$
  $sol = 7.53 + 0.18$
  -0.18×10^{-5} eV^2[20]

- $|\Delta m2$
  $31| \approx |\Delta m2$
  $32| \equiv \Delta m2$
  $atm = 2.44 + 0.06$
  -0.06×10^{-3} eV^2 (normal mass hierarchy)[20]

- δ, α_1, α_2, and the sign of $\Delta m2$
  32 are currently unknown

Solar neutrino experiments combined with KamLAND have measured the so-called solar parameters $\Delta m2$ sol and $\sin^2 \theta_{sol}$. Atmospheric neutrino experiments such as Super-Kamiokande together with the K2K and MINOS long baseline accelerator neutrino experiment have determined the so-called atmospheric parameters $\Delta m2$ atm and $\sin^2 \theta_{atm}$. The last mixing angle, θ_{13}, has been measured by the experiments Daya Bay, Double Chooz and RENO as $\sin^2 2\theta_{13}$.

For atmospheric neutrinos (where the relevant difference of masses is about $\Delta m^2 = 2.4 \times 10^{-3}$ eV^2 and the typical energies are ~1 GeV), oscillations become visible for neutrinos traveling several hundred km, which means neutrinos that reach the detector from below the horizon.

The mixing parameter θ_{13} is measured using electron antineutrinos from nuclear reactors. The rate of anti-neutrino interactions is measured in detectors sited near the reactors to determine the flux prior to any significant oscillations and then it is measured in far detectors (sited km from the reactors). The oscillation is observed as an apparent disappearance of electron anti-neutrinos in the far detectors (*i.e.* the interaction rate at the far site is lower than predicted from the observed rate at the near site).

From atmospheric and solar neutrino oscillation experiments, it is known that two mixing angles of the MNS matrix are large and the third is smaller. This is in sharp contrast to the CKM matrix in which all three angles are small and hierarchically decreasing. Nothing is known about the CP-violating phase of the MNS matrix.

If the neutrino mass proves to be of Majorana type (making the neutrino its own antiparticle), it is possible that the MNS matrix has more than one phase.

Since experiments observing neutrino oscillation measure the squared mass difference and not absolute mass, one can claim that the lightest neutrino mass is exactly zero, without contradicting observations. This is however regarded as unlikely by theorists.

1.3.5 Origins of neutrino mass

The question of how neutrino masses arise has not been answered conclusively. In the Standard Model of particle physics, fermions only have mass because of interactions with the Higgs field (see *Higgs boson*). These interactions involve both left- and right-handed versions of the fermion (see *chirality*). However, only left-handed neutrinos have been observed so far.

Neutrinos may have another source of mass through the Majorana mass term. This type of mass applies for electrically-neutral particles since otherwise it would allow particles to turn into anti-particles, which would violate conservation of electric charge.

The smallest modification to the Standard Model, which only has left-handed neutrinos, is to allow these left-handed neutrinos to have Majorana masses. The problem with this is that the neutrino masses are surprisingly smaller than the rest of the known particles (at least 500,000 times smaller than the mass of an electron), which, while it does not invalidate the theory, is widely regarded as unsatisfactory as this construction offers no insight into the origin of the neutrino mass scale.

The next simplest addition would be to add into the Standard Model right-handed neutrinos that interact with the left-handed neutrinos and the Higgs field in an analogous way to the rest of the fermions. These new neutrinos would interact with the other fermions solely in this way, so are not phenomenologically excluded. The problem of the disparity of the mass scales remains.

Seesaw mechanism

Main article: Seesaw mechanism

The most popular conjectured solution currently is the *seesaw mechanism*, where right-handed neutrinos with very large Majorana masses are added. If the right-handed neutrinos are very heavy, they induce a very small mass for the left-handed neutrinos, which is proportional to the inverse of the heavy mass.

If it is assumed that the neutrinos interact with the Higgs field with approximately the same strengths as the charged fermions do, the heavy mass should be close to the GUT scale. Note that, in the Standard Model there is just one fundamental mass scale (which can be taken as the scale of SU(2)L × U(1)Y breaking) and all masses (such as the electron or the mass of the Z boson) have to originate from this one.

There are other varieties of seesaw[22] and there is currently great interest in the so-called low-scale seesaw schemes, such as the inverse seesaw mechanism.[23]

The addition of right-handed neutrinos has the effect of adding new mass scales, unrelated to the mass scale of the Standard Model, hence the observation of heavy right-handed neutrinos would reveal physics beyond the Standard Model. Right-handed neutrinos would help to explain the origin of matter through a mechanism known as leptogenesis.

Other sources

There are alternative ways to modify the standard model that are similar to the addition of heavy right-handed neutrinos (e.g., the addition of new scalars or fermions in triplet states) and other modifications that are less similar (e.g., neutrino masses from loop effects and/or from suppressed couplings). One example of the last type of models is provided by certain versions supersymmetric extensions of the standard model of fundamental interactions, where R parity is not a symmetry. There, the exchange of supersymmetric particles such as squarks and sleptons can break the lepton number and lead to neutrino masses. These interactions are normally excluded from theories as they come from a class of interactions that lead to unacceptably rapid proton decay if they are all included. These models have little predictive power and are not able to provide a cold dark matter candidate.

1.3.6 Oscillations in the early universe

During the early universe when particle concentrations and temperatures were high, neutrino oscillations can behave differently.[124] Depending on neutrino mixing-angle parameters and masses, a broad spectrum of behavior may arise including vacuum-like neutrino oscillations, smooth evolution, or self-maintained coherence. The physics for this system is non-trivial and involves neutrino oscillations in a dense neutrino gas.

1.3.7 See also

- MSW effect
- Majoron
- Neutral kaon mixing
- Neutral particle oscillation
- Neutrino astronomy

1.3.8 Notes

[1] More formally, the neutrinos are emitted in an entangled state with the other bodies in the decay or reaction, and the mixed state is properly described by a density matrix. However, for all practical situations, the other particles in the decay may be well localized in time and space (e.g. to within a nuclear distance), leaving their momentum with a large spread. When these partner states are projected out, the neutrino is left in a state that for all intents and purposes behaves as the simple superposition of mass states described here. See [12] for more information.

1.3.9 References

[1] Barger, Vernon; Marfatia, Danny; Whisnant, Kerry Lewis (2012). *The Physics of Neutrinos*. Princeton University Press. ISBN 0-691-12853-7.

[2] B. Pontecorvo (1957). "Mesonium and anti-mesonium". *Zh. Eksp. Teor. Fiz.* **33**: 549–551. reproduced and translated in *Sov. Phys. JETP* **6**: 429. 1957. Missing or empty |title= (help) and B. Pontecorvo (1967). "Neutrino Experiments and the Problem of Conservation of Leptonic Charge". *Zh. Eksp. Teor. Fiz.* **53**: 1717. reproduced and translated in Pontecorvo, B. (1968). "Neutrino Experiments and the Problem of Conservation of Leptonic Charge". *Sov. Phys. JETP* **26**: 984. Bibcode:1968JETP...26..984P.

[3] Webb, Jonathan (6 October 2015). "Neutrino 'flip' wins physics Nobel Prize". BBC News. Retrieved 6 October 2015.

[4] M. C. Gonzalez-Garcia & Michele Maltoni (2008). "Phenomenology with Massive Neutrinos". *Physics Reports* **460**: 1–129. arXiv:0704.1800. Bibcode:2008PhR...460....1G. doi:10.1016/j.physrep.2007.12.004.

[5] Davis, Raymond; Harmer, Don S.; Hoffman, Kenneth C. (1968). "Search for Neutrinos from the Sun". *Physical Review Letters* **20** (21): 1205–1209. Bibcode:1968PhRvL..20.1205D. doi:10.1103/PhysRevLett.20.1205.

[6] Ahmad, Q. R.; et al. (SNO Collaboration) (2001). "Measurement of the Rate of $v_e + d \to p + p + e^-$ Interactions Produced by ^{8}B Solar Neutrinos at the Sudbury Neutrino Observatory". *Physical Review Letters* **87** (7). arXiv:nucl-ex/0106015. Bibcode:2001PhRvL..87g1301A. doi:10.1103/PhysRevLett.87.071301.

[7] Y. Fukudae; et al. (Super-Kamiokande Collaboration) (1998). "Evidence for Oscillation of Atmospheric Neutrinos". *Physical Review Letters* **81** (8): 1562–1567. arXiv:hep-ex/9807003. Bibcode:1998PhRvL..81.1562F. doi:10.1103/PhysRevLett.81.1562.

[8] F. P. An; et al. (Daya Bay Collaboration) (2012). "Observation of Electron-Antineutrino Disappearance at Daya Bay". *Physical Review Letters* **108** (17). arXiv:1203.1669. Bibcode:2012PhRvL.108q1803A. doi:10.1103/PhysRevLett.108.171803.

[9] Kim, Soo-Bong (2012). "Observation of Reactor Electron Antineutrino Disappearance in the RENO Experiment". arXiv:1204.0626v2 [hep-ex].

[10] N. Agafonova; et al. (OPERA Collaboration) (2010). "Observation of a first vτ candidate event in the OPERA experiment in the CNGS beam". *Physics Letters B* **691** (3): 138–145. arXiv:1006.1623. Bibcode:2010PhLB..691..138A. doi:10.1016/j.physletb.2010.06.022.

[11] K. Abe; et al. (T2K Collaboration) (2013-08-05). "Evidence of electron neutrino appearance in a muon neutrino beam". *Physical Review D* **88** (3). doi:10.1103/PhysRevD.88.032002. ISSN 1550-7998.

[12] Andrew G. Cohen; Sheldon L. Glashow & Zoltan Ligeti (2009). "Disentangling neutrino oscillations". *Physics Letters B* **678** (2): 191. arXiv:0810.4602. Bibcode:2009PhLB..678..191C. doi:10.1016/j.physletb.2009.06.020.

[13] Z. Maki; M. Nakagawa & S. Sakata (1962). "Remarks on the Unified Model of Elementary Particles". *Progress of Theoretical Physics* **28** (5): 870. Bibcode:1962PThPh..28..870M. doi:10.1143/PTP.28.870.

[14] B. Pontecorvo (1967). "Neutrino Experiments and the Problem of Conservation of Leptonic Charge". *Zh. Eksp. Teor. Fiz.* **53**: 1717. reproduced and translated in Pontecorvo, B. (1968). "Neutrino Experiments and the Problem of Conservation of Leptonic Charge". *Sov. Phys. JETP* **26**: 984. Bibcode:1968JETP...26..984P.

[15] Raymond   Davis   Jr.;   Don   S.   Harmer   &   Kenneth   C.   Hoffman   (1968).   "Search for Neutrinos from the Sun".   *Physical Review Letters* **20** (21):   1205.   Bibcode:1968PhRvL..20.1205D.   doi:10.1103/PhysRevLett.20.1205.

[16] V. Gribov & B. Pontecorvo (1969).   "Neutrino astronomy and lepton charge".   *Physics Letters B* **28** (7):   493.   Bibcode:1969PhLB...28..493G.   doi:10.1016/0370-2693(69)90525-5.

[17] .   J. Schechter, J.W.F. Valle; Valle (1980).   "Neutrino Masses in SU(2) x U(1) Theories".   *Physical Review D* **22** (9):   2227.   Bibcode:1980PhRvD..22.2227S.   doi:10.1103/PhysRevD.22.2227.

[18] S. Eidelman; et al. (2004).   "Particle Data Group - The Review of Particle Physics".   *Physics Letters B* **592** (1):   1.   arXiv:astro-ph/0406663.   Bibcode:2004PhLB..592....1P.   doi:10.1016/j.physletb.2004.06.001.   Chapter 15: *Neutrino mass, mixing, and flavor change*. Revised September 2005.

[19] Meszéna, Balázs.   "Neutrino Oscillations".   Wolfram Demonstrations Project. Retrieved 8 October 2015. Images in this section were created with Mathematica. The demonstration allows exploration of the parameters.

[20] K.A. Olive; et al. (Particle Data Group) (2014).   "2014 Review of Particle Physics".

[21] K. Nakamura; et al.   (2010).   "Review of Particle Physics".   *Journal of Physics G* **37** (7A):   1.   Bibcode:2010JPhG...37g5021N.   doi:10.1088/0954-3899/37/7a/075021.

[22] J. W. F. Valle (2006).   "Neutrino physics overview".   *Journal of Physics: Conference Series* **53** (1):   473.   arXiv:hep-ph/0608101.   Bibcode:2006JPhCS..53..473V.   doi:10.1088/1742-6596/53/1/031.

[23] R.N. Mohapatra & J. W. F. Valle (1986).   "Neutrino Mass and Baryon Number Nonconservation in Superstring Models".   *Physical Review D* **34** (5):   1642.   Bibcode:1986PhRvD..34.1642M.   doi:10.1103/PhysRevD.34.1642.

[24] Kostelecký, Alan; Samuel, Stuart (1994).   "Nonlinear neutrino oscillations in the expanding universe".   *Phys. Rev. D* **49** (4):   1740–1757.   Bibcode:1994PhRvD..49.1740K.   doi:10.1103/PhysRevD.49.1740.

1.3.10   Further reading

- Gonzalez-Garcia; Nir (2002).   "Neutrino Masses and Mixing: Evidence and Implications".   *Reviews of Modern Physics* **75** (2):   345–402.   arXiv:hep-ph/0202058.   Bibcode:2003RvMP...75..345G.   doi:10.1103/RevModPhys.75.345.

- Maltoni; Schwetz; Tortola; Valle (2004).   "Status of global fits to neutrino oscillations".   *New Journal of Physics* **6**:   122.   arXiv:hep-ph/0405172.   Bibcode:2004NJPh....6..122M.   doi:10.1088/1367-2630/6/1/122.

- Fogli; Lisi; Marrone; Montanino; Palazzo; Rotunno (2012).   "Global analysis of neutrino masses, mixings, and phases: Entering the era of leptonic CP violation searches".   *Physical Review D* **86** (1):   013012.   arXiv:1205.5254.   Bibcode:2012PhRvD..86a3012F.   doi:10.1103/PhysRevD.86.013012.

- Forero; Tortola; Valle (2012).   "Global status of neutrino oscillation parameters after Neutrino-2012".   *Physical Review D* **86** (7):   073012.   arXiv:1205.4018.   Bibcode:2012PhRvD..86g3012F.   doi:10.1103/PhysRevD.86.073012.

1.3.11   External links

- Maury Goodman. "The Neutrino Oscillation Industry" (2006). *(Provides links to many other neutrino oscillation websites.)*

- Review Articles on arxiv.org

1.4   Baryon asymmetry

The **baryon asymmetry** problem in physics refers to the fact that there is an imbalance in baryonic matter and antibaryonic matter in the observable universe. Neither the standard model of particle physics, nor the theory of general relativity provides an obvious explanation for why this should be so, and it is a natural assumption that the universe be neutral with all conserved charges.[1] The Big Bang should have produced equal amounts of matter and antimatter. Since this does not seem to be the case, it is likely some physical laws must have acted differently or did not exist for matter and antimatter. There are several competing hypotheses to explain the imbalance of matter and antimatter that resulted in baryogenesis, but there is as of yet no one consensus theory to explain the phenomenon.

1.4.1   Possible explanations

CP (charge parity) violations

Most explanations involve modifying the standard model of particle physics, to allow for some reactions (specifically involving the weak nuclear force) to proceed more easily than their opposite. This is called "violating CP symmetry" in weak interactions. Such a violation could allow matter to be produced more commonly than antimatter in conditions immediately after the Big Bang. In 2013 LHCb announced discovery of CP violation in B meson decays,[2] so did BaBar and Belle scientists in 2015.[3]

Regions of the universe where antimatter dominates

Another possible explanation of the apparent baryon asymmetry is that there are regions of the universe in which matter is dominant, and other regions of the universe in which antimatter is dominant, and these are widely separated. The problem then becomes a matter/antimatter separation problem, rather than a creation imbalance problem. Antimatter atoms would appear from a distance indistinguishable from matter atoms, as both matter and antimatter atoms would produce light (photons) in the same way. Only in the border between a matter dominated region and an antimatter dominated region would the antimatter's presence be detectable, as only there would matter/antimatter annihilation (and the subsequent production of gamma radiation) occur. How easy such a boundary would be to detect would depend on its distance and what the density of matter and antimatter is along it. Presumably such a boundary would lie (almost by necessity) in deep intergalactic space, and the density of matter in intergalactic space is reasonably well established at about one atom per cubic metre.[4][5] Assuming this is the typical density of both matter and antimatter near a boundary, the gamma ray luminosity of the boundary interaction zone is easily calculated. Approximately 30 years of scientific research have placed boundaries on how far away, at a minimum, any such boundary interaction zone would have to be, as no such zones have been detected. Hence, it is now considered very unlikely that any region within the observable universe is dominated by antimatter.[6]

The Alpha Magnetic Spectrometer at the International Space Station studies the amount of antimatter in cosmic rays to advance our capability of detecting very distant antimatter dominated regions.[7]

Yet another possibility is that antimatter repels ordinary matter rather than attracting it gravitationally. This would prevent observable interactions (see Motivations for antigravity). However, this idea is in conflict with general relativity. Einstein's field equations state that the energy–momentum tensor is the source of the gravitational field, which implies that gravity is attractive for antimatter. Furthermore, there are no astronomical observations that suggest the existence of a repulsive gravitational force between any two galaxies or galaxy clusters other than that caused by the overall accelerated expansion of the universe, and the vast majority of scientists believe that matter and antimatter attract each other gravitationally (see Antimatter

gravity debate).

Electric dipole moment

The presence of an electric dipole moment (EDM) in any fundamental particle would violate both parity (P) and time (T) symmetries. As such, an EDM would allow matter and antimatter to decay at different rates leading to a possible matter-antimatter asymmetry as observed today. Many experiments are currently being conducted to measure the EDM of various physical particles. All measurements are currently consistent with no dipole moment. However, the results do place rigorous constraints on the amount of symmetry violation that a physical model can permit. The most recent EDM limit, published in 2014, was that of the ACME Collaboration, which measured the EDM of the electron using a pulsed beam of thorium monoxide (ThO) molecules.[8]

1.4.2 See also

- Baryogenesis
- CP violation
- List of unsolved problems in physics

1.4.3 References

[1] Sarkar, Utpal (2007). *Particle and astroparticle physics.* CRC Press. p. 429. ISBN 1-58488-931-4.

[2] http://journals.aps.org/prl/pdf/10.1103/PhysRevLett.110.221601

[3] http://authors.library.caltech.edu/61145/2/1505.04147v2.pdf

[4] Davidson, Keay; Smoot, George (2008). *Wrinkles in Time.* New York: Avon. pp. 158–163. ISBN 0061344443.

[5] Silk, Joseph (1977). *Big Bang.* New York: Freeman. p. 299.

[6] Canetti, L.; Drewes, M.; Shaposhnikov, M. (2012). "Matter and Antimatter in the Universe". *New J.Phys.* **14**: 095012. arXiv:1204.4186. Bibcode:2012NJPh...14i5012C. doi:10.1088/1367-2630/14/9/095012.

[7] Barry, Patrick (12 May 2007). "The hunt for antihelium: finding a single heavy antimatter nucleus could revolutionize cosmology". *Science News.* Archived from the original on July 26, 2008.

[8] The ACME Collaboration; et al. (17 January 2014). "Order of Magnitude Smaller Limit on the Electric Dipole Moment of the Electron". *Science* **343** (269): 269–72. doi:10.1126/science.1248213.

1.5 Dark matter

Not to be confused with antimatter, dark energy, dark fluid, or dark flow. For other uses, see Dark Matter (disambiguation)

Dark matter is a hypothetical type of matter comprising approximately 27% of the mass and energy in the observable universe[1] that is not accounted for by dark energy, baryonic matter, and neutrinos.[2] The name refers to the fact that it does not emit or interact with electromagnetic radiation, such as light, and is thus invisible to the entire electromagnetic spectrum.[3] Although dark matter cannot be directly observed with conventional electromagnetic telescopes, its existence and properties are inferred from its various gravitational effects such as the motions of visible matter, via gravitational lensing, its influence on the universe's large-scale structure, and its effects in the cosmic microwave background. Dark matter is transparent to electromagnetic radiation and/or is so dense and small that it fails to absorb or emit enough radiation to be detectable with current imaging technology.

Estimates of masses for galaxies and larger structures via dynamical and general relativistic means are much greater than those based on the mass of the visible "luminous" matter.[4]

The standard model of cosmology indicates that the total mass–energy of the universe contains 4.9% ordinary matter, 26.8% dark matter and 68.3% dark energy.[5][6] Thus, dark matter constitutes 84.5%[note 1] of total mass, while dark energy plus dark matter constitute 95.1% of total mass–energy content.[7][8][9][10] The great majority of ordinary matter in the universe is also unseen, since visible stars and gas inside galaxies and clusters account for less than 10% of the ordinary matter contribution to the mass-energy density of the universe.[11]

The dark matter hypothesis plays a central role in current modeling of cosmic structure formation and galaxy formation and evolution and on explanations of the anisotropies observed in the cosmic microwave background (CMB). All these lines of evidence suggest that galaxies, clusters of galaxies, and the universe as a whole contain far more matter than that which is observable via electromagnetic signals.[12]

The most widely accepted hypothesis on the form for dark matter is that it is composed of weakly interacting massive particles (WIMPs) that interact only through gravity and the weak force.[13]

Although the existence of dark matter is generally accepted by most of the astronomical community, a minority of astronomers[14] argue for various modifications of the standard laws of general relativity, such as MOND, TeVeS, and Conformal gravity[15] that attempt to account for the observations without invoking additional matter.[16]

Many experiments to detect proposed dark matter particles through non-gravitational means are under way.[17]

1.5.1 History

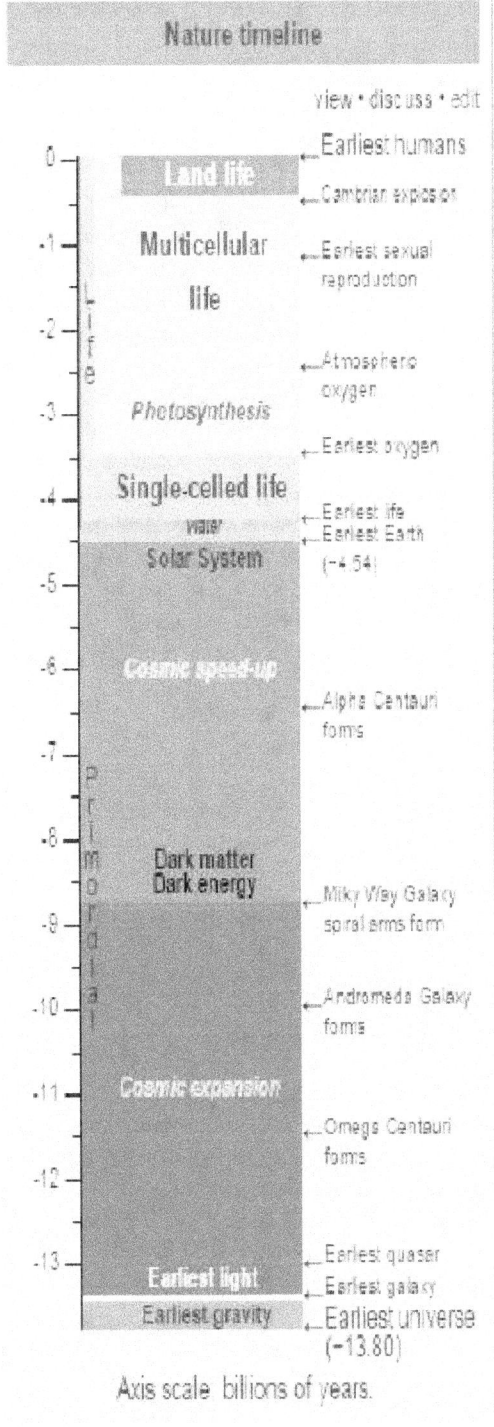

Axis scale billions of years.

galactic plane must be greater than what was observed, but this measurement was later determined to be erroneous.[22]

In 1933, Swiss astrophysicist Fritz Zwicky, who studied galactic clusters while working at the California Institute of Technology, made a similar inference.[23][24][25] Zwicky applied the virial theorem to the Coma galaxy cluster and obtained evidence of unseen mass that he called *dunkle Materie* 'dark matter'. Zwicky estimated its mass based on the motions of galaxies near its edge and compared that to an estimate based on its brightness and number of galaxies. He estimated that the cluster had about 400 times more mass than was visually observable. The gravity effect of the visible galaxies was far too small for such fast orbits, thus mass must be hidden from view. Based on these conclusions, Zwicky inferred that some unseen matter provided the mass and associated gravitation attraction to hold the cluster together. This was the first formal inference about the existence of dark matter.[26] Zwicky's estimates were off by more than an order of magnitude, mainly due to an obsolete value of the Hubble constant;[27] the same calculation today shows a smaller fraction, using greater values for luminous mass. However, Zwicky did correctly infer that the bulk of the matter was dark.[26]

The first robust indications that the mass to light ratio was anything other than unity came from measurements of galaxy rotation curves. In 1939, Horace W. Babcock reported the rotation curve for the Andromeda nebula, which suggested that the mass-to-luminosity ratio increases radially.[28] He attributed it to either light absorption within the galaxy or modified dynamics in the outer portions of the spiral and not to missing matter.

Vera Rubin and Kent Ford in the 1960s–1970s provided further strong evidence, also using galaxy rotation curves.[29][30] Rubin worked with a new spectrograph to measure the velocity curve of edge-on spiral galaxies with greater accuracy.[30] This result was independently confirmed in 1978.[31] An influential paper presented Rubin's results in 1980.[32] Rubin found that most galaxies must contain about six times as much dark as visible mass; thus, by around 1980 the apparent need for dark matter was widely recognized as a major unsolved problem in astronomy.

A stream of independent observations in the 1980s indicated its presence, including gravitational lensing of background objects by galaxy clusters, the temperature distribution of hot gas in galaxies and clusters, and the pattern of anisotropies in the cosmic microwave background. According to consensus among cosmologists, dark matter is composed primarily of a not yet characterized type of subatomic particle.[13][33] The search for this particle, by a variety of means, is one of the major efforts in particle physics.[17]

The first to suggest the existence of dark matter (using stellar velocities) was Dutch astronomer Jacobus Kapteyn in 1922.[18][19] Fellow Dutchman and radio astronomy pioneer Jan Oort also hypothesized the existence of dark matter in 1932.[19][20][21] Oort was studying stellar motions in the local galactic neighborhood and found that the mass in the

Cosmic microwave background radiation

In cosmology, the CMB is explained as relic radiation which has travelled freely since the era of recombination, around 375,000 years after the Big Bang. The CMB's anisotropies are explained as the result of small primordial density fluctuations, and subsequent acoustic oscillations in the photon-baryon plasma whose restoring force is gravity.[34]

The Cosmic Background Explorer (COBE) satellite found the CMB spectrum to be a very precise blackbody spectrum with a temperature of 2.726 K. In 1992, COBE detected CMB fluctuations (anisotropies) at a level of about one part in 10^5.[35]

In the following decade, CMB anisotropies were investigated by ground-based and balloon experiments. Their primary goal was to measure the angular scale of the first acoustic peak of the anisotropies' power spectrum, for which COBE had insufficient resolution. During the 1990s, the first peak was measured with increasing sensitivity, and in 2000 the BOOMERanG experiment[36] reported that the highest power fluctuations occur at scales of approximately one degree, showing that the Universe is close to flat. These measurements were able to rule out cosmic strings as the leading theory of cosmic structure formation, and suggested cosmic inflation was the correct theory.

Ground-based interferometers provided fluctuation measurements with higher accuracy, including the Very Small Array, the Degree Angular Scale Interferometer (DASI) and the Cosmic Background Imager (CBI). DASI first detected the CMB polarization,[37][38] and CBI provided the first E-mode polarization spectrum with compelling evidence that it is out of phase with the T-mode spectrum.[39] COBE's successor, the Wilkinson Microwave Anisotropy Probe (WMAP) provided the most detailed measurements of (large-scale) anisotropies in the CMB in 2003–2010.[40] ESA's Planck spacecraft returned more detailed results in 2013-2015.

WMAP's measurements played the key role in establishing the Standard Model of Cosmology, namely the Lambda-CDM model, which posits a dark energy-dominated flat universe, supplemented by dark matter and atoms with density fluctuations seeded by a Gaussian, adiabatic, nearly scale invariant process. Its basic properties are determined by six adjustable parameters: dark matter density, baryon (atom) density, the universe's age (or equivalently, the Hubble constant), the initial fluctuation amplitude and their scale dependence.

1.5.2 Observational evidence

Much of the evidence comes from the motions of galaxies.[42] Many of these appear to be fairly uniform, so

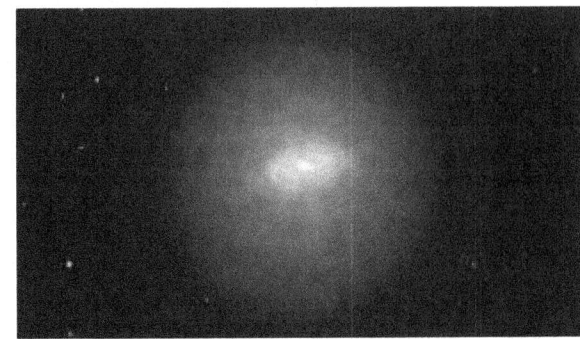

This artist's impression shows the expected distribution of dark matter in the Milky Way galaxy as a blue halo of material surrounding the galaxy.[41]

by the virial theorem, the total kinetic energy should be half the galaxies' total gravitational binding energy. Observationally, the total kinetic energy is much greater. In particular, assuming the gravitational mass is due to only visible matter, stars far from the center of galaxies have much higher velocities than predicted by the virial theorem. Galactic rotation curves, which illustrate the velocity of rotation versus the distance from the galactic center, show the "excess" velocity. Dark matter is the most straightforward way of accounting for this discrepancy.[43]

The distribution of dark matter in galaxies required to explain the motion of the observed matter suggests the presence of a roughly spherically symmetric, centrally concentrated halo of dark matter with the visible matter concentrated in a central disc.

Low surface brightness dwarf galaxies are important sources of information for studying dark matter. They have an uncommonly low ratio of visible to dark matter, and have few bright stars at the center that would otherwise impair observations of the rotation curve of outlying stars.

Gravitational lensing observations of galaxy clusters allow direct estimates of the gravitational mass based on its effect on light coming from background galaxies, since large collections of matter (dark or otherwise) gravitationally deflect light. In clusters such as Abell 1689, lensing observations confirm the presence of considerably more mass than is indicated by the clusters' light. In the Bullet Cluster, lensing observations show that much of the lensing mass is separated from the X-ray-emitting baryonic mass. In July 2012, lensing observations were used to identify a "filament" of dark matter between two clusters of galaxies, as cosmological simulations predicted.[44]

Galaxy rotation curves

Main article: Galaxy rotation curve

A galaxy rotation curve is a plot of the orbital velocities

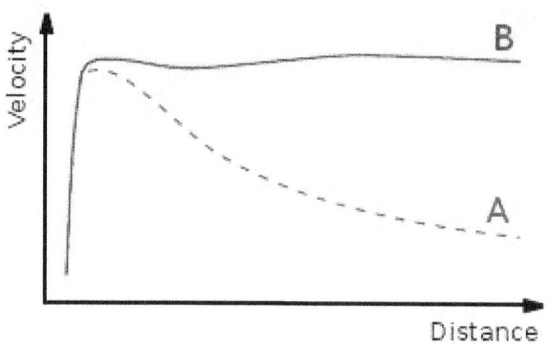

Rotation curve of a typical spiral galaxy: predicted (A) and ob-served (B). Dark matter can explain the 'flat' appearance of the velocity curve out to a large radius

(i.e., the speeds) of visible stars or gas in that galaxy versus their radial distance from that galaxy's center. The rotational/orbital speed of galaxies/stars does not decline with distance, unlike other orbital systems such as stars/planets and planets/moons that also have most of their mass at the centre. In the latter cases, this reflects the mass distributions within those systems. The mass observations for galaxies based on the light that they emit are far too low to explain the velocity observations. The dark matter hypothesis accounts for the missing mass, explaining the anomaly.[28]

A universal rotation curve can be expressed as the sum of an exponential distribution of visible matter that tapers to zero with distance from the center, and a spherical dark matter halo with a flat core of radius r_0 and density $\rho_0 = 4.5 \times 10^{-2}(r_0/\text{kpc})^{-2/3} \, M\odot\text{pc}^{-3}$.[45]

Low-surface-brightness (LSB) galaxies have a much larger visible mass deficit than others. This property simplifies the disentanglement of the dark and visible matter contributions to the rotation curves.[17]

Rotation curves for some elliptical galaxies do display low velocities for outlying stars (tracked for example by the motion of embedded planetary nebulae). A dark-matter compliant hypothesis proposes that some stars may have been torn by tidal forces from disk-galaxy mergers from their original galaxies during the first close passage and put on outgoing trajectories, explaining the low velocities of the remaining stars even in the presence of a halo.[17][46]

Velocity dispersions of galaxies

Velocity dispersion estimates of elliptical galaxies,[47] with some exceptions, generally indicate a relatively high dark matter content.

Diffuse interstellar gas measurements of galactic edges indicate missing ordinary matter beyond the visible boundary, but that galaxies are virialized (i.e., gravitationally bound and orbiting each other with velocities that correspond to predicted orbital velocities of general relativity) up to ten times their visible radii.[48] This has the effect of pushing up the dark matter as a fraction of the total matter from 50% as measured by Rubin to the now accepted value of nearly 95%.

Dark matter seems to be a small component or absent in some places. Globular clusters show little evidence of dark matter.[49] Star velocity profiles seemed to indicate a concentration of dark matter in the disk of the Milky Way. It now appears, however, that the high concentration of baryonic matter in the disk (especially in the interstellar medium) can account for this motion. Galaxy mass and light profiles appear to not match. The typical model for dark matter galaxies is a smooth, spherical distribution in virialized halos. This avoids small-scale (stellar) dynamical effects. A 2006 study explained the warp in the Milky Way's disk by the interaction of the Large and Small Magellanic Clouds and the 20-fold increase in predicted mass from dark matter.[50]

In 2005, astronomers claimed to have discovered a galaxy made almost entirely of dark matter, 50 million light years away in the Virgo Cluster, which was named VIRGOHI21.[51] Unusually, VIRGOHI21 does not appear to contain visible stars: it was discovered with radio frequency observations of hydrogen. Based on rotation profiles, scientists estimate that this object contains approximately 1000 times more dark matter than hydrogen and has a mass of about 1/10 that of the Milky Way. The Milky Way is estimated to have roughly 10 times as much dark matter as ordinary matter. Models of the Big Bang and structure formation suggested that such dark galaxies should be very common, but VIRGOHI21 was the first to be detected.

The velocity profiles of some galaxies such as NGC 3379 indicate an absence of dark matter.[52]

Galaxy clusters and gravitational lensing

Clusters of galaxies are particularly important for dark matter studies since their masses can be estimated in three independent ways:

- From the scatter in radial velocities of the galaxies within clusters

- From X-rays emitted by hot gas in the clusters. From the X-ray energy spectrum and flux, the gas temper-

Strong gravitational lensing as observed by the Hubble Space Telescope in Abell 1689 indicates the presence of dark matter—enlarge the image to see the lensing arcs.

Dark matter is invisible. Based on the effect of gravitational lensing, a ring of dark matter has been inferred in this image of a galaxy cluster (CL0024+17) and has been represented in blue.[54]

ature and density can be estimated, hence giving the pressure: assuming pressure and gravity balance determines the cluster's mass profile. Many Chandra X-ray Observatory experiments use this technique to independently determine cluster masses. These observations generally indicate that baryonic mass is approximately 12–15 percent, in reasonable agreement with the Planck spacecraft cosmic average of 15.5–16 percent.[53]

- Gravitational lensing (usually of more distant galaxies) can measure cluster masses without relying on observations of dynamics (e.g., velocity). There are two types of lensing: strong lensing produces multiple images or giant arcs near the cluster core, while weak lensing is observed as small shape distortions around the outer regions. Multiple Hubble projects have used this method to measure cluster masses.

Generally, these three methods are in reasonable agreement that dark matter outweighs visible matter by approximately 5 to 1.

Gravity acts as a lens to bend the light from a more distant source (such as a quasar) around a massive object (such as a cluster of galaxies) lying between the source and the observer in accordance with general relativity.

Strong lensing is the observed distortion of background galaxies into arcs when their light passes through such a

gravitational lens. It has been observed around many distant clusters including Abell 1689.[55] By measuring the distortion geometry, the mass of the intervening cluster can be obtained. In the dozens of cases where this has been done, the mass-to-light ratios obtained correspond to the dynamical dark matter measurements of clusters.[56]

Weak gravitational lensing investigates minute distortions of galaxies, using statistical analyses from vast galaxy surveys. By examining the apparent shear deformation of the adjacent background galaxies, astrophysicists can characterize the mean distribution of dark matter. The mass-to-light ratios correspond to dark matter densities predicted by other large-scale structure measurements.[57]

Galaxy cluster Abell 2029 comprises thousands of galaxies enveloped in a cloud of hot gas and dark matter equivalent to more than 10^{14} $M\odot$. At the center of this cluster is an enormous elliptical galaxy likely formed from many smaller galaxies.[58]

The most direct observational evidence comes from the Bullet Cluster. In most regions dark and visible matter are found together,[59] due to their gravitational attraction. In the Bullet Cluster however, the two matter types split apart, due to a past collision between two smaller clusters. Electromagnetic interactions between gas particles has caused the gas to slow and concentrate near the point of impact. The galaxies, stars and dark matter continued through with negligible collisions. Lensing observations show two dark matter peaks near the galaxy peaks, as expected in dark matter theory. Since the gas peaks contain more ordi-

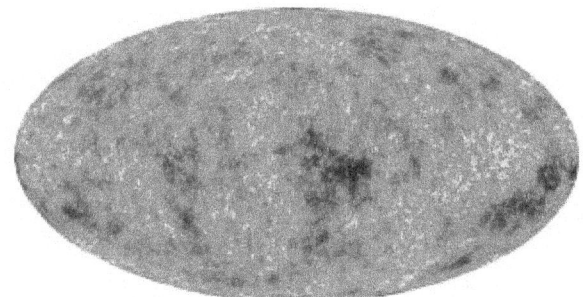

The cosmic microwave background by WMAP

The Bullet Cluster: HST image with overlays. The total projected mass distribution reconstructed from strong and weak gravitational lensing is shown in blue, while the X-ray emitting hot gas observed with Chandra is shown in red.

nary matter than the stars, modified-gravity theories should show the lensing peaks near the gas peaks, contrary to the observations.

X-ray observations show that much of the luminous matter (in the form of 10^7–10^8 Kelvin[60] gas or plasma) is concentrated in the cluster's center. Weak gravitational lensing observations show that much dark matter resides outside the central region. Unlike galactic rotation curves, this evidence is independent of the details of Newtonian gravity, directly supporting dark matter.[60] Dark matter's observed behavior constrains whether and how much it scatters off other dark matter particles, quantified as its self-interaction cross section. If dark matter has no pressure, it can be described as a perfect fluid that has no damping.[61] The distribution of mass in galaxy clusters has been used to argue both for[62][63] and against[64] the significance of self-interaction.

An ongoing survey using the Subaru Telescope uses weak lensing to analyze background light, bent by dark matter, to determine the statistical distribution of dark matter in the foreground. The survey studies galaxies more than a billion light-years distant, across an area greater than a thousand square degrees (about one fortieth of the entire sky).[65][66]

Cosmic microwave background

Main article: Cosmic microwave background
 Angular CMB fluctuations provide evidence for dark matter. The typical angular scales of CMB oscillations, measured as the power spectrum of the CMB anisotropies, reveal the different effects of baryonic and dark matter. Ordinary matter interacts strongly via radiation whereas dark matter particles (WIMPs) do not; both affect the oscilla-

tions by way of their gravity, so the two forms of matter have different effects.

The spectrum shows a large first peak and smaller successive peaks.[40] The first peak mostly shows the density of baryonic matter, while the third peak relates mostly to the density of dark matter, measuring the density of matter and the density of atoms.

Sky surveys and baryon acoustic oscillations

Main article: Baryon acoustic oscillations

The early universe's acoustic oscillations in the photon-baryon fluid are observed as the prominent acoustic peaks in the CMB spectrum. This set up a preferred length scale for baryons in the early universe which is determined as 147 megaparsec (comoving) by the Planck spacecraft. As the dark matter and baryons clumped together after recombination, the effect is much weaker in the galaxy distribution in the nearby universe, but is detectable as a subtle (~ 1 percent) preference for pairs of galaxies to be separated by 147 Mpc, rather than 130 or 160 Mpc, called the BAO feature. This feature was predicted theoretically in the 1990s and then discovered in 2005, in two large galaxy redshift surveys, the Sloan Digital Sky Survey and the 2dF Galaxy Redshift Survey.[67] Combining the CMB observations with BAO measurements from galaxy redshift surveys provides a precise estimate of the Hubble constant and the average matter density in the Universe.[34]

Redshift-space distortions

Large galaxy redshift surveys may be used to make a three-dimensional map of the galaxy distribution. These maps are slightly distorted because distances are estimated from observed redshifts; the redshift contains a contribution from the galaxy's so-called peculiar velocity in addition to the dominant Hubble expansion term. On average, superclusters are expanding but more slowly than the cosmic mean due to their gravity, while voids are expanding faster than average. In a redshift map, galaxies in front of a supercluster have excess radial velocities towards it and have redshifts slightly higher than their distance would imply, while galaxies behind the supercluster have redshifts slightly low for their distance. This effect causes superclusters to appear "squashed" in the radial direction, and likewise voids are "stretched"; angular positions are unaffected. The effect is not detectable for any one structure since the true shape is not known, but can be measured by averaging over many structures assuming we are not at a special location in the Universe.

The effect was predicted quantitatively by Nick Kaiser in 1987, and first decisively measured in 2001 by the 2dF Galaxy Redshift Survey.[68] Results are in agreement with the Lambda-CDM model.

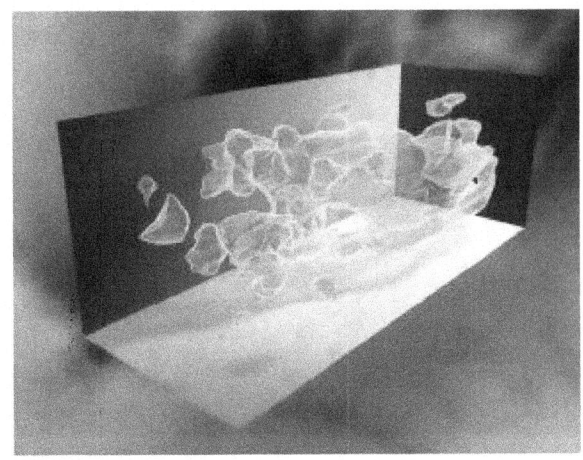

3D map of the large-scale distribution of dark matter, reconstructed from measurements of weak gravitational lensing with the Hubble Space Telescope.[71]

Type Ia supernova distance measurements

Main article: Type Ia supernova

Type Ia supernovae can be used as "standard candles" to measure extragalactic distances. Extensive data sets of these supernovae can be used to constrain cosmological models.[69] They constrain the dark energy density $\Omega\Lambda = $ ~0.713 for a flat, Lambda CDM universe and the parameter w for a quintessence model. The results are roughly consistent with those derived from the WMAP observations and further constrain the Lambda CDM model and (indirectly) dark matter.[34]

Lyman-alpha forest

Main article: Lyman-alpha forest

In astronomical spectroscopy, the Lyman-alpha forest is the sum of the absorption lines arising from the Lyman-alpha transition of neutral hydrogen in the spectra of distant galaxies and quasars. Lyman-alpha forest observations can also constrain cosmological models.[70] These constraints agree with those obtained from WMAP data.

Structure formation

Main article: Structure formation
Structure formation refers to the serial transformations of the universe following the Big Bang. Prior to structure formation, e.g., Friedmann cosmology solutions to general relativity describe a homogeneous universe. Later, small anisotropies gradually grew and condensed the homogeneous universe into stars, galaxies and larger structures.

Observations suggest that structure formation proceeds hierarchically, with the smallest structures collapsing first, followed by galaxies and then galaxy clusters. As the structures collapse in the evolving universe, they begin to "light up" as baryonic matter heats up through gravitational contraction and approaches hydrostatic pressure balance.

CMB anisotropy measurements fix models in which most matter is dark. Dark matter also closes gaps in models of large-scale structure. The dark matter hypothesis corresponds with statistical surveys of the visible structure and precisely to CMB predictions.

Initially, baryonic matter's post-Big Bang temperature and pressure were too high to collapse and form smaller structures, such as stars, via the Jeans instability. The gravity from dark matter increase the compaction force, allowing the formation of these structures.

Computer simulations of billions of dark matter particles[72] seem to confirm that the "cold" dark matter model of structure formation is consistent with the structures observed through galaxy surveys, such as the Sloan Digital Sky Survey and 2dF Galaxy Redshift Survey, as well as observations of the Lyman-alpha forest.

Tensions separate observations and simulations. Observations have turned up 90-99% fewer small galaxies than permitted by dark matter-based predictions.[73][74] In addition, simulations predict dark matter distributions with a dense cusp near galactic centers, but the observed halos are smoother than predicted.

1.5.3 Composition

The composition of dark matter remains uncertain. Possibilities include dense baryonic (interacts with electromagnetic force) matter and non-baryonic matter (interacts with its surroundings only through gravity).

Baryonic vs. nonbaryonic matter

Fermi-LAT observations of dwarf galaxies provide new insights on dark matter.

Baryonic matter   Baryonic matter is made of baryons (protons and neutrons) that make up stars and planets. It also encompasses less common black holes, neutron stars, faint old white dwarfs and brown dwarfs, collectively known as massive compact halo objects (MACHOs).

Multiple lines of evidence suggest the majority of dark matter is not made of baryons:

- Sufficient diffuse, baryonic gas or dust would be visible when backlit by stars.

- The theory of Big Bang nucleosynthesis predicts the observed abundance of the chemical elements;[75][76] agreement with observed abundances requires that baryonic matter makes up between 4–5% of the universe's critical density. In contrast, large-scale structure and other observations indicate that the total matter density is about 30% of the critical density (with dark energy providing the remaining 70%).

- Large astronomical searches for gravitational microlensing in the Milky Way found that at most a small fraction of the dark matter may be in dark, compact, conventional objects (MACHOs, etc.); the excluded range of object masses is from half the Earth's mass up to 30 solar masses, which covers nearly all the plausible candidates.[77][78][79][80][81][82]

- Detailed analysis of the small irregularities (anisotropies) in the cosmic microwave background observed by WMAP and Planck shows that around five-sixths of the total matter is in a form that interacts significantly with ordinary matter or photons only through gravitational effects.

Non-baryonic matter   Candidates for nonbaryonic dark matter are hypothetical particles such as axions or supersymmetric particles. The three neutrino types already observed are indeed abundant, and "dark", and matter, but because their individual masses – however uncertain they may be – are almost certainly tiny, they can only supply a small fraction of dark matter, due to limits derived from large-scale structure and high-redshift galaxies.[83]

Unlike baryonic matter, nonbaryonic matter did not contribute to the formation of the elements in the early universe ("Big Bang nucleosynthesis")[113] and so its presence is revealed only via its gravitational effects. In addition, if the particles of which it is composed are supersymmetric, they can undergo annihilation interactions with themselves, possibly resulting in observable by-products such as gamma rays and neutrinos ("indirect detection").[83]

Classification: cold/warm/hot

Dark matter can be divided into *cold*, *warm* and *hot* categories.[84] These categories refer to velocity rather than an actual temperature, indicating how far corresponding objects moved due to random motions in the early universe, before they slowed due to cosmic expansion – this is an important distance called the "free streaming length" (FSL). Primordial density fluctuations smaller than this length get washed out as particles spread from overdense to underdense regions, while larger fluctuations are unaffected; therefore this length sets a minimum scale for later structure formation. The categories are set with respect to the size of a protogalaxy (an object that later evolves into a dwarf galaxy): dark matter particles are classified as cold, warm, or hot according as their FSL: much smaller (cold), similar (warm), or much larger (hot) than a protogalaxy.[85][86]

Mixtures of the above are also possible: a theory of mixed dark matter was popular in the mid-1990s, but was rejected following the discovery of dark energy.

Cold dark matter leads to a "bottom-up" formation of structure while hot dark matter would result in a "top-down" formation scenario; the latter is excluded by high-redshift galaxy observations.[17]

Alternative definitions   These categories also correspond to fluctuation spectrum effects and the interval following the Big Bang at which each type became non-relativistic. Davis *et al.* wrote in 1985:

> Candidate particles can be grouped into three categories on the basis of their effect on the fluctuation spectrum (Bond *et al.* 1983). If the dark matter is composed of abundant light particles which remain relativistic until shortly before recombination, then it may be termed "hot". The best candidate for hot dark matter is a neutrino ... A second possibility is for the dark matter particles to interact more weakly than neutrinos, to be less abundant, and to have a mass of order 1 keV. Such particles are termed "warm dark matter", because they have lower thermal velocities than massive neutrinos ... there are at present few candidate particles which fit this description. Gravitinos and photinos have been suggested (Pagels and Primack 1982; Bond, Szalay and Turner 1982) ... Any particles which became nonrelativistic very early, and so were able to diffuse a negligible distance, are termed "cold" dark matter (CDM). There are many candidates for CDM including supersymmetric particles.[87]

Another approximate dividing line is that "warm" dark matter became non-relativistic when the universe was approximately 1 year old and 1 millionth of its present size and in the radiation-dominated era (photons and neutrinos), with a photon temperature 2.7 million K. Standard physical cosmology gives the particle horizon size as $2ct$ (speed of light multiplied by time) in the radiation-dominated era, thus 2 light-years. A region of this size would expand to 2 million light years today (absent structure formation). The actual FSL is roughly 5 times the above length, since it continues to grow slowly as particle velocities decrease inversely with the scale factor after they become non-relativistic. In this example the FSL would correspond to 10 million light-years or 3 Mpc today, around the size containing an average large galaxy.

The 2.7 million K photon temperature gives a typical photon energy of 250 electron-volts, thereby setting a typical mass scale for "warm" dark matter: particles much more massive than this, such as GeV – TeV mass WIMPs, would become non-relativistic much earlier than 1 year after the Big Bang and thus have FSLs much smaller than a protogalaxy, making them "cold". Conversely, much lighter particles, such as neutrinos with masses of only a few eV, have FSLs much larger than a protogalaxy, thus qualifying them as "hot".

Cold dark matter

Main article: Cold dark matter

"Cold" dark matter offers the simplest explanation for most cosmological observations. It is dark matter composed of constituents with an FSL much smaller than a protogalaxy. This is the focus for dark matter research, as hot dark matter does not seem to be capable of supporting galaxy or galaxy cluster formation, and most particle candidates slowed early.

The constituents of "cold" dark matter are unknown. Possibilities range from large objects like MACHOs (such as black holes[88]) or RAMBOs (such as clusters of brown dwarfs), to new particles such as WIMPs and axions.

Studies of Big Bang nucleosynthesis and gravitational lensing convinced most cosmologists[17][89][90][91][92][93] that MACHOs[89][91] cannot make up more than a small fraction of dark matter.[13][89] According to A. Peter: "... the only *really plausible* dark-matter candidates are new particles."[90]

The 1997 DAMA/NaI experiment and its successor DAMA/LIBRA in 2013, claimed to directly detect dark matter particles passing through the Earth, but many researchers remain skeptical, as negative results from similar experiments seem incompatible with the DAMA results.

Many supersymmetric models offer dark matter candidates in the form of the WIMPy Lightest Supersymmetric Particle (LSP).[94] Separately, heavy sterile neutrinos exist in non-supersymmetric extensions to the standard model that explain the small neutrino mass through the seesaw mechanism.

Warm dark matter

Main article: Warm dark matter

"Warm" dark matter refers to particles with an FSL comparable to the size of a protogalaxy. Predictions based on warm dark matter are similar to those for cold dark matter on large scales, but with less small-scale density perturbations. This reduces the predicted abundance of dwarf galaxies and may lead to lower density of dark matter in the central parts of large galaxies; some researchers consider this to be a better fit to observations. A challenge for this model is the lack of particle candidates with the required mass ~ 300 eV to 3000 eV.

No known particles can be categorized as "warm" dark matter. A postulated candidate is the sterile neutrino: a heavier, slower form of neutrino that does not interact through the weak force, unlike other neutrinos. Some modified gravity theories, such as scalar-tensor-vector gravity, require "warm" dark matter to make their equations work.

Hot dark matter

Main article: Hot dark matter

"Hot" dark matter consists of particles whose FSL is much larger than the size of a protogalaxy. The neutrino qualifies as such particle. They were discovered independently, long before the hunt for dark matter: they were postulated in 1930, and detected in 1956. Neutrinos' mass is less than 10^{-6} that of an electron. Neutrinos interact with normal matter only via gravity and the weak force, making them difficult to detect (the weak force only works over a small distance, thus a neutrino triggers a weak force event only if it hits a nucleus head-on). This makes them 'weakly interacting light particles' (WILPs), as opposed to WIMPs.

The three known flavours of neutrinos are the *electron*, *muon*, and *tau*. Their masses are slightly different. Neutrinos oscillate among the flavours as they move. It is hard to determine an exact upper bound on the collective average mass of the three neutrinos (or for any of the three individually). For example, if the average neutrino mass were over 50 eV/c^2 (less than 10^{-5} of the mass of an electron), the universe would collapse. CMB data and other methods indicate that their average mass probably does not exceed 0.3 eV/c^2. Thus, observed neutrinos cannot explain dark matter.[95]

Because galaxy-size density fluctuations get washed out by free-streaming, "hot" dark matter implies that the first objects that can form are huge supercluster-size pancakes, which then fragment into galaxies. Deep-field observations show instead that galaxies formed first, followed by clusters and superclusters as galaxies clump together.

1.5.4   Detection

If dark matter is made up of WIMPs, then millions, possibly billions, of WIMPs must pass through every square centimeter of the Earth each second.[96][97] Many experiments aim to test this hypothesis. Although WIMPs are popular search candidates,[17] the Axion Dark Matter eXperiment (ADMX) searches for axions. Another candidate is heavy hidden sector particles that only interact with ordinary matter via gravity.

These experiments can be divided into two classes: direct detection experiments, which search for the scattering of dark matter particles off atomic nuclei within a detector; and indirect detection, which look for the products of

WIMP annihilations.[83]

Direct detection

Direct detection experiments operate deep underground to reduce the interference from cosmic rays. Detectors include the Stawell mine, the Soudan mine, the SNOLAB underground laboratory at Sudbury, the Gran Sasso National Laboratory, the Canfranc Underground Laboratory, the Boulby Underground Laboratory, the Deep Underground Science and Engineering Laboratory and the Particle and Astrophysical Xenon Detector.

These experiments mostly use either cryogenic or noble liquid detector technologies. Cryogenic detectors operating at temperatures below 100mK, detect the heat produced when a particle hits an atom in a crystal absorber such as germanium. Noble liquid detectors detect scintillation produced by a particle collision in liquid xenon or argon. Cryogenic detector experiments include: CDMS, CRESST, EDELWEISS, EURECA. Noble liquid experiments include ZEPLIN, XENON, DEAP, ArDM, WARP, DarkSide, PandaX, and LUX, the Large Underground Xenon experiment. Both of these techniques distinguish background particles (that scatter off electrons) from dark matter particles (that scatter off nuclei). Other experiments include SIMPLE and PICASSO.

The DAMA/NaI, DAMA/LIBRA experiments detected an annual modulation in the event rate[98] that they claim is due to dark matter. (As the Earth orbits the Sun, the velocity of the detector relative to the dark matter halo will vary by a small amount). This claim is so far unconfirmed and unreconciled with negative results of other experiments.[99]

Directional detection is a search strategy based on the motion of the Solar System around the Galactic Center.[100][101][102][103]

A low pressure time projection chamber makes it possible to access information on recoiling tracks and constrain WIMP-nucleus kinematics. WIMPs coming from the direction in which the Sun is travelling (roughly towards Cygnus) may then be separated from background, which should be isotropic. Directional dark matter experiments include DMTPC, DRIFT, Newage and MIMAC.

Results   In 2009, CDMS researchers reported two possible WIMP candidate events. They estimate that the probability that these events are due to background (neutrons or misidentified beta or gamma events) is 23%, and conclude "this analysis cannot be interpreted as significant evidence for WIMP interactions, but we cannot reject either event as signal."[104]

In 2011, researchers using the CRESST detectors presented evidence of 67 collisions occurring in detector crystals from subatomic particles.[105] They calculated the probability that all were caused by known sources of interference/contamination was 1 in 10^5.

Indirect detection

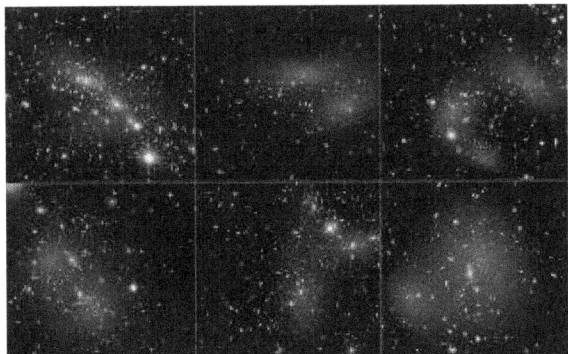

Collage of six cluster collisions with dark matter maps. The clusters were observed in a study of how dark matter in clusters of galaxies behaves when the clusters collide.[106]

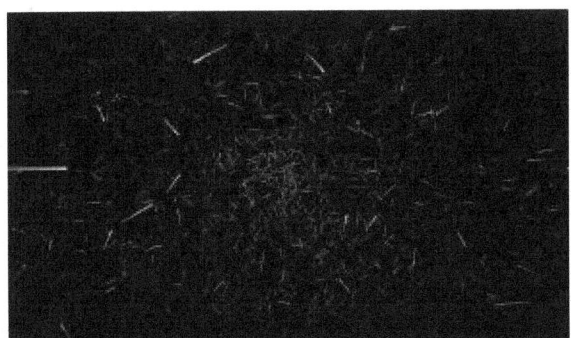

Video about the potential gamma-ray detection of dark matter annihilation around supermassive black holes. (Duration 3:13, also see file description.)

Indirect detection experiments search for the products of WIMP annihilation/decay. If WIMPs are Majorana particles (their own antiparticle) then two WIMPs could annihilate to produce gamma rays or Standard Model particle-antiparticle pairs. If the WIMP is unstable, WIMPs could decay into standard model (or other) particles. These processes could be detected indirectly through an excess of gamma rays, antiprotons or positrons emanating from high density regions. The detection of such a signal is not conclusive evidence, as the sources of gamma ray production are not fully understood.[17][83]

A few of the WIMPs passing through the Sun or Earth may scatter off atoms and lose energy. Thus WIMPs may accumulate at the center of these bodies, increasing the

chance of collision/annihilation. This could produce a distinctive signal in the form of high-energy neutrinos.[107] Such a signal would be strong indirect proof of WIMP dark matter.[17] High-energy neutrino telescopes such as AMANDA, IceCube and ANTARES are searching for this signal.

WIMP annihilation from the Milky Way galaxy as a whole may also be detected in the form of various annihilation products.[108] The Galactic Center is a particularly good place to look because the density of dark matter may be higher there.[109]

The recent detection by LIGO in February 2016, of gravity waves, opens the possibility of observing dark matter in a new way. Dark matter seems to have no effects except gravitational, and so the actual observation of gravitational waves provides scientists with a new way of observing the phenomenon.[110]

Results The EGRET gamma ray telescope observed more gamma rays in 2008 than expected from the Milky Way, but scientists concluded that this was most likely due to incorrect estimation of the telescope's sensitivity.[111]

The Fermi Gamma-ray Space Telescope is searching for similar gamma rays.[112] In April 2012, an analysis of previously available data from its Large Area Telescope instrument produced statistical evidence of a 130 GeV signal in the gamma radiation coming from the center of the Milky Way.[113] WIMP annihilation was seen as the most probable explanation.[114]

At higher energies, ground-based gamma-ray telescopes have set limits on the annihilation of dark matter in dwarf spheroidal galaxies[115] and in clusters of galaxies.[116]

The PAMELA experiment (launched 2006) detected excess positrons. They could be from dark matter annihilation or from pulsars. No excess antiprotons were observed.[117]

In 2013 results from the Alpha Magnetic Spectrometer on the International Space Station indicated excess high-energy cosmic rays that could be due to dark matter annihilation.[118][119][120][121][122][123]

1.5.5 Synthesis

An alternative approach to the detection of WIMPs in nature is to produce them in a laboratory. Experiments with the Large Hadron Collider (LHC) may be able to detect WIMPs produced in collisions of the LHC proton beams. Because a WIMP has negligible interaction with matter, it may be detected indirectly as (large amounts of) missing energy and momentum that escape the detectors, provided other (non-negligible) collision products are detected.[124]

These experiments could show that WIMPs can be formed, but a direct detection experiment must still show that they exist in sufficient numbers to account for dark matter.

1.5.6 Alternative theories

Mass in extra dimensions

In some multidimensional theories, the force of gravity is the only force with effect across all dimensions.[125] This explains the relative weakness of gravity compared to the other forces of nature that cannot cross into extra dimensions. In that case, dark matter could exist in a "Hidden Valley" in other dimensions that only interact with the matter in our dimensions through gravity. That dark matter could potentially aggregate in the same way as ordinary matter, forming other-dimensional galaxies.[12][126]

Topological defects

Dark matter could consist of primordial defects ("birth defects") in the topology of quantum fields, which would contain energy and therefore gravitate. This hypothesis may be investigated by the use of an orbital network of atomic clocks that would register the passage of topological defects by changes to clock synchronization. The Global Positioning System may be able to operate as such a network.[127]

Modified gravity

Some theories modify the laws of gravity. The earliest was Mordehai Milgrom's Modified Newtonian Dynamics (MOND) in 1983, which adjusts Newton's laws to increase gravitational field strength where gravitational acceleration becomes tiny (such as near the rim of a galaxy). It had some success explaining rotational velocity curves of elliptical and dwarf elliptical galaxies, but not galaxy cluster gravitational lensing. MOND was not relativistic: it was an adjustment of the Newtonian account. Attempts were made to bring MOND into conformity with general relativity; this spawned competing MOND-based hypotheses—including TeVeS, MOG or STV gravity, and the phenomenological covariant approach.[128]

In 2007, John Moffat proposed a modified gravity hypothesis based on nonsymmetric gravitational theory (NGT) that claims to account for the behavior of colliding galaxies.[129] This model requires the presence of non-relativistic neutrinos or other cold dark matter, to work.

Another proposal uses a gravitational backreaction from a theory that explains gravitational force between objects as an action, a reaction and then a back-reaction. Thus, an object A affects an object B, and the object B then re-affects object A, and so on, creating a feedback loop that strengthens gravity.[130]

In 2008, a group proposed "dark fluid", a modification of large-scale gravity. It hypothesized that attractive gravitational effects are instead a side-effect of dark energy. Dark fluid combines dark matter and dark energy in a single energy field that produces different effects at different scales. This treatment is a simplification of a previous fluid-like model called the generalized Chaplygin gas model in which the whole of spacetime is a compressible gas.[131] Dark fluid can be compared to an atmospheric system. Atmospheric pressure causes air to expand and air regions can collapse to form clouds. In the same way, the dark fluid might generally disperse, while collecting around galaxies.[131]

Spacetime fractality

Applying relativity to fractal, non-differentiable spacetime. Nottale suggests that potential energy may arise due to the fractality of spacetime, which would account for the missing mass-energy observed at cosmological scales.[132][133]

1.5.7 Popular culture

Main article: Dark matter in fiction

Mention of dark matter is made in some video games and other works of fiction. In such cases, it is usually attributed extraordinary physical or magical properties. Such descriptions are often inconsistent with the hypothesized properties of dark matter in physics and cosmology.

1.5.8 See also

- Chameleon particle
- Conformal gravity
- DEAP, a dark matter experiment
- DAMPE, a space mission
- General Antiparticle Spectrometer
- Illustris project, astrophysical simulations
- Light dark matter
- Mirror matter
- Multidark, a research program
- Scalar field dark matter

- Self-interacting dark matter

- SIMP, hypothetical particles of dark matter

- Unparticle physics

1.5.9   Notes

[1] Since dark energy, by convention, does not count as "matter", this is 26.8/(4.9 + 26.8)=0.845

1.5.10   References

[1] "Planck Mission Brings Universe Into Sharp Focus". *NASA Mission Pages*. 21 March 2013.

[2] "Dark Energy, Dark Matter". *NASA Science: Astrophysics*. 5 June 2015.

[3] "Dark Matter". *CERN Physics*. 20 January 2012.

[4] Trimble, V. (1987). "Existence and nature of dark matter in the universe". *Annual Review of Astronomy and Astrophysics* **25**: 425–472. Bibcode:1987ARA&A..25..425T. doi:10.1146/annurev.aa.25.090187.002233.

[5] Ade, P. A. R.; Aghanim, N.; Armitage-Caplan, C.; (Planck Collaboration); et al. (22 March 2013). "Planck 2013 results. I. Overview of products and scientific results – Table 9". *Astronomy and Astrophysics* **1303**: 5062. arXiv:1303.5062. Bibcode:2014A&A...571A...1P. doi:10.1051/0004-6361/201321529.

[6] Francis, Matthew (22 March 2013). "First Planck results: the Universe is still weird and interesting". *Arstechnica*.

[7] "Planck captures portrait of the young Universe, revealing earliest light". University of Cambridge. 21 March 2013. Retrieved 21 March 2013.

[8] Sean Carroll, Ph.D., Cal Tech, 2007, The Teaching Company, *Dark Matter, Dark Energy: The Dark Side of the Universe*, Guidebook Part 2 page 46, Accessed Oct. 7, 2013. "...dark matter: An invisible, essentially collisionless component of matter that makes up about 25 percent of the energy density of the universe... it's a different kind of particle... something not yet observed in the laboratory..."

[9] Ferris, Timothy. "Dark Matter". Retrieved 2015-06-10.

[10] Jarosik, N.; et al. (2011). "Seven-Year Wilson Microwave Anisotropy Probe (WMAP) Observations: Sky Maps, Systematic Errors, and Basic Results". *Astrophysical Journal Supplement* **192** (2): 14. arXiv:1001.4744. Bibcode:2011ApJS..192...14J. doi:10.1088/0067-0049/192/2/14.

[11] Persic, Massimo; Salucci, Paolo (1992-09-01). "The baryon content of the Universe". *Monthly Notices of the Royal Astronomical Society* **258** (1): 14P–18P. arXiv:astro-ph/0502178. Bibcode:1992MNRAS.258P..14P. doi:10.1093/mnras/258.1.14P. ISSN 0035-8711.

[12] Siegfried, T. (5 July 1999). "Hidden Space Dimensions May Permit Parallel Universes, Explain Cosmic Mysteries". *The Dallas Morning News*.

[13] Copi, C. J.; Schramm, D. N.; Turner, M. S. (1995). "Big-Bang Nucleosynthesis and the Baryon Density of the Universe". *Science* **267** (5195): 192–199. arXiv:astro-ph/9407006. Bibcode:1995Sci...267..192C. doi:10.1126/science.7809624. PMID 7809624.

[14] Kroupa, P.; et al. (2010). "Local-Group tests of dark-matter Concordance Cosmology: Towards a new paradigm for structure formation". *Astronomy and Astrophysics* **523**: 32–54. arXiv:1006.1647. Bibcode:2010A&A...523A..32K. doi:10.1051/0004-6361/201014892.

[15] Conformal theory: New light on dark matter, dark energy, and dark galactic halos." (PDF) Robert K. Nesbet. IBM Almaden Research Center, June 17, 2014.

[16] Angus, G. (2013). "Cosmological simulations in MOND: the cluster scale halo mass function with light sterile neutrinos". *Monthly Notices of the Royal Astronomical Society* **436**: 202–211. arXiv:1309.6094. Bibcode:2013MNRAS.436..202A. doi:10.1093/mnras/stt1564.

[17] Bertone, G.; Hooper, D.; Silk, J. (2005). "Particle dark matter: Evidence, candidates and constraints". *Physics Reports* **405** (5–6): 279–390. arXiv:hep-ph/0404175. Bibcode:2005PhR...405..279B. doi:10.1016/j.physrep.2004.08.031.

[18] Kapteyn, Jacobus Cornelius (1922). "First attempt at a theory of the arrangement and motion of the sidereal system". *Astrophysical Journal* **55**: 302–327. Bibcode:1922ApJ....55..302K. doi:10.1086/142670. It is incidentally suggested that when the theory is perfected it may be possible to determine *the amount of dark matter* from its gravitational effect. (emphasis in original)

[19] Rosenberg, Leslie J (30 June 2014). *Status of the Axion Dark-Matter Experiment (ADMX)* (PDF). 10th PATRAS Workshop on Axions, WIMPs and WISPs. p. 2.

[20] Oort, J.H. (1932) "The force exerted by the stellar system in the direction perpendicular to the galactic plane and some related problems." *Bulletin of the Astronomical Institutes of the Netherlands*. **6** : 249-287.

[21] "The Hidden Lives of Galaxies: Hidden Mass". *Imagine the Universe!*. NASA/GSFC.

[22] Kuijken, K.; Gilmore, G. (July 1989). "The Mass Distribution in the Galactic Disc - Part III - the Local Volume Mass Density" (PDF). *Monthly Notices of the Royal Astronomical Society* **239** (2): 651–664. Bibcode:1989MNRAS.239..651K. doi:10.1093/mnras/239.2.651.

[23] Zwicky, F. (1933). "Die Rotverschiebung von extragalaktischen Nebeln". *Helvetica Physica Acta* **6**: 110–127. Bibcode:1933AcHPh...6..110Z.

[24] Zwicky, F. (1937). "On the Masses of Nebulae and of Clusters of Nebulae". *The Astrophysical Journal* **86**: 217. Bibcode:1937ApJ....86..217Z. doi:10.1086/143864.

[25] Zwicky, F. (1933). "Die Rotverschiebung von extragalaktischen Nebeln". *Helvetica Physica Acta* **6**: 110–127. Bibcode:1933AcHPh...6..110Z See also Zwicky, F. (1937). "On the Masses of Nebulae and of Clusters of Nebulae", *Astrophysical Journal* **86**: 217. Bibcode:1937ApJ....86..217Z. doi:10.1086/143864

[26] Some details of Zwicky's calculation and of more modern values are given in Richmond, M., *Using the virial theorem: the mass of a cluster of galaxies*, retrieved 2007-07-10

[27] Freese, Katherine (4 May 2014). *The Cosmic Cocktail: Three Parts Dark Matter*. Princeton University Press. ISBN 978-1-4008-5007-5.

[28] Babcock, H, 1939, "The rotation of the Andromeda Nebula", Lick Observatory bulletin ; no. 498

[29] First observational evidence of dark matter. Darkmatterphysics.com. Retrieved 6 August 2013.

[30] Rubin, Vera C.; Ford, W. Kent, Jr. (February 1970). "Rotation of the Andromeda Nebula from a Spectroscopic Survey of Emission Regions". *The Astrophysical Journal* **159**: 379–403. Bibcode:1970ApJ...159..379R. doi:10.1086/150317.

[31] Bosma, A. (1978). "The distribution and kinematics of neutral hydrogen in spiral galaxies of various morphological types" (Ph.D. Thesis). Rijksuniversiteit Groningen.

[32] Rubin, V.; Thonnard, W. K. Jr.; Ford, N. (1980). "Rotational Properties of 21 Sc Galaxies with a Large Range of Luminosities and Radii from NGC 4605 ($R = 4$kpc) to UGC 2885 ($R = 122$kpc)". *The Astrophysical Journal* **238**: 471. Bibcode:1980ApJ...238..471R. doi:10.1086/158003.

[33] Bergstrom, L. (2000). "Non-baryonic dark matter: Observational evidence and detection methods". *Reports on Progress in Physics* **63** (5): 793–841. arXiv:hep-ph/0002126. Bibcode:2000RPPh...63..793B. doi:10.1088/0034-4885/63/5/2r3.

[34] Komatsu, E.; et al. (2009). "Five-Year Wilkinson Microwave Anisotropy Probe Observations: Cosmological Interpretation". *The Astrophysical Journal Supplement* **180** (2): 330–376. arXiv:0803.0547. Bibcode:2009ApJS..180..330K. doi:10.1088/0067-0049/180/2/330.

[35] Boggess, N. W.; et al. (1992). "The COBE Mission: Its Design and Performance Two Years after the launch". *The Astrophysical Journal* **397**: 420. Bibcode:1992ApJ...397..420B. doi:10.1086/171797.

[36] Melchiorri, A.; et al. (2000). "A Measurement of Ω from the North American Test Flight of Boomerang". *The Astrophysical Journal Letters* **536** (2): L63–L66. arXiv:astro-ph/9911445. Bibcode:2000ApJ...536L..63M. doi:10.1086/312744.

[37] Leitch, E. M.; et al. (2002). "Measurement of polarization with the Degree Angular Scale Interferometer". *Nature* **420** (6917): 763–771. arXiv:astro-ph/0209476. Bibcode:2002Natur.420..763L. doi:10.1038/nature01271. PMID 12490940.

[38] Leitch, E. M.; et al. (2005). "Degree Angular Scale Interferometer 3 Year Cosmic Microwave Background Polarization Results". *The Astrophysical Journal* **624** (1): 10–20. arXiv:astro-ph/0409357. Bibcode:2005ApJ...624...10L. doi:10.1086/428825.

[39] Readhead, A. C. S.; et al. (2004). "Polarization Observations with the Cosmic Background Imager". *Science* **306** (5697): 836–844. arXiv:astro-ph/0409569. Bibcode:2004Sci...306..836R. doi:10.1126/science.1105598. PMID 15472038.

[40] Hinshaw, G.; et al. (2009). "Five-Year Wilkinson Microwave Anisotropy Probe Observations: Data Processing, Sky Maps, and Basic Results". *The Astrophysical Journal Supplement* **180** (2): 225–245. arXiv:0803.0732. Bibcode:2009ApJS..180..225H. doi:10.1088/0067-0049/180/2/225.

[41] "Serious Blow to Dark Matter Theories?" (Press release). European Southern Observatory. 18 April 2012.

[42] Freeman, K.; McNamara, G. (2006). *In Search of Dark Matter*. Birkhäuser. p. 37. ISBN 0-387-27616-5.

[43] Randall, Lisa (2015). *Dark matter and the dinosaurs: The astounding interconnectedness of the universe*. Harper Collins Publishers. ISBN 978-0-06-232847-2.

[44] Jörg, D.; et al. (2012). "A filament of dark matter between two clusters of galaxies". *Nature* **487** (7406): 202–204. arXiv:1207.0809. Bibcode:2012Natur.487..202D. doi:10.1038/nature11224.

[45] Salucci, P.; Borriello, A. (2003). "The Intriguing Distribution of Dark Matter in Galaxies". *Lecture Notes in Physics*. Lecture Notes in Physics **616**: 66–77. arXiv:astro-ph/0203457. Bibcode:2003LNP...616...66S. doi:10.1007/3-540-36539-7_5. ISBN 978-3-540-00711-1.

[46] Dekel, A.; et al. (2005). "Lost and found dark matter in elliptical galaxies". *Nature* **437** (7059): 707–710. arXiv:astro-ph/0501622. Bibcode:2005Natur.437..707D. doi:10.1038/nature03970. PMID 16193046.

[47] Faber, S. M.; Jackson, R. E. (1976). "Velocity dispersions and mass-to-light ratios for elliptical galaxies". *The Astrophysical Journal* **204**: 668–683. Bibcode:1976ApJ...204..668F. doi:10.1086/154215.

[48] Collins, G. W. (1978). "The Virial Theorem in Stellar Astrophysics". Pachart Press.

[49] Rejkuba, M.; Dubath, P.; Minniti, D.; Meylan, G. (2008). "Masses and M/L Ratios of Bright Globular Clusters in NGC 5128". *Proceedings of the International Astronomical Union* **246**: 418–422. Bibcode:2008IAUS..246..418R. doi:10.1017/S1743921308016074.

[50] Weinberg, M. D.; Blitz, L. (2006). "A Magellanic Origin for the Warp of the Galaxy". *The Astrophysical Journal Letters* **641** (1): L33–L36. arXiv:astro-ph/0601694. Bibcode:2006ApJ...641L..33W. doi:10.1086/503607.

[51] Minchin, R.; et al. (2005). "A Dark Hydrogen Cloud in the Virgo Cluster". *The Astrophysical Journal Letters* **622**: L21–L24. arXiv:astro-ph/0502312. Bibcode:2005ApJ...622L..21M. doi:10.1086/429538.

[52] Ciardullo, R.; Jacoby, G. H.; Dejonghe, H. B. (1993). "The radial velocities of planetary nebulae in NGC 3379". *The Astrophysical Journal* **414**: 454–462. Bibcode:1993ApJ...414..454C. doi:10.1086/173092.

[53] Vikhlinin, A.; et al. (2006). "Chandra Sample of Nearby Relaxed Galaxy Clusters: Mass, Gas Fraction, and Mass–Temperature Relation". *The Astrophysical Journal* **640** (2): 691–709. arXiv:astro-ph/0507092. Bibcode:2006ApJ...640..691V. doi:10.1086/500288.

[54] "Hubble Finds Dark Matter Ring in Galaxy Cluster".

[55] Taylor, A. N.; et al. (1998). "Gravitational Lens Magnification and the Mass of Abell 1689". *The Astrophysical Journal* **501** (2): 539–553. arXiv:astro-ph/9801158. Bibcode:1998ApJ...501..539T. doi:10.1086/305827.

[56] Wu, X.; Chiueh, T.; Fang, L.; Xue, Y. (1998). "A comparison of different cluster mass estimates: consistency or discrepancy?". *Monthly Notices of the Royal Astronomical Society* **301** (3): 861–871. arXiv:astro-ph/9808179. Bibcode:1998MNRAS.301..861W. doi:10.1046/j.1365-8711.1998.02055.x.

[57] Refregier, A. (2003). "Weak gravitational lensing by large-scale structure". *Annual Review of Astronomy and Astrophysics* **41** (1): 645–668. arXiv:astro-ph/0307212. Bibcode:2003ARA&A..41..645R. doi:10.1146/annurev.astro.41.111302.102207.

[58] "Abell 2029: Hot News for Cold Dark Matter". Chandra X-ray Observatory. 11 June 2003.

[59] Massey, R.; et al. (2007). "Dark matter maps reveal cosmic scaffolding". *Nature* **445** (7125): 286–290. arXiv:astro-ph/0701594. Bibcode:2007Natur.445..286M. doi:10.1038/nature05497. PMID 17206154.

[60] Clowe, D.; et al. (2006). "A direct empirical proof of the existence of dark matter". *The Astrophysical Journal* **648** (2): 109–113. arXiv:astro-ph/0608407. Bibcode:2006ApJ...648L.109C. doi:10.1086/508162.

[61] Tiberiu, H.; Lobo, F. S. N. (2011). "Two-fluid dark matter models". *Physical Review D* **83** (12): 124051. arXiv:1106.2642. Bibcode:2011PhRvD..83d4051H. doi:10.1103/PhysRevD.83.124051.

[62] Spergel, D. N.; Steinhardt, P. J. (2000). "Observational evidence for self-interacting cold dark matter". *Physical Review Letters* **84** (17): 3760–3763. arXiv:astro-ph/9909386. Bibcode:2000PhRvL..84.3760S. doi:10.1103/PhysRevLett.84.3760.

[63] Markevitch, M.; et al. (2004). "Direct Constraints on the Dark Matter Self-Interaction Cross Section from the Merging Galaxy Cluster 1E 0657-56". *The Astrophysical Journal* **606** (2): 819–824. arXiv:astro-ph/0309303. Bibcode:2004ApJ...606..819M. doi:10.1086/383178.

[64] Allen, S. W.; Evrard, A. E.; Mantz, A. B. (2011). "Cosmological Parameters from Observations of Galaxy Clusters". *Annual Review of Astronomy & Astrophysics* **49**: 409–470. arXiv:1103.4829. Bibcode:2011ARA&A..49..409A. doi:10.1146/annurev-astro-081710-102514.

[65] "Press Release - Dark Matter Map Begins to Reveal the Universe's Early History - Subaru Telescope". *www.subarutelescope.org*. Retrieved 2015-07-03.

[66] Miyazaki, Satoshi; Oguri, Masamune; Hamana, Takashi; Tanaka, Masayuki; Miller, Lance; Utsumi, Yousuke; Komiyama, Yutaka; Furusawa, Hisanori; Sakurai, Junya (2015-07-01). "Properties of Weak Lensing Clusters Detected on Hyper Suprime-Cam's 2.3 deg2 field". *The Astrophysical Journal* **807** (1): 22. arXiv:1504.06974. Bibcode:2015ApJ...807...22M. doi:10.1088/0004-637X/807/1/22. ISSN 0004-637X.

[67] Percival, W. J.; et al. (2007). "Measuring the Baryon Acoustic Oscillation scale using the Sloan Digital Sky Survey and 2dF Galaxy Redshift Survey". *Monthly Notices of the Royal Astronomical Society* **381** (3): 1053–1066. arXiv:0705.3323. Bibcode:2007MNRAS.381.1053P. doi:10.1111/j.1365-2966.2007.12268.x.

[68] Peacock, J.; et al. (2001). "A measurement of the cosmological mass density from clustering in the 2dF Galaxy Redshift Survey". *Nature* **410**: 169. arXiv:astro-ph/0103143. Bibcode:2001Natur.410..169P.

[69] Kowalski, M.; et al. (2008). "Improved Cosmological Constraints from New, Old, and Combined Supernova Data Sets". *The Astrophysical Journal* **686** (2): 749–778. arXiv:0804.4142. Bibcode:2008ApJ...686..749K. doi:10.1086/589937.

[70] Viel, M.; Bolton, J. S.; Haehnelt, M. G. (2009). "Cosmological and astrophysical constraints from the Lyman α forest flux probability distribution function". *Monthly Notices of the Royal Astronomical Society* **399** (1): L39–L43. arXiv:0907.2927. Bibcode:2009MNRAS.399L..39V. doi:10.1111/j.1745-3933.2009.00720.x.

[71] "Hubble Maps the Cosmic Web of "Clumpy" Dark Matter in 3-D" (Press release). NASA. 7 January 2007.

[72] Springel, V.; et al. (2005). "Simulations of the formation, evolution and clustering of galaxies and quasars". *Nature* **435** (7042): 629–636. arXiv:astro-ph/0504097. Bibcode:2005Natur.435..629S. doi:10.1038/nature03597. PMID 15931216.

[73] Mateo, M. L. (1998). "Dwarf Galaxies of the Local Group". *Annual Review of Astronomy*

and Astrophysics **36** (1): 435–506. arXiv:astro-ph/9810070. Bibcode:1998ARA&A..36..435M. doi:10.1146/annurev.astro.36.1.435.

[74] Moore, B.; et al. (1999). "Dark Matter Substructure within Galactic Halos". *The Astrophysical Journal Letters* **524** (1): L19–L22. arXiv:astro-ph/9907411. Bibcode:1999ApJ...524L..19M. doi:10.1086/312287.

[75] Achim Weiss, "Big Bang Nucleosynthesis: Cooking up the first light elements" in: Einstein Online Vol. 2 (2006), 1017

[76] Raine, D.; Thomas, T. (2001). *An Introduction to the Science of Cosmology*. IOP Publishing. p. 30. ISBN 0-7503-0405-7.

[77] Tisserand, P.; Le Guillou, L.; Afonso, C.; Albert, J. N.; Andersen, J.; Ansari, R.; Aubourg, É.; Bareyre, P.; Beaulieu, J. P.; Charlot, X.; Coutures, C.; Ferlet, R.; Fouqué, P.; Glicenstein, J. F.; Goldman, B.; Gould, A.; Graff, D.; Gros, M.; Haissinski, J.; Hamadache, C.; De Kat, J.; Lasserre, T.; Lesquoy, É.; Loup, C.; Magneville, C.; Marquette, J. B.; Maurice, É.; Maury, A.; Milsztajn, A.; Moniez, M. (2007). "Limits on the Macho content of the Galactic Halo from the EROS-2 Survey of the Magellanic Clouds". *Astronomy and Astrophysics* **469** (2): 387–404. arXiv:astro-ph/0607207. Bibcode:2007A&A...469..387T. doi:10.1051/0004-6361:20066017.

[78] Graff, D. S.; Freese, K. (1996). "Analysis of a *Hubble Space Telescope* Search for Red Dwarfs: Limits on Baryonic Matter in the Galactic Halo". *The Astrophysical Journal* **456**. arXiv:astro-ph/9507097. Bibcode:1996ApJ...456L..49G. doi:10.1086/309850.

[79] Najita, J. R.; Tiede, G. P.; Carr, J. S. (2000). "From Stars to Superplanets: The Low-Mass Initial Mass Function in the Young Cluster IC 348". *The Astrophysical Journal* **541** (2): 977–1003. arXiv:astro-ph/0005290. Bibcode:2000ApJ...541..977N. doi:10.1086/309477.

[80] Wyrzykowski, Lukasz et al. (2011) The OGLE view of microlensing towards the Magellanic Clouds – IV. OGLE-III SMC data and final conclusions on MACHOs. MNRAS, 416, 2949

[81] Freese, Katherine; Fields, Brian; Graff, David (2000). "Death of Stellar Baryonic Dark Matter Candidates". arXiv:astro-ph/0007444 [astro-ph].

[82] Freese, Katherine; Fields, Brian; Graff, David (2000). "Death of Stellar Baryonic Dark Matter". *The First Stars*. ESO Astrophysics Symposia. p. 18. arXiv:astro-ph/0002058. Bibcode:2000fist.conf...18F. doi:10.1007/10719504_3. ISBN 3-540-67222-2.

[83] Bertone, G.; Merritt, D. (2005). "Dark Matter Dynamics and Indirect Detection". *Modern Physics Letters A* **20** (14): 1021–1036. arXiv:astro-ph/0504422. Bibcode:2005MPLA...20.1021B. doi:10.1142/S0217732305017391.

[84] Silk, Joseph (6 December 2000). "IX". *The Big Bang: Third Edition*. Henry Holt and Company. ISBN 978-0-8050-7256-3.

[85] Vittorio, N.; J. Silk (1984). "Fine-scale anisotropy of the cosmic microwave background in a universe dominated by cold dark matter". *Astrophysical Journal Letters* **285**: L39–L43. Bibcode:1984ApJ...285L..39V. doi:10.1086/184361.

[86] Umemura, Masayuki; Satoru Ikeuchi (1985). "Formation of Subgalactic Objects within Two-Component Dark Matter". *Astrophysical Journal* **299**: 583–592. Bibcode:1985ApJ...299..583U. doi:10.1086/163726.

[87] Davis, M.; Efstathiou, G., Frenk, C. S., & White, S. D. M. (May 15, 1985). "The evolution of large-scale structure in a universe dominated by cold dark matter". *Astrophysical Journal* **292**: 371–394. Bibcode:1985ApJ...292..371D. doi:10.1086/163168.

[88] Hawkins, M. R. S. (2011). "The case for primordial black holes as dark matter". *Monthly Notices of the Royal Astronomical Society* **415** (3): 2744–2757. arXiv:1106.3875. Bibcode:2011MNRAS.415.2744H. doi:10.1111/j.1365-2966.2011.18890.x.

[89] Carr, B. J.; et al. (May 2010). "New cosmological constraints on primordial black holes" (PDF). *Physical Review D* **81** (10): 104019. arXiv:0912.5297. Bibcode:2010PhRvD..81j4019C. doi:10.1103/PhysRevD.81.104019.

[90] Peter. A. H. G. (2012). "Dark Matter: A Brief Review". arXiv:1201.3942 [astro-ph.CO].

[91] Garrett, Katherine; Dūda, Gintaras (2011). "Dark Matter: A Primer". *Advances in Astronomy* **2011**: 1–22. arXiv:1006.2483. Bibcode:2011AdAst2011E...8G. doi:10.1155/2011/968283. MACHOs can only account for a very small percentage of the nonluminous mass in our galaxy, revealing that most dark matter cannot be strongly concentrated or exist in the form of baryonic astrophysical objects. Although microlensing surveys rule out baryonic objects like brown dwarfs, black holes, and neutron stars in our galactic halo, can other forms of baryonic matter make up the bulk of dark matter? The answer, surprisingly, is no...

[92] Bertone, G. (2010). "The moment of truth for WIMP dark matter". *Nature* **468** (7322): 389–393. arXiv:1011.3532. Bibcode:2010Natur.468..389B. doi:10.1038/nature09509. PMID 21085174.

[93] Olive, Keith A. (2003). "TASI Lectures on Dark Matter". p. 21

[94] Jungman, Gerard; Kamionkowski, Marc; Griest, Kim (1996-03-01). "Supersymmetric dark matter". *Physics Reports* **267** (5–6): 195–373. arXiv:hep-ph/9506380. Bibcode:1996PhR...267..195J. doi:10.1016/0370-1573(95)00058-5.

[95] "Neutrinos as Dark Matter". Astro.ucla.edu. 21 September 1998. Retrieved 6 January 2011.

[96] Gaitskell, Richard J. (2004). "Direct Detection of Dark Matter". *Annual Review of Nuclear and Particle Science* **54**: 315–359. Bibcode:2004ARNPS..54..315G. doi:10.1146/annurev.nucl.54.070103.181244.

[97] "NEUTRALINO DARK MATTER". Retrieved 26 December 2011. Griest, Kim. "WIMPs and MACHOs" (PDF). Retrieved 26 December 2011.

[98] Drukier, A.; Freese, K.; Spergel, D. (1986). "Detecting Cold Dark Matter Candidates". *Physical Review D* **33** (12): 3495–3508. Bibcode:1986PhRvD..33.3495D. doi:10.1103/PhysRevD.33.3495.

[99] Bernabei, R.; Belli, P.; Cappella, F.; Cerulli, R.; Dai, C. J.; d'Angelo, A.; He, H. L.; Incicchitti, A.; Kuang, H. H.; Ma, J. M.; Montecchia, F.; Nozzoli, F.; Prosperi, D.; Sheng, X. D.; Ye, Z. P. (2008). "First results from DAMA/LIBRA and the combined results with DAMA/NaI". *Eur. Phys. J. C* **56** (3): 333–355. arXiv:0804.2741. doi:10.1140/epjc/s10052-008-0662-y.

[100] Stonebraker, Alan (2014-01-03). "Synopsis: Dark-Matter Wind Sways through the Seasons". *Physics - Synopses* (American Physical Society). Retrieved 6 January 2014.

[101] Lee, Samuel K.; Mariangela Lisanti, Annika H. G. Peter, and Benjamin R. Safdi (2014-01-03). "Effect of Gravitational Focusing on Annual Modulation in Dark-Matter Direct-Detection Experiments". *Phys. Rev. Lett.* (American Physical Society) **112** (1): 011301 (2014) [5 pages]. arXiv:1308.1953. Bibcode:2014PhRvL.112a1301L. doi:10.1103/PhysRevLett.112.011301.

[102] The Dark Matter Group. "An Introduction to Dark Matter". *Dark Matter Research* (Sheffield, UK: University of Sheffield). Retrieved 7 January 2014.

[103] "Blowing in the Wind". *Kavli News* (Sheffield, UK: Kavli Foundation). Retrieved 7 January 2014. Scientists at Kavli MIT are working on...a tool to track the movement of dark matter.

[104] The CDMS II Collaboration; Ahmed, Z.; Akerib, D. S.; Arrenberg, S.; Bailey, C. N.; Balakishiyeva, D.; Baudis, L.; Bauer, D. A.; Brink, P. L.; Bruch, T.; Bunker, R.; Cabrera, B.; Caldwell, D. O.; Cooley, J.; Cushman, P.; Daal, M.; Dejongh, F.; Dragowsky, M. R.; Duong, L.; Fallows, S.; Figueroa-Feliciano, E.; Filippini, J.; Fritts, M.; Golwala, S. R.; Grant, D. R.; Hall, J.; Hennings-Yeomans, R.; Hertel, S. A.; Holmgren, D.; Hsu, L. (2010). "Dark Matter Search Results from the CDMS II Experiment". *Science* **327** (5973): 1619–1621. arXiv:0912.3592. Bibcode:2010Sci...327.1619C. doi:10.1126/science.1186112. PMID 20150446.

[105] Angloher, G.; Bauer; Bavykina; Bento; Bucci; Ciemniak; Deuter; von Feilitzsch; Hauff; Huff (2011). "Results from 730kg days of the CRESST-II Dark Matter Search". arXiv:1109.0702v1 [astro-ph.CO].

[106] "Dark matter even darker than once thought". Retrieved 16 June 2015.

[107] Freese, K. (1986). "Can Scalar Neutrinos or Massive Dirac Neutrinos be the Missing Mass?". *Physics Letters B* **167** (3): 295–300. Bibcode:1986PhLB..167..295F. doi:10.1016/0370-2693(86)90349-7.

[108] Ellis, J.; Flores, R. A.; Freese, K.; Ritz, S.; Seckel, D.; Silk, J. (1988). "Cosmic ray constraints on the annihilations of relic particles in the galactic halo". *Physics Letters B* **214** (3): 403–412. Bibcode:1988PhLB..214..403E. doi:10.1016/0370-2693(88)91385-8.

[109] Bertone, Gianfranco (2010). "Dark Matter at the Centers of Galaxies". *Particle Dark Matter: Observations, Models and Searches.* Cambridge University Press. pp. 83–104. arXiv:1001.3706. ISBN 978-0-521-76368-4.

[110] Sokol, Joshua et al (2016). "Surfing gravity's waves" (*New Scientist* 20 February)

[111] Stecker, F.W.; Hunter, S; Kniffen, D (2008). "The likely cause of the EGRET GeV anomaly and its implications". *Astroparticle Physics* **29** (1): 25–29. arXiv:0705.4311. Bibcode:2008APh....29...25S. doi:10.1016/j.astropartphys.2007.11.002.

[112] Atwood, W.B.; Abdo, A. A.; Ackermann, M.; Althouse, W.; Anderson, B.; Axelsson, M.; Baldini, L.; Ballet, J.; et al. (2009). "The large area telescope on the Fermi Gamma-ray Space Telescope Mission". *Astrophysical Journal* **697** (2): 1071–1102. arXiv:0902.1089. Bibcode:2009ApJ...697.1071A. doi:10.1088/0004-637X/697/2/1071.

[113] Weniger, Christoph (2012). "A Tentative Gamma-Ray Line from Dark Matter Annihilation at the Fermi Large Area Telescope". *Journal of Cosmology and Astroparticle Physics* **2012** (8): 7. arXiv:1204.2797v2. Bibcode:2012JCAP...08..007W. doi:10.1088/1475-7516/2012/08/007.

[114] Cartlidge, Edwin (24 April 2012). "Gamma rays hint at dark matter". Institute Of Physics. Retrieved 23 April 2013.

[115] Albert, J.; Aliu, E.; Anderhub, H.; Antoranz, P.; Backes, M.; Baixeras, C.; Barrio, J. A.; Bartko, H.; Bastieri, D.; Becker, J. K.; Bednarek, W.; Berger, K.; Bigongiari, C.; Biland, A.; Bock, R. K.; Bordas, P.; Bosch-Ramon, V.; Bretz, T.; Britvitch, I.; Camara, M.; Carmona, E.; Chilingarian, A.; Commichau, S.; Contreras, J. L.; Cortina, J.; Costado, M. T.; Curtef, V.; Danielyan, V.; Dazzi, F.; De Angelis, A. (2008). "Upper Limit for γ-Ray Emission above 140 GeV from the Dwarf Spheroidal Galaxy Draco". *The Astrophysical Journal* **679**: 428–431. arXiv:0711.2574. Bibcode:2008ApJ...679..428A. doi:10.1086/529135.

[116] Aleksić, J.; Antonelli, L. A.; Antoranz, P.; Backes, M.; Baixeras, C.; Balestra, S.; Barrio, J. A.; Bastieri, D.; González, J. B.; Bednarek, W.; Berdyugin, A.; Berger, K.; Bernardini, E.; Biland, A.; Bock, R. K.; Bonnoli,

G.; Bordas, P.; Tridon, D. B.; Bosch-Ramon, V.; Bose, D.; Braun, I.; Bretz, T.; Britzger, D.; Camara, M.; Carmona, E.; Carosi, A.; Colin, P.; Commichau, S.; Contreras, J. L.; Cortina, J. (2010). "Magic Gamma-Ray Telescope Observation of the Perseus Cluster of Galaxies: Implications for Cosmic Rays, Dark Matter, and Ngc 1275". *The Astrophysical Journal* **710**: 634–647. arXiv:0909.3267. Bibcode:2010ApJ...710..634A. doi:10.1088/0004-637X/710/1/634.

[117] Adriani, O.; Barbarino, G. C.; Bazilevskaya, G. A.; Bellotti, R.; Boezio, M.; Bogomolov, E. A.; Bonechi, L.; Bongi, M.; Bonvicini, V.; Bottai, S.; Bruno, A.; Cafagna, F.; Campana, D.; Carlson, P.; Casolino, M.; Castellini, G.; De Pascale, M. P.; De Rosa, G.; De Simone, N.; Di Felice, V.; Galper, A. M.; Grishantseva, L.; Hofverberg, P.; Koldashov, S. V.; Krutkov, S. Y.; Kvashnin, A. N.; Leonov, A.; Malvezzi, V.; Marcelli, L.; Menn, W. (2009). "An anomalous positron abundance in cosmic rays with energies 1.5–100 GeV". *Nature* **458** (7238): 607–609. arXiv:0810.4995. Bibcode:2009Natur.458..607A. doi:10.1038/nature07942. PMID 19340076.

[118] Aguilar, M. (AMS Collaboration); et al. (3 April 2013). "First Result from the Alpha Magnetic Spectrometer on the International Space Station: Precision Measurement of the Positron Fraction in Primary Cosmic Rays of 0.5–350 GeV". *Physical Review Letters* **110**. Bibcode:2013PhRvL.110n1102A. doi:10.1103/PhysRevLett.110.141102. Retrieved 3 April 2013.

[119] "First Result from the Alpha Magnetic Spectrometer Experiment". *AMS Collaboration*. 3 April 2013. Retrieved 3 April 2013.

[120] Heilprin, John; Borenstein, Seth (3 April 2013). "Scientists find hint of dark matter from cosmos". Associated Press. Retrieved 3 April 2013.

[121] Amos, Jonathan (3 April 2013). "Alpha Magnetic Spectrometer zeroes in on dark matter". *BBC*. Retrieved 3 April 2013.

[122] Perrotto, Trent J.; Byerly, Josh (2 April 2013). "NASA TV Briefing Discusses Alpha Magnetic Spectrometer Results". *NASA*. Retrieved 3 April 2013.

[123] Overbye, Dennis (3 April 2013). "New Clues to the Mystery of Dark Matter". *New York Times*. Retrieved 3 April 2013.

[124] Kane, G.; Watson, S. (2008). "Dark Matter and LHC: what is the Connection?". *Modern Physics Letters A* **23** (26): 2103–2123. arXiv:0807.2244. Bibcode:2008MPLA...23.2103K. doi:10.1142/S0217732308028314.

[125] Extra dimensions, gravitons, and tiny black holes. CERN. Retrieved on 17 November 2014.

[126] Dark matter. CERN. Retrieved on 17 November 2014.

[127] Rzetelny, Xaq (19 November 2014). "Looking for a different sort of dark matter with GPS satellites". *Ars Technica*. Retrieved 24 November 2014.

[128] Exirifard, Q. (2010). "Phenomenological covariant approach to gravity". *General Relativity and Gravitation* **43** (1): 93–106. arXiv:0808.1962. Bibcode:2011GReGr..43...93E. doi:10.1007/s10714-010-1073-6.

[129] Brownstein, J.R.; Moffat, J. W. (2007). "The Bullet Cluster 1E0657-558 evidence shows modified gravity in the absence of dark matter". *Monthly Notices of the Royal Astronomical Society* **382** (1): 29–47. arXiv:astro-ph/0702146. Bibcode:2007MNRAS.382...29B. doi:10.1111/j.1365-2966.2007.12275.x.

[130] Anastopoulos, C. (2009). "Gravitational backreaction in cosmological space-times". *Physical Review D* **79** (8): 084029. arXiv:0902.0159. Bibcode:2009PhRvD..79h4029A. doi:10.1103/PhysRevD.79.084029.

[131] "New Cosmic Theory Unites Dark Forces". SPACE.com. 11 February 2008. Retrieved 6 January 2011.

[132] Nottale, Laurent (May 29, 2009). "Scale relativity and fractal space-time: theory and applications" (PDF).

[133] Nottale, Laurent (17 June 2011). *Scale Relativity and Fractal Space-Time: A New Approach to Unifying Relativity and Quantum Mechanics*. World Scientific. p. 516. ISBN 978-1-908977-87-8.

1.5.11 External links

- Dark matter at DMOZ

- Dark matter (Astronomy) at *Encyclopædia Britannica*

- What is dark matter? at cosmosmagazine.com

- The Dark Matter Crisis 18 August 2010 by Pavel Kroupa, posted in General

- The European astroparticle physics network

- Helmholtz Alliance for Astroparticle Physics

- "NASA Finds Direct Proof of Dark Matter" (Press release). NASA. 21 August 2006.

- Tuttle, Kelen (22 August 2006). "Dark Matter Observed". SLAC (Stanford Linear Accelerator Center) Today.

- "Astronomers claim first 'dark galaxy' find". New Scientist. 23 February 2005.

- Sample, Ian (17 December 2009). "Dark Matter Detected". London: Guardian. Retrieved 1 May 2010.

- Video lecture on dark matter by Scott Tremaine, IAS professor

- Science Daily story "Astronomers' Doubts About the Dark Side ..."

- Gray, Meghan; Merrifield, Mike; Copeland, Ed (2010). "Dark Matter". *Sixty Symbols*. Brady Haran for the University of Nottingham.

1.6   Dark energy

Not to be confused with dark fluid, dark flow, or dark matter.

In physical cosmology and astronomy, **dark energy** is an unknown form of energy which is hypothesized to permeate all of space, tending to accelerate the expansion of the universe.[1] Dark energy is the most accepted hypothesis to explain the observations since the 1990s indicating that the universe is expanding at an accelerating rate.

Assuming that the standard model of cosmology is correct, the best current measurements indicate that dark energy contributes 68.3% of the total energy in the present-day observable universe. The mass–energy of dark matter and ordinary (baryonic) matter contribute 26.8% and 4.9%, respectively, and other components such as neutrinos and photons contribute a very small amount.[2][3][4][5] Again, on a mass–energy equivalence basis, the density of dark energy ($\sim 7 \times 10^{-30}$ g/cm^3) is very low, much less than the density of ordinary matter or dark matter within galaxies. However, it comes to dominate the mass–energy of the universe because it is uniform across space.[6][7][8]

Two proposed forms for dark energy are the cosmological constant,[9] a *constant* energy density filling space homogeneously,[10] and scalar fields such as quintessence or moduli, *dynamic* quantities whose energy density can vary in time and space. Contributions from scalar fields that are constant in space are usually also included in the cosmological constant. The cosmological constant can be formulated to be equivalent to vacuum energy. Scalar fields that do change in space can be difficult to distinguish from a cosmological constant because the change may be extremely slow.

High-precision measurements of the expansion of the universe are required to understand how the expansion rate changes over time and space. In general relativity, the evolution of the expansion rate is parameterized by the cos-

mological equation of state (the relationship between temperature, pressure, and combined matter, energy, and vacuum energy density for any region of space). Measuring the equation of state for dark energy is one of the biggest efforts in observational cosmology today. Adding the cosmological constant to cosmology's standard FLRW metric leads to the Lambda-CDM model, which has been referred to as the *"standard model of cosmology"* because of its precise agreement with observations. Dark energy has been used as a crucial ingredient in a recent attempt to formulate a cyclic model for the universe.[11]

Many things about the nature of dark energy remain matters of speculation.[12] The evidence for dark energy is indirect but comes from three independent sources:

- Distance measurements and their relation to redshift, which suggest the universe has expanded more in the last half of its life.[13]

- The theoretical need for a type of additional energy that is not matter or dark matter to form the observationally flat universe (absence of any detectable global curvature).

- It can be inferred from measures of large scale wave-patterns of mass density in the universe.

Dark energy is thought to be very homogeneous, not very dense and is not known to interact through any of the fundamental forces other than gravity. Since it is quite rarefied — roughly 10^{-27} kg/m^3 — it is unlikely to be detectable in laboratory experiments. Dark energy can have such a profound effect on the universe, making up 68% of universal density, only because it uniformly fills otherwise empty space. The two leading models are a cosmological constant and quintessence. Both models include the common characteristic that dark energy must have negative pressure.

Effect of dark energy: a small constant negative pressure of vacuum

Independently of its actual nature, dark energy would need to have a strong negative pressure (acting repulsively) like radiation pressure in a metamaterial[14] to explain the observed acceleration of the expansion of the universe. According to general relativity, the pressure within a substance contributes to its gravitational attraction for other things just as its mass density does. This happens because the physical quantity that causes matter to generate gravitational effects is the stress–energy tensor, which contains both the energy (or matter) density of a substance and its pressure and viscosity. In the Friedmann–Lemaître–Robertson–Walker metric, it can be shown that a strong constant negative pressure in all the universe causes an acceleration in universe expansion if the universe is already expanding, or a deceleration in universe contraction if the universe is already contracting. This accelerating expansion effect is sometimes labeled "gravitational repulsion", which is a colorful but possibly confusing expression. In fact, a negative pressure does not influence the gravitational interaction between masses—which remains attractive—but rather alters the overall evolution of the universe at the cosmological scale, typically resulting in the accelerating expansion of the universe despite the attraction among the masses present in the universe. The acceleration is simply a function of dark energy density. Dark energy is persistent: its density remains constant (experimentally, within a factor of 1:10), i.e. it does not get diluted when space expands.

1.6.2   Evidence of existence

Supernovae

A Type Ia supernova (bright spot on the bottom-left) near a galaxy

In 1998, the High-Z Supernova Search Team[15] published observations of Type Ia ("one-A") supernovae. In 1999, the Supernova Cosmology Project[16] followed by suggesting that the expansion of the universe is accelerating.[17] The 2011 Nobel Prize in Physics was awarded to Saul Perlmutter, Brian P. Schmidt and Adam G. Riess for their leadership in the discovery.[18][19]

Since then, these observations have been corroborated by several independent sources. Measurements of the cosmic microwave background, gravitational lensing, and the large-scale structure of the cosmos as well as improved measurements of supernovae have been consistent with the Lambda-CDM model.[20] Some people argue that the only indication for the existence of dark energy is observations of distance measurements and associated redshifts. Cosmic microwave background anisotropies and baryon acoustic oscillations are only observations that distances to a given redshift are larger than expected from a "dusty" Friedmann–Lemaître universe and the local measured Hubble constant.[21]

Supernovae are useful for cosmology because they are excellent standard candles across cosmological distances. They allow the expansion history of the universe to be mea-

sured by looking at the relationship between the distance to an object and its redshift, which gives how fast it is receding from us. The relationship is roughly linear, according to Hubble's law. It is relatively easy to measure redshift, but finding the distance to an object is more difficult. Usually, astronomers use standard candles: objects for which the intrinsic brightness, the absolute magnitude, is known. This allows the object's distance to be measured from its actual observed brightness, or apparent magnitude. Type Ia supernovae are the best-known standard candles across cosmological distances because of their extreme and consistent luminosity.

Recent observations of supernovae are consistent with a universe made up 71.3% of dark energy and 27.4% of a combination of dark matter and baryonic matter.[22]

Cosmic microwave background

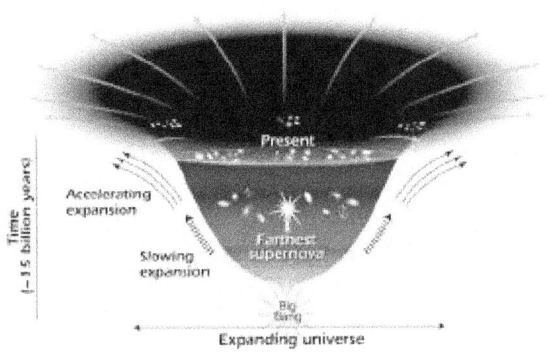

This diagram reveals changes in the rate of expansion since the universe's birth 15 billion years ago. The more shallow the curve, the faster the rate of expansion. The curve changes noticeably about 7.5 billion years ago, when objects in the universe began flying apart at a faster rate. Astronomers theorize that the faster expansion rate is due to a mysterious, dark force that is pulling galaxies apart.

Diagram representing the accelerated expansion of the universe due to dark energy.

The existence of dark energy, in whatever form, is needed to reconcile the measured geometry of space with the total amount of matter in the universe. Measurements of cosmic microwave background (CMB) anisotropies indicate that the universe is close to flat. For the shape of the universe to be flat, the mass/energy density of the universe must be equal to the critical density. The total amount of matter in the universe (including baryons and dark matter), as measured from the CMB spectrum, accounts for only about 30% of the critical density. This implies the existence of an additional form of energy to account for the remaining 70%.[20] The Wilkinson Microwave Anisotropy Probe (WMAP) spacecraft seven-year analysis estimated a universe made up of 72.8% dark energy, 22.7% dark matter and 4.5% ordinary matter.[4] Work done in 2013 based on the Planck spacecraft observations of the CMB gave a more accurate estimate of 68.3% of dark energy, 26.8% of dark matter and 4.9% of ordinary matter.[23]

Large-scale structure

The theory of large-scale structure, which governs the formation of structures in the universe (stars, quasars, galaxies and galaxy groups and clusters), also suggests that the density of matter in the universe is only 30% of the critical density.

A 2011 survey, the WiggleZ galaxy survey of more than 200,000 galaxies, provided further evidence towards the existence of dark energy, although the exact physics behind it remains unknown.[24][25] The WiggleZ survey from the Australian Astronomical Observatory scanned the galaxies to determine their redshift. Then, by exploiting the fact that baryon acoustic oscillations have left voids regularly of ~150 Mpc diameter, surrounded by the galaxies, the voids were used as standard rulers to determine distances to galaxies as far as 2,000 Mpc (redshift 0.6), which allowed astronomers to determine more accurately the speeds of the galaxies from their redshift and distance. The data confirmed cosmic acceleration up to half of the age of the universe (7 billion years) and constrain its inhomogeneity to 1 part in 10.[25] This provides a confirmation to cosmic acceleration independent of supernovae.

Late-time integrated Sachs-Wolfe effect

Accelerated cosmic expansion causes gravitational potential wells and hills to flatten as photons pass through them, producing cold spots and hot spots on the CMB aligned with vast supervoids and superclusters. This so-called late-time Integrated Sachs–Wolfe effect (ISW) is a direct signal of dark energy in a flat universe.[26] It was reported at high significance in 2008 by Ho et al.[27] and Giannantonio et al.[28]

Observational Hubble constant data

A new approach to test evidence of dark energy through observational Hubble constant (H(z)) data (OHD) has gained significant attention in recent years.[29][30][31][32] The Hubble constant is measured as a function of cosmological redshift. OHD directly tracks the expansion history of the universe by taking passively evolving early-type galaxies as "cosmic chronometers".[33] From this point, this approach provides standard clocks in the universe. The core of this idea is the measurement of the differential age evolution as a function of redshift of these cosmic chronometers.

Thus, it provides a direct estimate of the Hubble parameter $H(z) = -1/(1+z)dz/dt \approx -1/(1+z)\Delta z/\Delta t$. The merit of this approach is clear: the reliance on a differential quantity, $\Delta z/\Delta t$, can minimize many common issues and systematic effects; and as a direct measurement of the Hubble parameter instead of its integral, like supernovae and baryon acoustic oscillations (BAO), it brings more information and is appealing in computation. For these reasons, it has been widely used to examine the accelerated cosmic expansion and study properties of dark energy.

1.6.3   Theories of explanation

Cosmological constant

Main article: Cosmological constant
For more details on this topic, see Equation of state (cosmology).

The simplest explanation for dark energy is that it is simply

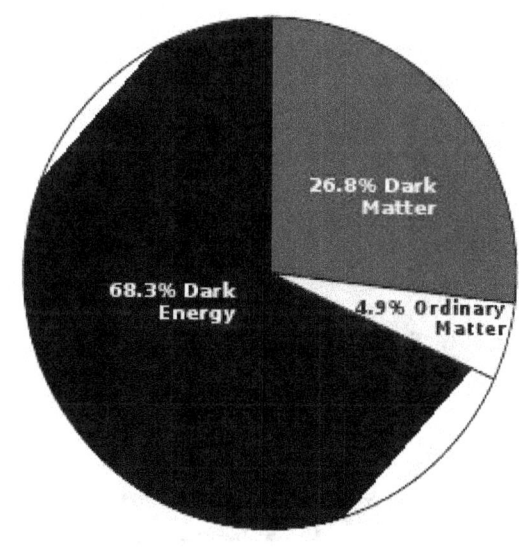

Estimated distribution of matter and energy in the universe[34]

the "cost of having space": that is, a volume of space has some intrinsic, fundamental energy. This is the cosmological constant, sometimes called Lambda (hence Lambda-CDM model) after the Greek letter Λ, the symbol used to represent this quantity mathematically. Since energy and mass are related by $E = mc^2$, Einstein's theory of general relativity predicts that this energy will have a gravitational effect. It is sometimes called a vacuum energy because it is the energy density of empty vacuum. In fact, most theories of particle physics predict vacuum fluctuations that would give the vacuum this sort of energy. This is related to the Casimir effect, in which there is a small suction into regions where virtual particles are geometrically inhibited from forming (e.g. between plates with tiny separation). The cosmological constant is estimated by cosmologists to be on the order of 10^{-29} g/cm^3, or about 10^{-120} in reduced Planck units. Particle physics predicts a natural value of 1 in reduced Planck units, leading to a large discrepancy.

Lambda, the letter that represents the cosmological constant

The cosmological constant has negative pressure equal to its energy density and so causes the expansion of the universe to accelerate. The reason why a cosmological constant has negative pressure can be seen from classical thermodynamics: Energy must be lost from inside a container to do work on the container. A change in volume dV requires work done equal to a change of energy $-P\,dV$, where P is the pressure. But the amount of energy in a container full of vacuum actually increases when the volume increases (dV is positive), because the energy is equal to ϱV, where ϱ (rho) is the energy density of the cosmological constant. Therefore, P is negative and, in fact, $P = -\varrho$.

A major outstanding problem is that most quantum field theories predict a huge cosmological constant from the energy of the quantum vacuum, more than 100 orders of magnitude too large.[10] This would need to be cancelled almost, but not exactly, by an equally large term of the opposite sign. Some supersymmetric theories require a cosmological constant that is exactly zero,[35] which does not help because supersymmetry must be broken. The present scientific consensus amounts to extrapolating the empirical evidence where it is relevant to predictions, and fine-tuning theories until a more elegant solution is found. Technically, this amounts to checking theories against macroscopic observations. Unfortunately, as the known error-margin in the constant predicts the fate of the universe more than

its present state, many such "deeper" questions remain unknown.

In spite of its problems, the cosmological constant is in many respects the most economical solution to the problem of cosmic acceleration. One number successfully explains a multitude of observations. Thus, the current standard model of cosmology, the Lambda-CDM model, includes the cosmological constant as an essential feature.

Quintessence

Main article: Quintessence (physics)

In quintessence models of dark energy, the observed acceleration of the scale factor is caused by the potential energy of a dynamical field, referred to as quintessence field. Quintessence differs from the cosmological constant in that it can vary in space and time. In order for it not to clump and form structure like matter, the field must be very light so that it has a large Compton wavelength.

No evidence of quintessence is yet available, but it has not been ruled out either. It generally predicts a slightly slower acceleration of the expansion of the universe than the cosmological constant. Some scientists think that the best evidence for quintessence would come from violations of Einstein's equivalence principle and variation of the fundamental constants in space or time.[36] Scalar fields are predicted by the *Standard Model of particle physics* and string theory, but an analogous problem to the cosmological constant problem (or the problem of constructing models of cosmological inflation) occurs: renormalization theory predicts that scalar fields should acquire large masses.

The coincidence problem asks why the acceleration of the Universe began when it did. If acceleration began earlier in the universe, structures such as galaxies would never have had time to form and life, at least as we know it, would never have had a chance to exist. Proponents of the anthropic principle view this as support for their arguments. However, many models of quintessence have a so-called **tracker** behavior, which solves this problem. In these models, the quintessence field has a density which closely tracks (but is less than) the radiation density until matter-radiation equality, which triggers quintessence to start behaving as dark energy, eventually dominating the universe. This naturally sets the low energy scale of the dark energy.).[37][38]

In 2004, when scientists fit the evolution of dark energy with the cosmological data, they found that the equation of state had possibly crossed the cosmological constant boundary (w=−1) from above to below. A No-Go theorem has been proved that gives this scenario at least two degrees of freedom as required for dark energy models. This scenario is so-called Quintom scenario.

Some special cases of quintessence are phantom energy, in which the energy density of quintessence actually increases with time, and k-essence (short for kinetic quintessence) which has a non-standard form of kinetic energy such as a negative kinetic energy.[39] They can have unusual properties: phantom energy, for example, can cause a Big Rip.

1.6.4 Alternative ideas

Some alternatives to dark energy aim to explain the observational data by a more refined use of established theories, focusing, for example, on the gravitational effects of density inhomogeneities, or on consequences of electroweak symmetry breaking in the early universe. If we are located in an emptier-than-average region of space, the observed cosmic expansion rate could be mistaken for a variation in time, or acceleration.[40][41][42][43] A different approach uses a cosmological extension of the equivalence principle to show how space might appear to be expanding more rapidly in the voids surrounding our local cluster. While weak, such effects considered cumulatively over billions of years could become significant, creating the illusion of cosmic acceleration, and making it appear as if we live in a Hubble bubble.[44][45][46]

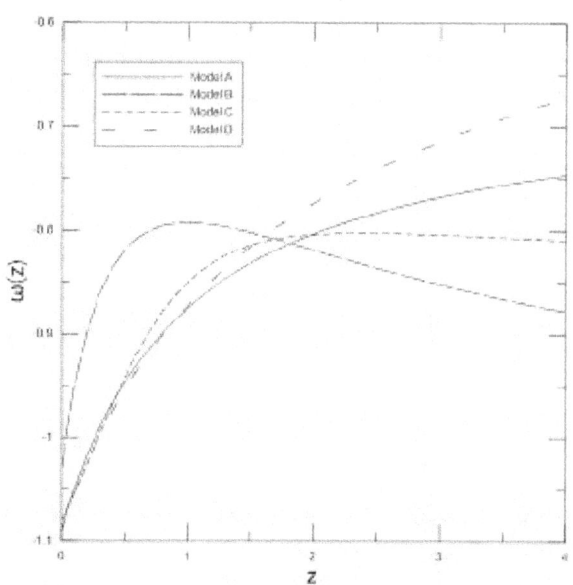

The equation of state of Dark Energy for 4 common models by Redshift.[47]

A: CPL Model,

B: Jassal Model,

C: Barboza & Alcaniz Model,

D: Wetterich Model

Another class of theories attempts to come up with an all-encompassing theory of both dark matter and dark energy as a single phenomenon that modifies the laws of gravity at various scales. An example of this type of theory is the theory of dark fluid. Another class of theories that unifies dark matter and dark energy are suggested to be covariant theories of modified gravities. These theories alter the dynamics of the space-time such that the modified dynamic stems what have been assigned to the presence of dark energy and dark matter.[48]

A 2011 paper in the journal *Physical Review D* by Christos Tsagas, a cosmologist at Aristotle University of Thessaloniki in Greece, argued that it is likely that the accelerated expansion of the universe is an illusion caused by the relative motion of us to the rest of the universe. The paper cites data showing that the 2.5 billion light-years-wide region of space we are inside of is moving very quickly relative to everything around it. If the theory is confirmed, then dark energy would not exist (but the "dark flow" still might).[49][50]

Some theorists think that dark energy and cosmic acceleration are a failure of general relativity on very large scales, larger than superclusters. However most attempts at modifying general relativity have turned out to be either equivalent to theories of quintessence, or inconsistent with observations. Other ideas for dark energy have come from string theory, brane cosmology and the holographic principle, but have not yet proved as compelling as quintessence and the cosmological constant. In other hand, M.R. Khoshbin-e-Khoshnazar believes that a model discretization of the universe could explain the origin of dark energy.[51]

On string theory, an article in the journal *Nature* described:[52]

> *String theories, popular with many particle physicists, make it possible, even desirable, to think that the observable universe is just one of 10^{500} universes in a grander multiverse, says Leonard Susskind, a cosmologist at Stanford University in California. The vacuum energy will have different values in different universes, and in many or most it might indeed be vast. But it must be small in ours because it is only in such a universe that observers such as ourselves can evolve.*

Paul Steinhardt in the same article criticizes string theory's explanation of dark energy stating "...Anthropics and randomness don't explain anything... I am disappointed with what most theorists are willing to accept".[52]

Another set of proposals is based on the possibility of a double metric tensor for space-time.[53][54] It has been argued that time reversed solutions in general relativity require such double metric for consistency, and that both dark matter and dark energy can be understood in terms of time reversed solutions of general relativity.[55]

It has been shown that if inertia is assumed to be due to the effect of horizons on Unruh radiation, then this predicts galaxy rotation and a cosmic acceleration similar to that observed.[56]

Variable Dark Energy models

In general, the dark energy can be variable. Modern observational data have determined the density of dark energy in the present. Using baryon acoustic oscillations, we can investigate the effect of dark energy in the history of the Universe and we can constrain parameters of the equation of state of dark energy. One of the proposed solutions to get closer to answering the question of dark energy, is to assume that it is variable. To that end, several models have been proposed. One of their most popular models is Chevallier–Polarski–Linder model (CPL).[57][58] Some other common models are, (Barboza & Alcaniz, 2008),[59] (Jassal et al. 2005),[60] (Wetterich, 2004).[61]

1.6.5   Implications for the fate of the universe

Cosmologists estimate that the acceleration began roughly 5 billion years ago. Before that, it is thought that the expansion was decelerating, due to the attractive influence of dark matter and baryons. The density of dark matter in an expanding universe decreases more quickly than dark energy, and eventually the dark energy dominates. Specifically, when the volume of the universe doubles, the density of dark matter is halved, but the density of dark energy is nearly unchanged (it is exactly constant in the case of a cosmological constant).

If the acceleration continues indefinitely, the ultimate result will be that galaxies outside the local supercluster will have a line-of-sight velocity that continually increases with time, eventually far exceeding the speed of light.[62] This is not a violation of special relativity because the notion of "velocity" used here is different from that of velocity in a local inertial frame of reference, which is still constrained to be less than the speed of light for any massive object (see Uses of the proper distance for a discussion of the subtleties of defining any notion of relative velocity in cosmology). Because the Hubble parameter is decreasing with time, there can actually be cases where a galaxy that is receding from us faster than light does manage to emit a signal which reaches us eventually.[63][64] However, because of the accelerating expansion, it is projected that most galaxies will eventually cross a type of cosmological event horizon where any light they emit past that point will never be able to reach us at any time in the infinite future[65] because the light never reaches a point where its "peculiar velocity" toward us exceeds the

expansion velocity away from us (these two notions of velocity are also discussed in Uses of the proper distance). Assuming the dark energy is constant (a cosmological constant), the current distance to this cosmological event horizon is about 16 billion light years, meaning that a signal from an event happening *at present* would eventually be able to reach us in the future if the event were less than 16 billion light years away, but the signal would never reach us if the event were more than 16 billion light years away.[64]

As galaxies approach the point of crossing this cosmological event horizon, the light from them will become more and more redshifted, to the point where the wavelength becomes too large to detect in practice and the galaxies appear to vanish completely[66][67] (*see* Future of an expanding universe). The Earth, the Milky Way, and the Virgo Supercluster, however, would remain virtually undisturbed while the rest of the universe recedes and disappears from view. In this scenario, the local supercluster would ultimately suffer heat death, just as was thought for the flat, matter-dominated universe before measurements of cosmic acceleration.

There are some very speculative ideas about the future of the universe. One suggests that phantom energy causes *divergent* expansion, which would imply that the effective force of dark energy continues growing until it dominates all other forces in the universe. Under this scenario, dark energy would ultimately tear apart all gravitationally bound structures, including galaxies and solar systems, and eventually overcome the electrical and nuclear forces to tear apart atoms themselves, ending the universe in a "Big Rip". On the other hand, dark energy might dissipate with time or even become attractive. Such uncertainties leave open the possibility that gravity might yet rule the day and lead to a universe that contracts in on itself in a "Big Crunch".[68] Some scenarios, such as the cyclic model, suggest this could be the case. It is also possible the universe may never have an end and continue in its present state forever (see The Second Law as a law of disorder). While these ideas are not supported by observations, they are not ruled out.

1.6.6   History of discovery and previous speculation

The cosmological constant was first proposed by Einstein as a mechanism to obtain a solution of the gravitational field equation that would lead to a static universe, effectively using dark energy to balance gravity.[69] Not only was the mechanism an inelegant example of fine-tuning but it was also later realized that Einstein's static universe would actually be unstable because local inhomogeneities would ultimately lead to either the runaway expansion or contraction of the universe. The equilibrium is unstable: If the universe expands slightly, then the expansion releases vacuum energy, which causes yet more expansion. Likewise, a universe which contracts slightly will continue contracting. These sorts of disturbances are inevitable, due to the uneven distribution of matter throughout the universe. More importantly, observations made by Edwin Hubble in 1929 showed that the universe appears to be expanding and not static at all. Einstein reportedly referred to his failure to predict the idea of a dynamic universe, in contrast to a static universe, as his greatest blunder.[70]

Alan Guth and Alexei Starobinsky proposed in 1980 that a negative pressure field, similar in concept to dark energy, could drive cosmic inflation in the very early universe. Inflation postulates that some repulsive force, qualitatively similar to dark energy, resulted in an enormous and exponential expansion of the universe slightly after the Big Bang. Such expansion is an essential feature of most current models of the Big Bang. However, inflation must have occurred at a much higher energy density than the dark energy we observe today and is thought to have completely ended when the universe was just a fraction of a second old. It is unclear what relation, if any, exists between dark energy and inflation. Even after inflationary models became accepted, the cosmological constant was thought to be irrelevant to the current universe.

Nearly all inflation models predict that the total (matter+energy) density of the universe should be very close to the critical density. During the 1980s, most cosmological research focused on models with critical density in matter only, usually 95% cold dark matter and 5% ordinary matter (baryons). These models were found to be successful at forming realistic galaxies and clusters, but some problems appeared in the late 1980s: notably, the model required a value for the Hubble constant lower than preferred by observations, and the model under-predicted observations of large-scale galaxy clustering. These difficulties became stronger after the discovery of anisotropy in the cosmic microwave background by the COBE spacecraft in 1992, and several modified CDM models came under active study through the mid-1990s: these included the Lambda-CDM model and a mixed cold/hot dark matter model. The first direct evidence for dark energy came from supernova observations in 1998 of accelerated expansion in Riess et al.[15] and in Perlmutter et al.[16] and the Lambda-CDM model then became the leading model. Soon after, dark energy was supported by independent observations: in 2000, the BOOMERanG and Maxima cosmic microwave background experiments observed the first acoustic peak in the CMB, showing that the total (matter+energy) density is close to 100% of critical density. Then in 2001, the 2dF Galaxy Redshift Survey gave strong evidence that the matter density is around 30% of critical. The large difference between these two supports a smooth component of dark energy making up the difference. Much more precise measurements from WMAP in 2003–2010 have continued to support the standard model and give more accurate measurements of the key parameters.

The term "dark energy", echoing Fritz Zwicky's "dark matter" from the 1930s, was coined by Michael Turner in 1998.[71]

As of 2013, the Lambda-CDM model is consistent with a series of increasingly rigorous cosmological observations, including the Planck spacecraft and the Supernova Legacy Survey. First results from the SNLS reveal that the average behavior (i.e., equation of state) of dark energy behaves like Einstein's cosmological constant to a precision of 10%.[72] Recent results from the Hubble Space Telescope Higher-Z Team indicate that dark energy has been present for at least 9 billion years and during the period preceding cosmic acceleration.

1.6.7 See also

- Conformal gravity

- De Sitter relativity

- Illustris project

- *The Dark Energy Survey*

- *Quintessence: The Search for Missing Mass in the Universe*

- Vacuum state

1.6.8 References

[1] Peebles, P. J. E.; Ratra, Bharat (2003). "The cosmological constant and dark energy". *Reviews of Modern Physics* **75** (2): 559–606. arXiv:astro-ph/0207347. Bibcode:2003RvMP...75..559P. doi:10.1103/RevModPhys.75.559.

[2] Ade, P. A. R.; Aghanim, N.; Armitage-Caplan, C.; et al. (Planck Collaboration), C.; Arnaud, M.; Ashdown, M.; Atrio-Barandela, F.; Aumont, J.; Aussel, H.; Baccigalupi, C.; Banday, A. J.; Barreiro, R. B.; Barrena, R.; Bartelmann, M.; Bartlett, J. G.; Bartolo, N.; Basak, S.; Battaner, E.; Battye, R.; Benabed, K.; Benoît, A.; Benoit-Lévy, A.; Bernard, J.-P.; Bersanelli, M.; Bertincourt, B.; Bethermin, M.; Bielewicz, P.; Bikmaev, I.; Blanchard, A.; et al. (22 March 2013). "Planck 2013 results. I. Overview of products and scientific results – Table 9". *Astronomy and Astrophysics* **571**: A1. arXiv:1303.5062. Bibcode:2014A&A...571A...1P. doi:10.1051/0004-6361/201321529.

[3] Ade, P. A. R.; Aghanim, N.; Armitage-Caplan, C.; et al. (Planck Collaboration). C.; Arnaud, M.; Ashdown, M.; Atrio-Barandela, F.; Aumont, J.; Aussel, H.; Baccigalupi, C.; Banday, A. J.; Barreiro, R. B.; Barrena, R.; Bartelmann, M.; Bartlett, J. G.; Bartolo, N.; Basak, S.; Battaner, E.; Battye, R.; Benabed, K.; Benoît, A.; Benoit-Lévy, A.; Bernard, J.-P.; Bersanelli, M.; Bertincourt, B.; Bethermin, M.; Bielewicz, P.; Bikmaev, I.; Blanchard, A.; et al. (31 March 2013). "Planck 2013 Results Papers". *Astronomy and Astrophysics* **571**: A1. arXiv:1303.5062. Bibcode:2014A&A...571A...1P. doi:10.1051/0004-6361/201321529.

[4] "First Planck results: the Universe is still weird and interesting".

[5] Sean Carroll, Ph.D., Caltech, 2007, The Teaching Company, *Dark Matter, Dark Energy: The Dark Side of the Universe*, Guidebook Part 2 page 46. Retrieved Oct. 7, 2013. "...dark energy: A smooth, persistent component of invisible energy, thought to make up about 70 percent of the current energy density of the universe. Dark energy is known to be smooth because it doesn't accumulate preferentially in galaxies and clusters..."

[6] Paul J. Steinhardt, Neil Turok (2006). "Why the cosmological constant is small and positive". *Science* **312** (5777): 1180–1183. arXiv:astro-ph/0605173. Bibcode:2006Sci...312.1180S. doi:10.1126/science.1126231.

[7] "Dark Energy". *Hyperphysics*. Retrieved January 4, 2014.

[8] Ferris, Timothy. "Dark Matter(Dark Energy)". Retrieved 2015-06-10.

[9] http://www.ft.com/intl/cms/s/2/493de45a-8bef-11e0-854c-00144feab49a.html#axzz3m9WSVVkC

[10] Carroll, Sean (2001). "The cosmological constant". *Living Reviews in Relativity* **4**. arXiv:astro-ph/0004075. Bibcode:2001LRR.....4....1C. doi:10.12942/lrr-2001-1. Retrieved 2006-09-28.

[11] Baum, L.; Frampton, P.H. (2007). "Turnaround in Cyclic Cosmology". *Physical Review Letters* **98** (7): 071301. arXiv:hep-th/0610213. Bibcode:2007PhRvL..98g1301B. doi:10.1103/PhysRevLett.98.071301. PMID 17359014.

[12] Overbye, Dennis. "Astronomers Report Evidence of 'Dark Energy' Splitting the Universe". The New York Times. Retrieved August 5, 2015.

[13] Durrer, R. (2011). "What do we really know about Dark Energy?". *Philosophical Transactions of the Royal Society A: Mathematical, Physical and Engineering Sciences* **369** (1957): 5102. arXiv:1103.5331. Bibcode:2011RSPTA.369.5102D. doi:10.1098/rsta.2011.0285.

[14] Zhong-Yue Wang (2016). "Modern Theory for Electromagnetic Metamaterials". *Plasmonics* **11** (2): 503–508. doi:10.1007/s11468-015-0071-7.

[15] Riess, Adam G.; Filippenko; Challis; Clocchiatti; Diercks; Garnavich; Gilliland; Hogan; Jha; Kirshner; Leibundgut; Phillips; Reiss; Schmidt; Schommer; Smith; Spyromilio; Stubbs; Suntzeff; Tonry (1998). "Observational evidence from supernovae for an accelerating universe and a cosmological constant". *Astronomical Journal* **116** (3): 1009–38. arXiv:astro-ph/9805201. Bibcode:1998AJ....116.1009R. doi:10.1086/300499.

[16] Perlmutter, S.; Aldering; Goldhaber; Knop; Nugent; Castro; Deustua; Fabbro; Goobar; Groom; Hook; Kim; Kim; Lee; Nunes; Pain; Pennypacker; Quimby; Lidman; Ellis; Irwin; McMahon; Ruiz-Lapuente; Walton; Schaefer; Boyle; Filippenko; Matheson; Fruchter; et al. (1999). "Measurements of Omega and Lambda from 42 high redshift supernovae". *Astrophysical Journal* **517** (2): 565–86. arXiv:astro-ph/9812133. Bibcode:1999ApJ...517..565P. doi:10.1086/307221.

[17] The first paper, using observed data, which claimed a positive Lambda term was Paál, G.; et al. (1992). "Inflation and compactification from galaxy redshifts?". *Astrophysics and Space Science* **191**: 107–24. Bibcode:1992Ap&SS.191..107P. doi:10.1007/BF00644200.

[18] "The Nobel Prize in Physics 2011". Nobel Foundation. Retrieved 2011-10-04.

[19] The Nobel Prize in Physics 2011. Perlmutter got half the prize, and the other half was shared between Schmidt and Riess.

[20] Spergel, D. N. (WMAP collaboration); et al. (March 2006). "Wilkinson Microwave Anisotropy Probe (WMAP) three year results: implications for cosmology".

[21] Durrer, R. (2011). "What do we really know about dark energy?". *Philosophical Transactions of the Royal Society A* **369** (1957): 5102–5114. arXiv:1103.5331. Bibcode:2011RSPTA.369.5102D. doi:10.1098/rsta.2011.0285.

[22] Kowalski, Marek; Rubin, David; Aldering, G.; Agostinho, R. J.; Amadon, A.; Amanullah, R.; Balland, C.; Barbary, K.; Blanc, G.; Challis, P. J.; Conley, A.; Connolly, N. V.; Covarrubias, R.; Dawson, K. S.; Deustua, S. E.; Ellis, R.; Fabbro, S.; Fadeyev, V.; Fan, X.; Farris, B.; Folatelli, G.; Frye, B. L.; Garavini, G.; Gates, E. L.; Germany, L.; Goldhaber, G.; Goldman, B.; Goobar, A.; Groom, D. E.; et al. (October 27, 2008). "Improved Cosmological Constraints from New, Old and Combined Supernova Datasets". *The Astrophysical Journal* (Chicago: University of Chicago Press) **686** (2): 749–778. arXiv:0804.4142. Bibcode:2008ApJ...686..749K. doi:10.1086/589937.. They find a best fit value of the dark energy density, $\Omega\Lambda$ of 0.713+0.027−0.029(stat)+0.036−0.039(sys), of the total

matter density, ΩM, of 0.274+0.016−0.016(stat)+0.013−0.012(sys) with an equation of state parameter w of −0.969+0.059−0.063(stat)+0.063−0.066(sys).

[23] "Big Bang's afterglow shows universe is 80 million years older than scientists first thought". *The Washington Post*. Retrieved 22 March 2013.

[24] "New method 'confirms dark energy'". BBC News. 2011-05-19.

[25] Dark energy is real. Swinburne University of Technology. 19 May 2011

[26] Crittenden; Neil Turok (1995). "Looking for Λ with the Rees-Sciama Effect". *Physical Review Letters* **76** (4): 575–578. arXiv:astro-ph/9510072. Bibcode:1996PhRvL..76..575C. doi:10.1103/PhysRevLett.76.575. PMID 10061494.

[27] Shirley Ho; Hirata; Nikhil Padmanabhan; Uros Seljak; Neta Bahcall (2008). "Correlation of CMB with large-scale structure: I. ISW Tomography and Cosmological Implications". *Physical Review D* **78** (4): 043519. arXiv:0801.0642. Bibcode:2008PhRvD..78d3519H. doi:10.1103/PhysRevD.78.043519.

[28] Tommaso Giannantonio; Ryan Scranton; Crittenden; Nichol; Boughn; Myers; Richards (2008). "Combined analysis of the integrated Sachs-Wolfe effect and cosmological implications". *Physical Review D* **77** (12): 123520. arXiv:0801.4380. Bibcode:2008PhRvD..77l3520G. doi:10.1103/PhysRevD.77.123520.

[29] Zelong Yi; Tongjie Zhang (2007). "Constraints on holographic dark energy models using the differential ages of passively evolving galaxies". *Modern Physics Letters A* **22** (1): 41. arXiv:astro-ph/0605596. Bibcode:2007MPLA...22...41Y. doi:10.1142/S0217732307020889.

[30] Haoyi Wan; Zelong Yi; Tongjie Zhang; Jie Zhou (2007). "Constraints on the DGP Universe Using Observational Hubble parameter". *Physics Letters B* **651** (5): 352. arXiv:0706.2723. Bibcode:2007PhLB..651..352W. doi:10.1016/j.physletb.2007.06.053.

[31] Cong Ma; Tongjie Zhang (2010). "Power of Observational Hubble Parameter Data: a Figure of Merit Exploration". *Astrophysical Journal* **730** (2): 74. arXiv:1007.3787. Bibcode:2011ApJ...730...74M. doi:10.1088/0004-637X/730/2/74.

[32] Tongjie Zhang; Cong Ma; Tian Lan (2010). "Constraints on the Dark Side of the Universe and Observational Hubble Parameter Data". *Advances in Astronomy* **2010** (1): 1. arXiv:1010.1307. Bibcode:2010AdAst2010E..81Z. doi:10.1155/2010/184284.

[33] Joan Simon; Licia Verde; Raul Jimenez (2005). "Constraints on the redshift dependence of the dark energy potential". *Physical Review D* **71** (12): 123001.

arXiv:astro-ph/0412269. Bibcode:2005PhRvD..71l3001S. doi:10.1103/PhysRevD.71.123001.

[34] "Planck reveals an almost perfect universe". *Planck*. ESA. 2013-03-21. Retrieved 2013-03-21.

[35] Wess, Julius; Bagger, Jonathan. *Supersymmetry and Supergravity*. ISBN 978-0691025308.

[36] Carroll, Sean M. (1998). "Quintessence and the Rest of the World: Suppressing Long-Range Interactions". *Physical Review Letters* **81** (15): 3067–3070. arXiv:astro-ph/9806099. Bibcode:1998PhRvL..81.3067C. doi:10.1103/PhysRevLett.81.3067. ISSN 0031-9007.

[37] Ratra, Bharat; Peebles, P.J.E. "Cosmological consequences of a rolling homogeneous scalar field". *Phys. Rev.* **D37**: 3406. Bibcode:1988PhRvD..37.3406R. doi:10.1103/PhysRevD.37.3406.

[38] Steinhardt, Paul J.; Wang, Li-Min; Zlatev, Ivaylo. "Cosmological tracking solutions". *Phys. Rev.* **D59**: 123504. arXiv:astro-ph/9812313. Bibcode:1999PhRvD..5913504S. doi:10.1103/PhysRevD.59.123504.

[39] R.R.Caldwell (2002). "A phantom menace? Cosmological consequences of a dark energy component with super-negative equation of state". *Physics Letters B* **545** (1-2): 23–29. arXiv:astro-ph/9908168. Bibcode:2002PhLB..545...23C. doi:10.1016/S0370-2693(02)02589-3.

[40] Wiltshire, David L. (2007). "Exact Solution to the Averaging Problem in Cosmology". *Physical Review Letters* **99** (25): 251101. arXiv:0709.0732. Bibcode:2007PhRvL..99y1101W. doi:10.1103/PhysRevLett.99.251101. PMID 18233512.

[41] Ishak, Mustapha; Richardson, James; Garred, David; Whittington, Delilah; Nwankwo, Anthony; Sussman, Roberto (2007). "Dark Energy or Apparent Acceleration Due to a Relativistic Cosmological Model More Complex than FLRW?". *Physical Review D* **78** (12): 123531. arXiv:0708.2943. Bibcode:2008PhRvD..78l3531I. doi:10.1103/PhysRevD.78.123531.

[42] Mattsson, Teppo (2007). "Dark energy as a mirage". *Gen. Rel. Grav.* **42** (3): 567–599. arXiv:0711.4264. Bibcode:2010GReGr..42..567M. doi:10.1007/s10714-009-0873-z.

[43] Clifton, Timothy; Ferreira, Pedro (April 2009). "Does Dark Energy Really Exist?". *Scientific American* **300** (4): 48–55. doi:10.1038/scientificamerican0409-48. PMID 19363920. Retrieved April 30, 2009.

[44] Wiltshire, D. (2008). "Cosmological equivalence principle and the weak-field limit". *Physical Review D* **78** (8): 084032. arXiv:0809.1183. Bibcode:2008PhRvD..78h4032W. doi:10.1103/PhysRevD.78.084032.

[45] Gray, Stuart. "Dark questions remain over dark energy". ABC Science Australia. Retrieved 27 January 2013.

[46] Merali, Zeeya (March 2012). "Is Einstein's Greatest Work All Wrong—Because He Didn't Go Far Enough?". *Discover magazine*. Retrieved 27 January 2013.

[47] by Ehsan Sadri M.A Ap

[48] Exirifard. Q. (2010). "Phenomenological covariant approach to gravity". *General Relativity and Gravitation* **43**: 93–106. arXiv:0808.1962. Bibcode:2011GReGr..43...93E. doi:10.1007/s10714-010-1073-6.

[49] Wolchover, Natalie (27 September 2011) 'Accelerating universe' could be just an illusion, MSNBC

[50] Tsagas. Christos G. (2011). "Peculiar motions, accelerated expansion, and the cosmological axis". *Physical Review D* **84** (6): 063503. arXiv:1107.4045. Bibcode:2011PhRvD..84f3503T. doi:10.1103/PhysRevD.84.063503.

[51] Khoshbin-e-Khoshnazar, M.R. (2013). "Binding Energy of the Very Early Universe: Abandoning Einstein for a Discretized Three–Torus Poset.A Proposal on the Origin of Dark Energy". *Gravitation and Cosmology* **19** (2): 106–113. doi:10.1134/s0202289313020059.

[52] Hogan, Jenny (2007). "Unseen Universe: Welcome to the dark side". *Nature* **448** (7151): 240–245. Bibcode:2007Natur.448..240H. doi:10.1038/448240a. PMID 17637630.

[53] Hossenfelder. S. (2008). "A Bi-Metric Theory with Exchange Symmetry". *Physical Review D* **78** (4): 044015. arXiv:0807.2838. Bibcode:2008PhRvD..78d4015H. doi:10.1103/PhysRevD.78.044015.

[54] Henry-Couannier. F. (2005). "Discrete Symmetries and General Relativity, the Dark Side of Gravity". *International Journal of Modern Physics A* **20** (11): 2341. arXiv:gr-qc/0410055. Bibcode:2005IJMPA..20.2341H. doi:10.1142/S0217751X05024602.

[55] Ripalda, Jose M. (1999). "Time reversal and negative energies in general relativity". arXiv:gr-qc/9906012.

[56] McCulloch. M.E. (2010). "Minimum accelerations from quantised inertia". *EPL* **90** (2): 29001. arXiv:1004.3303. Bibcode:2010EL.....9029001M. doi:10.1209/0295-5075/90/29001.

[57] Chevallier. M; Polarski. D (2001). "Accelerating Universes with Scaling Dark Matter". *International Journal of Modern Physics D* **10**: 213–224. arXiv:gr-qc/0009008. Bibcode:2001IJMPD..10..213C. doi:10.1142/S0218271801000822.

[58] Linder. Eric V. (3 March 2003). "Exploring the Expansion History of the Universe". *Physical Review Letters* **90** (9). arXiv:astro-ph/0208512v1. Bibcode:2003PhRvL..90i1301L. doi:10.1103/PhysRevLett.90.091301.

[59] Alcaniz. E.M.; Alcaniz. J.S. (2008). "A parametric model for dark energy". *Physics Letters B* **666**: 415–419. arXiv:0805.1713. Bibcode:2008PhLB..666..415B. doi:10.1016/j.physletb.2008.08.012.

[60] Jassal, H.K; Bagla. J.S (2010). "Understanding the origin of CMB constraints on Dark Energy". *Monthly Notices of the Royal Astronomical Society* **405**: 2639–2650. arXiv:astro-ph/0601389. Bibcode:2010MNRAS.405.2639J. doi:10.1111/j.1365-2966.2010.16647.x.

[61] Wetterich. C. (2004). "Phenomenological parameterization of quintessence". arXiv:astro-ph/0403289v1.

[62] Krauss. Lawrence M.; Scherrer, Robert J. (March 2008). "The End of Cosmology?". *Scientific American* **82**. Retrieved 2011-01-06.

[63] Is the universe expanding faster than the speed of light? (see the last two paragraphs)

[64] Lineweaver. Charles; Tamara M. Davis (2005). "Misconceptions about the Big Bang" (PDF). *Scientific American*. Retrieved 2008-11-06.

[65] Loeb. Abraham (2002). "The Long-Term Future of Extragalactic Astronomy". *Physical Review D* **65** (4): 047301. arXiv:astro-ph/0107568. Bibcode:2002PhRvD..65d7301L. doi:10.1103/PhysRevD.65.047301.

[66] Krauss. Lawrence M.; Robert J. Scherrer (2007). "The Return of a Static Universe and the End of Cosmology". *General Relativity and Gravitation* **39** (10): 1545–1550. arXiv:0704.0221. Bibcode:2007GReGr..39.1545K. doi:10.1007/s10714-007-0472-9.

[67] Using Tiny Particles To Answer Giant Questions. Science Friday. 3 Apr 2009. According to the transcript. Brian Greene makes the comment "And actually, in the far future. everything we now see, except for our local galaxy and a region of galaxies will have disappeared. The entire universe will disappear before our very eyes, and it's one of my arguments for actually funding cosmology. We've got to do it while we have a chance."

[68] *How the Universe Works 3*. End of the Universe. Discovery Channel. 2014.

[69] Harvey, Alex (2012). "How Einstein Discovered Dark Energy". arXiv:1211.6338.

[70] Gamow, George (1970) *My World Line: An Informal Autobiography*. p. 44: "Much later. when I was discussing cosmological problems with Einstein. he remarked that the introduction of the cosmological term was the biggest blunder he ever made in his life." – Here the "cosmological term" refers to the cosmological constant in the equations of general relativity, whose value Einstein initially picked to ensure that his model of the universe would neither expand nor contract; if he hadn't done this he might have theoretically predicted the universal expansion that was first observed by Edwin Hubble.

[71] The first appearance of the term "dark energy" is in the article with another cosmologist and Turner's student at the time, Dragan Huterer, "Prospects for Probing the Dark Energy via Supernova Distance Measurements", which was posted to the ArXiv.org e-print archive in August 1998 and published in Huterer, D.; Turner, M. (1999). "Prospects for probing the dark energy via supernova distance measurements". *Physical Review D* **60** (8). arXiv:astro-ph/9808133. Bibcode:1999PhRvD..60h1301H. doi:10.1103/PhysRevD.60.081301., although the manner in which the term is treated there suggests it was already in general use. Cosmologist Saul Perlmutter has credited Turner with coining the term in an article they wrote together with Martin White, where it is introduced in quotation marks as if it were a neologism. Perlmutter, S.; Turner, M.; White, M. (1999). "Constraining Dark Energy with Type Ia Supernovae and Large-Scale Structure". *Physical Review Letters* **83** (4): 670. arXiv:astro-ph/9901052. Bibcode:1999PhRvL..83..670P. doi:10.1103/PhysRevLett.83.670.

[72] Astier, Pierre (Supernova Legacy Survey); Guy; Regnault; Pain; Aubourg; Balam; Basa; Carlberg; Fabbro; Fouchez; Hook; Howell; Lafoux; Neill; Palanque-Delabrouille; Perrett; Pritchet; Rich; Sullivan; Taillet; Aldering; Antilogus; Arsenijevic; Balland; Baumont; Bronder; Courtois; Ellis; Filiol; et al. (2006). "The Supernova legacy survey: Measurement of ΩM, ΩΛ and W from the first year data set". *Astronomy and Astrophysics* **447**: 31–48. arXiv:astro-ph/0510447. Bibcode:2006A&A...447...31A. doi:10.1051/0004-6361:20054185.

1.6.9 External links

*

- Dark Energy on *In Our Time* at the BBC. (listen now)

- Dark energy Eric Linder Scholarpedia 3(2):4900. doi: 10.4249/scholarpedia.4900

- Dark energy: how the paradigm shifted Physicsworld.com

- Dennis Overbye (November 2006). "9 Billion-Year-Old 'Dark Energy' Reported". *The New York Times.*

- "Mysterious force's long presence" BBC News online (2006) More evidence for dark energy being the cosmological constant

- "Astronomy Picture of the Day" one of the images of the Cosmic Microwave Background which confirmed the presence of dark energy and dark matter

- SuperNova Legacy Survey home page The Canada-France-Hawaii Telescope Legacy Survey Supernova Program aims primarily at measuring the equation of

state of Dark Energy. It is designed to precisely measure several hundred high-redshift supernovae.

- "Report of the Dark Energy Task Force"

- "HubbleSite.org – Dark Energy Website" Multimedia presentation explores the science of dark energy and Hubble's role in its discovery.

- "Surveying the dark side"

- "Dark energy and 3-manifold topology" Acta Physica Polonica 38 (2007), p. 3633–3639

- The Dark Energy Survey

- The Joint Dark Energy Mission

- Harvard: Dark Energy Found Stifling Growth in Universe, primary source

- April 2010 Smithsonian Magazine Article

- HETDEX Dark energy experiment

- Dark Energy FAQ

- "The Hunt for Dark Energy" George FR Ellis, Peter Cameron and David Tong discuss the presence of dark energy in the Universe

- Euclid ESA Satellite, a mission to map the geometry of the dark universe

1.7 General relativity

For the book by Robert Wald, see General Relativity (book).

For a more accessible and less technical introduction to this topic, see Introduction to general relativity.

General relativity (**GR**, also known as the **general theory of relativity** or **GTR**) is the geometric theory of gravitation published by Albert Einstein in 1915[2] and the current description of gravitation in modern physics. General relativity generalizes special relativity and Newton's law of universal gravitation, providing a unified description of gravity as a geometric property of space and time, or spacetime. In particular, the curvature of spacetime is directly related to the energy and momentum of whatever matter and radiation are present. The relation is specified by the Einstein field equations, a system of partial differential equations.

Some predictions of general relativity differ significantly from those of classical physics, especially concerning the passage of time, the geometry of space, the motion of bodies in free fall, and the propagation of light. Examples of such differences include gravitational time dilation,

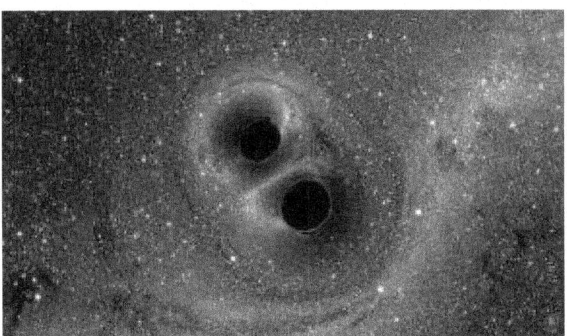

Slow motion computer simulation of the black hole binary system GW150914 as seen by a nearby observer, during 0.33 s of its final inspiral, merge, and ringdown. The star field behind the black holes is being heavily distorted and appears to rotate and move, due to extreme gravitational lensing, as space-time itself is distorted and dragged around by the rotating black holes.[1]

Albert Einstein developed the theories of special and general relativity. Picture from 1921.

gravitational lensing, the gravitational redshift of light, and the gravitational time delay. The predictions of general relativity have been confirmed in all observations and experiments to date. Although general relativity is not the only relativistic theory of gravity, it is the simplest theory that is consistent with experimental data. However, unanswered questions remain, the most fundamental being how general relativity can be reconciled with the laws of quantum physics to produce a complete and self-consistent theory of quantum gravity.

Einstein's theory has important astrophysical implications. For example, it implies the existence of black holes—regions of space in which space and time are distorted in such a way that nothing, not even light, can escape—as an end-state for massive stars. There is ample evidence that the intense radiation emitted by certain kinds of astronomical objects is due to black holes; for example, microquasars and active galactic nuclei result from the presence of stellar black holes and black holes of a much more massive type, respectively. The bending of light by gravity can lead to the phenomenon of gravitational lensing, in which multiple images of the same distant astronomical object are visible in the sky. General relativity also predicts the existence of gravitational waves, which have since been observed directly by physics collaboration LIGO. In addition, general relativity is the basis of current cosmological models of a consistently expanding universe.

1.7.1   History

Main articles: History of general relativity and Classical theories of gravitation

Soon after publishing the special theory of relativity in 1905, Einstein started thinking about how to incorporate gravity into his new relativistic framework. In 1907, beginning with a simple thought experiment involving an observer in free fall, he embarked on what would be an eight-year search for a relativistic theory of gravity. After numerous detours and false starts, his work culminated in the presentation to the Prussian Academy of Science in November 1915 of what are now known as the Einstein field equations. These equations specify how the geometry of space and time is influenced by whatever matter and radiation are present, and form the core of Einstein's general theory of relativity.[3]

The Einstein field equations are nonlinear and very difficult to solve. Einstein used approximation methods in working out initial predictions of the theory. But as early as 1916, the astrophysicist Karl Schwarzschild found the first non-trivial exact solution to the Einstein field equations, the so-called Schwarzschild metric. This solution laid the groundwork for the description of the final stages of gravitational collapse, and the objects known today as black holes. In the same year, the first steps towards generalizing Schwarzschild's solution to electrically charged objects were taken, which eventually resulted in the Reissner–Nordström solution, now associated with electrically charged black holes.[4] In 1917, Einstein applied his theory to the universe as a whole, initiating the field of relativistic cosmology. In line with contemporary thinking, he assumed a static universe, adding a new parameter to his original field equations—the cosmological constant—to match that observational presumption.[5] By 1929, however, the work of Hubble and others had shown that our universe is expanding. This is readily described by the expanding cosmological solutions found by Friedmann in 1922, which do not require a cosmological constant. Lemaître used these solutions to formulate the earliest version of the Big Bang models, in which our universe has evolved from an extremely hot and dense earlier state.[6] Einstein later declared the cosmological constant the biggest blunder of his life.[7]

During that period, general relativity remained something of a curiosity among physical theories. It was clearly superior to Newtonian gravity, being consistent with special relativity and accounting for several effects unexplained by the Newtonian theory. Einstein himself had shown in 1915 how his theory explained the anomalous perihelion advance of the planet Mercury without any arbitrary parameters ("fudge factors").[8] Similarly, a 1919 expedition led by Eddington confirmed general relativity's prediction for the deflection of starlight by the Sun during the total solar eclipse of May 29, 1919,[9] making Einstein instantly famous.[10] Yet the theory entered the mainstream of theoretical physics and astrophysics only with the developments between approximately 1960 and 1975, now known as the golden age of general relativity.[11] Physicists began to understand the concept of a black hole, and to identify quasars as one of these objects' astrophysical manifestations.[12] Ever more precise solar system tests confirmed the theory's predictive power,[13] and relativistic cosmology, too, became amenable to direct observational tests.[14]

1.7.2   From classical mechanics to general relativity

General relativity can be understood by examining its similarities with and departures from classical physics. The first step is the realization that classical mechanics and Newton's law of gravity admit a geometric description. The combination of this description with the laws of special relativity results in a heuristic derivation of general relativity.[15]

Geometry of Newtonian gravity

At the base of classical mechanics is the notion that a body's motion can be described as a combination of free (or inertial) motion, and deviations from this free motion. Such deviations are caused by external forces acting on a

According to general relativity, objects in a gravitational field behave similarly to objects within an accelerating enclosure. For example, an observer will see a ball fall the same way in a rocket (left) as it does on Earth (right), provided that the acceleration of the rocket is equal to 9.8 m/s² (the acceleration due to gravity at the surface of the Earth).

body in accordance with Newton's second law of motion, which states that the net force acting on a body is equal to that body's (inertial) mass multiplied by its acceleration.[16] The preferred inertial motions are related to the geometry of space and time: in the standard reference frames of classical mechanics, objects in free motion move along straight lines at constant speed. In modern parlance, their paths are geodesics, straight world lines in curved spacetime.[17]

Conversely, one might expect that inertial motions, once identified by observing the actual motions of bodies and making allowances for the external forces (such as electromagnetism or friction), can be used to define the geometry of space, as well as a time coordinate. However, there is an ambiguity once gravity comes into play. According to Newton's law of gravity, and independently verified by experiments such as that of Eötvös and its successors (see Eötvös experiment), there is a universality of free fall (also known as the weak equivalence principle, or the universal equality of inertial and passive-gravitational mass): the trajectory of a test body in free fall depends only on its position and initial speed, but not on any of its material properties.[18] A simplified version of this is embodied in Einstein's elevator experiment, illustrated in the figure on the right: for an observer in a small enclosed room, it is impossible to decide, by mapping the trajectory of bodies such as a dropped ball, whether the room is at rest in a gravitational field, or in free space aboard a rocket that is accelerating at a rate equal to that of the gravitational field.[19]

Given the universality of free fall, there is no observable distinction between inertial motion and motion under the influence of the gravitational force. This suggests the definition of a new class of inertial motion, namely that of objects in free fall under the influence of gravity. This new class of preferred motions, too, defines a geometry of space and time—in mathematical terms, it is the geodesic motion associated with a specific connection which depends on the gradient of the gravitational potential. Space, in this construction, still has the ordinary Euclidean geometry. However, space*time* as a whole is more complicated. As can be shown using simple thought experiments following the free-fall trajectories of different test particles, the result of transporting spacetime vectors that can denote a particle's velocity (time-like vectors) will vary with the particle's trajectory; mathematically speaking, the Newtonian connection is not integrable. From this, one can deduce that spacetime is curved. The resulting Newton–Cartan theory is a geometric formulation of Newtonian gravity using only covariant concepts, i.e. a description which is valid in any desired coordinate system.[20] In this geometric description, tidal effects—the relative acceleration of bodies in free fall—are related to the derivative of the connection, showing how the modified geometry is caused by the presence of mass.[21]

Relativistic generalization

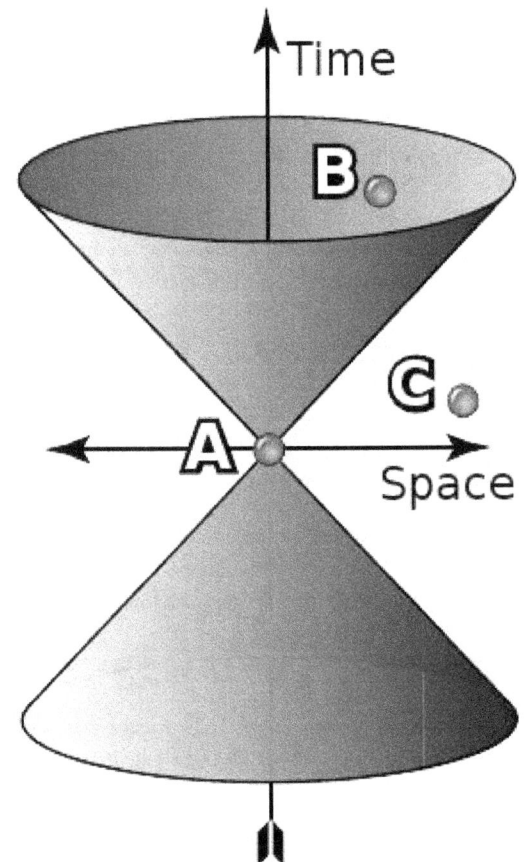

Light cone

As intriguing as geometric Newtonian gravity may be, its basis, classical mechanics, is merely a limiting case of (special) relativistic mechanics.[22] In the language of symmetry: where gravity can be neglected, physics is Lorentz invariant as in special relativity rather than Galilei invariant as in classical mechanics. (The defining symmetry of special relativity is the Poincaré group, which includes translations and rotations.) The differences between the two become significant when dealing with speeds approaching the speed of light, and with high-energy phenomena.[23]

With Lorentz symmetry, additional structures come into play. They are defined by the set of light cones (see image). The light-cones define a causal structure: for each event A, there is a set of events that can, in principle, either influence or be influenced by A via signals or interactions that do not need to travel faster than light (such as event B in the image), and a set of events for which such an influence is impossible (such as event C in the image). These sets are observer-independent.[24] In conjunction with the world-lines of freely falling particles, the light-cones can be used to reconstruct the space–time's semi-Riemannian metric, at least up to a positive scalar factor. In mathematical terms, this defines a Conformal structure[25] or conformal geometry.

Special relativity is defined in the absence of gravity, so for practical applications, it is a suitable model whenever gravity can be neglected. Bringing gravity into play, and assuming the universality of free fall, an analogous reasoning as in the previous section applies: there are no global inertial frames. Instead there are approximate inertial frames moving alongside freely falling particles. Translated into the language of spacetime: the straight time-like lines that define a gravity-free inertial frame are deformed to lines that are curved relative to each other, suggesting that the inclusion of gravity necessitates a change in spacetime geometry.[26]

A priori, it is not clear whether the new local frames in free fall coincide with the reference frames in which the laws of special relativity hold—that theory is based on the propagation of light, and thus on electromagnetism, which could have a different set of preferred frames. But using different assumptions about the special-relativistic frames (such as their being earth-fixed, or in free fall), one can derive different predictions for the gravitational redshift, that is, the way in which the frequency of light shifts as the light propagates through a gravitational field (cf. below). The actual measurements show that free-falling frames are the ones in which light propagates as it does in special relativity.[27] The generalization of this statement, namely that the laws of special relativity hold to good approximation in freely falling (and non-rotating) reference frames, is known as the Einstein equivalence principle, a crucial guiding principle for generalizing special-relativistic physics to include gravity.[28]

The same experimental data shows that time as measured by clocks in a gravitational field—proper time, to give the technical term—does not follow the rules of special relativity. In the language of spacetime geometry, it is not measured by the Minkowski metric. As in the Newtonian case, this is suggestive of a more general geometry. At small scales, all reference frames that are in free fall are equivalent, and approximately Minkowskian. Consequently, we are now dealing with a curved generalization of Minkowski space. The metric tensor that defines the geometry—in particular, how lengths and angles are measured—is not the Minkowski metric of special relativity, it is a generalization known as a semi- or pseudo-Riemannian metric. Furthermore, each Riemannian metric is naturally associated with one particular kind of connection, the Levi-Civita connection, and this is, in fact, the connection that satisfies the equivalence principle and makes space locally Minkowskian (that is, in suitable locally inertial coordinates, the metric is Minkowskian, and its first partial derivatives and the connection coefficients vanish).[29]

Einstein's equations

Main articles: Einstein field equations and Mathematics of general relativity

Having formulated the relativistic, geometric version of the effects of gravity, the question of gravity's source remains. In Newtonian gravity, the source is mass. In special relativity, mass turns out to be part of a more general quantity called the energy–momentum tensor, which includes both energy and momentum densities as well as stress (that is, pressure and shear).[30] Using the equivalence principle, this tensor is readily generalized to curved space-time. Drawing further upon the analogy with geometric Newtonian gravity, it is natural to assume that the field equation for gravity relates this tensor and the Ricci tensor, which describes a particular class of tidal effects: the change in volume for a small cloud of test particles that are initially at rest, and then fall freely. In special relativity, conservation of energy–momentum corresponds to the statement that the energy–momentum tensor is divergence-free. This formula, too, is readily generalized to curved spacetime by replacing partial derivatives with their curved-manifold counterparts, covariant derivatives studied in differential geometry. With this additional condition—the covariant divergence of the energy–momentum tensor, and hence of whatever is on the other side of the equation, is zero— the simplest set of equations are what are called Einstein's (field) equations:

On the left-hand side is the Einstein tensor, a specific divergence-free combination of the Ricci tensor $R_{\mu\nu}$ and the metric. Where $G_{\mu\nu}$ is symmetric. In particular,

$$R = g^{\mu\nu} R_{\mu\nu}$$

is the curvature scalar. The Ricci tensor itself is related to the more general Riemann curvature tensor as

$$R_{\mu\nu} = R^\alpha{}_{\mu\alpha\nu}.$$

On the right-hand side, $T_{\mu\nu}$ is the energy–momentum tensor. All tensors are written in abstract index notation.[31] Matching the theory's prediction to observational results for planetary orbits (or, equivalently, assuring that the weak-gravity, low-speed limit is Newtonian mechanics), the proportionality constant can be fixed as $\kappa = 8\pi G/c^4$, with G the gravitational constant and c the speed of light.[32] When there is no matter present, so that the energy–momentum tensor vanishes, the results are the vacuum Einstein equations,

$$R_{\mu\nu} = 0.$$

Alternatives to general relativity

Main article: Alternatives to general relativity

There are alternatives to general relativity built upon the same premises, which include additional rules and/or constraints, leading to different field equations. Examples are Brans–Dicke theory, teleparallelism, f(R) gravity and Einstein–Cartan theory.[33]

1.7.3 Definition and basic applications

See also: Mathematics of general relativity and Physical theories modified by general relativity

The derivation outlined in the previous section contains all the information needed to define general relativity, describe its key properties, and address a question of crucial importance in physics, namely how the theory can be used for model-building.

Definition and basic properties

General relativity is a metric theory of gravitation. At its core are Einstein's equations, which describe the relation between the geometry of a four-dimensional, pseudo-Riemannian manifold representing spacetime, and the energy–momentum contained in that spacetime.[34] Phenomena that in classical mechanics are ascribed to the action of the force of gravity (such as free-fall, orbital motion, and spacecraft trajectories), correspond to inertial motion within a curved geometry of spacetime in general relativity; there is no gravitational force deflecting objects from their natural, straight paths. Instead, gravity corresponds to changes in the properties of space and time, which in turn changes the straightest-possible paths that objects will naturally follow.[35] The curvature is, in turn, caused by the energy–momentum of matter. Paraphrasing the relativist John Archibald Wheeler, spacetime tells matter how to move; matter tells spacetime how to curve.[36]

While general relativity replaces the scalar gravitational potential of classical physics by a symmetric rank-two tensor, the latter reduces to the former in certain limiting cases. For weak gravitational fields and slow speed relative to the speed of light, the theory's predictions converge on those of Newton's law of universal gravitation.[37]

As it is constructed using tensors, general relativity exhibits general covariance: its laws—and further laws formulated within the general relativistic framework—take on the same form in all coordinate systems.[38] Furthermore, the theory does not contain any invariant geometric background structures, i.e. it is background independent. It thus satisfies a more stringent general principle of relativity, namely that the laws of physics are the same for all observers.[39] Locally, as expressed in the equivalence principle, spacetime is Minkowskian, and the laws of physics exhibit local Lorentz invariance.[40]

Model-building

The core concept of general-relativistic model-building is that of a solution of Einstein's equations. Given both Einstein's equations and suitable equations for the properties of matter, such a solution consists of a specific semi-Riemannian manifold (usually defined by giving the metric in specific coordinates), and specific matter fields defined on that manifold. Matter and geometry must satisfy Einstein's equations, so in particular, the matter's energy–momentum tensor must be divergence-free. The matter must, of course, also satisfy whatever additional equations were imposed on its properties. In short, such a solution is a model universe that satisfies the laws of general relativity, and possibly additional laws governing whatever matter might be present.[41]

Einstein's equations are nonlinear partial differential equations and, as such, difficult to solve exactly.[42] Nevertheless, a number of exact solutions are known, although only a few have direct physical applications.[43] The best-known exact solutions, and also those most interesting from a physics point of view, are the Schwarzschild solution, the Reissner–Nordström solution and the Kerr metric, each corresponding to a certain type of black hole in an otherwise empty universe,[44] and the Friedmann–Lemaître–Robertson–Walker and de Sitter universes, each describing an expanding cosmos.[45] Exact solutions of great theoretical interest include the Gödel universe (which opens up the intriguing possibility of time travel in curved spacetimes), the Taub-NUT solution (a model universe that is homogeneous, but anisotropic), and anti-de Sitter space (which has recently come to prominence in the context of what is called the Maldacena conjecture).[46]

Given the difficulty of finding exact solutions, Einstein's field equations are also solved frequently by numerical integration on a computer, or by considering small perturbations of exact solutions. In the field of numerical relativity, powerful computers are employed to simulate the geometry of spacetime and to solve Einstein's equations for interesting situations such as two colliding black holes.[47] In principle, such methods may be applied to any system, given sufficient computer resources, and may address fundamental questions such as naked singularities. Approximate solutions may also be found by perturbation theories such as linearized gravity[48] and its generalization, the post-Newtonian expansion, both of which were developed by Einstein. The latter provides a systematic approach to solving for the geometry of a spacetime that contains a distribution of matter that moves slowly compared with the speed of light. The expansion involves a series of terms; the first terms represent Newtonian gravity, whereas the later terms represent ever smaller corrections to Newton's theory due to general relativity.[49] An extension of this expansion is the parametrized post-Newtonian (PPN) formalism, which allows quantitative comparisons between the predictions of general relativity and alternative theories.[50]

1.7.4   Consequences of Einstein's theory

General relativity has a number of physical consequences. Some follow directly from the theory's axioms, whereas others have become clear only in the course of many years of research that followed Einstein's initial publication.

Gravitational time dilation and frequency shift

Main article: Gravitational time dilation

Assuming that the equivalence principle holds,[51] gravity influences the passage of time. Light sent down into a gravity well is blueshifted, whereas light sent in the opposite direction (i.e., climbing out of the gravity well) is redshifted; collectively, these two effects are known as the

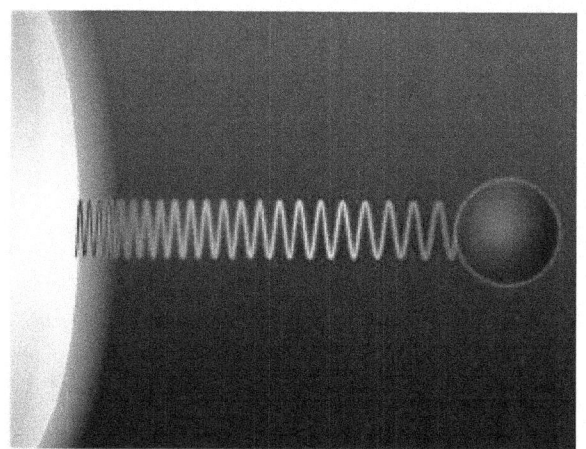

Schematic representation of the gravitational redshift of a light wave escaping from the surface of a massive body

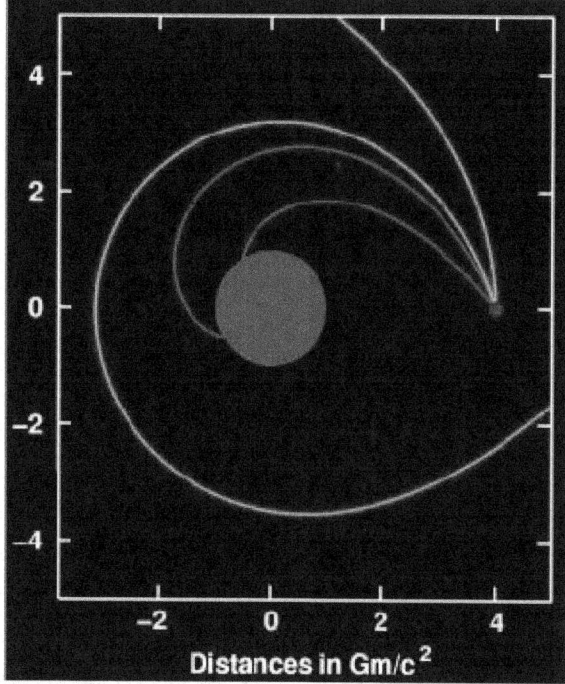

Distances in Gm/c²

Deflection of light (sent out from the location shown in blue) near a compact body (shown in gray)

gravitational frequency shift. More generally, processes close to a massive body run more slowly when compared with processes taking place farther away; this effect is known as gravitational time dilation.[52]

Gravitational redshift has been measured in the laboratory[53] and using astronomical observations.[54] Gravitational time dilation in the Earth's gravitational field has been measured numerous times using atomic clocks,[55] while ongoing validation is provided as a side effect of the operation of the Global Positioning System (GPS).[56] Tests in stronger gravitational fields are provided by the observation of binary pulsars.[57] All results are in agreement with general relativity.[58] However, at the current level of accuracy, these observations cannot distinguish between general relativity and other theories in which the equivalence principle is valid.[59]

Light deflection and gravitational time delay

Main articles: Kepler problem in general relativity, Gravitational lens and Shapiro delay

General relativity predicts that the path of light is bent in a gravitational field; light passing a massive body is deflected towards that body. This effect has been confirmed by observing the light of stars or distant quasars being deflected as it passes the Sun.[60]

This and related predictions follow from the fact that light follows what is called a light-like or null geodesic—a generalization of the straight lines along which light travels in classical physics. Such geodesics are the generalization of the invariance of lightspeed in special relativity.[61] As one examines suitable model spacetimes (either the exterior Schwarzschild solution or, for more than a single mass, the post-Newtonian expansion),[62] several effects of gravity on

light propagation emerge. Although the bending of light can also be derived by extending the universality of free fall to light,[63] the angle of deflection resulting from such calculations is only half the value given by general relativity.[64]

Closely related to light deflection is the gravitational time delay (or Shapiro delay), the phenomenon that light signals take longer to move through a gravitational field than they would in the absence of that field. There have been numerous successful tests of this prediction.[65] In the parameterized post-Newtonian formalism (PPN), measurements of both the deflection of light and the gravitational time delay determine a parameter called γ, which encodes the influence of gravity on the geometry of space.[66]

Gravitational waves

Main article: Gravitational wave

Predicted in 1916[67][68] by Albert Einstein, there are gravitational waves: ripples in the metric of spacetime that propagate at the speed of light. These are one of several analogies between weak-field gravity and electromagnetism in that, they are analogous to electromagnetic waves. On February 11, 2016, the Advanced LIGO team announced that they had directly detected gravitational waves from a pair of black holes merging.[69][70][71]

The simplest type of such a wave can be visualized by its action on a ring of freely floating particles. A sine wave prop-

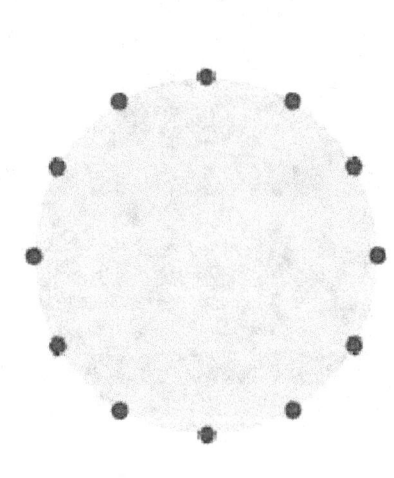

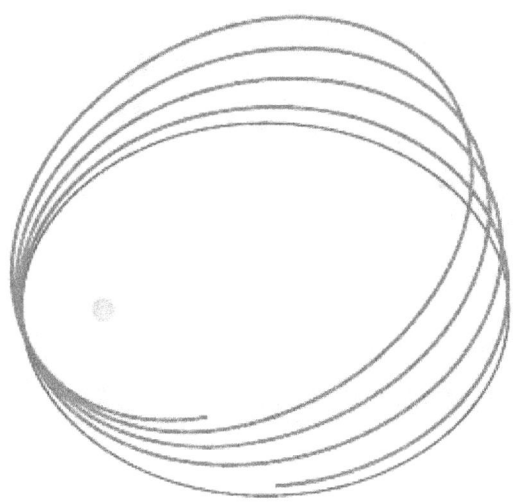

Newtonian (red) vs. Einsteinian orbit (blue) of a lone planet orbiting a star

Ring of test particles influenced by gravitational wave

agating through such a ring towards the reader distorts the ring in a characteristic, rhythmic fashion (animated image to the right).[72] Since Einstein's equations are non-linear, arbitrarily strong gravitational waves do not obey linear superposition, making their description difficult. However, for weak fields, a linear approximation can be made. Such linearized gravitational waves are sufficiently accurate to describe the exceedingly weak waves that are expected to arrive here on Earth from far-off cosmic events, which typically result in relative distances increasing and decreasing by 10^{-21} or less. Data analysis methods routinely make use of the fact that these linearized waves can be Fourier decomposed.[73]

Some exact solutions describe gravitational waves without any approximation, e.g., a wave train traveling through empty space[74] or so-called Gowdy universes, varieties of an expanding cosmos filled with gravitational waves.[75] But for gravitational waves produced in astrophysically relevant situations, such as the merger of two black holes, numerical methods are presently the only way to construct appropriate models.[76]

Orbital effects and the relativity of direction

Main article: Kepler problem in general relativity

General relativity differs from classical mechanics in a number of predictions concerning orbiting bodies. It predicts an overall rotation (precession) of planetary orbits, as well as orbital decay caused by the emission of gravitational waves and effects related to the relativity of direction.

Precession of apsides   In general relativity, the apsides of any orbit (the point of the orbiting body's closest approach to the system's center of mass) will precess—the orbit is not an ellipse, but akin to an ellipse that rotates on its focus, resulting in a rose curve-like shape (see image). Einstein first derived this result by using an approximate metric representing the Newtonian limit and treating the orbiting body as a test particle. For him, the fact that his theory gave a straightforward explanation of the anomalous perihelion shift of the planet Mercury, discovered earlier by Urbain Le Verrier in 1859, was important evidence that he had at last identified the correct form of the gravitational field equations.[77]

The effect can also be derived by using either the exact Schwarzschild metric (describing spacetime around a spherical mass)[78] or the much more general post-Newtonian formalism.[79] It is due to the influence of gravity on the geometry of space and to the contribution of self-energy to a body's gravity (encoded in the nonlinearity of Einstein's equations).[80] Relativistic precession has been observed for all planets that allow for accurate precession measurements (Mercury, Venus, and Earth),[81] as well as in binary pulsar systems, where it is larger by five orders of magnitude.[82]

In general relativity the perihelion shift σ, expressed in radians per revolution, is approximately given by:[83]

$$\sigma = \frac{24\pi^3 L^2}{T^2 c^2 (1 - e^2)},$$

where L is the semi-major axis, T is the orbital period, c is the speed of light, and e is the orbital eccentricity.

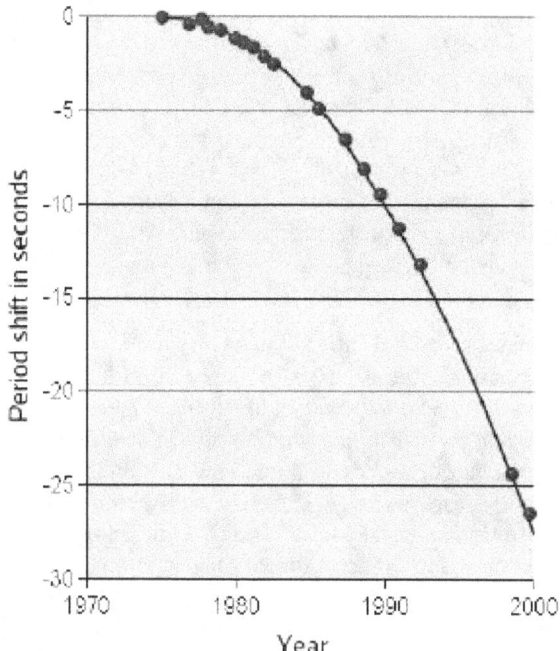

Orbital decay for PSR1913+16: time shift in seconds, tracked over three decades.[84]

Orbital decay According to general relativity, a binary system will emit gravitational waves, thereby losing energy. Due to this loss, the distance between the two orbiting bodies decreases, and so does their orbital period. Within the Solar System or for ordinary double stars, the effect is too small to be observable. This is not the case for a close binary pulsar, a system of two orbiting neutron stars, one of which is a pulsar: from the pulsar, observers on Earth receive a regular series of radio pulses that can serve as a highly accurate clock, which allows precise measurements of the orbital period. Because neutron stars are immensely compact, significant amounts of energy are emitted in the form of gravitational radiation.[85]

The first observation of a decrease in orbital period due to the emission of gravitational waves was made by Hulse and Taylor, using the binary pulsar PSR1913+16 they had discovered in 1974. This was the first detection of gravitational waves, albeit indirect, for which they were awarded the 1993 Nobel Prize in physics.[86] Since then, several other binary pulsars have been found, in particular the double pulsar PSR J0737-3039, in which both stars are pulsars.[87]

Geodetic precession and frame-dragging Main articles: Geodetic precession and Frame dragging

Several relativistic effects are directly related to the relativity of direction.[88] One is geodetic precession: the axis

direction of a gyroscope in free fall in curved spacetime will change when compared, for instance, with the direction of light received from distant stars—even though such a gyroscope represents the way of keeping a direction as stable as possible ("parallel transport").[89] For the Moon–Earth system, this effect has been measured with the help of lunar laser ranging.[90] More recently, it has been measured for test masses aboard the satellite Gravity Probe B to a precision of better than 0.3%.[91][92]

Near a rotating mass, there are so-called gravitomagnetic or frame-dragging effects. A distant observer will determine that objects close to the mass get "dragged around". This is most extreme for rotating black holes where, for any object entering a zone known as the ergosphere, rotation is inevitable.[93] Such effects can again be tested through their influence on the orientation of gyroscopes in free fall.[94] Somewhat controversial tests have been performed using the LAGEOS satellites, confirming the relativistic prediction.[95] Also the Mars Global Surveyor probe around Mars has been used.[96][97]

1.7.5 Astrophysical applications

Gravitational lensing

Main article: Gravitational lensing
The deflection of light by gravity is responsible for a new

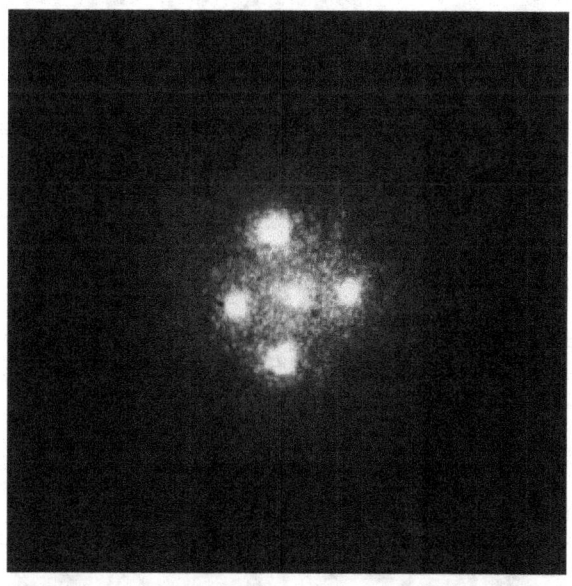

Einstein cross: four images of the same astronomical object, produced by a gravitational lens

class of astronomical phenomena. If a massive object is situated between the astronomer and a distant target object with appropriate mass and relative distances, the astronomer will see multiple distorted images of the target.

Such effects are known as gravitational lensing.[98] Depending on the configuration, scale, and mass distribution, there can be two or more images, a bright ring known as an Einstein ring, or partial rings called arcs.[99] The earliest example was discovered in 1979;[100] since then, more than a hundred gravitational lenses have been observed.[101] Even if the multiple images are too close to each other to be resolved, the effect can still be measured, e.g., as an overall brightening of the target object; a number of such "microlensing events" have been observed.[102]

Gravitational lensing has developed into a tool of observational astronomy. It is used to detect the presence and distribution of dark matter, provide a "natural telescope" for observing distant galaxies, and to obtain an independent estimate of the Hubble constant. Statistical evaluations of lensing data provide valuable insight into the structural evolution of galaxies.[103]

Gravitational wave astronomy

Main articles: Gravitational wave and Gravitational wave astronomy

Observations of binary pulsars provide strong indirect evidence for the existence of gravitational waves (see Orbital decay, above). Detection of these waves is a major goal of current relativity-related research.[104] Several land-based gravitational wave detectors are currently in operation, most notably the interferometric detectors GEO 600, LIGO (two detectors), TAMA 300 and VIRGO.[105] Various pulsar timing arrays are using millisecond pulsars to detect gravitational waves in the 10^{-9} to 10^{-6} Hertz frequency range, which originate from binary supermassive blackholes.[106] A European space-based detector, eLISA / NGO, is currently under development,[107] with a precursor mission (LISA Pathfinder) having launched in December 2015.[108]

Observations of gravitational waves promise to complement observations in the electromagnetic spectrum.[109] They are expected to yield information about black holes and other dense objects such as neutron stars and white dwarfs, about certain kinds of supernova implosions, and about processes in the very early universe, including the signature of certain types of hypothetical cosmic string.[110] In February 2016, the Advanced LIGO team announced that they had detected gravitational waves from a black hole merger.[69][70][111]

Black holes and other compact objects

Main article: Black hole

Whenever the ratio of an object's mass to its radius becomes sufficiently large, general relativity predicts the formation of a black hole, a region of space from which nothing, not even light, can escape. In the currently accepted models of stellar evolution, neutron stars of around 1.4 solar masses, and stellar black holes with a few to a few dozen solar masses, are thought to be the final state for the evolution of massive stars.[112] Usually a galaxy has one supermassive black hole with a few million to a few billion solar masses in its center,[113] and its presence is thought to have played an important role in the formation of the galaxy and larger cosmic structures.[114]

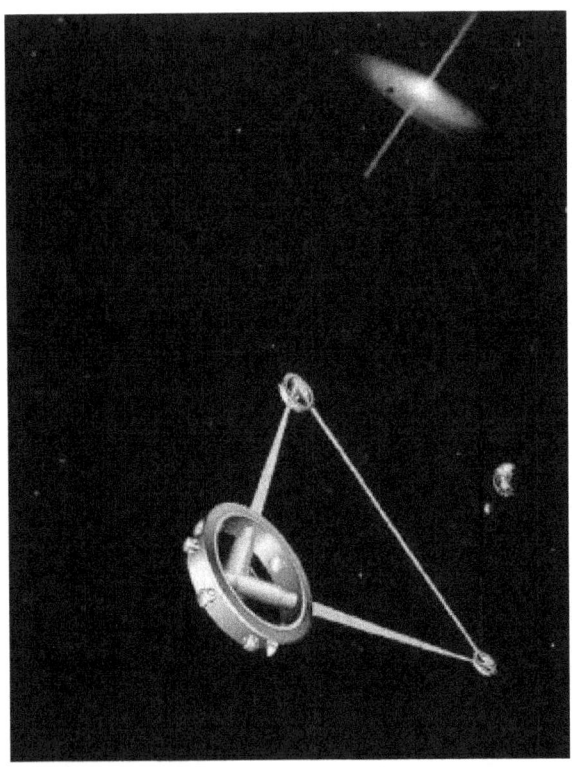

Artist's impression of the space-borne gravitational wave detector LISA

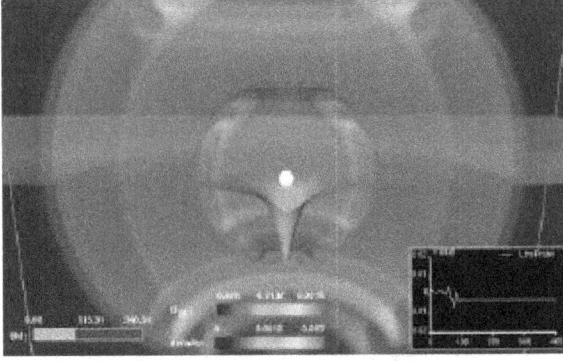

Simulation based on the equations of general relativity: a star collapsing to form a black hole while emitting gravitational waves

Astronomically, the most important property of compact

objects is that they provide a supremely efficient mechanism for converting gravitational energy into electromagnetic radiation.[115] Accretion, the falling of dust or gaseous matter onto stellar or supermassive black holes, is thought to be responsible for some spectacularly luminous astronomical objects, notably diverse kinds of active galactic nuclei on galactic scales and stellar-size objects such as microquasars.[116] In particular, accretion can lead to relativistic jets, focused beams of highly energetic particles that are being flung into space at almost light speed.[117] General relativity plays a central role in modelling all these phenomena,[118] and observations provide strong evidence for the existence of black holes with the properties predicted by the theory.[119]

Black holes are also sought-after targets in the search for gravitational waves (cf. Gravitational waves, above). Merging black hole binaries should lead to some of the strongest gravitational wave signals reaching detectors here on Earth, and the phase directly before the merger ("chirp") could be used as a "standard candle" to deduce the distance to the merger events—and hence serve as a probe of cosmic expansion at large distances.[120] The gravitational waves produced as a stellar black hole plunges into a supermassive one should provide direct information about the supermassive black hole's geometry.[121]

Cosmology

This blue horseshoe is a distant galaxy that has been magnified and warped into a nearly complete ring by the strong gravitational pull of the massive foreground luminous red galaxy.

Main article: Physical cosmology

The current models of cosmology are based on Einstein's field equations, which include the cosmological constant Λ since it has important influence on the large-scale dynamics of the cosmos,

$$R_{\mu\nu} - \frac{1}{2} R\, g_{\mu\nu} + \Lambda\, g_{\mu\nu} = \frac{8\pi G}{c^4}\, T_{\mu\nu}$$

where $g_{\mu\nu}$ is the spacetime metric.[122] Isotropic and homogeneous solutions of these enhanced equations, the Friedmann–Lemaître–Robertson–Walker solutions,[123] allow physicists to model a universe that has evolved over the past 14 billion years from a hot, early Big Bang phase.[124] Once a small number of parameters (for example the universe's mean matter density) have been fixed by astronomical observation,[125] further observational data can be used to put the models to the test.[126] Predictions, all successful, include the initial abundance of chemical elements formed in a period of primordial nucleosynthesis,[127] the large-scale structure of the universe,[128] and the existence and properties of a "thermal echo" from the early cosmos, the cosmic background radiation.[129]

Astronomical observations of the cosmological expansion rate allow the total amount of matter in the universe to be estimated, although the nature of that matter remains mysterious in part. About 90% of all matter appears to be so-called dark matter, which has mass (or, equivalently, gravitational influence), but does not interact electromagnetically and, hence, cannot be observed directly.[130] There is no generally accepted description of this new kind of matter, within the framework of known particle physics[131] or otherwise.[132] Observational evidence from redshift surveys of distant supernovae and measurements of the cosmic background radiation also show that the evolution of our universe is significantly influenced by a cosmological constant resulting in an acceleration of cosmic expansion or, equivalently, by a form of energy with an unusual equation of state, known as dark energy, the nature of which remains unclear.[133]

A so-called inflationary phase,[134] an additional phase of strongly accelerated expansion at cosmic times of around 10^{-33} seconds, was hypothesized in 1980 to account for several puzzling observations that were unexplained by classical cosmological models, such as the nearly perfect homogeneity of the cosmic background radiation.[135] Recent measurements of the cosmic background radiation have resulted in the first evidence for this scenario.[136] However, there is a bewildering variety of possible inflationary scenarios, which cannot be restricted by current observations.[137] An even larger question is the physics of the earliest universe, prior to the inflationary phase and close to where the classical models predict the big bang singularity. An authoritative answer would require a complete theory of quantum gravity, which has not yet been developed[138] (cf. the section on quantum gravity, below).

Time travel

Kurt Gödel showed[139] that solutions to Einstein's equations exist that contain closed timelike curves (CTCs), which allow for loops in time. The solutions require extreme physical conditions unlikely ever to occur in practice, and it remains an open question whether further laws of physics will eliminate them completely. Since then other—similarly impractical—GR solutions containing CTCs have been found, such as the Tipler cylinder and traversable wormholes.

1.7.6 Advanced concepts

Causal structure and global geometry

Main article: Causal structure

In general relativity, no material body can catch up with

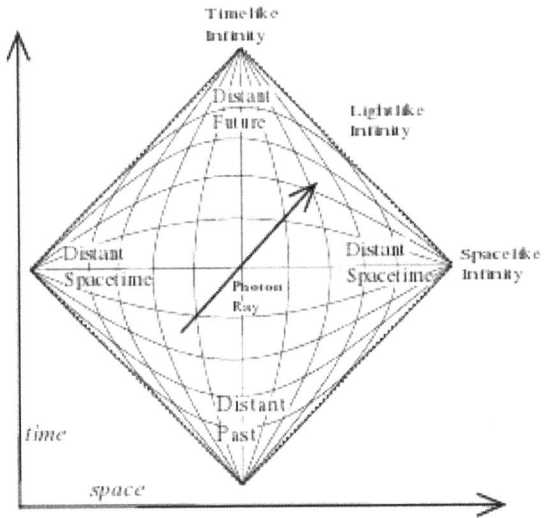

Penrose–Carter diagram of an infinite Minkowski universe

or overtake a light pulse. No influence from an event A can reach any other location X before light sent out at A to X. In consequence, an exploration of all light worldlines (null geodesics) yields key information about the spacetime's causal structure. This structure can be displayed using Penrose–Carter diagrams in which infinitely large regions of space and infinite time intervals are shrunk ("compactified") so as to fit onto a finite map, while light still travels along diagonals as in standard spacetime diagrams.[140]

Aware of the importance of causal structure, Roger Penrose and others developed what is known as global geometry. In global geometry, the object of study is not one particular solution (or family of solutions) to Einstein's equations. Rather, relations that hold true for all geodesics, such as

the Raychaudhuri equation, and additional non-specific assumptions about the nature of matter (usually in the form of so-called energy conditions) are used to derive general results.[141]

Horizons

Main articles: Horizon (general relativity), No hair theorem and Black hole mechanics

Using global geometry, some spacetimes can be shown to contain boundaries called horizons, which demarcate one region from the rest of spacetime. The best-known examples are black holes: if mass is compressed into a sufficiently compact region of space (as specified in the hoop conjecture[142], the relevant length scale is the Schwarzschild radius[142]), no light from inside can escape to the outside. Since no object can overtake a light pulse, all interior matter is imprisoned as well. Passage from the exterior to the interior is still possible, showing that the boundary, the black hole's *horizon*, is not a physical barrier.[143]

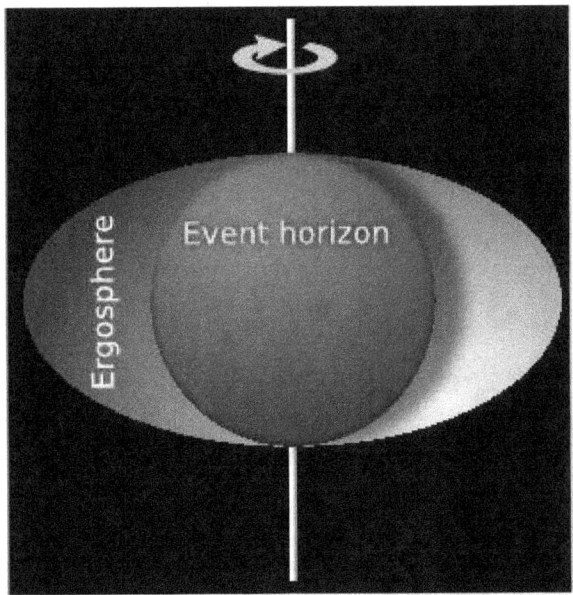

The ergosphere of a rotating black hole, which plays a key role when it comes to extracting energy from such a black hole

Early studies of black holes relied on explicit solutions of Einstein's equations, notably the spherically symmetric Schwarzschild solution (used to describe a static black hole) and the axisymmetric Kerr solution (used to describe a rotating, stationary black hole, and introducing interesting features such as the ergosphere). Using global geometry, later studies have revealed more general properties of black holes. In the long run, they are rather simple objects characterized by eleven parameters specifying energy, linear momentum, angular momentum, location at a specified time and electric charge. This is stated by the black hole uniqueness theorems: "black holes have no hair", that is, no distinguishing marks like the hairstyles of humans. Irrespective of the complexity of a gravitating object collapsing to form a black hole, the object that results (having emitted gravitational waves) is very simple.[144]

Even more remarkably, there is a general set of laws known as black hole mechanics, which is analogous to the laws of thermodynamics. For instance, by the second law of black hole mechanics, the area of the event horizon of a general black hole will never decrease with time, analogous to the entropy of a thermodynamic system. This limits the energy that can be extracted by classical means from a rotating black hole (e.g. by the Penrose process).[145] There is strong evidence that the laws of black hole mechanics are, in fact, a subset of the laws of thermodynamics, and that the black hole area is proportional to its entropy.[146] This leads to a modification of the original laws of black hole mechanics: for instance, as the second law of black hole mechanics becomes part of the second law of thermodynamics, it is possible for black hole area to decrease—as long as other processes ensure that, overall, entropy increases. As thermodynamical objects with non-zero temperature, black holes should emit thermal radiation. Semi-classical calculations indicate that indeed they do, with the surface gravity playing the role of temperature in Planck's law. This radiation is known as Hawking radiation (cf. the quantum theory section, below).[147]

There are other types of horizons. In an expanding universe, an observer may find that some regions of the past cannot be observed ("particle horizon"), and some regions of the future cannot be influenced (event horizon).[148] Even in flat Minkowski space, when described by an accelerated observer (Rindler space), there will be horizons associated with a semi-classical radiation known as Unruh radiation.[149]

Singularities

Main article: Spacetime singularity

Another general feature of general relativity is the appearance of spacetime boundaries known as singularities. Spacetime can be explored by following up on timelike and lightlike geodesics—all possible ways that light and particles in free fall can travel. But some solutions of Einstein's equations have "ragged edges"—regions known as spacetime singularities, where the paths of light and falling particles come to an abrupt end, and geometry becomes ill-defined. In the more interesting cases, these are "curvature singularities", where geometrical quantities char-

acterizing spacetime curvature, such as the Ricci scalar, take on infinite values.[150] Well-known examples of spacetimes with future singularities—where worldlines end—are the Schwarzschild solution, which describes a singularity inside an eternal static black hole,[151] or the Kerr solution with its ring-shaped singularity inside an eternal rotating black hole.[152] The Friedmann–Lemaître–Robertson–Walker solutions and other spacetimes describing universes have past singularities on which worldlines begin, namely Big Bang singularities, and some have future singularities (Big Crunch) as well.[153]

Given that these examples are all highly symmetric—and thus simplified—it is tempting to conclude that the occurrence of singularities is an artifact of idealization.[154] The famous singularity theorems, proved using the methods of global geometry, say otherwise: singularities are a generic feature of general relativity, and unavoidable once the collapse of an object with realistic matter properties has proceeded beyond a certain stage[155] and also at the beginning of a wide class of expanding universes.[156] However, the theorems say little about the properties of singularities, and much of current research is devoted to characterizing these entities' generic structure (hypothesized e.g. by the so-called BKL conjecture).[157] The cosmic censorship hypothesis states that all realistic future singularities (no perfect symmetries, matter with realistic properties) are safely hidden away behind a horizon, and thus invisible to all distant observers. While no formal proof yet exists, numerical simulations offer supporting evidence of its validity.[158]

Evolution equations

Main article: Initial value formulation (general relativity)

Each solution of Einstein's equation encompasses the whole history of a universe — it is not just some snapshot of how things are, but a whole, possibly matter-filled, spacetime. It describes the state of matter and geometry everywhere and at every moment in that particular universe. Due to its general covariance, Einstein's theory is not sufficient by itself to determine the time evolution of the metric tensor. It must be combined with a coordinate condition, which is analogous to gauge fixing in other field theories.[159]

To understand Einstein's equations as partial differential equations, it is helpful to formulate them in a way that describes the evolution of the universe over time. This is done in so-called "3+1" formulations, where spacetime is split into three space dimensions and one time dimension. The best-known example is the ADM formalism.[160] These decompositions show that the spacetime evolution equations of general relativity are well-behaved: solutions always exist, and are uniquely defined, once suitable initial conditions have been specified.[161] Such formulations of Einstein's field equations are the basis of numerical relativity.[162]

Global and quasi-local quantities

Main article: Mass in general relativity

The notion of evolution equations is intimately tied in with another aspect of general relativistic physics. In Einstein's theory, it turns out to be impossible to find a general definition for a seemingly simple property such as a system's total mass (or energy). The main reason is that the gravitational field—like any physical field—must be ascribed a certain energy, but that it proves to be fundamentally impossible to localize that energy.[163]

Nevertheless, there are possibilities to define a system's total mass, either using a hypothetical "infinitely distant observer" (ADM mass)[164] or suitable symmetries (Komar mass).[165] If one excludes from the system's total mass the energy being carried away to infinity by gravitational waves, the result is the so-called Bondi mass at null infinity.[166] Just as in classical physics, it can be shown that these masses are positive.[167] Corresponding global definitions exist for momentum and angular momentum.[168] There have also been a number of attempts to define *quasi-local* quantities, such as the mass of an isolated system formulated using only quantities defined within a finite region of space containing that system. The hope is to obtain a quantity useful for general statements about isolated systems, such as a more precise formulation of the hoop conjecture.[169]

1.7.7   Relationship with quantum theory

If general relativity were considered to be one of the two pillars of modern physics, then quantum theory, the basis of understanding matter from elementary particles to solid state physics, would be the other.[170] However, how to reconcile quantum theory with general relativity is still an open question.

Quantum field theory in curved spacetime

Main article: Quantum field theory in curved spacetime

Ordinary quantum field theories, which form the basis of modern elementary particle physics, are defined in flat Minkowski space, which is an excellent approximation when it comes to describing the behavior of microscopic particles in weak gravitational fields like those found on Earth.[171] In order to describe situations in which gravity is strong enough to influence (quantum) matter, yet not strong enough to require quantization itself, physicists have formulated quantum field theories in curved spacetime. These theories rely on general relativity to describe a curved background spacetime, and define a generalized quantum field theory to describe the behavior of quantum matter within that spacetime.[172] Using this formalism, it can be shown that black holes emit a blackbody spectrum of particles known as Hawking radiation, leading to the possibility that they evaporate over time.[173] As briefly mentioned above, this radiation plays an important role for the thermodynamics of black holes.[174]

Quantum gravity

Main article: Quantum gravity
See also: String theory, Canonical general relativity, Loop quantum gravity, Causal Dynamical Triangulations, and Causal sets

The demand for consistency between a quantum description of matter and a geometric description of spacetime,[175] as well as the appearance of singularities (where curvature length scales become microscopic), indicate the need for a full theory of quantum gravity: for an adequate description of the interior of black holes, and of the very early universe, a theory is required in which gravity and the associated geometry of spacetime are described in the language of quantum physics.[176] Despite major efforts, no complete and consistent theory of quantum gravity is currently known, even though a number of promising candidates exist.[177][178]

Attempts to generalize ordinary quantum field theories, used in elementary particle physics to describe fundamental interactions, so as to include gravity have led to serious problems.[179] Some have argued that at low energies, this approach proves successful, in that it results in an acceptable effective (quantum) field theory of gravity.[180] At very high energies, however, the perturbative results are badly divergent and lead to models devoid of predictive power ("perturbative non-renormalizability").[181]

One attempt to overcome these limitations is string theory, a quantum theory not of point particles, but of minute one-dimensional extended objects.[182] The theory promises to be a unified description of all particles and interactions, including gravity;[183] the price to pay is unusual features such as six extra dimensions of space in addition to the usual three.[184] In what is called the second superstring revolution, it was conjectured that both string theory and a unification of general relativity and supersymmetry known as supergravity[185] form part of a hypothesized eleven-dimensional model known as M-theory, which would con-

Projection of a Calabi–Yau manifold, one of the ways of compactifying the extra dimensions posited by string theory

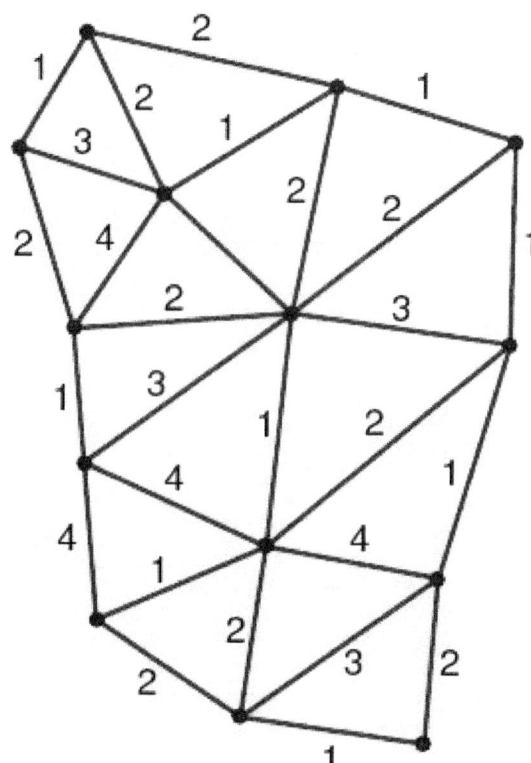

Simple spin network of the type used in loop quantum gravity

stitute a uniquely defined and consistent theory of quantum gravity.[186]

Another approach starts with the canonical quantization procedures of quantum theory. Using the initial-value-formulation of general relativity (cf. evolution equations above), the result is the Wheeler–deWitt equation (an analogue of the Schrödinger equation) which, regrettably, turns out to be ill-defined without a proper ultraviolet (lattice) cutoff.[187] However, with the introduction of what are now known as Ashtekar variables,[188] this leads to a promising model known as loop quantum gravity. Space is represented by a web-like structure called a spin network, evolving over time in discrete steps.[189]

Depending on which features of general relativity and quantum theory are accepted unchanged, and on what level changes are introduced,[190] there are numerous other attempts to arrive at a viable theory of quantum gravity, some examples being the lattice theory of gravity based on the Feynman Path Integral approach and Regge Calculus,[177] dynamical triangulations,[191] causal sets,[192] twistor models[193] or the path-integral based models of quantum cosmology.[194]

All candidate theories still have major formal and conceptual problems to overcome. They also face the common problem that, as yet, there is no way to put quantum gravity predictions to experimental tests (and thus to decide between the candidates where their predictions vary), although there is hope for this to change as future data from cosmological observations and particle physics experiments becomes available.[195]

1.7.8   Current status

General relativity has emerged as a highly successful model of gravitation and cosmology, which has so far passed many unambiguous observational and experimental tests. However, there are strong indications the theory is incomplete.[196] The problem of quantum gravity and the question of the reality of spacetime singularities remain open.[197] Observational data that is taken as evidence for dark energy and dark matter could indicate the need for new physics.[198] Even taken as is, general relativity is rich with possibilities for further exploration. Mathematical relativists seek to understand the nature of singularities and the fundamental properties of Einstein's equations,[199] and increasingly powerful computer simulations (such as those describing merging black holes) are run.[200] In February 2016, it was announced that the existence of gravitational waves was directly detected by the Advanced LIGO team on September 14, 2015.[71][201][202] A century after its publication, general relativity remains a highly active area of research.[203]

1.7.9   See also

- Alcubierre drive (warp drive)

- Center of mass (relativistic)

- Contributors to general relativity

- Derivations of the Lorentz transformations

- Ehrenfest paradox

- Einstein–Hilbert action

- Introduction to mathematics of general relativity

- Relativity priority dispute

- Ricci calculus

- Tests of general relativity

- Timeline of gravitational physics and relativity

- Two-body problem in general relativity

- Weak Gravity Conjecture

1.7.10   Notes

[1] "GW150914: LIGO Detects Gravitational Waves". *Blackholes.org*. Retrieved 18 April 2016.

[2] O'Connor, J.J. and Robertson, E.F. (1996), *General relativity. Mathematical Physics index*, School of Mathematics and Statistics, University of St. Andrews, Scotland. Retrieved 2015-02-04.

[3] Pais 1982, ch. 9 to 15, Janssen 2005; an up-to-date collection of current research, including reprints of many of the original articles, is Renn 2007; an accessible overview can be found in Renn 2005, pp. 110ff. Einstein's original papers are found in Digital Einstein, volumes 4 and 6. An early key article is Einstein 1907, cf. Pais 1982, ch. 9. The publication featuring the field equations is Einstein 1915, cf. Pais 1982, ch. 11–15

[4] Schwarzschild 1916a, Schwarzschild 1916b and Reissner 1916 (later complemented in Nordström 1918)

[5] Einstein 1917, cf. Pais 1982, ch. 15e

[6] Hubble's original article is Hubble 1929; an accessible overview is given in Singh 2004, ch. 2–4

[7] As reported in Gamow 1970. Einstein's condemnation would prove to be premature, cf. the section Cosmology, below

[8] Pais 1982, pp. 253–254

[9] Kennefick 2005, Kennefick 2007

[10] Pais 1982, ch. 16

[11] Thorne, Kip (2003). "Warping spacetime". *The future of theoretical physics and cosmology: celebrating Stephen Hawking's 60th birthday*. Cambridge University Press. p. 74. ISBN 0-521-82081-2. Extract of page 74

[12] Israel 1987, ch. 7.8–7.10, Thorne 1994, ch. 3–9

[13] Sections Orbital effects and the relativity of direction, Gravitational time dilation and frequency shift and Light deflection and gravitational time delay, and references therein

[14] Section Cosmology and references therein; the historical development is in Overbye 1999

[15] The following exposition re-traces that of Ehlers 1973, sec. 1

[16] Arnold 1989, ch. 1

[17] Ehlers 1973, pp. 5f

[18] Will 1993, sec. 2.4, Will 2006, sec. 2

[19] Wheeler 1990, ch. 2

[20] Ehlers 1973, sec. 1.2, Havas 1964, Künzle 1972. The simple thought experiment in question was first described in Heckmann & Schücking 1959

[21] Ehlers 1973, pp. 10f

[22] Good introductions are, in order of increasing presupposed knowledge of mathematics, Giulini 2005, Mermin 2005, and Rindler 1991; for accounts of precision experiments, cf. part IV of Ehlers & Lämmerzahl 2006

[23] An in-depth comparison between the two symmetry groups can be found in Giulini 2006a

[24] Rindler 1991, sec. 22, Synge 1972, ch. 1 and 2

[25] Ehlers 1973, sec. 2.3

[26] Ehlers 1973, sec. 1.4, Schutz 1985, sec. 5.1

[27] Ehlers 1973, pp. 17ff; a derivation can be found in Mermin 2005, ch. 12. For the experimental evidence, cf. the section Gravitational time dilation and frequency shift, below

[28] Rindler 2001, sec. 1.13; for an elementary account, see Wheeler 1990, ch. 2; there are, however, some differences between the modern version and Einstein's original concept used in the historical derivation of general relativity, cf. Norton 1985

[29] Ehlers 1973, sec. 1.4 for the experimental evidence, see once more section Gravitational time dilation and frequency shift. Choosing a different connection with non-zero torsion leads to a modified theory known as Einstein–Cartan theory

[30] Ehlers 1973, p. 16, Kenyon 1990, sec. 7.2, Weinberg 1972, sec. 2.8

[31] Ehlers 1973, pp. 19–22; for similar derivations, see sections 1 and 2 of ch. 7 in Weinberg 1972. The Einstein tensor is the only divergence-free tensor that is a function of the metric coefficients, their first and second derivatives at most, and allows the spacetime of special relativity as a solution in the absence of sources of gravity, cf. Lovelock 1972. The tensors on both side are of second rank, that is, they can each be thought of as 4×4 matrices, each of which contains ten independent terms; hence, the above represents ten coupled equations. The fact that, as a consequence of geometric relations known as Bianchi identities, the Einstein tensor satisfies a further four identities reduces these to six independent equations, e.g. Schutz 1985, sec. 8.3

[32] Kenyon 1990, sec. 7.4

[33] Brans & Dicke 1961, Weinberg 1972, sec. 3 in ch. 7, Goenner 2004, sec. 7.2, and Trautman 2006, respectively

[34] Wald 1984, ch. 4, Weinberg 1972, ch. 7 or, in fact, any other textbook on general relativity

[35] At least approximately, cf. Poisson 2004

[36] Wheeler 1990, p. xi

[37] Wald 1984, sec. 4.4

[38] Wald 1984, sec. 4.1

[39] For the (conceptual and historical) difficulties in defining a general principle of relativity and separating it from the notion of general covariance, see Giulini 2006b

[40] section 5 in ch. 12 of Weinberg 1972

[41] Introductory chapters of Stephani et al. 2003

[42] A review showing Einstein's equation in the broader context of other PDEs with physical significance is Geroch 1996

[43] For background information and a list of solutions, cf. Stephani et al. 2003; a more recent review can be found in MacCallum 2006

[44] Chandrasekhar 1983, ch. 3,5,6

[45] Narlikar 1993, ch. 4, sec. 3.3

[46] Brief descriptions of these and further interesting solutions can be found in Hawking & Ellis 1973, ch. 5

[47] Lehner 2002

[48] For instance Wald 1984, sec. 4.4

[49] Will 1993, sec. 4.1 and 4.2

[50] Will 2006, sec. 3.2, Will 1993, ch. 4

[51] Rindler 2001, pp. 24–26 vs. pp. 236–237 and Ohanian & Ruffini 1994, pp. 164–172. Einstein derived these effects using the equivalence principle as early as 1907, cf. Einstein 1907 and the description in Pais 1982, pp. 196–198

[52] Rindler 2001, pp. 24–26; Misner, Thorne & Wheeler 1973, § 38.5

[53] Pound–Rebka experiment, see Pound & Rebka 1959, Pound & Rebka 1960; Pound & Snider 1964; a list of further experiments is given in Ohanian & Ruffini 1994, table 4.1 on p. 186

[54] Greenstein, Oke & Shipman 1971; the most recent and most accurate Sirius B measurements are published in Barstow, Bond et al. 2005.

[55] Starting with the Hafele–Keating experiment, Hafele & Keating 1972a and Hafele & Keating 1972b, and culminating in the Gravity Probe A experiment; an overview of experiments can be found in Ohanian & Ruffini 1994, table 4.1 on p. 186

[56] GPS is continually tested by comparing atomic clocks on the ground and aboard orbiting satellites; for an account of relativistic effects, see Ashby 2002 and Ashby 2003

[57] Stairs 2003 and Kramer 2004

[58] General overviews can be found in section 2.1. of Will 2006; Will 2003, pp. 32–36; Ohanian & Ruffini 1994, sec. 4.2

[59] Ohanian & Ruffini 1994, pp. 164–172

[60] Cf. Kennefick 2005 for the classic early measurements by the Eddington expeditions; for an overview of more recent measurements, see Ohanian & Ruffini 1994, ch. 4.3. For the most precise direct modern observations using quasars, cf. Shapiro et al. 2004

[61] This is not an independent axiom; it can be derived from Einstein's equations and the Maxwell Lagrangian using a WKB approximation, cf. Ehlers 1973, sec. 5

[62] Blanchet 2006, sec. 1.3

[63] Rindler 2001, sec. 1.16; for the historical examples, Israel 1987, pp. 202–204; in fact, Einstein published one such derivation as Einstein 1907. Such calculations tacitly assume that the geometry of space is Euclidean, cf. Ehlers & Rindler 1997

[64] From the standpoint of Einstein's theory, these derivations take into account the effect of gravity on time, but not its consequences for the warping of space, cf. Rindler 2001, sec. 11.11

[65] For the Sun's gravitational field using radar signals reflected from planets such as Venus and Mercury, cf. Shapiro 1964, Weinberg 1972, ch. 8, sec. 7; for signals actively sent back by space probes (transponder measurements), cf. Bertotti, Iess & Tortora 2003; for an overview, see Ohanian & Ruffini 1994, table 4.4 on p. 200; for more recent measurements using signals received from a pulsar that is part of a binary system, the gravitational field causing the time delay being that of the other pulsar, cf. Stairs 2003, sec. 4.4

[66] Will 1993, sec. 7.1 and 7.2

[67] Einstein. A (June 1916). "Näherungsweise Integration der Feldgleichungen der Gravitation". *Sitzungsberichte der Königlich Preussischen Akademie der Wissenschaften Berlin*. part 1: 688–696.

[68] Einstein. A (1918). "Über Gravitationswellen". *Sitzungsberichte der Königlich Preussischen Akademie der Wissenschaften Berlin*. part 1: 154–167.

[69] Castelvecchi. Davide: Witze. Witze (February 11, 2016). "Einstein's gravitational waves found at last". *Nature News*. doi:10.1038/nature.2016.19361. Retrieved 2016-02-11.

[70] B. P. Abbott et al. (LIGO Scientific Collaboration and Virgo Collaboration) (2016). "Observation of Gravitational Waves from a Binary Black Hole Merger". *Physical Review Letters* **116** (6). doi:10.1103/PhysRevLett.116.061102.

[71] "Gravitational waves detected 100 years after Einstein's prediction | NSF - National Science Foundation". *www.nsf.gov*. Retrieved 2016-02-11.

[72] Most advanced textbooks on general relativity contain a description of these properties. e.g. Schutz 1985. ch. 9

[73] For example Jaranowski & Królak 2005

[74] Rindler 2001. ch. 13

[75] Gowdy 1971. Gowdy 1974

[76] See Lehner 2002 for a brief introduction to the methods of numerical relativity. and Seidel 1998 for the connection with gravitational wave astronomy

[77] Schutz 2003. pp. 48–49. Pais 1982. pp. 253–254

[78] Rindler 2001. sec. 11.9

[79] Will 1993. pp. 177–181

[80] In consequence. in the parameterized post-Newtonian formalism (PPN). measurements of this effect determine a linear combination of the terms β and γ. cf. Will 2006. sec. 3.5 and Will 1993. sec. 7.3

[81] The most precise measurements are VLBI measurements of planetary positions: see Will 1993. ch. 5, Will 2006. sec. 3.5. Anderson et al. 1992: for an overview, Ohanian & Ruffini 1994. pp. 406–407

[82] Kramer et al. 2006

[83] Dediu. Adrian-Horia: Magdalena. Luis: Martín-Vide. Carlos (2015). *Theory and Practice of Natural Computing: Fourth International Conference, TPNC 2015, Mieres, Spain, December 15-16, 2015. Proceedings* (illustrated ed.). Springer. p. 141. ISBN 978-3-319-26841-5. Extract of page 141

[84] A figure that includes error bars is fig. 7 in Will 2006. sec. 5.1

[85] Stairs 2003. Schutz 2003. pp. 317–321. Bartusiak 2000. pp. 70–86

[86] Weisberg & Taylor 2003: for the pulsar discovery, see Hulse & Taylor 1975: for the initial evidence for gravitational radiation. see Taylor 1994

[87] Kramer 2004

[88] Penrose 2004. §14.5. Misner. Thorne & Wheeler 1973. §11.4

[89] Weinberg 1972. sec. 9.6. Ohanian & Ruffini 1994. sec. 7.8

[90] Bertotti. Ciufolini & Bender 1987. Nordtvedt 2003

[91] Kahn 2007

[92] A mission description can be found in Everitt et al. 2001: a first post-flight evaluation is given in Everitt. Parkinson & Kahn 2007: further updates will be available on the mission website Kahn 1996–2012.

[93] Townsend 1997. sec. 4.2.1. Ohanian & Ruffini 1994. pp. 469–471

[94] Ohanian & Ruffini 1994. sec. 4.7. Weinberg 1972. sec. 9.7: for a more recent review. see Schäfer 2004

[95] Ciufolini & Pavlis 2004. Ciufolini. Pavlis & Peron 2006. Iorio 2009

[96] Iorio L. (August 2006). "COMMENTS. REPLIES AND NOTES: A note on the evidence of the gravitomagnetic field of Mars". *Classical Quantum Gravity* **23** (17): 5451–5454. arXiv:gr-qc/0606092. Bibcode:2006CQGra..23.5451I. doi:10.1088/0264-9381/23/17/N01

[97] Iorio L. (June 2010). "On the Lense–Thirring test with the Mars Global Surveyor in the gravitational field of Mars". *Central European Journal of Physics* **8** (3): 509–513. arXiv:gr-qc/0701146. Bibcode:2010CEJPh...8..509I. doi:10.2478/s11534-009-0117-6

[98] For overviews of gravitational lensing and its applications. see Ehlers. Falco & Schneider 1992 and Wambsganss 1998

[99] For a simple derivation. see Schutz 2003. ch. 23: cf. Narayan & Bartelmann 1997. sec. 3

[100] Walsh. Carswell & Weymann 1979

[101] Images of all the known lenses can be found on the pages of the CASTLES project. Kochanek et al. 2007

[102] Roulet & Mollerach 1997

[103] Narayan & Bartelmann 1997. sec. 3.7

[104] Barish 2005. Bartusiak 2000. Blair & McNamara 1997

[105] Hough & Rowan 2000

[106] Hobbs. George: Archibald. A.: Arzoumanian. Z.: Backer. D.: Bailes. M.: Bhat. N. D. R.: Burgay. M.: Burke-Spolaor. S.: et al. (2010). "The international pulsar timing array project: using pulsars as a gravitational wave detector". *Classical and Quantum Gravity* **27** (8): 084013. arXiv:0911.5206. Bibcode:2010CQGra..27h4013H. doi:10.1088/0264-9381/27/8/084013

[107] Danzmann & Rüdiger 2003

[108] "LISA pathfinder overview". ESA. Retrieved 2012-04-23.

[109] Thorne 1995

[110] Cutler & Thorne 2002

[111] "Gravitational waves detected 100 years after Einstein's prediction | NSF - National Science Foundation". *www.nsf.gov*. Retrieved 2016-02-11.

[112] Miller 2002, lectures 19 and 21

[113] Celotti, Miller & Sciama 1999, sec. 3

[114] Springel et al. 2005 and the accompanying summary Gnedin 2005

[115] Blandford 1987, sec. 8.2.4

[116] For the basic mechanism, see Carroll & Ostlie 1996, sec. 17.2; for more about the different types of astronomical objects associated with this, cf. Robson 1996

[117] For a review, see Begelman, Blandford & Rees 1984. To a distant observer, some of these jets even appear to move faster than light; this, however, can be explained as an optical illusion that does not violate the tenets of relativity, see Rees 1966

[118] For stellar end states, cf. Oppenheimer & Snyder 1939 or, for more recent numerical work, Font 2003, sec. 4.1; for supernovae, there are still major problems to be solved, cf. Buras et al. 2003; for simulating accretion and the formation of jets, cf. Font 2003, sec. 4.2. Also, relativistic lensing effects are thought to play a role for the signals received from X-ray pulsars, cf. Kraus 1998

[119] The evidence includes limits on compactness from the observation of accretion-driven phenomena ("Eddington luminosity"), see Celotti, Miller & Sciama 1999, observations of stellar dynamics in the center of our own Milky Way galaxy, cf. Schödel et al. 2003, and indications that at least some of the compact objects in question appear to have no solid surface, which can be deduced from the examination of X-ray bursts for which the central compact object is either a neutron star or a black hole; cf. Remillard et al. 2006 for an overview, Narayan 2006, sec. 5. Observations of the "shadow" of the Milky Way galaxy's central black hole horizon are eagerly sought for, cf. Falcke, Melia & Agol 2000

[120] Dalal et al. 2006

[121] Barack & Cutler 2004

[122] Originally Einstein 1917; cf. Pais 1982, pp. 285–288

[123] Carroll 2001, ch. 2

[124] Bergström & Goobar 2003, ch. 9–11; use of these models is justified by the fact that, at large scales of around hundred million light-years and more, our own universe indeed appears to be isotropic and homogeneous, cf. Peebles et al. 1991

[125] E.g. with WMAP data, see Spergel et al. 2003

[126] These tests involve the separate observations detailed further on, see, e.g., fig. 2 in Bridle et al. 2003

[127] Peebles 1966; for a recent account of predictions, see Coc, Vangioni-Flam et al. 2004; an accessible account can be found in Weiss 2006; compare with the observations in Olive & Skillman 2004, Bania, Rood & Balser 2002, O'Meara et al. 2001, and Charbonnel & Primas 2005

[128] Lahav & Suto 2004, Bertschinger 1998, Springel et al. 2005

[129] Alpher & Herman 1948, for a pedagogical introduction, see Bergström & Goobar 2003, ch. 11; for the initial detection, see Penzias & Wilson 1965 and, for precision measurements by satellite observatories, Mather et al. 1994 (COBE) and Bennett et al. 2003 (WMAP). Future measurements could also reveal evidence about gravitational waves in the early universe; this additional information is contained in the background radiation's polarization, cf. Kamionkowski, Kosowsky & Stebbins 1997 and Seljak & Zaldarriaga 1997

[130] Evidence for this comes from the determination of cosmological parameters and additional observations involving the dynamics of galaxies and galaxy clusters cf. Peebles 1993, ch. 18, evidence from gravitational lensing, cf. Peacock 1999, sec. 4.6, and simulations of large-scale structure formation, see Springel et al. 2005

[131] Peacock 1999, ch. 12, Peskin 2007; in particular, observations indicate that all but a negligible portion of that matter is not in the form of the usual elementary particles ("nonbaryonic matter"), cf. Peacock 1999, ch. 12

[132] Namely, some physicists have questioned whether or not the evidence for dark matter is, in fact, evidence for deviations from the Einsteinian (and the Newtonian) description of gravity cf. the overview in Mannheim 2006, sec. 9

[133] Carroll 2001; an accessible overview is given in Caldwell 2004. Here, too, scientists have argued that the evidence indicates not a new form of energy, but the need for modifications in our cosmological models, cf. Mannheim 2006, sec. 10; aforementioned modifications need not be modifications of general relativity, they could, for example, be modifications in the way we treat the inhomogeneities in the universe, cf. Buchert 2007

[134] A good introduction is Linde 1990; for a more recent review, see Linde 2005

[135] More precisely, these are the flatness problem, the horizon problem, and the monopole problem; a pedagogical introduction can be found in Narlikar 1993, sec. 6.4, see also Börner 1993, sec. 9.1

[136] Spergel et al. 2007, sec. 5.6

[137] More concretely, the potential function that is crucial to determining the dynamics of the inflaton is simply postulated, but not derived from an underlying physical theory

[138] Brandenberger 2007, sec. 2

[139] Gödel 1949

[140] Frauendiener 2004, Wald 1984, sec. 11.1, Hawking & Ellis 1973, sec. 6.8, 6.9

[141] Wald 1984, sec. 9.2–9.4 and Hawking & Ellis 1973, ch. 6

[142] Thorne 1972; for more recent numerical studies, see Berger 2002, sec. 2.1

[143] Israel 1987. A more exact mathematical description distinguishes several kinds of horizon, notably event horizons and apparent horizons cf. Hawking & Ellis 1973, pp. 312–320 or Wald 1984, sec. 12.2; there are also more intuitive definitions for isolated systems that do not require knowledge of spacetime properties at infinity, cf. Ashtekar & Krishnan 2004

[144] For first steps, cf. Israel 1971; see Hawking & Ellis 1973, sec. 9.3 or Heusler 1996, ch. 9 and 10 for a derivation, and Heusler 1998 as well as Beig & Chruściel 2006 as overviews of more recent results

[145] The laws of black hole mechanics were first described in Bardeen, Carter & Hawking 1973; a more pedagogical presentation can be found in Carter 1979; for a more recent review, see Wald 2001, ch. 2. A thorough, book-length introduction including an introduction to the necessary mathematics Poisson 2004. For the Penrose process, see Penrose 1969

[146] Bekenstein 1973, Bekenstein 1974

[147] The fact that black holes radiate, quantum mechanically, was first derived in Hawking 1975; a more thorough derivation can be found in Wald 1975. A review is given in Wald 2001, ch. 3

[148] Narlikar 1993, sec. 4.4.4, 4.4.5

[149] Horizons: cf. Rindler 2001, sec. 12.4. Unruh effect: Unruh 1976, cf. Wald 2001, ch. 3

[150] Hawking & Ellis 1973, sec. 8.1, Wald 1984, sec. 9.1

[151] Townsend 1997, ch. 2; a more extensive treatment of this solution can be found in Chandrasekhar 1983, ch. 3

[152] Townsend 1997, ch. 4; for a more extensive treatment, cf. Chandrasekhar 1983, ch. 6

[153] Ellis & Van Elst 1999; a closer look at the singularity itself is taken in Börner 1993, sec. 1.2

[154] Here one should remind to the well-known fact that the important "quasi-optical" singularities of the so-called eikonal approximations of many wave-equations, namely the "caustics", are resolved into finite peaks beyond that approximation.

[155] Namely when there are trapped null surfaces, cf. Penrose 1965

[156] Hawking 1966

[157] The conjecture was made in Belinskii, Khalatnikov & Lifschitz 1971; for a more recent review, see Berger 2002. An accessible exposition is given by Garfinkle 2007

[158] The restriction to future singularities naturally excludes initial singularities such as the big bang singularity, which in principle be visible to observers at later cosmic time. The cosmic censorship conjecture was first presented in Penrose 1969; a textbook-level account is given in Wald 1984, pp. 302–305. For numerical results, see the review Berger 2002, sec. 2.1

[159] Hawking & Ellis 1973, sec. 7.1

[160] Arnowitt, Deser & Misner 1962; for a pedagogical introduction, see Misner, Thorne & Wheeler 1973, §21.4–§21.7

[161] Fourès-Bruhat 1952 and Bruhat 1962; for a pedagogical introduction, see Wald 1984, ch. 10; an online review can be found in Reula 1998

[162] Gourgoulhon 2007; for a review of the basics of numerical relativity, including the problems arising from the peculiarities of Einstein's equations, see Lehner 2001

[163] Misner, Thorne & Wheeler 1973, §20.4

[164] Arnowitt, Deser & Misner 1962

[165] Komar 1959; for a pedagogical introduction, see Wald 1984, sec. 11.2; although defined in a totally different way, it can be shown to be equivalent to the ADM mass for stationary spacetimes, cf. Ashtekar & Magnon-Ashtekar 1979

[166] For a pedagogical introduction, see Wald 1984, sec. 11.2

[167] Wald 1984, p. 295 and refs therein; this is important for questions of stability—if there were negative mass states, then flat, empty Minkowski space, which has mass zero, could evolve into these states

[168] Townsend 1997, ch. 5

[169] Such quasi-local mass–energy definitions are the Hawking energy, Geroch energy, or Penrose's quasi-local energy-momentum based on twistor methods; cf. the review article Szabados 2004

[170] An overview of quantum theory can be found in standard textbooks such as Messiah 1999; a more elementary account is given in Hey & Walters 2003

[171] Ramond 1990, Weinberg 1995, Peskin & Schroeder 1995; a more accessible overview is Auyang 1995

[172] Wald 1994, Birrell & Davies 1984

[173] For Hawking radiation Hawking 1975, Wald 1975; an accessible introduction to black hole evaporation can be found in Traschen 2000

[174] Wald 2001, ch. 3

[175] Put simply, matter is the source of spacetime curvature, and once matter has quantum properties, we can expect space-time to have them as well. Cf. Carlip 2001, sec. 2

[176] Schutz 2003, p. 407

[177] Hamber 2009

[178] A timeline and overview can be found in Rovelli 2000

[179] t'Hooft 1974

[180] Donoghue 1995

[181] In particular, a perturbative technique known as renormalization, an integral part of deriving predictions which take into account higher-energy contributions, cf. Weinberg 1996, ch. 17, 18, fails in this case; cf. Veltman 1975, Goroff & Sagnotti 1985; for a recent comprehensive review of the failure of perturbative renormalizability for quantum gravity see Hamber 2009

[182] An accessible introduction at the undergraduate level can be found in Zwiebach 2004; more complete overviews can be found in Polchinski 1998a and Polchinski 1998b

[183] At the energies reached in current experiments, these strings are indistinguishable from point-like particles, but, crucially, different modes of oscillation of one and the same type of fundamental string appear as particles with different (electric and other) charges, e.g. Ibanez 2000. The theory is successful in that one mode will always correspond to a graviton, the messenger particle of gravity, e.g. Green, Schwarz & Witten 1987, sec. 2.3, 5.3

[184] Green, Schwarz & Witten 1987, sec. 4.2

[185] Weinberg 2000, ch. 31

[186] Townsend 1996, Duff 1996

[187] Kuchař 1973, sec. 3

[188] These variables represent geometric gravity using mathematical analogues of electric and magnetic fields; cf. Ashtekar 1986, Ashtekar 1987

[189] For a review, see Thiemann 2006; more extensive accounts can be found in Rovelli 1998, Ashtekar & Lewandowski 2004 as well as in the lecture notes Thiemann 2003

[190] Isham 1994, Sorkin 1997

[191] Loll 1998

[192] Sorkin 2005

[193] Penrose 2004, ch. 33 and refs therein

[194] Hawking 1987

[195] Ashtekar 2007, Schwarz 2007

[196] Maddox 1998, pp. 52–59, 98–122; Penrose 2004, sec. 34.1, ch. 30

[197] section Quantum gravity, above

[198] section Cosmology, above

[199] Friedrich 2005

[200] A review of the various problems and the techniques being developed to overcome them, see Lehner 2002

[201] See Bartusiak 2000 for an account up to that year; up-to-date news can be found on the websites of major detector collaborations such as GEO 600 and LIGO

[202] For the most recent papers on gravitational wave polarizations of inspiralling compact binaries, see Blanchet et al. 2008, and Arun et al. 2007; for a review of work on compact binaries, see Blanchet 2006 and Futamase & Itoh 2006; for a general review of experimental tests of general relativity, see Will 2006

[203] See, e.g., the electronic review journal Living Reviews in Relativity

1.7.11   References

- Alpher, R. A.; Herman, R. C. (1948), "Evolution of the universe", *Nature* **162** (4124): 774–775, Bibcode:1948Natur.162..774A, doi:10.1038/162774b0

- Anderson, J. D.; Campbell, J. K.; Jurgens, R. F.; Lau, E. L. (1992), "Recent developments in solar-system tests of general relativity", in Sato, H.; Nakamura, T., *Proceedings of the Sixth Marcel Großmann Meeting on General Relativity*, World Scientific, pp. 353–355, ISBN 981-02-0950-9

- Arnold, V. I. (1989). *Mathematical Methods of Classical Mechanics*, Springer, ISBN 3-540-96890-3

- Arnowitt, Richard; Deser, Stanley; Misner, Charles W. (1962). "The dynamics of general relativity", in Witten, Louis, *Gravitation: An Introduction to Current Research*, Wiley, pp. 227–265

- Arun, K.G.; Blanchet, L.; Iyer, B. R.; Qusailah, M. S. S. (2007), "Inspiralling compact binaries in quasi-elliptical orbits: The complete 3PN energy flux", *Physical Review D* **77** (6), arXiv:0711.0302, Bibcode:2008PhRvD..77f4035A, doi:10.1103/PhysRevD.77.064035

- Ashby, Neil (2002), "Relativity and the Global Positioning System" (PDF), *Physics Today* **55** (5): 41–47, Bibcode:2002PhT....55e..41A, doi:10.1063/1.1485583

- Ashby, Neil (2003), "Relativity in the Global Positioning System", *Living Reviews in Relativity* **6**, doi:10.12942/lrr-2003-1, retrieved 2007-07-06

- Ashtekar, Abhay (1986), "New variables for classical and quantum gravity", *Phys. Rev. Lett.* **57** (18): 2244–2247, Bibcode:1986PhRvL..57.2244A, doi:10.1103/PhysRevLett.57.2244, PMID 10033673

- Ashtekar, Abhay (1987), "New Hamiltonian formulation of general relativity", *Phys. Rev.* **D36** (6): 1587–1602, Bibcode:1987PhRvD..36.1587A, doi:10.1103/PhysRevD.36.1587

- Ashtekar, Abhay (2007), "LOOP QUANTUM GRAVITY: FOUR RECENT ADVANCES AND A DOZEN FREQUENTLY ASKED QUESTIONS", *The Eleventh Marcel Grossmann Meeting - on Recent Developments in Theoretical and Experimental General Relativity, Gravitation and Relativistic Field Theories - Proceedings of the MG11 Meeting on General Relativity*, p. 126, arXiv:0705.2222, Bibcode:2008mgm..conf..126A, doi:10.1142/9789812834300_0008, ISBN 9789812834263

- Ashtekar, Abhay; Krishnan, Badri (2004), "Isolated and Dynamical Horizons and Their Applications", *Living Reviews in Relativity* **7**, arXiv:gr-qc/0407042, Bibcode:2004LRR.....7...10A, doi:10.12942/lrr-2004-10, retrieved 2007-08-28

- Ashtekar, Abhay; Lewandowski, Jerzy (2004), "Background Independent Quantum Gravity: A Status Report", *Class. Quant. Grav.* **21** (15): R53–R152, arXiv:gr-qc/0404018, Bibcode:2004CQGra..21R..53A, doi:10.1088/0264-9381/21/15/R01

- Ashtekar, Abhay; Magnon-Ashtekar, Anne (1979), "On conserved quantities in general relativity", *Journal of Mathematical Physics* **20** (5): 793–800, Bibcode:1979JMP....20..793A, doi:10.1063/1.524151

- Auyang, Sunny Y. (1995), *How is Quantum Field Theory Possible?*, Oxford University Press, ISBN 0-19-509345-3

- Bania, T. M.; Rood, R. T.; Balser, D. S. (2002), "The cosmological density of baryons from observations of 3He+ in the Milky Way", *Nature* **415** (6867): 54–57, Bibcode:2002Natur.415...54B, doi:10.1038/415054a, PMID 11780112

- Barack, Leor; Cutler, Curt (2004), "LISA Capture Sources: Approximate Waveforms, Signal-to-Noise Ratios, and Parameter Estimation Accuracy", *Phys. Rev.* **D69** (8): 082005, arXiv:gr-qc/0310125, Bibcode:2004PhRvD..69h2005B, doi:10.1103/PhysRevD.69.082005

- Bardeen, J. M.; Carter, B.; Hawking, S. W. (1973), "The Four Laws of Black Hole Mechanics", *Comm. Math. Phys.* **31** (2): 161–170, Bibcode:1973CMaPh..31..161B, doi:10.1007/BF01645742

- Barish, Barry (2005), "Towards detection of gravitational waves", in Florides, P.; Nolan, B.; Ottewil, A., *General Relativity and Gravitation. Proceedings of the 17th International Conference*, World Scientific, pp. 24–34, ISBN 981-256-424-1

- Barstow, M; Bond, Howard E.; Holberg, J. B.; Burleigh, M. R.; Hubeny, I.; Koester, D. (2005), "Hubble Space Telescope Spectroscopy of the Balmer lines in Sirius B", *Mon. Not. Roy. Astron. Soc.* **362** (4): 1134–1142, arXiv:astro-ph/0506600, Bibcode:2005MNRAS.362.1134B, doi:10.1111/j.1365-2966.2005.09359.x

- Bartusiak, Marcia (2000), *Einstein's Unfinished Symphony: Listening to the Sounds of Space-Time*, Berkley, ISBN 978-0-425-18620-6

- Begelman, Mitchell C.; Blandford, Roger D.; Rees, Martin J. (1984), "Theory of extragalactic radio sources", *Rev. Mod. Phys.* **56** (2): 255–351, Bibcode:1984RvMP...56..255B, doi:10.1103/RevModPhys.56.255

- Beig, Robert; Chruściel, Piotr T. (2006), "Stationary black holes", in Françoise, J.-P.; Naber, G.; Tsou, T.S., *Encyclopedia of Mathematical Physics, Volume 2*, Elsevier, p. 2041, arXiv:gr-qc/0502041, Bibcode:2005gr.qc.....2041B, ISBN 0-12-512660-3

- Bekenstein, Jacob D. (1973), "Black Holes and Entropy", *Phys. Rev.* **D7** (8): 2333–2346, Bibcode:1973PhRvD...7.2333B, doi:10.1103/PhysRevD.7.2333

- Bekenstein, Jacob D. (1974), "Generalized Second Law of Thermodynamics in Black-Hole Physics", *Phys. Rev.* **D9** (12): 3292–3300, Bibcode:1974PhRvD...9.3292B, doi:10.1103/PhysRevD.9.3292

- Belinskii, V. A.; Khalatnikov, I. M.; Lifschitz, E. M. (1971), "Oscillatory approach to the singular point in relativistic cosmology", *Advances in Physics* **19** (80): 525–573, Bibcode:1970AdPhy..19..525B, doi:10.1080/00018737000101171; original paper in Russian: Belinsky, V. A.; Lifshits, I. M.; Khalatnikov, E. M. (1970), "Колебательный Режим Приближения К Особой Точке В Релятивистской Космологии". *Uspekhi Fizicheskikh Nauk (Успехи Физических Наук)*, 102(3) (11): 463–500, Bibcode:1970UsFiN.102..463B

- Bennett, C. L.; Halpern, M.; Hinshaw, G.; Jarosik, N.; Kogut, A.; Limon, M.; Meyer, S. S.; Page, L.; et al. (2003), "First Year Wilkinson Microwave Anisotropy Probe (WMAP) Observations: Preliminary Maps and Basic Results", *Astrophys. J. Suppl.* **148** (1): 1–27, arXiv:astro-ph/0302207, Bibcode:2003ApJS..148....1B, doi:10.1086/377253

- Berger, Beverly K. (2002), "Numerical Approaches to Spacetime Singularities", *Living Reviews in Relativity* **5**, arXiv:gr-qc/0201056, Bibcode:2002LRR.....5....1B, doi:10.12942/lrr-2002-1, retrieved 2007-08-04

- Bergström, Lars; Goobar, Ariel (2003), *Cosmology and Particle Astrophysics* (2nd ed.), Wiley & Sons, ISBN 3-540-43128-4

- Bertotti, Bruno; Ciufolini, Ignazio; Bender, Peter L. (1987), "New test of general relativity: Measurement of de Sitter geodetic precession rate for lunar perigee", *Physical Review Letters* **58** (11): 1062–1065, Bibcode:1987PhRvL..58.1062B, doi:10.1103/PhysRevLett.58.1062, PMID 10034329

- Bertotti, Bruno; Iess, L.; Tortora, P. (2003), "A test of general relativity using radio links with the Cassini spacecraft", *Nature* **425** (6956): 374–376, Bibcode:2003Natur.425..374B, doi:10.1038/nature01997, PMID 14508481

- Bertschinger, Edmund (1998), "Simulations of structure formation in the universe", *Annu. Rev. Astron. Astrophys.* **36** (1): 599–654, Bibcode:1998ARA&A..36..599B, doi:10.1146/annurev.astro.36.1.599

- Birrell, N. D.; Davies, P. C. (1984), *Quantum Fields in Curved Space*, Cambridge University Press, ISBN 0-521-27858-9

- Blair, David; McNamara, Geoff (1997), *Ripples on a Cosmic Sea. The Search for Gravitational Waves*, Perseus, ISBN 0-7382-0137-5

- Blanchet, L.; Faye, G.; Iyer, B. R.; Sinha, S. (2008), "The third post-Newtonian gravitational wave polarisations and associated spherical harmonic modes for inspiralling compact binaries in quasi-circular orbits", *Classical and Quantum Gravity* **25** (16): 165003, arXiv:0802.1249, Bibcode:2008CQGra..25p5003B, doi:10.1088/0264-9381/25/16/165003

- Blanchet, Luc (2006), "Gravitational Radiation from Post-Newtonian Sources and Inspiralling Compact Binaries", *Living Reviews in Relativity* **9**, Bibcode:2006LRR.....9....4B, doi:10.12942/lrr-2006-4, retrieved 2007-08-07

- Blandford, R. D. (1987), "Astrophysical Black Holes", in Hawking, Stephen W.; Israel, Werner, *300 Years of Gravitation*, Cambridge University Press, pp. 277–329, ISBN 0-521-37976-8

- Börner, Gerhard (1993), *The Early Universe. Facts and Fiction*, Springer, ISBN 0-387-56729-1

- Brandenberger, Robert H. (2007), "Conceptual Problems of Inflationary Cosmology and a New Approach to Cosmological Structure Formation", *Inflationary Cosmology*, Lecture Notes in Physics **738**, p. 393, arXiv:hep-th/0701111, Bibcode:2008LNP...738..393B, doi:10.1007/978-3-540-74353-8_11, ISBN 978-3-540-74352-1

- Brans, C. H.; Dicke, R. H. (1961), "Mach's Principle and a Relativistic Theory of Gravitation", *Physical Review* **124** (3): 925–935, Bibcode:1961PhRv..124..925B, doi:10.1103/PhysRev.124.925

- Bridle, Sarah L.; Lahav, Ofer; Ostriker, Jeremiah P.; Steinhardt, Paul J. (2003), "Precision Cosmology? Not Just Yet", *Science* **299** (5612): 1532–1533, arXiv:astro-ph/0303180, Bibcode:2003Sci...299.1532B, doi:10.1126/science.1082158, PMID 12624255

- Bruhat, Yvonne (1962), "The Cauchy Problem", in Witten, Louis, *Gravitation: An Introduction to Current Research*, Wiley, p. 130, ISBN 978-1-114-29166-9

- Buchert, Thomas (2007), "Dark Energy from Structure—A Status Report", *General Relativity and Gravitation* **40** (2–3): 467–527, arXiv:0707.2153, Bibcode:2008GReGr..40..467B, doi:10.1007/s10714-007-0554-8

- Buras, R.; Rampp, M.; Janka, H.-Th.; Kifonidis, K. (2003), "Improved Models of Stellar Core Collapse and Still no Explosions: What is Missing?", *Phys. Rev. Lett.* **90** (24): 241101, arXiv:astro-ph/0303171, Bibcode:2003PhRvL..90x1101B, doi:10.1103/PhysRevLett.90.241101, PMID 12857181

- Caldwell, Robert R. (2004), "Dark Energy", *Physics World* **17** (5): 37–42

- Carlip, Steven (2001), "Quantum Gravity: a Progress Report", *Rept. Prog. Phys.* **64** (8): 885–942, arXiv:gr-qc/0108040, Bibcode:2001RPPh...64..885C, doi:10.1088/0034-4885/64/8/301

- Carroll, Bradley W.; Ostlie, Dale A. (1996), *An Introduction to Modern Astrophysics*, Addison-Wesley, ISBN 0-201-54730-9

- Carroll, Sean M. (2001), "The Cosmological Constant", *Living Reviews in Relativity* **4**, arXiv:astro-ph/0004075, Bibcode:2001LRR.....4....1C, doi:10.12942/lrr-2001-1, retrieved 2007-07-21

- Carter, Brandon (1979), "The general theory of the mechanical, electromagnetic and thermodynamic properties of black holes", in Hawking, S. W.; Israel, W., *General Relativity, an Einstein Centenary Survey*, Cambridge University Press, pp. 294–369 and 860–863, ISBN 0-521-29928-4

- Celotti, Annalisa; Miller, John C.; Sciama, Dennis W. (1999), "Astrophysical evidence for the existence of black holes", *Class. Quant. Grav.* **16** (12A): A3–A21, arXiv:astro-ph/9912186, doi:10.1088/0264-9381/16/12A/301

- Chandrasekhar, Subrahmanyan (1983), *The Mathematical Theory of Black Holes*, Oxford University Press, ISBN 0-19-850370-9

- Charbonnel, C.; Primas, F. (2005), "The Lithium Content of the Galactic Halo Stars", *Astronomy & Astrophysics* **442** (3): 961–992, arXiv:astro-ph/0505247, Bibcode:2005A&A...442..961C, doi:10.1051/0004-6361:20042491

- Ciufolini, Ignazio; Pavlis, Erricos C. (2004), "A confirmation of the general relativistic prediction of the Lense-Thirring effect", *Nature* **431** (7011): 958–960, Bibcode:2004Natur.431..958C, doi:10.1038/nature03007, PMID 15496915

- Ciufolini, Ignazio; Pavlis, Erricos C.; Peron, R. (2006), "Determination of frame-dragging using Earth gravity models from CHAMP and GRACE", *New Astron.* **11** (8): 527–550, Bibcode:2006NewA...11..527C, doi:10.1016/j.newast.2006.02.001

- Coc, A.; Vangioni-Flam, Elisabeth; Descouvemont, Pierre; Adahchour, Abderrahim; Angulo, Carmen (2004), "Updated Big Bang Nucleosynthesis confronted to WMAP observations and to the Abundance of Light Elements", *Astrophysical Journal* **600** (2): 544–552, arXiv:astro-ph/0309480, Bibcode:2004ApJ...600..544C, doi:10.1086/380121

- Cutler, Curt; Thorne, Kip S. (2002), "An overview of gravitational wave sources", in Bishop, Nigel; Maharaj, Sunil D., *Proceedings of 16th International Conference on General Relativity and Gravitation (GR16)*, World Scientific, p. 4090, arXiv:gr-qc/0204090, Bibcode:2002gr.qc.....4090C, ISBN 981-238-171-6

- Dalal, Neal; Holz, Daniel E.; Hughes, Scott A.; Jain, Bhuvnesh (2006), "Short GRB and binary black hole standard sirens as a probe of dark energy", *Phys.Rev.* **D74** (6): 063006, arXiv:astro-ph/0601275, Bibcode:2006PhRvD..74f3006D, doi:10.1103/PhysRevD.74.063006

- Danzmann, Karsten; Rüdiger, Albrecht (2003), "LISA Technology—Concepts, Status, Prospects" (PDF), *Class. Quant. Grav.* **20** (10): S1–S9, Bibcode:2003CQGra..20S...1D, doi:10.1088/0264-9381/20/10/301

- Dirac, Paul (1996), *General Theory of Relativity*, Princeton University Press, ISBN 0-691-01146-X

- Donoghue, John F. (1995), "Introduction to the Effective Field Theory Description of Gravity", in Cornet, Fernando, *Effective Theories: Proceedings of the Advanced School, Almunecar, Spain, 26 June–1 July 1995*, Singapore: World Scientific, p. 12024, arXiv:gr-qc/9512024, Bibcode:1995gr.qc....12024D, ISBN 981-02-2908-9

- Duff, Michael (1996), "M-Theory (the Theory Formerly Known as Strings)", *Int. J. Mod. Phys.* **A11** (32): 5623–5641, arXiv:hep-th/9608117, Bibcode:1996IJMPA..11.5623D, doi:10.1142/S0217751X96002583

- Ehlers, Jürgen (1973), "Survey of general relativity theory", in Israel, Werner, *Relativity, Astrophysics and Cosmology*, D. Reidel, pp. 1–125, ISBN 90-277-0369-8

- Ehlers, Jürgen; Falco, Emilio E.; Schneider, Peter (1992), *Gravitational lenses*, Springer, ISBN 3-540-66506-4

- Ehlers, Jürgen; Lämmerzahl, Claus, eds. (2006), *Special Relativity—Will it Survive the Next 101 Years?*, Springer, ISBN 3-540-34522-1

- Ehlers, Jürgen; Rindler, Wolfgang (1997), "Local and Global Light Bending in Einstein's and other Gravitational Theories", *General Relativity and Gravitation* **29** (4): 519–529, Bibcode:1997GReGr..29..519E, doi:10.1023/A:1018843001842

- Einstein, Albert (1907), "Über das Relativitätsprinzip und die aus demselben gezogene Folgerungen" (PDF), *Jahrbuch der Radioaktivität und Elektronik* **4**: 411, retrieved 2008-05-05

- Einstein, Albert (1915), "Die Feldgleichungen der Gravitation", *Sitzungsberichte der Preussischen Akademie der Wissenschaften zu Berlin*: 844–847, retrieved 2006-09-12

- Einstein, Albert (1916), "Die Grundlage der allgemeinen Relativitätstheorie", *Annalen der Physik* **49**: 769–822, Bibcode:1916AnP...354..769E, doi:10.1002/andp.19163540702, archived from the original (PDF) on 2006-08-29, retrieved 2016-02-14

- Einstein, Albert (1917), "Kosmologische Betrachtungen zur allgemeinen Relativitätstheorie", *Sitzungsberichte der Preußischen Akademie der Wissenschaften*: 142

- Ellis, George F R; Van Elst, Henk (1999), Lachièze-Rey, Marc, ed., "Theoretical and Observational Cosmology: Cosmological models (Cargèse lectures 1998)", *Theoretical and observational cosmology : proceedings of the NATO Advanced Study Institute on Theoretical and Observational Cosmology* (Kluwer): 1–116, arXiv:gr-qc/9812046, Bibcode:1999toc..conf....1E, doi:10.1007/978-94-011-4455-1_1, ISBN 978-0-7923-5946-3

- Everitt, C. W. F.; Buchman, S.; DeBra, D. B.; Keiser, G. M. (2001), "Gravity Probe B: Countdown to launch", in Lämmerzahl, C.; Everitt, C. W. F.; Hehl, F. W., *Gyros, Clocks, and Interferometers: Testing Relativistic Gravity in Space (Lecture Notes in Physics 562)*, Springer, pp. 52–82, ISBN 3-540-41236-0

- Everitt, C. W. F.; Parkinson, Bradford; Kahn, Bob (2007), *The Gravity Probe B experiment. Post Flight Analysis—Final Report (Preface and Executive Summary)* (PDF), Project Report: NASA, Stanford University and Lockheed Martin, retrieved 2007-08-05

- Falcke, Heino; Melia, Fulvio; Agol, Eric (2000), "Viewing the Shadow of the Black Hole at the Galactic Center", *Astrophysical Journal* **528** (1): L13–L16, arXiv:astro-ph/9912263, Bibcode:2000ApJ...528L..13F, doi:10.1086/312423, PMID 10587484

- Flanagan, Éanna É.; Hughes, Scott A. (2005), "The basics of gravitational wave theory", *New J.Phys.* **7**: 204, arXiv:gr-qc/0501041, Bibcode:2005NJPh....7..204F, doi:10.1088/1367-2630/7/1/204

- Font, José A. (2003), "Numerical Hydrodynamics in General Relativity", *Living Reviews in Relativity* **6**, doi:10.12942/lrr-2003-4, retrieved 2007-08-19

- Fourès-Bruhat, Yvonne (1952), "Théoréme d'existence pour certains systémes d'équations aux derivées partielles non linéaires", *Acta Mathematica* **88** (1): 141–225, Bibcode:1952AcM....88..141F, doi:10.1007/BF02392131

- Frauendiener, Jörg (2004), "Conformal Infinity", *Living Reviews in Relativity* **7**, Bibcode:2004LRR.....7....1F, doi:10.12942/lrr-2004-1, retrieved 2007-07-21

- Friedrich, Helmut (2005), "Is general relativity 'essentially understood'?", *Annalen der Physik* **15** (1–2): 84–108, arXiv:gr-qc/0508016, Bibcode:2006AnP...518...84F, doi:10.1002/andp.200510173

- Futamase, T.; Itoh, Y. (2006), "The Post-Newtonian Approximation for Relativistic Compact Binaries", *Living Reviews in Relativity* **10**, doi:10.12942/lrr-2007-2, retrieved 2008-02-29

- Gamow, George (1970), *My World Line*, Viking Press, ISBN 0-670-50376-2

- Garfinkle, David (2007), "Of singularities and breadmaking", *Einstein Online*, retrieved 2007-08-03

- Geroch, Robert (1996), "Partial Differential Equations of Physics", arXiv:gr-qc/9602055 [gr-qc].

- Giulini, Domenico (2005), *Special Relativity: A First Encounter*, Oxford University Press, ISBN 0-19-856746-4

- Giulini, Domenico (2006a), "Algebraic and Geometric Structures in Special Relativity", in Ehlers, Jürgen; Lämmerzahl, Claus, *Special Relativity—Will it Survive the Next 101 Years?*, Springer, pp. 45–111, arXiv:math-ph/0602018, Bibcode:2006math.ph...2018G, ISBN 3-540-34522-1

- Giulini, Domenico (2006b), Stamatescu, I. O., ed., "An assessment of current paradigms in the physics of fundamental interactions: Some remarks on the notions of general covariance and background independence", *Approaches to Fundamental Physics*, Lecture Notes in Physics (Springer) **721**: 105, arXiv:gr-qc/0603087, Bibcode:2007LNP...721..105G, doi:10.1007/978-3-540-71117-9_6, ISBN 978-3-540-71115-5

- Gnedin, Nickolay Y. (2005), "Digitizing the Universe", *Nature* **435** (7042): 572–573, Bibcode:2005Natur.435..572G, doi:10.1038/435572a, PMID 15931201

- Goenner, Hubert F. M. (2004), "On the History of Unified Field Theories", *Living Reviews in Relativity* **7**, Bibcode:2004LRR.....7....2G, doi:10.12942/lrr-2004-2, retrieved 2008-02-28

- Goroff, Marc H.; Sagnotti, Augusto (1985), "Quantum gravity at two loops", *Phys. Lett.* **160B** (1–3): 81–86, Bibcode:1985PhLB..160...81G, doi:10.1016/0370-2693(85)91470-4

- Gourgoulhon, Eric (2007). "3+1 Formalism and Bases of Numerical Relativity". arXiv:gr-qc/0703035 [gr-qc].

- Gowdy, Robert H. (1971), "Gravitational Waves in Closed Universes", *Phys. Rev. Lett.* **27** (12): 826–829, Bibcode:1971PhRvL..27..826G, doi:10.1103/PhysRevLett.27.826

- Gowdy, Robert H. (1974), "Vacuum spacetimes with two-parameter spacelike isometry groups and compact invariant hypersurfaces: Topologies and boundary conditions", *Annals of Physics* **83** (1): 203–241, Bibcode:1974AnPhy..83..203G, doi:10.1016/0003-4916(74)90384-4

- Green, M. B.; Schwarz, J. H.; Witten, E. (1987), *Superstring theory. Volume 1: Introduction*, Cambridge University Press, ISBN 0-521-35752-7

- Greenstein, J. L.; Oke, J. B.; Shipman, H. L. (1971), "Effective Temperature, Radius, and Gravitational Redshift of Sirius B", *Astrophysical Journal* **169**: 563, Bibcode:1971ApJ...169..563G, doi:10.1086/151174

- Hamber, Herbert W. (2009), *Quantum Gravitation - The Feynman Path Integral Approach*, Springer Publishing, doi:10.1007/978-3-540-85293-3, ISBN 978-3-540-85292-6

- Gödel, Kurt (1949), "An Example of a New Type of Cosmological Solution of Einstein's Field Equations of Gravitation", *Rev. Mod. Phys.* **21** (3): 447. Bibcode:1949RvMP...21..447G, doi:10.1103/RevModPhys.21.447.

- Hafele, J. C.; Keating, R. E. (July 14, 1972). "Around-the-World Atomic Clocks: Predicted Relativistic Time Gains". *Science* **177** (4044): 166–168. Bibcode:1972Sci...177..166H. doi:10.1126/science.177.4044.166. PMID 17779917.

- Hafele, J. C.; Keating, R. E. (July 14, 1972). "Around-the-World Atomic Clocks: Observed Relativistic Time Gains". *Science* **177** (4044): 168–170. Bibcode:1972Sci...177..168H. doi:10.1126/science.177.4044.168. PMID 17779918.

- Havas, P. (1964). "Four-Dimensional Formulation of Newtonian Mechanics and Their Relation to the Special and the General Theory of Relativity". *Rev. Mod. Phys.* **36** (4): 938–965. Bibcode:1964RvMP...36..938H. doi:10.1103/RevModPhys.36.938

- Hawking, Stephen W. (1966), "The occurrence of singularities in cosmology", *Proceedings of the Royal Society* **A294** (1439): 511–521. Bibcode:1966RSPSA.294..511H. doi:10.1098/rspa.1966.0221

- Hawking, S. W. (1975), "Particle Creation by Black Holes", *Communications in Mathematical Physics* **43** (3): 199–220, Bibcode:1975CMaPh..43..199H, doi:10.1007/BF02345020

- Hawking, Stephen W. (1987), "Quantum cosmology", in Hawking, Stephen W.; Israel, Werner, *300 Years of Gravitation*, Cambridge University Press, pp. 631–651, ISBN 0-521-37976-8

- Hawking, Stephen W.; Ellis, George F. R. (1973), *The large scale structure of space-time*, Cambridge University Press, ISBN 0-521-09906-4

- Heckmann, O. H. L.; Schücking, E. (1959), "Newtonsche und Einsteinsche Kosmologie", in Flügge, S., *Encyclopedia of Physics* **53**, p. 489

- Heusler, Markus (1998), "Stationary Black Holes: Uniqueness and Beyond", *Living Reviews in Relativity* **1**, doi:10.12942/lrr-1998-6, retrieved 2007-08-04

- Heusler, Markus (1996), *Black Hole Uniqueness Theorems*, Cambridge University Press, ISBN 0-521-56735-1

- Hey, Tony; Walters, Patrick (2003), *The new quantum universe*, Cambridge University Press, ISBN 0-521-56457-3

- Hough, Jim; Rowan, Sheila (2000), "Gravitational Wave Detection by Interferometry (Ground and Space)", *Living Reviews in Relativity* **3**, retrieved 2007-07-21

- Hubble, Edwin (1929), "A Relation between Distance and Radial Velocity among Extra-Galactic Nebulae" (PDF). *Proc. Natl. Acad. Sci.* **15** (3): 168–173, Bibcode:1929PNAS...15..168H, doi:10.1073/pnas.15.3.168. PMC 522427. PMID 16577160

- Hulse, Russell A.; Taylor, Joseph H. (1975), "Discovery of a pulsar in a binary system", *Astrophys. J.* **195**: L51–L55, Bibcode:1975ApJ...195L..51H, doi:10.1086/181708

- Ibanez, L. E. (2000), "The second string (phenomenology) revolution", *Class. Quant. Grav.* **17** (5): 1117–1128, arXiv:hep-ph/9911499, Bibcode:2000CQGra..17.1117I, doi:10.1088/0264-9381/17/5/321

- Iorio, L. (2009), "An Assessment of the Systematic Uncertainty in Present and Future Tests of the Lense-Thirring Effect with Satellite Laser Ranging", *Space Sci. Rev.* **148** (1–4): 363, arXiv:0809.1373, Bibcode:2009SSRv..148..363I, doi:10.1007/s11214-008-9478-1

- Isham, Christopher J. (1994), "Prima facie questions in quantum gravity", in Ehlers, Jürgen; Friedrich, Helmut, *Canonical Gravity: From Classical to Quantum*, Springer, ISBN 3-540-58339-4

- Israel, Werner (1971), "Event Horizons and Gravitational Collapse", *General Relativity and Gravitation* **2** (1): 53–59, Bibcode:1971GReGr...2...53I, doi:10.1007/BF02450518

- Israel, Werner (1987), "Dark stars: the evolution of an idea", in Hawking, Stephen W.; Israel, Werner, *300 Years of Gravitation*, Cambridge University Press, pp. 199–276, ISBN 0-521-37976-8

- Janssen, Michel (2005), "Of pots and holes: Einstein's bumpy road to general relativity" (PDF), *Annalen der Physik* **14** (S1): 58–85, Bibcode:2005AnP...517S..58J, doi:10.1002/andp.200410130

- Jaranowski, Piotr; Królak, Andrzej (2005), "Gravitational-Wave Data Analysis. Formalism and Sample Applications: The Gaussian Case", *Living Reviews in Relativity* **8**, doi:10.12942/lrr-2005-3, retrieved 2007-07-30

- Kahn, Bob (1996–2012), *Gravity Probe B Website*, Stanford University, retrieved 2012-04-20

- Kahn, Bob (April 14, 2007), *Was Einstein right? Scientists provide first public peek at Gravity Probe B results (Stanford University Press Release)* (PDF), Stanford University News Service

- Kamionkowski, Marc; Kosowsky, Arthur; Stebbins, Albert (1997), "Statistics of Cosmic Microwave Background Polarization", *Phys. Rev.* **D55** (12): 7368–7388, arXiv:astro-ph/9611125, Bibcode:1997PhRvD..55.7368K, doi:10.1103/PhysRevD.55.7368

- Kennefick, Daniel (2005), "Astronomers Test General Relativity: Light-bending and the Solar Redshift", in Renn, Jürgen, *One hundred authors for Einstein*, Wiley-VCH, pp. 178–181, ISBN 3-527-40574-7

- Kennefick, Daniel (2007), "Not Only Because of Theory: Dyson, Eddington and the Competing Myths of the 1919 Eclipse Expedition", *Proceedings of the 7th Conference on the History of General Relativity, Tenerife, 2005* **0709**, p. 685, arXiv:0709.0685, Bibcode:2007arXiv0709.0685K

- Kenyon, I. R. (1990), *General Relativity*, Oxford University Press, ISBN 0-19-851996-6

- Kochanek, C.S.; Falco, E.E.; Impey, C.; Lehar, J. (2007), *CASTLES Survey Website*, Harvard-Smithsonian Center for Astrophysics, retrieved 2007-08-21

- Komar, Arthur (1959), "Covariant Conservation Laws in General Relativity", *Phys. Rev.* **113** (3): 934–936, Bibcode:1959PhRv..113..934K, doi:10.1103/PhysRev.113.934

- Kramer, Michael (2004), Karshenboim, S. G.; Peik, E., eds., "Astrophysics, Clocks and Fundamental Constants: Millisecond Pulsars as Tools of Fundamental Physics", *Lecture Notes in Physics* (Springer) **648**: 33–54, arXiv:astro-ph/0405178, Bibcode:2004LNP...648...33K, doi:10.1007/978-3-540-40991-5_3, ISBN 978-3-540-21967-5

- Kramer, M.; Stairs, I. H.; Manchester, R. N.; McLaughlin, M. A.; Lyne, A. G.; Ferdman, R. D.; Burgay, M.; Lorimer, D. R.; et al. (2006), "Tests of general relativity from timing the double pulsar", *Science* **314** (5796): 97–102, arXiv:astro-ph/0609417, Bibcode:2006Sci...314...97K, doi:10.1126/science.1132305, PMID 16973838

- Kraus, Ute (1998), "Light Deflection Near Neutron Stars", *Relativistic Astrophysics*, Vieweg, pp. 66–81, ISBN 3-528-06909-0

- Kuchař, Karel (1973), "Canonical Quantization of Gravity", in Israel, Werner, *Relativity, Astrophysics and Cosmology*, D. Reidel, pp. 237–288, ISBN 90-277-0369-8

- Künzle, H. P. (1972), "Galilei and Lorentz Structures on spacetime: comparison of the corresponding geometry and physics", *Annales de l'Institut Henri Poincaré A* **17**: 337–362

- Lahav, Ofer; Suto, Yasushi (2004), "Measuring our Universe from Galaxy Redshift Surveys", *Living Reviews in Relativity* **7**, arXiv:astro-ph/0310642, Bibcode:2004LRR.....7....8L, doi:10.12942/lrr-2004-8, retrieved 2007-08-19

- Landgraf, M.; Hechler, M.; Kemble, S. (2005), "Mission design for LISA Pathfinder", *Class. Quant. Grav.* **22** (10): S487–S492, arXiv:gr-qc/0411071, Bibcode:2005CQGra..22S.487L, doi:10.1088/0264-9381/22/10/048

- Lehner, Luis (2001), "Numerical Relativity: A review", *Class. Quant. Grav.* **18** (17): R25–R86, arXiv:gr-qc/0106072, Bibcode:2001CQGra..18R..25L, doi:10.1088/0264-9381/18/17/202

- Lehner, Luis (2002), "NUMERICAL RELATIVITY: STATUS AND PROSPECTS", *General Relativity and Gravitation - Proceedings of the 16th International Conference*, p. 210, arXiv:gr-qc/0202055, Bibcode:2002grg..conf..210L, doi:10.1142/9789812776556_0010, ISBN 9789812381712

- Linde, Andrei (1990), *Particle Physics and Inflationary Cosmology*, Harwood, p. 3203, arXiv:hep-th/0503203, Bibcode:2005hep.th....3203L, ISBN 3-7186-0489-2

- Linde, Andrei (2005), "Towards inflation in string theory", *J. Phys. Conf. Ser.* **24**: 151–160, arXiv:hep-th/0503195, Bibcode:2005JPhCS..24..151L, doi:10.1088/1742-6596/24/1/018

- Loll, Renate (1998), "Discrete Approaches to Quantum Gravity in Four Dimensions", *Living Reviews in Relativity* **1**, arXiv:gr-qc/9805049, Bibcode:1998LRR.....1...13L, doi:10.12942/lrr-1998-13, retrieved 2008-03-09

- Lovelock, David (1972), "The Four-Dimensionality of Space and the Einstein Tensor", *J. Math. Phys.* **13** (6): 874–876, Bibcode:1972JMP....13..874L, doi:10.1063/1.1666069

- Ludyk, Günter (2013). *Einstein in Matrix Form* (1st ed.). Berlin: Springer. ISBN 9783642357978.

- MacCallum, M. (2006), "Finding and using exact solutions of the Einstein equations", in Mornas, L.; Alonso, J. D., *A Century of Relativity Physics (ERE05, the XXVIII Spanish Relativity Meeting)* **841**, American Institute of Physics, p. 129, arXiv:gr-qc/0601102, Bibcode:2006AIPC..841..129M, doi:10.1063/1.2218172

- Maddox, John (1998), *What Remains To Be Discovered*, Macmillan, ISBN 0-684-82292-X

- Mannheim, Philip D. (2006), "Alternatives to Dark Matter and Dark Energy", *Prog. Part. Nucl. Phys.* **56** (2): 340–445, arXiv:astro-ph/0505266, Bibcode:2006PrPNP..56..340M, doi:10.1016/j.ppnp.2005.08.001

- Mather, J. C.; Cheng, E. S.; Cottingham, D. A.; Eplee, R. E.; Fixsen, D. J.; Hewagama, T.; Isaacman, R. B.; Jensen, K. A.; et al. (1994), "Measurement of the cosmic microwave spectrum by the COBE FIRAS instrument", *Astrophysical Journal* **420**: 439–444, Bibcode:1994ApJ...420..439M, doi:10.1086/173574

- Mermin, N. David (2005), *It's About Time. Understanding Einstein's Relativity*, Princeton University Press, ISBN 0-691-12201-6

- Messiah, Albert (1999), *Quantum Mechanics*, Dover Publications, ISBN 0-486-40924-4

- Miller, Cole (2002), *Stellar Structure and Evolution (Lecture notes for Astronomy 606)*, University of Maryland, retrieved 2007-07-25

- Misner, Charles W.; Thorne, Kip. S.; Wheeler, John A. (1973), *Gravitation*, W. H. Freeman, ISBN 0-7167-0344-0

- Møller, Christian (1952), *The Theory of Relativity* (3rd ed.), Oxford University Press

- Narayan, Ramesh (2006), "Black holes in astrophysics", *New Journal of Physics* **7**: 199, arXiv:gr-qc/0506078, Bibcode:2005NJPh....7..199N, doi:10.1088/1367-2630/7/1/199

- Narayan, Ramesh; Bartelmann, Matthias (1997), "Lectures on Gravitational Lensing", arXiv:astro-ph/9606001 [astro-ph].

- Narlikar, Jayant V. (1993), *Introduction to Cosmology*, Cambridge University Press, ISBN 0-521-41250-1

- Nieto, Michael Martin (2006), "The quest to understand the Pioneer anomaly" (PDF), *EurophysicsNews* **37** (6): 30–34, Bibcode:2006ENews..37...30N, doi:10.1051/epn:2006604

- Nordström, Gunnar (1918), "On the Energy of the Gravitational Field in Einstein's Theory", *Verhandl. Koninkl. Ned. Akad. Wetenschap.* **26**: 1238–1245

- Nordtvedt, Kenneth (2003). "Lunar Laser Ranging—a comprehensive probe of post-Newtonian gravity". arXiv:gr-qc/0301024 [gr-qc].

- Norton, John D. (1985), "What was Einstein's principle of equivalence?" (PDF), *Studies in History and Philosophy of Science* **16** (3): 203–246, doi:10.1016/0039-3681(85)90002-0, retrieved 2007-06-11

- Ohanian, Hans C.; Ruffini, Remo (1994), *Gravitation and Spacetime*, W. W. Norton & Company, ISBN 0-393-96501-5

- Olive, K. A.; Skillman, E. A. (2004), "A Realistic Determination of the Error on the Primordial Helium Abundance", *Astrophysical Journal* **617** (1): 29–49, arXiv:astro-ph/0405588, Bibcode:2004ApJ...617...29O, doi:10.1086/425170

- O'Meara, John M.; Tytler, David; Kirkman, David; Suzuki, Nao; Prochaska, Jason X.; Lubin, Dan; Wolfe, Arthur M. (2001), "The Deuterium to Hydrogen Abundance Ratio Towards a Fourth QSO: HS0105+1619", *Astrophysical Journal* **552** (2): 718–730, arXiv:astro-ph/0011179, Bibcode:2001ApJ...552..718O, doi:10.1086/320579

- Oppenheimer, J. Robert; Snyder, H. (1939), "On continued gravitational contraction", *Physical Review* **56** (5): 455–459, Bibcode:1939PhRv...56..455O, doi:10.1103/PhysRev.56.455

- Overbye, Dennis (1999), *Lonely Hearts of the Cosmos: the story of the scientific quest for the secret of the Universe*, Back Bay, ISBN 0-316-64896-5

- Pais, Abraham (1982), *'Subtle is the Lord...' The Science and life of Albert Einstein*, Oxford University Press, ISBN 0-19-853907-X

- Peacock, John A. (1999), *Cosmological Physics*, Cambridge University Press, ISBN 0-521-41072-X

- Peebles, P. J. E. (1966), "Primordial Helium abundance and primordial fireball II", *Astrophysical Journal* **146**: 542–552, Bibcode:1966ApJ...146..542P, doi:10.1086/148918

- Peebles, P. J. E. (1993), *Principles of physical cosmology*, Princeton University Press, ISBN 0-691-01933-9

- Peebles, P.J.E.; Schramm, D.N.; Turner, E.L.; Kron, R.G. (1991), "The case for the relativistic hot Big Bang cosmology", *Nature* **352** (6338): 769–776, Bibcode:1991Natur.352..769P, doi:10.1038/352769a0

- Penrose, Roger (1965), "Gravitational collapse and spacetime singularities", *Physical Review Letters* **14** (3): 57–59, Bibcode:1965PhRvL..14...57P, doi:10.1103/PhysRevLett.14.57

- Penrose, Roger (1969), "Gravitational collapse: the role of general relativity", *Rivista del Nuovo Cimento* **1**: 252–276, Bibcode:1969NCimR...1..252P

- Penrose, Roger (2004), *The Road to Reality*, A. A. Knopf, ISBN 0-679-45443-8

- Penzias, A. A.; Wilson, R. W. (1965), "A measurement of excess antenna temperature at 4080 Mc/s", *Astrophysical Journal* **142**: 419–421, Bibcode:1965ApJ...142..419P, doi:10.1086/148307

- Peskin, Michael E.; Schroeder, Daniel V. (1995), *An Introduction to Quantum Field Theory*, Addison-Wesley, ISBN 0-201-50397-2

- Peskin, Michael E. (2007), "Dark Matter and Particle Physics", *Journal of the Physical Society of Japan* **76** (11): 111017, arXiv:0707.1536, Bibcode:2007JPSJ...76k1017P, doi:10.1143/JPSJ.76.111017

- Poisson, Eric (2004), "The Motion of Point Particles in Curved Spacetime", *Living Reviews in Relativity* **7**, doi:10.12942/lrr-2004-6, retrieved 2007-06-13

- Poisson, Eric (2004), *A Relativist's Toolkit. The Mathematics of Black-Hole Mechanics*, Cambridge University Press, ISBN 0-521-83091-5

- Polchinski, Joseph (1998a), *String Theory Vol. I: An Introduction to the Bosonic String*, Cambridge University Press, ISBN 0-521-63303-6

- Polchinski, Joseph (1998b), *String Theory Vol. II: Superstring Theory and Beyond*, Cambridge University Press, ISBN 0-521-63304-4

- Pound, R. V.; Rebka, G. A. (1959), "Gravitational Red-Shift in Nuclear Resonance", *Physical Review Letters* **3** (9): 439–441, Bibcode:1959PhRvL...3..439P, doi:10.1103/PhysRevLett.3.439

- Pound, R. V.; Rebka, G. A. (1960), "Apparent weight of photons", *Phys. Rev. Lett.* **4** (7): 337–341, Bibcode:1960PhRvL...4..337P, doi:10.1103/PhysRevLett.4.337

- Pound, R. V.; Snider, J. L. (1964), "Effect of Gravity on Nuclear Resonance", *Phys. Rev. Lett.* **13** (18): 539–540, Bibcode:1964PhRvL..13..539P, doi:10.1103/PhysRevLett.13.539

- Ramond, Pierre (1990), *Field Theory: A Modern Primer*, Addison-Wesley, ISBN 0-201-54611-6

- Rees, Martin (1966), "Appearance of Relativistically Expanding Radio Sources", *Nature* **211** (5048): 468–470, Bibcode:1966Natur.211..468R, doi:10.1038/211468a0

- Reissner, H. (1916), "Über die Eigengravitation des elektrischen Feldes nach der Einsteinschen Theorie", *Annalen der Physik* **355**

(9): 106–120. Bibcode:1916AnP...355..106R. doi:10.1002/andp.19163550905

- Remillard, Ronald A.; Lin, Dacheng; Cooper, Randall L.; Narayan, Ramesh (2006). "The Rates of Type I X-Ray Bursts from Transients Observed with RXTE: Evidence for Black Hole Event Horizons", *Astrophysical Journal* **646** (1): 407–419, arXiv:astro-ph/0509758, Bibcode:2006ApJ...646..407R, doi:10.1086/504862

- Renn, Jürgen, ed. (2007), *The Genesis of General Relativity (4 Volumes)*, Dordrecht: Springer, ISBN 1-4020-3999-9

- Renn, Jürgen, ed. (2005), *Albert Einstein—Chief Engineer of the Universe: Einstein's Life and Work in Context*, Berlin: Wiley-VCH, ISBN 3-527-40571-2

- Reula, Oscar A. (1998). "Hyperbolic Methods for Einstein's Equations", *Living Reviews in Relativity* **1**. Bibcode:1998LRR.....1....3R, doi:10.12942/lrr-1998-3, retrieved 2007-08-29

- Rindler, Wolfgang (2001), *Relativity. Special, General and Cosmological*, Oxford University Press, ISBN 0-19-850836-0

- Rindler, Wolfgang (1991), *Introduction to Special Relativity*, Clarendon Press, Oxford, ISBN 0-19-853952-5

- Robson, Ian (1996), *Active galactic nuclei*, John Wiley, ISBN 0-471-95853-0

- Roulet, E.; Mollerach, S. (1997). "Microlensing", *Physics Reports* **279** (2): 67–118, arXiv:astro-ph/9603119, Bibcode:1997PhR...279...67R, doi:10.1016/S0370-1573(96)00020-8

- Rovelli, Carlo (2000). "Notes for a brief history of quantum gravity". arXiv:gr-qc/0006061 [gr-qc].

- Rovelli, Carlo (1998). "Loop Quantum Gravity", *Living Reviews in Relativity* **1**, doi:10.12942/lrr-1998-1, retrieved 2008-03-13

- Schäfer, Gerhard (2004). "Gravitomagnetic Effects", *General Relativity and Gravitation* **36** (10): 2223–2235, arXiv:gr-qc/0407116, Bibcode:2004GReGr..36.2223S, doi:10.1023/B:GERG.0000046180.97877.32

- Schödel, R.; Ott, T.; Genzel, R.; Eckart, A.; Mouawad, N.; Alexander, T. (2003), "Stellar Dynamics in the Central Arcsecond of Our Galaxy", *Astrophysical Journal* **596** (2): 1015–1034, arXiv:astro-ph/0306214, Bibcode:2003ApJ...596.1015S, doi:10.1086/378122

- Schutz, Bernard F. (1985), *A first course in general relativity*, Cambridge University Press, ISBN 0-521-27703-5

- Schutz, Bernard F. (2001), "Gravitational radiation", in Murdin, Paul, *Encyclopedia of Astronomy and Astrophysics*, Grove's Dictionaries, ISBN 1-56159-268-4

- Schutz, Bernard F. (2003), *Gravity from the ground up*, Cambridge University Press, ISBN 0-521-45506-5

- Schwarz, John H. (2007), "String Theory: Progress and Problems". *Progress of Theoretical Physics Supplement* **170**: 214, arXiv:hep-th/0702219, Bibcode:2007PThPS.170..214S, doi:10.1143/PTPS.170.214

- Schwarzschild, Karl (1916a). "Über das Gravitationsfeld eines Massenpunktes nach der Einsteinschen Theorie", *Sitzungsber. Preuss. Akad. D. Wiss.*: 189–196

- Schwarzschild, Karl (1916b). "Über das Gravitationsfeld einer Kugel aus inkompressibler Flüssigkeit nach der Einsteinschen Theorie", *Sitzungsber. Preuss. Akad. D. Wiss.*: 424–434

- Seidel, Edward (1998). "Numerical Relativity: Towards Simulations of 3D Black Hole Coalescence", in Narlikar, J. V.; Dadhich, N., *Gravitation and Relativity: At the turn of the millennium (Proceedings of the GR-15 Conference, held at IUCAA, Pune, India, December 16–21, 1997)*, IUCAA, p. 6088, arXiv:gr-qc/9806088, Bibcode:1998gr.qc.....6088S, ISBN 81-900378-3-8

- Seljak, Uroš; Zaldarriaga, Matias (1997). "Signature of Gravity Waves in the Polarization of the Microwave Background". *Phys. Rev. Lett.* **78** (11): 2054–2057, arXiv:astro-ph/9609169, Bibcode:1997PhRvL..78.2054S, doi:10.1103/PhysRevLett.78.2054

- Shapiro, S. S.; Davis, J. L.; Lebach, D. E.; Gregory, J. S. (2004), "Measurement of the solar gravitational deflection of radio waves using geodetic very-long-baseline interferometry data, 1979–1999", *Phys. Rev. Lett.* **92** (12): 121101, Bibcode:2004PhRvL..92l1101S, doi:10.1103/PhysRevLett.92.121101, PMID 15089661

- Shapiro, Irwin I. (1964). "Fourth test of general relativity", *Phys. Rev. Lett.* **13** (26): 789–791, Bibcode:1964PhRvL..13..789S, doi:10.1103/PhysRevLett.13.789

- Shapiro, I. I.; Pettengill, Gordon; Ash, Michael; Stone, Melvin; Smith, William; Ingalls, Richard; Brockelman, Richard (1968), "Fourth test of general relativity: preliminary results", *Phys. Rev. Lett.* **20** (22): 1265–1269, Bibcode:1968PhRvL..20.1265S, doi:10.1103/PhysRevLett.20.1265

- Singh, Simon (2004), *Big Bang: The Origin of the Universe*, Fourth Estate, ISBN 0-00-715251-5

- Sorkin, Rafael D. (2005), "Causal Sets: Discrete Gravity", in Gomberoff, Andres; Marolf, Donald, *Lectures on Quantum Gravity*, Springer, p. 9009, arXiv:gr-qc/0309009, Bibcode:2003gr.qc.....9009S, ISBN 0-387-23995-2

- Sorkin, Rafael D. (1997), "Forks in the Road, on the Way to Quantum Gravity", *Int. J. Theor. Phys.* **36** (12): 2759–2781, arXiv:gr-qc/9706002, Bibcode:1997IJTP...36.2759S, doi:10.1007/BF02435709

- Spergel, D. N.; Verde, L.; Peiris, H. V.; Komatsu, E.; Nolta, M. R.; Bennett, C. L.; Halpern, M.; Hinshaw, G.; et al. (2003), "First Year Wilkinson Microwave Anisotropy Probe (WMAP) Observations: Determination of Cosmological Parameters", *Astrophys. J. Suppl.* **148** (1): 175–194, arXiv:astro-ph/0302209, Bibcode:2003ApJS..148..175S, doi:10.1086/377226

- Spergel, D. N.; Bean, R.; Doré, O.; Nolta, M. R.; Bennett, C. L.; Dunkley, J.; Hinshaw, G.; Jarosik, N.; et al. (2007), "Wilkinson Microwave Anisotropy Probe (WMAP) Three Year Results: Implications for Cosmology", *Astrophysical Journal Supplement* **170** (2): 377–408, arXiv:astro-ph/0603449, Bibcode:2007ApJS..170..377S, doi:10.1086/513700

- Springel, Volker; White, Simon D. M.; Jenkins, Adrian; Frenk, Carlos S.; Yoshida, Naoki; Gao, Liang; Navarro, Julio; Thacker, Robert; et al. (2005), "Simulations of the formation, evolution and clustering of galaxies and quasars", *Nature* **435** (7042): 629–636, arXiv:astro-ph/0504097, Bibcode:2005Natur.435..629S, doi:10.1038/nature03597, PMID 15931216

- Stairs, Ingrid H. (2003), "Testing General Relativity with Pulsar Timing", *Living Reviews in Relativity* **6**, arXiv:astro-ph/0307536, Bibcode:2003LRR.....6....5S, doi:10.12942/lrr-2003-5, retrieved 2007-07-21

- Stephani, H.; Kramer, D.; MacCallum, M.; Hoenselaers, C.; Herlt, E. (2003), *Exact Solutions of Einstein's Field Equations* (2 ed.), Cambridge University Press, ISBN 0-521-46136-7

- Synge, J. L. (1972), *Relativity: The Special Theory*, North-Holland Publishing Company, ISBN 0-7204-0064-3

- Szabados, László B. (2004), "Quasi-Local Energy-Momentum and Angular Momentum in GR", *Living Reviews in Relativity* **7**, doi:10.12942/lrr-2004-4, retrieved 2007-08-23

- Taylor, Joseph H. (1994), "Binary pulsars and relativistic gravity", *Rev. Mod. Phys.* **66** (3): 711–719, Bibcode:1994RvMP...66..711T, doi:10.1103/RevModPhys.66.711

- Thiemann, Thomas (2006), "Approaches to Fundamental Physics: Loop Quantum Gravity: An Inside View", *Lecture Notes in Physics* **721**: 185–263, arXiv:hep-th/0608210, Bibcode:2007LNP...721..185T, doi:10.1007/978-3-540-71117-9_10, ISBN 978-3-540-71115-5

- Thiemann, Thomas (2003), "Lectures on Loop Quantum Gravity", *Lecture Notes in Physics* **631**: 41–135, arXiv:gr-qc/0210094, doi:10.1007/978-3-540-45230-0_3, ISBN 978-3-540-40810-9

- 't Hooft, Gerard; Veltman, Martinus (1974), "One Loop Divergencies in the Theory of Gravitation", *Ann. Inst. Poincare* **20**: 69

- Thorne, Kip S. (1972), "Nonspherical Gravitational Collapse—A Short Review", in Klauder, J., *Magic without Magic*, W. H. Freeman, pp. 231–258

- Thorne, Kip S. (1994), *Black Holes and Time Warps: Einstein's Outrageous Legacy*, W W Norton & Company, ISBN 0-393-31276-3

- Thorne, Kip S. (1995), "Gravitational radiation", *Particle and Nuclear Astrophysics and Cosmology in the Next Millenium*: 160, arXiv:gr-qc/9506086, Bibcode:1995pnac.conf..160T, ISBN 0-521-36853-7

- Townsend, Paul K. (1997), "Black Holes (Lecture notes)". arXiv:gr-qc/9707012 [gr-qc].

- Townsend, Paul K. (1996), "Four Lectures on M-Theory". arXiv:hep-th/9612121 [hep-th].

- Traschen, Jenny (2000), Bytsenko, A.; Williams, F., eds., "An Introduction to Black Hole Evaporation", *Mathematical Methods of Physics (Proceedings of the 1999 Londrina Winter School)* (World Scientific): 180, arXiv:gr-qc/0010055, Bibcode:2000mmp..conf..180T

- Trautman, Andrzej (2006), "Einstein–Cartan theory", in Françoise, J.-P.; Naber, G. L.; Tsou,

S. T., *Encyclopedia of Mathematical Physics, Vol. 2*, Elsevier, pp. 189–195, arXiv:gr-qc/0606062, Bibcode:2006gr.qc.....6062T

- Unruh, W. G. (1976), "Notes on Black Hole Evaporation", *Phys. Rev. D* **14** (4): 870–892, Bibcode:1976PhRvD..14..870U, doi:10.1103/PhysRevD.14.870

- Valtonen, M. J.; Lehto, H. J.; Nilsson, K.; Heidt, J.; Takalo, L. O.; Sillanpää, A.; Villforth, C.; Kidger, M.; et al. (2008), "A massive binary black-hole system in OJ 287 and a test of general relativity", *Nature* **452** (7189): 851–853, arXiv:0809.1280, Bibcode:2008Natur.452..851V, doi:10.1038/nature06896, PMID 18421348

- Veltman, Martinus (1975), "Quantum Theory of Gravitation", in Balian, Roger; Zinn-Justin, Jean, *Methods in Field Theory - Les Houches Summer School in Theoretical Physics.* **77**, North Holland

- Wald, Robert M. (1975), "On Particle Creation by Black Holes", *Commun. Math. Phys.* **45** (3): 9–34, Bibcode:1975CMaPh..45....9W, doi:10.1007/BF01609863

- Wald, Robert M. (1984), *General Relativity*, University of Chicago Press, ISBN 0-226-87033-2

- Wald, Robert M. (1994), *Quantum field theory in curved spacetime and black hole thermodynamics*, University of Chicago Press, ISBN 0-226-87027-8

- Wald, Robert M. (2001), "The Thermodynamics of Black Holes", *Living Reviews in Relativity* **4**. Bibcode:2001LRR.....4....6W, doi:10.12942/lrr-2001-6, retrieved 2007-08-08

- Walsh, D.; Carswell, R. F.; Weymann, R. J. (1979), "0957 + 561 A, B: twin quasistellar objects or gravitational lens?", *Nature* **279** (5712): 381–4, Bibcode:1979Natur.279..381W, doi:10.1038/279381a0, PMID 16068158

- Wambsganss, Joachim (1998), "Gravitational Lensing in Astronomy", *Living Reviews in Relativity* **1**, arXiv:astro-ph/9812021, Bibcode:1998LRR.....1...12W, doi:10.12942/lrr-1998-12, retrieved 2007-07-20

- Weinberg, Steven (1972), *Gravitation and Cosmology*, John Wiley, ISBN 0-471-92567-5

- Weinberg, Steven (1995), *The Quantum Theory of Fields I: Foundations*, Cambridge University Press, ISBN 0-521-55001-7

- Weinberg, Steven (1996), *The Quantum Theory of Fields II: Modern Applications*, Cambridge University Press, ISBN 0-521-55002-5

- Weinberg, Steven (2000), *The Quantum Theory of Fields III: Supersymmetry*, Cambridge University Press, ISBN 0-521-66000-9

- Weisberg, Joel M.; Taylor, Joseph H. (2003), "The Relativistic Binary Pulsar B1913+16"", in Bailes, M.; Nice, D. J.; Thorsett, S. E., *Proceedings of "Radio Pulsars," Chania, Crete, August, 2002*, ASP Conference Series

- Weiss, Achim (2006), "Elements of the past: Big Bang Nucleosynthesis and observation", *Einstein Online* (Max Planck Institute for Gravitational Physics), retrieved 2007-02-24

- Wheeler, John A. (1990), *A Journey Into Gravity and Spacetime*, Scientific American Library, San Francisco: W. H. Freeman, ISBN 0-7167-6034-7

- Will, Clifford M. (1993), *Theory and experiment in gravitational physics*, Cambridge University Press, ISBN 0-521-43973-6

- Will, Clifford M. (2006), "The Confrontation between General Relativity and Experiment", *Living Reviews in Relativity* **9**, arXiv:gr-qc/0510072, Bibcode:2006LRR.....9....3W, doi:10.12942/lrr-2006-3, retrieved 2007-06-12

- Zwiebach, Barton (2004), *A First Course in String Theory*, Cambridge University Press, ISBN 0-521-83143-1

1.7.12 Further reading

Popular books

- Geroch, R (1981), *General Relativity from A to B*, Chicago: University of Chicago Press, ISBN 0-226-28864-1

- Lieber, Lillian (2008), *The Einstein Theory of Relativity: A Trip to the Fourth Dimension*, Philadelphia: Paul Dry Books, Inc., ISBN 978-1-58988-044-3

- Wald, Robert M. (1992), *Space, Time, and Gravity: the Theory of the Big Bang and Black Holes*, Chicago: University of Chicago Press, ISBN 0-226-87029-4

- Wheeler, John; Ford, Kenneth (1998), *Geons, Black Holes, & Quantum Foam: a life in physics*, New York: W. W. Norton, ISBN 0-393-31991-1

Beginning undergraduate textbooks

- Callahan, James J. (2000), *The Geometry of Spacetime: an Introduction to Special and General Relativity*, New York: Springer, ISBN 0-387-98641-3

- Taylor, Edwin F.; Wheeler, John Archibald (2000), *Exploring Black Holes: Introduction to General Relativity*, Addison Wesley, ISBN 0-201-38423-X

Advanced undergraduate textbooks

- B. F. Schutz (2009), *A First Course in General Relativity (Second Edition)*, Cambridge University Press, ISBN 978-0-521-88705-2

- Cheng, Ta-Pei (2005), *Relativity, Gravitation and Cosmology: a Basic Introduction*, Oxford and New York: Oxford University Press, ISBN 0-19-852957-0

- Gron, O.; Hervik, S. (2007), *Einstein's General theory of Relativity*, Springer, ISBN 978-0-387-69199-2

- Hartle, James B. (2003), *Gravity: an Introduction to Einstein's General Relativity*, San Francisco: Addison-Wesley, ISBN 0-8053-8662-9

- Hughston, L. P. & Tod, K. P. (1991), *Introduction to General Relativity*, Cambridge: Cambridge University Press, ISBN 0-521-33943-X

- d'Inverno, Ray (1992), *Introducing Einstein's Relativity*, Oxford: Oxford University Press, ISBN 0-19-859686-3

- Ludyk, Günter (2013). *Einstein in Matrix Form* (1st ed.). Berlin: Springer. ISBN 9783642357978.

Graduate-level textbooks

- Carroll, Sean M. (2004), *Spacetime and Geometry: An Introduction to General Relativity*, San Francisco: Addison-Wesley, ISBN 0-8053-8732-3

- Grøn, Øyvind; Hervik, Sigbjørn (2007), *Einstein's General Theory of Relativity*, New York: Springer, ISBN 978-0-387-69199-2

- Landau, Lev D.; Lifshitz, Evgeny F. (1980), *The Classical Theory of Fields (4th ed.)*, London: Butterworth-Heinemann, ISBN 0-7506-2768-9

- Misner, Charles W.; Thorne, Kip. S.; Wheeler, John A. (1973), *Gravitation*, W. H. Freeman, ISBN 0-7167-0344-0

- Stephani, Hans (1990), *General Relativity: An Introduction to the Theory of the Gravitational Field*, Cambridge: Cambridge University Press, ISBN 0-521-37941-5

- Wald, Robert M. (1984), *General Relativity*, University of Chicago Press, ISBN 0-226-87033-2

1.7.13 External links

- Einstein Online – Articles on a variety of aspects of relativistic physics for a general audience; hosted by the Max Planck Institute for Gravitational Physics

- NCSA Spacetime Wrinkles – produced by the numerical relativity group at the NCSA, with an elementary introduction to general relativity

- **Courses**

- **Lectures**

- **Tutorials**

- Einstein's General Theory of Relativity on YouTube (lecture by Leonard Susskind recorded September 22, 2008 at Stanford University).

- Series of lectures on General Relativity given in 2006 at the Institut Henri Poincaré (introductory/advanced).

- General Relativity Tutorials by John Baez.

- Brown, Kevin. "Reflections on relativity". *Mathpages.com*. Retrieved May 29, 2005.

- Carroll, Sean M. "Lecture Notes on General Relativity". Retrieved January 5, 2014.

- Moor, Rafi. "Understanding General Relativity". Retrieved July 11, 2006.

- Waner, Stefan. "Introduction to Differential Geometry and General Relativity" (PDF). Retrieved 2015-04-05.

1.8 Big Bang

"Big Bang theory" redirects here. For the American TV sitcom, see The Big Bang Theory. For other uses, see Big Bang (disambiguation) and Big Bang Theory (disambiguation).

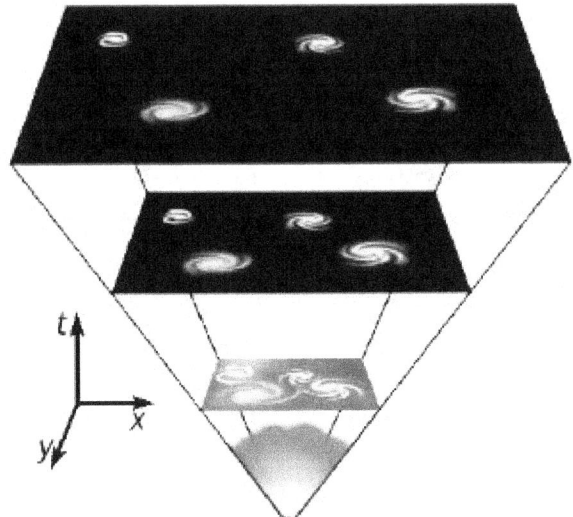

According to the Big Bang model, the universe expanded from an extremely dense and hot state and continues to expand.

The **Big Bang** theory is the prevailing cosmological model for the universe from the earliest known periods through its subsequent large-scale evolution.[1][2][3] The model accounts for the fact that the universe expanded from a very high density and high temperature state,[4][5] and offers a comprehensive explanation for a broad range of phenomena, including the abundance of light elements, the cosmic microwave background, large scale structure and Hubble's Law.[6] If the known laws of physics are extrapolated beyond where they have been verified, there is a singularity. Some estimates place this moment at approximately 13.8 billion years ago, which is thus considered the age of the universe.[7] After the initial expansion, the universe cooled sufficiently to allow the formation of subatomic particles, and later simple atoms. Giant clouds of these primordial elements later coalesced through gravity to form stars and galaxies.

Since Georges Lemaître first noted, in 1927, that an expanding universe might be traced back in time to an originating single point, scientists have built on his idea of cosmic expansion. While the scientific community was once divided between supporters of two different expanding universe theories, the Big Bang and the Steady State theory, accumulated empirical evidence provides strong support for the former.[8] In 1929, from analysis of galactic redshifts, Edwin Hubble concluded that galaxies are drifting apart; this is important observational evidence consistent with the hypothesis of an expanding universe. In 1965, the cosmic microwave background radiation was discovered, which was crucial evidence in favor of the Big Bang model, since that theory predicted the existence of background radiation throughout the universe before it was discovered. More recently, measurements of the redshifts of supernovae indicate that the expansion of the universe is accelerating, an observation attributed to dark energy's existence.[9] The known physical laws of nature can be used to calculate the characteristics of the universe in detail back in time to an initial state of extreme density and temperature.[10][11][12]

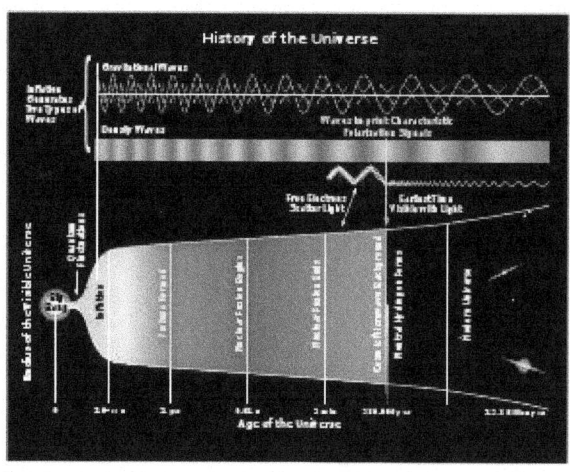

History of the Universe - gravitational waves are hypothesized to arise from cosmic inflation, an expansion just after the Big Bang.[13][14][15][16]

universe), the only remaining interpretation is that all observable regions of the universe are receding from all others. Since we know that the distance between galaxies increases today, it must mean that in the past galaxies were closer together. The continuous expansion of the universe implies that the universe was denser and hotter in the past.

Large particle accelerators can replicate the conditions that prevailed after the early moments of the universe, resulting in confirmation and refinement of the details of the Big Bang model. However, these accelerators can only probe so far into high energy regimes. Consequently, the state of the universe in the earliest instants of the Big Bang expansion is still poorly understood and an area of open investigation and speculation.

The first subatomic particles included protons, neutrons, and electrons. Though simple atomic nuclei formed within the first three minutes after the Big Bang, thousands of years passed before the first electrically neutral atoms formed. The majority of atoms produced by the Big Bang were hydrogen, along with helium and traces of lithium. Giant clouds of these primordial elements later coalesced through gravity to form stars and galaxies, and the heavier elements were synthesized either within stars or during supernovae.

American astronomer Edwin Hubble observed that the distances to faraway galaxies were strongly correlated with their redshifts. This was interpreted to mean that all dis-
tant galaxies and clusters are receding away from our vantage point with an apparent velocity proportional to their distance: that is,the farther they are,the faster they move away from us, regardless of direction.[17] Assuming the Copernican principle(that the Earth is not the center of the

The Big Bang theory offers a comprehensive explanation for a broad range of observed phenomena, including the abundance of light elements, the cosmic microwave background, large scale structure, and Hubble's Law.[6] The framework for the Big Bang model relies on Albert Einstein's theory of general relativity and on simplifying assumptions such as homogeneity and isotropy of space. The governing equations were formulated by Alexander Friedmann, and similar solutions were worked on by Willem de Sitter. Since then, astrophysicists have incorporated observational and theoretical additions into the Big Bang model, and its parametrization as the Lambda-CDM model serves as the framework for current investigations of theoretical cosmology. The Lambda-CDM model is the standard model of Big Bang cosmology, the simplest model that provides a reasonably good account of various observations about the universe.

1.8.2   Timeline

Main article: Chronology of the universe

Singularity

See also: Gravitational singularity and Planck epoch

Extrapolation of the expansion of the universe backwards in time using general relativity yields an infinite density and temperature at a finite time in the past.[18] This singularity signals the breakdown of general relativity and thus, all the laws of physics. How closely this can be extrapolated toward the singularity is debated—certainly no closer than the end of the Planck epoch. This singularity is sometimes called "the Big Bang",[19] but the term can also refer to the early hot, dense phase itself,[20][notes 1] which can be considered the "birth" of our universe. Based on measurements of the expansion using Type Ia supernovae and measurements of temperature fluctuations in the cosmic microwave background, the universe has an estimated age of 13.799 ± 0.021 billion years.[21] The agreement of these three independent measurements strongly supports the ΛCDM model that describes in detail the contents of the universe.

Inflation and baryogenesis

Main articles: Cosmic inflation and baryogenesis

The earliest phases of the Big Bang are subject to much speculation. In the most common models the universe was filled homogeneously and isotropically with a very high energy density and huge temperatures and pressures and was very rapidly expanding and cooling. Approximately 10^{-37} seconds into the expansion, a phase transition caused a cosmic inflation, during which the universe grew exponentially.[22] After inflation stopped, the universe consisted of a quark–gluon plasma, as well as all other elementary particles.[23] Temperatures were so high that the random motions of particles were at relativistic speeds, and particle–antiparticle pairs of all kinds were being continuously created and destroyed in collisions.[4] At some point

an unknown reaction called baryogenesis violated the conservation of baryon number, leading to a very small excess of quarks and leptons over antiquarks and antileptons— of the order of one part in 30 million. This resulted in the predominance of matter over antimatter in the present universe.[24]

Cooling

Main articles: Big Bang nucleosynthesis and cosmic microwave background radiation

The universe continued to decrease in density and fall

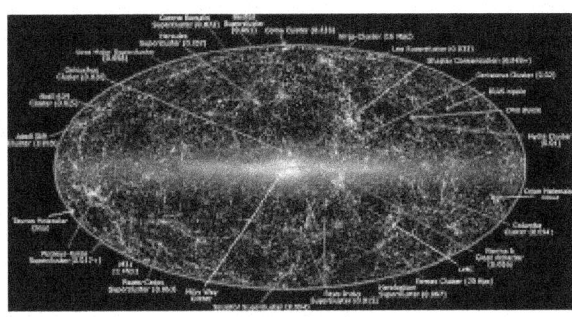

Panoramic view of the entire near-infrared sky reveals the distribution of galaxies beyond the Milky Way. Galaxies are color-coded by redshift.

in temperature, hence the typical energy of each particle was decreasing. Symmetry breaking phase transitions put the fundamental forces of physics and the parameters of elementary particles into their present form.[25] After about 10^{-11} seconds, the picture becomes less speculative, since particle energies drop to values that can be attained in particle physics experiments. At about 10^{-6} seconds, quarks and gluons combined to form baryons such as protons and neutrons. The small excess of quarks over antiquarks led to a small excess of baryons over antibaryons. The temperature was now no longer high enough to create new proton–antiproton pairs (similarly for neutrons– antineutrons), so a mass annihilation immediately followed, leaving just one in 10^{10} of the original protons and neutrons, and none of their antiparticles. A similar process happened at about 1 second for electrons and positrons. After these annihilations, the remaining protons, neutrons and electrons were no longer moving relativistically and the energy density of the universe was dominated by photons (with a minor contribution from neutrinos).

A few minutes into the expansion, when the temperature was about a billion (one thousand million; 10^9; SI prefix giga-) kelvin and the density was about that of air, neutrons combined with protons to form the universe's deuterium and helium nuclei in a process called Big Bang nucleosynthesis.[26] Most protons remained uncombined as hydrogen nuclei. As the universe cooled, the rest mass energy density of matter came to gravitationally dominate that of the photon radiation. After about 379,000 years the electrons and nuclei combined into atoms (mostly hydrogen); hence the radiation decoupled from matter and continued through space largely unimpeded. This relic radiation is known as the cosmic microwave background radiation.[27] The chemistry of life may have begun shortly after the Big Bang, 13.8 billion years ago, during a habitable epoch when the universe was only 10–17 million years old.[28][29]

Structure formation

Main article: Structure formation
Over a long period of time, the slightly denser regions of

Abell 2744 galaxy cluster - Hubble Frontier Fields view.[30]

the nearly uniformly distributed matter gravitationally attracted nearby matter and thus grew even denser, forming gas clouds, stars, galaxies, and the other astronomical structures observable today.[4] The details of this process depend on the amount and type of matter in the universe. The four possible types of matter are known as cold dark matter, warm dark matter, hot dark matter, and baryonic matter. The best measurements available (from WMAP) show that the data is well-fit by a Lambda-CDM model in which dark matter is assumed to be cold (warm dark matter is ruled out by early reionization),[31] and is estimated to make up about 23% of the matter/energy of the universe, while baryonic matter makes up about 4.6%.[32] In an "extended model" which includes hot dark matter in the form of neutrinos,

then if the "physical baryon density" $\Omega_b h^2$ is estimated at about 0.023 (this is different from the 'baryon density' Ω_b expressed as a fraction of the total matter/energy density, which as noted above is about 0.046), and the corresponding cold dark matter density $\Omega_c h^2$ is about 0.11, the corresponding neutrino density $\Omega_v h^2$ is estimated to be less than 0.0062.[32]

Cosmic acceleration

Main article: Accelerating universe

Independent lines of evidence from Type Ia supernovae and the CMB imply that the universe today is dominated by a mysterious form of energy known as dark energy, which apparently permeates all of space. The observations suggest 73% of the total energy density of today's universe is in this form. When the universe was very young, it was likely infused with dark energy, but with less space and everything closer together, gravity predominated, and it was slowly braking the expansion. But eventually, after numerous billion years of expansion, the growing abundance of dark energy caused the expansion of the universe to slowly begin to accelerate. Dark energy in its simplest formulation takes the form of the cosmological constant term in Einstein's field equations of general relativity, but its composition and mechanism are unknown and, more generally, the details of its equation of state and relationship with the Standard Model of particle physics continue to be investigated both through observation and theoretically.[9]

All of this cosmic evolution after the inflationary epoch can be rigorously described and modeled by the ΛCDM model of cosmology, which uses the independent frameworks of quantum mechanics and Einstein's General Relativity. There is no well-supported model describing the action prior to 10^{-15} seconds or so. Apparently a new unified theory of quantum gravitation is needed to break this barrier. Understanding this earliest of eras in the history of the universe is currently one of the greatest unsolved problems in physics.

1.8.3 Underlying assumptions

The Big Bang theory depends on two major assumptions: the universality of physical laws and the cosmological principle. The cosmological principle states that on large scales the universe is homogeneous and isotropic.

These ideas were initially taken as postulates, but today there are efforts to test each of them. For example, the first assumption has been tested by observations showing that largest possible deviation of the fine structure constant over much of the age of the universe is of order 10^{-5}.[33] Also, general relativity has passed stringent tests on the scale of the Solar System and binary stars.[notes 2]

If the large-scale universe appears isotropic as viewed from Earth, the cosmological principle can be derived from the simpler Copernican principle, which states that there is no preferred (or special) observer or vantage point. To this end, the cosmological principle has been confirmed to a level of 10^{-5} via observations of the CMB. The universe has been measured to be homogeneous on the largest scales at the 10% level.[34]

Expansion of space

Main articles: Friedmann–Lemaître–Robertson–Walker metric and Metric expansion of space

General relativity describes spacetime by a metric, which determines the distances that separate nearby points. The points, which can be galaxies, stars, or other objects, themselves are specified using a coordinate chart or "grid" that is laid down over all spacetime. The cosmological principle implies that the metric should be homogeneous and isotropic on large scales, which uniquely singles out the Friedmann–Lemaître–Robertson–Walker metric (FLRW metric). This metric contains a scale factor, which describes how the size of the universe changes with time. This enables a convenient choice of a coordinate system to be made, called comoving coordinates. In this coordinate system the grid expands along with the universe, and objects that are moving only because of the expansion of the universe remain at fixed points on the grid. While their *coordinate* distance (comoving distance) remains constant, the *physical* distance between two such comoving points expands proportionally with the scale factor of the universe.[35]

The Big Bang is not an explosion of matter moving outward to fill an empty universe. Instead, space itself expands with time everywhere and increases the physical distance between two comoving points. In other words, the Big Bang is not an explosion *in space*, but rather an expansion *of space*.[4] Because the FLRW metric assumes a uniform distribution of mass and energy, it applies to our universe only on large scales—local concentrations of matter such as our galaxy are gravitationally bound and as such do not experience the large-scale expansion of space.[36]

Horizons

Main article: Cosmological horizon

An important feature of the Big Bang spacetime is the presence of horizons. Since the universe has a finite age, and light travels at a finite speed, there may be events in the past whose light has not had time to reach us. This places a limit or a *past horizon* on the most distant objects that can be observed. Conversely, because space is expanding, and more distant objects are receding ever more quickly, light emitted by us today may never "catch up" to very distant objects. This defines a *future horizon*, which limits the events in the future that we will be able to influence. The presence of either type of horizon depends on the details of the FLRW model that describes our universe. Our understanding of the universe back to very early times suggests that there is a past horizon, though in practice our view is also limited by the opacity of the universe at early times. So our view cannot extend further backward in time, though the horizon recedes in space. If the expansion of the universe continues to accelerate, there is a future horizon as well.[37]

1.8.4   History

Main article: History of the Big Bang theory
See also: Timeline of cosmology

Etymology

English astronomer Fred Hoyle is credited with coining the term "Big Bang" during a 1949 BBC radio broadcast. It is popularly reported that Hoyle, who favored an alternative "steady state" cosmological model, intended this to be pejorative, but Hoyle explicitly denied this and said it was just a striking image meant to highlight the difference between the two models.[38][39][40]:129

Development

Hubble eXtreme Deep Field (XDF)

XDF size compared to the size of the moon - several thousand galaxies, each

consisting of billions of stars, are in this small view.

XDF (2012) view - each light speck is a galaxy - some of these are as old as 13.2 billion years[41] - the universe is estimated to contain 200 billion galaxies.

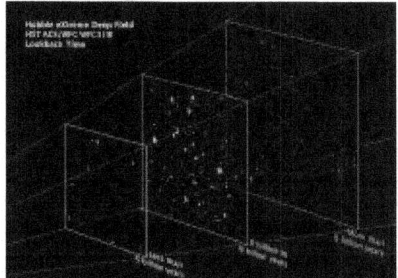

XDF image shows fully mature galaxies in the foreground plane - nearly mature galaxies from 5 to 9 billion years ago - protogalaxies, blazing with young stars, beyond 9 billion years.

The Big Bang theory developed from observations of the structure of the universe and from theoretical considerations. In 1912 Vesto Slipher measured the first Doppler shift of a "spiral nebula" (spiral nebula is the obsolete term for spiral galaxies), and soon discovered that almost all such nebulae were receding from Earth. He did not grasp the cosmological implications of this fact, and indeed at the time it was highly controversial whether or not these nebulae were "island universes" outside our Milky Way.[42][43] Ten years later, Alexander Friedmann, a Russian cosmologist and mathematician, derived the Friedmann equations from Albert Einstein's equations of general relativity, showing that the universe might be expanding in contrast to the static universe model advocated by Einstein at that time.[44] In 1924 Edwin Hubble's measurement of the great distance to the nearest spiral nebulae showed that these systems were indeed other galaxies. Independently deriving Friedmann's equations in 1927, Georges Lemaître, a Belgian physicist and Roman Catholic priest, proposed that the inferred recession of the nebulae was due to the expansion of the universe.[45]

In 1931 Lemaître went further and suggested that the evident expansion of the universe, if projected back in time, meant that the further in the past the smaller the universe was, until at some finite time in the past all the mass of the universe was concentrated into a single point, a "primeval atom" where and when the fabric of time and space came into existence.[46]

Starting in 1924, Hubble painstakingly developed a series of distance indicators, the forerunner of the cosmic distance ladder, using the 100-inch (2.5 m) Hooker telescope at Mount Wilson Observatory. This allowed him to estimate distances to galaxies whose redshifts had already been measured, mostly by Slipher. In 1929 Hubble discovered a correlation between distance and recession velocity—now known as Hubble's law.[17][47] Lemaître had already shown that this was expected, given the cosmological principle.[9]

In the 1920s and 1930s almost every major cosmologist preferred an eternal steady state universe, and several complained that the beginning of time implied by the Big Bang imported religious concepts into physics; this objection was later repeated by supporters of the steady state theory.[48] This perception was enhanced by the fact that the originator of the Big Bang theory, Monsignor Georges Lemaître, was a Roman Catholic priest.[49] Arthur Eddington agreed with Aristotle that the universe did not have a beginning in time, viz., that matter is eternal. A beginning in time was "repugnant" to him.[50][51] Lemaître, however, thought that

> If the world has begun with a single quantum, the notions of space and time would altogether fail to have any meaning at the beginning; they would only begin to have a sensible meaning when the original quantum had been divided into a sufficient number of quanta. If this suggestion is correct, the beginning of the world happened a little before the beginning of space and time.[52]

During the 1930s other ideas were proposed as nonstandard cosmologies to explain Hubble's observations, including the Milne model,[53] the oscillatory universe (originally suggested by Friedmann, but advocated by Albert Einstein and Richard Tolman)[54] and Fritz Zwicky's tired light hypothesis.[55]

After World War II, two distinct possibilities emerged. One was Fred Hoyle's steady state model, whereby new matter would be created as the universe seemed to expand. In this model the universe is roughly the same at any point in time.[56] The other was Lemaître's Big Bang theory, advocated and developed by George Gamow, who introduced big bang nucleosynthesis (BBN)[57] and whose associates, Ralph Alpher and Robert Herman, predicted the cosmic microwave background radiation (CMB).[58] Ironically, it was Hoyle who coined the phrase that came to be applied to Lemaître's theory, referring to it as "this *big bang* idea" during a BBC Radio broadcast in March 1949.[40][notes 3] For a while, support was split between these two theories. Eventually, the observational evidence, most notably from radio source counts, began to favor Big Bang over Steady State. The discovery and confirmation of the cosmic microwave background radiation in 1965 secured the Big Bang as the best theory of the origin and evolution of the universe.[60] Much of the current work in cosmology includes understanding how galaxies form in the context of the Big Bang, understanding the physics of the universe at earlier and earlier times, and reconciling observations with the basic theory.

In 1968 and 1970, Roger Penrose, Stephen Hawking, and George F. R. Ellis published papers where they showed that mathematical singularities were an inevitable initial condition of general relativistic models of the Big Bang.[61][62] Then, from the 1970s to the 1990s, cosmologists worked on characterizing the features of the Big Bang universe and resolving outstanding problems. In 1981, Alan Guth made a breakthrough in theoretical work on resolving certain outstanding theoretical problems in the Big Bang theory with the introduction of an epoch of rapid expansion in the early universe he called "inflation".[63] Meanwhile, during these decades, two questions in observational cosmology that generated much discussion and disagreement were over the precise values of the Hubble Constant[64] and the matter-density of the universe (before the discovery of dark energy, thought to be the key predictor for the eventual fate of the universe).[65] In the mid-1990s observations of certain globular clusters appeared to indicate that they were about 15 billion years old, which conflicted with most then-current estimates of the age of the universe (and indeed with the age measured today). This issue was later resolved when new computer simulations, which included the effects of mass loss due to stellar winds, indicated a much younger age for globular clusters.[66] While there still remain some questions as to how accurately the ages of the clusters are measured, globular clusters are of interest to cosmology as some of the oldest objects in the universe.

Significant progress in Big Bang cosmology have been made since the late 1990s as a result of advances in telescope technology as well as the analysis of data from satellites such as COBE,[67] the Hubble Space Telescope and WMAP.[68] Cosmologists now have fairly precise and accurate measurements of many of the parameters of the Big Bang model, and have made the unexpected discovery that the expansion of the universe appears to be accelerating.

1.8.5   Observational evidence

"[The] big bang picture is too firmly grounded in data from every area to be proved invalid in its general features."

Lawrence Krauss[69]

The earliest and most direct observational evidence of the

Artist's depiction of the WMAP satellite gathering data to help scientists understand the Big Bang

validity of the theory are the expansion of the universe according to Hubble's law (as indicated by the redshifts of galaxies), discovery and measurement of the cosmic microwave background and the relative abundances of light elements produced by Big Bang nucleosynthesis. More recent evidence includes observations of galaxy formation and evolution, and the distribution of large-scale cosmic structures.[70] These are sometimes called the "four pillars" of the Big Bang theory.[71]

Precise modern models of the Big Bang appeal to various exotic physical phenomena that have not been observed in terrestrial laboratory experiments or incorporated into the Standard Model of particle physics. Of these features, dark matter is currently subjected to the most active laboratory investigations.[72] Remaining issues include the cuspy halo problem and the dwarf galaxy problem of cold dark matter. Dark energy is also an area of intense interest for scientists, but it is not clear whether direct detection of dark energy will be possible.[73] Inflation and baryogenesis remain more speculative features of current Big Bang models. Viable, quantitative explanations for such phenomena are still being sought. These are currently unsolved problems in physics.

Hubble's law and the expansion of space

Main articles: Hubble's law and Metric expansion of space
See also: Distance measures (cosmology) and Scale factor (universe)

Observations of distant galaxies and quasars show that these objects are redshifted—the light emitted from them has been shifted to longer wavelengths. This can be seen by taking a frequency spectrum of an object and matching the spectroscopic pattern of emission lines or absorption lines corresponding to atoms of the chemical elements interacting with the light. These redshifts are uniformly isotropic, distributed evenly among the observed objects in all directions. If the redshift is interpreted as a Doppler shift, the recessional velocity of the object can be calculated. For some galaxies, it is possible to estimate distances via the cosmic distance ladder. When the recessional velocities are plotted against these distances, a linear relationship known as Hubble's law is observed:[17]

$$v = H_0 D.$$

where

- v is the recessional velocity of the galaxy or other distant object,

- D is the comoving distance to the object, and

- H_0 is Hubble's constant, measured to be 70.4+1.3 −1.4 km/s/Mpc by the WMAP probe.[32]

Hubble's law has two possible explanations. Either we are at the center of an explosion of galaxies—which is untenable given the Copernican principle—or the universe is uniformly expanding everywhere. This universal expansion was predicted from general relativity by Alexander Friedmann in 1922[44] and Georges Lemaître in 1927,[45] well before Hubble made his 1929 analysis and observations, and it remains the cornerstone of the Big Bang theory as developed by Friedmann, Lemaître, Robertson, and Walker.

The theory requires the relation $v = HD$ to hold at all times, where D is the comoving distance, v is the recessional velocity, and v, H, and D vary as the universe expands (hence we write H_0 to denote the present-day Hubble "constant"). For distances much smaller than the size of the observable universe, the Hubble redshift can be thought of as the Doppler shift corresponding to the recession velocity v. However, the redshift is not a true Doppler shift, but rather the result of the expansion of the universe between the time the light was emitted and the time that it was detected.[74]

That space is undergoing metric expansion is shown by direct observational evidence of the Cosmological principle and the Copernican principle, which together with Hubble's law have no other explanation. Astronomical redshifts are extremely isotropic and homogeneous,[17] supporting the Cosmological principle that the universe looks the same in all directions, along with much other evidence. If the redshifts were the result of an explosion from a center distant from us, they would not be so similar in different directions.

Measurements of the effects of the cosmic microwave background radiation on the dynamics of distant astrophysical systems in 2000 proved the Copernican principle, that, on a cosmological scale, the Earth is not in a central position.[75] Radiation from the Big Bang was demonstrably

warmer at earlier times throughout the universe. Uniform cooling of the cosmic microwave background over billions of years is explainable only if the universe is experiencing a metric expansion, and excludes the possibility that we are near the unique center of an explosion.

Cosmic microwave background radiation

Main article: Cosmic microwave background radiation
In 1965, Arno Penzias and Robert Wilson serendipitously

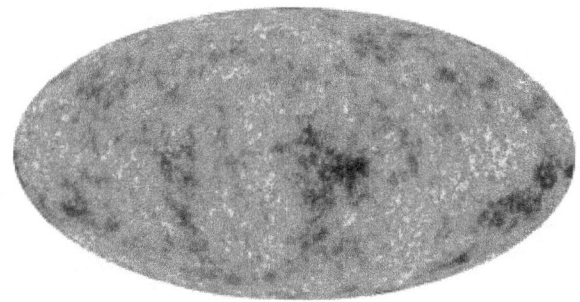

9 year WMAP image of the cosmic microwave background radiation (2012).[76][77] The radiation is isotropic to roughly one part in 100,000.[78]

discovered the cosmic background radiation, an omnidirectional signal in the microwave band.[60] Their discovery provided substantial confirmation of the big-bang predictions by Alpher, Herman and Gamow around 1950. Through the 1970s the radiation was found to be approximately consistent with a black body spectrum in all directions; this spectrum has been redshifted by the expansion of the universe, and today corresponds to approximately 2.725 K. This tipped the balance of evidence in favor of the Big Bang model, and Penzias and Wilson were awarded a Nobel Prize in 1978.

The *surface of last scattering* corresponding to emission of the CMB occurs shortly after *recombination*, the epoch when neutral hydrogen becomes stable. Prior to this, the universe comprised a hot dense photon-baryon plasma sea where photons were quickly scattered from free charged particles. Peaking at around 372±14 kyr,[31] the mean free path for a photon becomes long enough to reach the present day and the universe becomes transparent.

In 1989 NASA launched the Cosmic Background Explorer satellite (COBE) which made two major advances: in 1990, high-precision spectrum measurements showed the CMB frequency spectrum is an almost perfect blackbody with no deviations at a level of 1 part in 10^4, and measured a residual temperature of 2.726 K (more recent measurements have revised this figure down slightly to 2.7255 K); then in 1992 further COBE measurements discovered tiny fluctuations (anisotropies) in the CMB temperature across

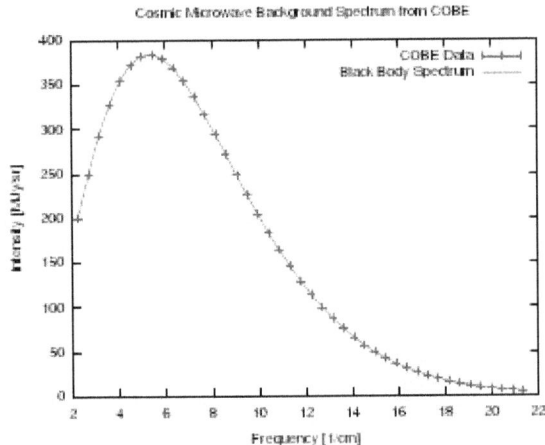

The cosmic microwave background spectrum measured by the FI-RAS instrument on the COBE satellite is the most-precisely measured black body spectrum in nature.[79] The data points and error bars on this graph are obscured by the theoretical curve.

the sky, at a level of about one part in 10^5.[67] John C. Mather and George Smoot were awarded the 2006 Nobel Prize in Physics for their leadership in these results. During the following decade, CMB anisotropies were further investigated by a large number of ground-based and balloon experiments. In 2000–2001 several experiments, most notably BOOMERanG, found the shape of the universe to be spatially almost flat by measuring the typical angular size (the size on the sky) of the anisotropies.[80][81][82]

In early 2003 the first results of the Wilkinson Microwave Anisotropy Probe (WMAP) were released, yielding what were at the time the most accurate values for some of the cosmological parameters. The results disproved several specific cosmic inflation models, but are consistent with the inflation theory in general.[68] The Planck space probe was launched in May 2009. Other ground and balloon based cosmic microwave background experiments are ongoing.

Abundance of primordial elements

Main article: Big Bang nucleosynthesis

Using the Big Bang model it is possible to calculate the concentration of helium-4, helium-3, deuterium, and lithium-7 in the universe as ratios to the amount of ordinary hydrogen.[26] The relative abundances depend on a single parameter, the ratio of photons to baryons. This value can be calculated independently from the detailed structure of CMB fluctuations. The ratios predicted (by mass, not by number) are about 0.25 for 4He/H, about 10^{-3} for 2H/H, about 10^{-4} for 3He/H and about 10^{-9} for 7Li/H.[26]

The measured abundances all agree at least roughly with those predicted from a single value of the baryon-to-photon ratio. The agreement is excellent for deuterium, close but formally discrepant for 4He, and off by a factor of two for 7Li; in the latter two cases there are substantial systematic uncertainties. Nonetheless, the general consistency with abundances predicted by Big Bang nucleosynthesis is strong evidence for the Big Bang, as the theory is the only known explanation for the relative abundances of light elements, and it is virtually impossible to "tune" the Big Bang to produce much more or less than 20–30% helium.[83] Indeed, there is no obvious reason outside of the Big Bang that, for example, the young universe (i.e., before star formation, as determined by studying matter supposedly free of stellar nucleosynthesis products) should have more helium than deuterium or more deuterium than 3He, and in constant ratios, too.[84]:182–185

Galactic evolution and distribution

Main articles: Galaxy formation and evolution and Structure formation

Detailed observations of the morphology and distribution of galaxies and quasars are in agreement with the current state of the Big Bang theory. A combination of observations and theory suggest that the first quasars and galaxies formed about a billion years after the Big Bang, and since then larger structures have been forming, such as galaxy clusters and superclusters. Populations of stars have been aging and evolving, so that distant galaxies (which are observed as they were in the early universe) appear very different from nearby galaxies (observed in a more recent state). Moreover, galaxies that formed relatively recently appear markedly different from galaxies formed at similar distances but shortly after the Big Bang. These observations are strong arguments against the steady-state model. Observations of star formation, galaxy and quasar distributions and larger structures agree well with Big Bang simulations of the formation of structure in the universe and are helping to complete details of the theory.[185][186]

Primordial gas clouds

In 2011 astronomers found what they believe to be pristine clouds of primordial gas, by analyzing absorption lines in the spectra of distant quasars. Before this discovery, all other astronomical objects have been observed to contain heavy elements that are formed in stars. These two clouds of gas contain no elements heavier than hydrogen and deuterium.[87][88] Since the clouds of gas have no heavy elements, they likely formed in the first few minutes after

Focal plane of BICEP2 telescope under a microscope - used to search for polarization in the CMB.[13][14][15][16]

the Big Bang, during Big Bang nucleosynthesis.

Other lines of evidence

The age of the universe as estimated from the Hubble expansion and the CMB is now in good agreement with other estimates using the ages of the oldest stars, both as measured by applying the theory of stellar evolution to globular clusters and through radiometric dating of individual Population II stars.[89]

The prediction that the CMB temperature was higher in the past has been experimentally supported by observations of very low temperature absorption lines in gas clouds at high redshift.[90] This prediction also implies that the amplitude of the Sunyaev–Zel'dovich effect in clusters of galaxies does not depend directly on redshift. Observations have found this to be roughly true, but this effect depends on cluster properties that do change with cosmic time, making precise measurements difficult.[91][92]

On 17 March 2014, astronomers at the Harvard-Smithsonian Center for Astrophysics announced the apparent detection of primordial gravitational waves, which, was shown to be due to galactic dust.[13][14][15][16] On February 11, 2016, the LIGO Scientific Collaboration and Virgo Collaboration teams announced that they had made first observation of gravitational waves, originating from a pair of merging black holes using the Advanced LIGO detectors.[93][94][95]

Future observations

Future gravitational waves observatories might see primordial gravitational waves, relics of the early universe, up to less than a second of the Big Bang.[96][97]

1.8.6 Problems and related issues in physics

See also: List of unsolved problems in physics

As with any theory, a number of mysteries and problems have arisen as a result of the development of the Big Bang theory. Some of these mysteries and problems have been resolved while others are still outstanding. Proposed solutions to some of the problems in the Big Bang model have revealed new mysteries of their own. For example, the horizon problem, the magnetic monopole problem, and the flatness problem are most commonly resolved with inflationary theory, but the details of the inflationary universe are still left unresolved and many, including some founders of the theory, say it has been disproven.[98][99][100][101] What follows are a list of the mysterious aspects of the Big Bang theory still under intense investigation by cosmologists and astrophysicists.

Baryon asymmetry

Main article: Baryon asymmetry

It is not yet understood why the universe has more matter than antimatter.[102] It is generally assumed that when the universe was young and very hot, it was in statistical equilibrium and contained equal numbers of baryons and antibaryons. However, observations suggest that the universe, including its most distant parts, is made almost entirely of matter. A process called baryogenesis was hypothesized to account for the asymmetry. For baryogenesis to occur, the Sakharov conditions must be satisfied. These require that baryon number is not conserved, that C-symmetry and CP-symmetry are violated and that the universe depart from thermodynamic equilibrium.[103] All these conditions occur in the Standard Model, but the effect is not strong enough to explain the present baryon asymmetry.

Dark energy

Main article: Dark energy

Measurements of the redshift–magnitude relation for type Ia supernovae indicate that the expansion of the universe has been accelerating since the universe was about half its present age. To explain this acceleration, general relativity requires that much of the energy in the universe consists of a component with large negative pressure, dubbed "dark energy".[9] Dark energy, though speculative, solves numerous problems. Measurements of the cosmic microwave background indicate that the universe is very nearly spatially flat, and therefore according to general relativity the universe must have almost exactly the critical density of mass/energy. But the mass density of the universe can be measured from its gravitational clustering, and is found to have only about 30% of the critical density.[9] Since theory suggests that dark energy does not cluster in the usual way it is the best explanation for the "missing" energy density. Dark energy also helps to explain two geometrical measures of the overall curvature of the universe, one using the frequency of gravitational lenses, and the other using the characteristic pattern of the large-scale structure as a cosmic ruler.

Negative pressure is believed to be a property of vacuum energy, but the exact nature and existence of dark energy remains one of the great mysteries of the Big Bang. Results from the WMAP team in 2008 are in accordance with a universe that consists of 73% dark energy, 23% dark matter, 4.6% regular matter and less than 1% neutrinos.[32] According to theory, the energy density in matter decreases with the expansion of the universe, but the dark energy density remains constant (or nearly so) as the universe expands. Therefore, matter made up a larger fraction of the total energy of the universe in the past than it does today, but its fractional contribution will fall in the far future as dark energy becomes even more dominant.

The dark energy component of the universe has been explained by theorists using a variety of competing theories including Einstein's cosmological constant but also extending to more exotic forms of quintessence or other modified gravity schemes.[104] A cosmological constant problem sometimes called the "most embarrassing problem in physics" results from the apparent discrepancy between the measured energy density of dark energy and the one naively predicted from Planck units.[105]

Dark matter

Main article: Dark matter

During the 1970s and 80s, various observations showed that there is not sufficient visible matter in the universe to account for the apparent strength of gravitational forces within and between galaxies. This led to the idea that up to 90% of the matter in the universe is dark matter that does not emit light or interact with normal baryonic matter. In addition, the assumption that the universe is mostly normal matter led to predictions that were strongly incon-

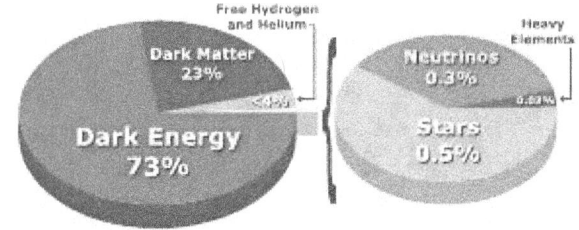

Chart shows the proportion of different components of the universe – about 95% is dark matter and dark energy.

sistent with observations. In particular, the universe today is far more lumpy and contains far less deuterium than can be accounted for without dark matter. While dark matter has always been controversial, it is inferred by various observations: the anisotropies in the CMB, galaxy cluster velocity dispersions, large-scale structure distributions, gravitational lensing studies, and X-ray measurements of galaxy clusters.[106]

Indirect evidence for dark matter comes from its gravitational influence on other matter, as no dark matter particles have been observed in laboratories. Many particle physics candidates for dark matter have been proposed, and several projects to detect them directly are underway.[107]

Additionally, there are outstanding problems associated with the currently favored cold dark matter model which include the dwarf galaxy problem[108] and the cuspy halo problem.[109] Alternative theories have been proposed that do not require a large amount of undetected matter but instead modify the laws of gravity established by Newton and Einstein, but no alternative theory as been as successful as the cold dark matter proposal in explaining all extant observations.[110]

Horizon problem

The horizon problem results from the premise that information cannot travel faster than light. In a universe of finite age this sets a limit—the particle horizon—on the separation of any two regions of space that are in causal contact.[111] The observed isotropy of the CMB is problematic in this regard: if the universe had been dominated by radiation or matter at all times up to the epoch of last scattering, the particle horizon at that time would correspond to about 2 degrees on the sky. There would then be no mechanism to cause wider regions to have the same temperature.[84]:191–202

A resolution to this apparent inconsistency is offered by inflationary theory in which a homogeneous and isotropic scalar energy field dominates the universe at some very early period (before baryogenesis). During inflation, the universe undergoes exponential expansion, and the particle horizon expands much more rapidly than previously assumed, so that regions presently on opposite sides of the observable universe are well inside each other's particle horizon. The observed isotropy of the CMB then follows from the fact that this larger region was in causal contact before the beginning of inflation.[22]:180–186

Heisenberg's uncertainty principle predicts that during the inflationary phase there would be quantum thermal fluctuations, which would be magnified to cosmic scale. These fluctuations serve as the seeds of all current structure in the universe.[84]:207 Inflation predicts that the primordial fluctuations are nearly scale invariant and Gaussian, which has been accurately confirmed by measurements of the CMB.[112]:sec 6

If inflation occurred, exponential expansion would push large regions of space well beyond our observable horizon.[22]:180–186

A related issue to the classic horizon problem arises because in most standard cosmological inflation models, inflation ceases well before electroweak symmetry breaking occurs, so inflation should not be able to prevent large-scale discontinuities in the electroweak vacuum since distant parts of the observable universe were causally separate when the electroweak epoch ended.[113]

Magnetic monopoles

The magnetic monopole objection was raised in the late 1970s. Grand unified theories predicted topological defects in space that would manifest as magnetic monopoles. These objects would be produced efficiently in the hot early universe, resulting in a density much higher than is consistent with observations, given that no monopoles have been found. This problem is also resolved by cosmic inflation, which removes all point defects from the observable universe, in the same way that it drives the geometry to flatness.[111]

Flatness problem

The flatness problem (also known as the oldness problem) is an observational problem associated with a Friedmann–Lemaître–Robertson–Walker metric.[111] The universe may have positive, negative, or zero spatial curvature depending on its total energy density. Curvature is negative if its density is less than the critical density, positive if greater, and zero at the critical density, in which case space is said to be *flat*. The problem is that any small departure from the critical density grows with time, and yet the universe today remains very close to flat.[notes 4] Given that a natural timescale for departure from flatness might be the Planck time, 10^{-43} seconds,[4] the fact that the universe has reached neither a

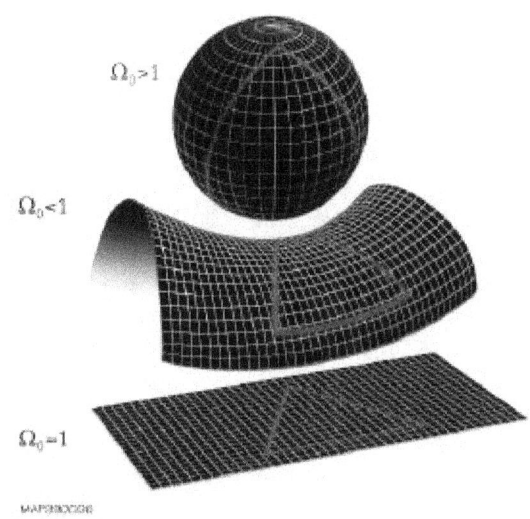

$\Omega_0 > 1$

$\Omega_0 < 1$

$\Omega_0 = 1$

MAP990006

The overall geometry of the universe is determined by whether the Omega cosmological parameter is less than, equal to or greater than 1. Shown from top to bottom are a closed universe with positive curvature, a hyperbolic universe with negative curvature and a flat universe with zero curvature.

heat death nor a Big Crunch after billions of years requires an explanation. For instance, even at the relatively late age of a few minutes (the time of nucleosynthesis), the universe density must have been within one part in 10^{14} of its critical value, or it would not exist as it does today.[114]

1.8.7 Ultimate fate of the universe

Main article: Ultimate fate of the universe

Before observations of dark energy, cosmologists considered two scenarios for the future of the universe. If the mass density of the universe were greater than the critical density, then the universe would reach a maximum size and then begin to collapse. It would become denser and hotter again, ending with a state similar to that in which it started—a Big Crunch.[37] Alternatively, if the density in the universe were equal to or below the critical density, the expansion would slow down but never stop. Star formation would cease with the consumption of interstellar gas in each galaxy; stars would burn out leaving white dwarfs, neutron stars, and black holes. Very gradually, collisions between these would result in mass accumulating into larger and larger black holes. The average temperature of the universe would asymptotically approach absolute zero—a Big Freeze.[115] Moreover, if the proton were unstable, then baryonic matter would disappear, leaving only radiation and black holes. Eventually, black holes would evaporate by emitting Hawking radiation. The entropy of the universe would increase to the point where no organized form of energy could be extracted from it, a scenario known as heat death.[116];sec VI.D

Modern observations of accelerating expansion imply that more and more of the currently visible universe will pass beyond our event horizon and out of contact with us. The eventual result is not known. The ΛCDM model of the universe contains dark energy in the form of a cosmological constant. This theory suggests that only gravitationally bound systems, such as galaxies, will remain together, and they too will be subject to heat death as the universe expands and cools. Other explanations of dark energy, called phantom energy theories, suggest that ultimately galaxy clusters, stars, planets, atoms, nuclei, and matter itself will be torn apart by the ever-increasing expansion in a so-called Big Rip.[117]

1.8.8   Speculations

Main article: Cosmogony

While the Big Bang model is well established in cosmol-

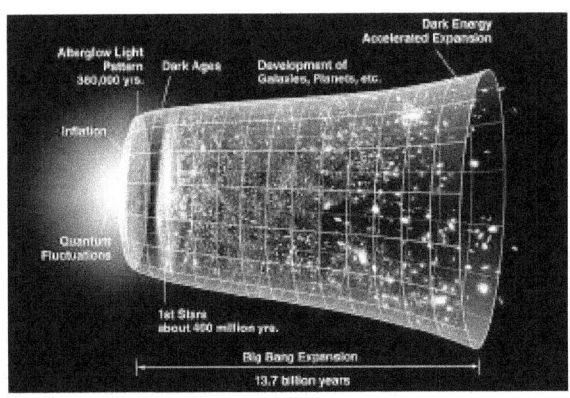

Timeline of the metric expansion of space, where space (including hypothetical non-observable portions of the universe) is represented at each time by the circular sections. On the left the dramatic expansion occurs in the inflationary epoch, and at the center the expansion accelerates (artist's concept; not to scale).

ogy, it is likely to be refined. The Big Bang theory, built upon the equations of classical general relativity, indicates a singularity at the origin of cosmic time; this infinite energy density is regarded as impossible in physics. Still, it is known that the equations are not applicable before the time when the universe cooled down to the Planck temperature, and this conclusion depends on various assumptions, of which some could never be experimentally verified. *(Also see Planck epoch.)*

One proposed refinement to avoid this would-be singularity is to develop a correct treatment of quantum gravity.[118]

It is not known what could have preceded the hot dense state

of the early universe or how and why it originated, though speculation abounds in the field of cosmogony.

Some proposals, each of which entails untested hypotheses, are:

- Models including the Hartle–Hawking no-boundary condition, in which the whole of space-time is finite; the Big Bang does represent the limit of time but without any singularity.[119]

- Big Bang lattice model, states that the universe at the moment of the Big Bang consists of an infinite lattice of fermions, which is smeared over the fundamental domain so it has rotational, translational and gauge symmetry. The symmetry is the largest symmetry possible and hence the lowest entropy of any state.[120]

- Brane cosmology models, in which inflation is due to the movement of branes in string theory; the pre-Big Bang model; the ekpyrotic model, in which the Big Bang is the result of a collision between branes; and the cyclic model, a variant of the ekpyrotic model in which collisions occur periodically. In the latter model the Big Bang was preceded by a Big Crunch and the universe cycles from one process to the other.[121][122][123][124]

- Eternal inflation, in which universal inflation ends locally here and there in a random fashion, each endpoint leading to a *bubble universe*, expanding from its own big bang.[125][126]

Proposals in the last two categories, see the Big Bang as an event in either a much larger and older universe or in a multiverse.

1.8.9   Religious and philosophical interpretations

Main article: Religious interpretations of the Big Bang theory

As a description of the origin of the universe, the Big Bang has significant bearing on religion and philosophy.[127][128] As a result, it has become one of the liveliest areas in the discourse between science and religion.[129] Some believe the Big Bang implies a creator,[130][131] and some see its mention in their holy books,[132] while others argue that Big Bang cosmology makes the notion of a creator superfluous.[128][133]

1.8.10 See also

- Big Crunch

- Cosmic Calendar

- Shape of the universe

1.8.11 Notes

[1] There is no consensus about how long the Big Bang phase lasted. For some writers this denotes only the initial singularity, for others the whole history of the universe. Usually, at least the first few minutes (during which helium is synthesized) are said to occur "during the Big Bang".

[2] Detailed information of and references for tests of general relativity are given in the article tests of general relativity.

[3] It is commonly reported that Hoyle intended this to be pejorative. However, Hoyle later denied that, saying that it was just a striking image meant to emphasize the difference between the two theories for radio listeners.[59]

[4] Strictly, dark energy in the form of a cosmological constant drives the universe towards a flat state; however, our universe remained close to flat for several billion years, before the dark energy density became significant.

1.8.12 References

[1] Joseph Silk (2009). *Horizons of Cosmology*. Templeton Press. p. 208.

[2] Simon Singh (2005). *Big Bang: The Origin of the Universe*. Harper Perennial. p. 560.

[3] Wollack, E. J. (10 December 2010). "Cosmology: The Study of the Universe". *Universe 101: Big Bang Theory*. NASA. Archived from the original on 14 May 2011. Retrieved 27 April 2011. The second section discusses the classic tests of the Big Bang theory that make it so compelling as the likely valid description of our universe.

[4] "First Second of the Big Bang". *How The Universe Works 3*. 2014. Discovery Science.

[5] "Big-bang model". *Encyclopedia Britannica*. Retrieved 11 February 2015.

[6] Wright, E. L. (9 May 2009). "What is the evidence for the Big Bang?". *Frequently Asked Questions in Cosmology*. UCLA, Division of Astronomy and Astrophysics. Retrieved 16 October 2009.

[7] "Planck reveals an almost perfect universe". *Planck*. ESA. 2013-03-21. Retrieved 2013-03-21.

[8] Kragh, H. (1996). *Cosmology and Controversy*. Princeton University Press. p. 318. ISBN 0-691-02623-8.

[9] Peebles, P. J. E.; Ratra, Bharat (2003). "The cosmological constant and dark energy". *Reviews of Modern Physics* **75** (2): 559–606. arXiv:astro-ph/0207347. Bibcode:2003RvMP...75..559P. doi:10.1103/RevModPhys.75.559.

[10] Gibson, C. H. (2001). "The First Turbulent Mixing and Combustion" (PDF). *IUTAM Turbulent Mixing and Combustion*.

[11] Gibson, C. H. (2001). "Turbulence And Mixing In The Early Universe". arXiv:astro-ph/0110012 [astro-ph].

[12] Gibson, C. H. (2005). "The First Turbulent Combustion". arXiv:astro-ph/0501416 [astro-ph].

[13] Staff (17 March 2014). "BICEP2 2014 Results Release". *National Science Foundation*. Retrieved 18 March 2014.

[14] Clavin, Whitney (17 March 2014). "NASA Technology Views Birth of the Universe". NASA. Retrieved 17 March 2014.

[15] Overbye, Dennis (17 March 2014). "Detection of Waves in Space Buttresses Landmark Theory of Big Bang". *The New York Times*. Retrieved 17 March 2014.

[16] Overbye, Dennis (24 March 2014). "Ripples From the Big Bang". *New York Times*. Retrieved 24 March 2014.

[17] Hubble, E. (1929). "A Relation Between Distance and Radial Velocity Among Extra-Galactic Nebulae". *Proceedings of the National Academy of Sciences* **15** (3): 168–73. Bibcode:1929PNAS...15..168H. doi:10.1073/pnas.15.3.168. PMC 522427. PMID 16577160.

[18] Hawking, S. W.; Ellis, G. F. R. (1973). *The Large-Scale Structure of Space-Time*. Cambridge University Press. ISBN 0-521-20016-4.

[19] Roos, M. (2008). "Expansion of the Universe – Standard Big Bang Model". In Engvold, O.; Stabell, R.; Czerny, B.; Lattanzio, J. *Astronomy and Astrophysics*. Encyclopedia of Life Support Systems. UNESCO. arXiv:0802.2005. This singularity is termed the *Big Bang*.

[20] Drees, W. B. (1990). *Beyond the big bang: quantum cosmologies and God*. Open Court Publishing. pp. 223–224. ISBN 978-0-8126-9118-4.

[21] Planck Collaboration (2015). "Planck 2015 results. XIII. Cosmological parameters (See Table 4 on page 31 of pdf).". arXiv:1502.01589.

[22] Guth, A. H. (1998). *The Inflationary Universe: Quest for a New Theory of Cosmic Origins*. Vintage Books. ISBN 978-0-09-995950-2.

[23] Schewe, P. (2005). "An Ocean of Quarks". *Physics News Update* (American Institute of Physics) **728** (1).

[24] Kolb and Turner (1988), chapter 6

[25] Kolb and Turner (1988), chapter 7

[26] Kolb and Turner (1988), chapter 4

[27] Peacock (1999), chapter 9

[28] Loeb, Abraham (October 2014). "The Habitable Epoch of the Early Universe". *International Journal of Astrobiology* **13** (4): 337–339. arXiv:1312.0613. Bibcode:2014IJAsB..13..337L. doi:10.1017/S1473550414000196.

[29] Dreitus, Claudia (2 December 2014). "Much-Discussed Views That Go Way Back - Avi Loeb Ponders the Early Universe, Nature and Life". *New York Times*. Retrieved 3 December 2014.

[30] Clavin, Whitney; Jenkins, Ann; Villard, Ray (7 January 2014). "NASA's Hubble and Spitzer Team up to Probe Faraway Galaxies". NASA. Retrieved 8 January 2014.

[31] Spergel, D. N.; et al. (2003). "First year Wilkinson Microwave Anisotropy Probe (WMAP) observations: determination of cosmological parameters". *The Astrophysical Journal Supplement* **148** (1): 175–194. arXiv:astro-ph/0302209. Bibcode:2003ApJS..148..175S. doi:10.1086/377226.

[32] Jarosik, N.; et al. (WMAP Collaboration) (2011). "Seven-Year Wilkinson Microwave Anisotropy Probe (WMAP) Observations: Sky Maps, Systematic Errors, and Basic Results" (PDF). NASA/GSFC: 39. Table 8. Retrieved 4 December 2010.

[33] Ivanchik, A. V.; Potekhin, A. Y.; Varshalovich, D. A. (1999). "The Fine-Structure Constant: A New Observational Limit on Its Cosmological Variation and Some Theoretical Consequences". *Astronomy and Astrophysics* **343**: 459. arXiv:astro-ph/9810166. Bibcode:1999A&A...343..439I.

[34] Goodman, J. (1995). "Geocentrism Reexamined". *Physical Review D* **52** (4): 1821–1827. arXiv:astro-ph/9506068. Bibcode:1995PhRvD..52.1821G. doi:10.1103/PhysRevD.52.1821.

[35] d'Inverno, R. (1992). "Chapter 23". *Introducing Einstein's Relativity*. Oxford University Press. ISBN 0-19-859686-3.

[36] Tamara M. Davis and Charles H. Lineweaver, *Expanding Confusion: common misconceptions of cosmological horizons and the superluminal expansion of the Universe*. astro-ph/0310808

[37] Kolb and Turner (1988), chapter 3

[38] "'Big bang' astronomer dies". BBC News. 22 August 2001. Archived from the original on 8 December 2008. Retrieved 7 December 2008.

[39] Croswell, K. (1995). "Chapter 9". *The Alchemy of the Heavens*. Anchor Books.

[40] Mitton. *Fred Hoyle: A Life in Science*. Cambridge University Press. ISBN 978-1-139-49595-0. "To create a picture in the mind of the listener, Hoyle had likened the explosive theory of the universe's origin to a 'big bang'"

[41] Moskowitz, C. (25 September 2012). "Hubble Telescope Reveals Farthest View Into Universe Ever". Space.com. Retrieved 26 September 2012.

[42] Slipher, V. M. (1913). "The Radial Velocity of the Andromeda Nebula". *Lowell Observatory Bulletin* **1**: 56–57. Bibcode:1913LowOB...2...56S.

[43] Slipher, V. M. (1915). "Spectrographic Observations of Nebulae". *Popular Astronomy* **23**: 21–24. Bibcode:1915PA.....23Q..21S.

[44] Friedman, A. A. (1922). "Über die Krümmung des Raumes". *Zeitschrift für Physik* (in German) **10** (1): 377–386. Bibcode:1922ZPhy...10..377F. doi:10.1007/BF01332580.

(English translation in: Friedman, A. (1999). "On the Curvature of Space". *General Relativity and Gravitation* **31** (12): 1991–2000. Bibcode:1999GReGr..31.1991F. doi:10.1023/A:1026751225741.)

[45] Lemaître, G. (1927). "Un univers homogène de masse constante et de rayon croissant rendant compte de la vitesse radiale des nébuleuses extragalactiques". *Annals of the Scientific Society of Brussels* (in French) **47A**: 41.

(Translated in: Lemaître, G. (1931). "A Homogeneous Universe of Constant Mass and Growing Radius Accounting for the Radial Velocity of Extragalactic Nebulae". *Monthly Notices of the Royal Astronomical Society* **91**: 483–490. Bibcode:1931MNRAS..91..483L. doi:10.1093/mnras/91.5.483.)

[46] Lemaître, G. (1931). "The Evolution of the Universe: Discussion". *Nature* **128** (3234): 699–701. Bibcode:1931Natur.128..704L. doi:10.1038/128704a0.

[47] Christianson, E. (1995). *Edwin Hubble: Mariner of the Nebulae*. Farrar, Straus and Giroux. ISBN 0-374-14660-8.

[48] Kragh, H. (1996). *Cosmology and Controversy*. Princeton University Press. ISBN 0-691-02623-8.

[49] "People and Discoveries: Big Bang Theory". *A Science Odyssey*. PBS. Retrieved 9 March 2012.

[50] Eddington, A. (1931). "The End of the World: from the Standpoint of Mathematical Physics". *Nature* **127** (3203): 447–453. Bibcode:1931Natur.127..447E. doi:10.1038/127447a0.

[51] Appolloni, S. (2011). ""Repugnant", "Not Repugnant at All": How the Respective Epistemic Attitudes of Georges Lemaitre and Sir Arthur Eddington Influenced How Each Approached the Idea of a Beginning of the Universe". *IBSU Scientific Journal* **5** (1): 19–44.

[52] Lemaitre, G. (1931). "The Beginning of the World from the Point of View of Quantum Theory". *Nature* **127** (3210): 706. Bibcode:1931Natur.127..706L. doi:10.1038/127706b0.

[53] Milne, E. A. (1935). *Relativity, Gravitation and World Structure*. Oxford University Press. LCCN 35019093.

[54] Tolman, R. C. (1934). *Relativity, Thermodynamics, and Cosmology*. Clarendon Press. ISBN 0-486-65383-8. LCCN 34032023.

[55] Zwicky, F. (1929). "On the Red Shift of Spectral Lines through Interstellar Space". *Proceedings of the National Academy of Sciences* **15** (10): 773–779. Bibcode:1929PNAS...15..773Z. doi:10.1073/pnas.15.10.773. PMC 522555. PMID 16577237.

[56] Hoyle, F. (1948). "A New Model for the Expanding Universe". *Monthly Notices of the Royal Astronomical Society* **108**: 372–382. Bibcode:1948MNRAS.108..372H. doi:10.1093/mnras/108.5.372.

[57] Alpher, R. A.; Bethe, H.; Gamow, G. (1948). "The Origin of Chemical Elements". *Physical Review* **73** (7): 803–804. Bibcode:1948PhRv...73..803A. doi:10.1103/PhysRev.73.803.

[58] Alpher, R. A.; Herman, R. (1948). "Evolution of the Universe". *Nature* **162** (4124): 774–775. Bibcode:1948Natur.162..774A. doi:10.1038/162774b0.

[59] Croswell, K. (1995). *The Alchemy of the Heavens*. Anchor Books. chapter 9. ISBN 978-0-385-47213-5.

[60] Penzias, A. A.; Wilson, R. W. (1965). "A Measurement of Excess Antenna Temperature at 4080 Mc/s". *The Astrophysical Journal* **142**: 419. Bibcode:1965ApJ...142..419P. doi:10.1086/148307.

[61] Hawking, S.; Ellis, G. F. (1968). "The Cosmic Black-Body Radiation and the Existence of Singularities in our Universe". *The Astrophysical Journal* **152**: 25. Bibcode:1968ApJ...152...25H. doi:10.1086/149520.

[62] Hawking, S.; Penrose, R. (27 January 1970). "The Singularities of Gravitational Collapse and Cosmology". *Proceedings of the Royal Society A: Mathematical, Physical & Engineering Sciences* (The Royal Society) **314** (1519): 529–548. Bibcode:1970RSPSA.314..529H. doi:10.1098/rspa.1970.0021. Retrieved 27 March 2015.

[63] Guth, Alan (15 January 1981). "Inflationary universe: A possible solution to the horizon and flatness problems". *Phys. Rev. D* **23** (2): 347–356. Bibcode:1981PhRvD..23..347G. doi:10.1103/PhysRevD.23.347.

[64] Huchra, John (2008). "The Hubble Constant". Center for Astrophysics, Harvard University.

[65] Livio, Mario (2001). *The Accelerating Universe: Infinite Expansion, the Cosmological Constant, and the Beauty of the Cosmos*. John Wiley & Sons. p. 160. ISBN 047143714X.

[66] Navabi, A. A.; Riazi, N. (2003). "Is the Age Problem Resolved?". *Journal of Astrophysics and Astronomy* **24** (1–2): 3–10. Bibcode:2003JApA...24....3N. doi:10.1007/BF03012187.

[67] Boggess, N. W.; et al. (1992). "The COBE Mission: Its Design and Performance Two Years after the launch". *The Astrophysical Journal* **397**: 420. Bibcode:1992ApJ...397..420B. doi:10.1086/171797.

[68] Spergel, D. N.; et al. (2006). "Wilkinson Microwave Anisotropy Probe (WMAP) Three Year Results: Implications for Cosmology". *Astrophysical Journal Supplement* **170** (2): 377–408. arXiv:astro-ph/0603449. Bibcode:2007ApJS..170..377S. doi:10.1086/513700.

[69] Krauss, L. (2012). *A Universe From Nothing: Why there is Something Rather than Nothing*. Free Press. p. 118. ISBN 978-1-4516-2445-8.

[70] Gladders, M. D.; et al. (2007). "Cosmological Constraints from the Red-Sequence Cluster Survey". *The Astrophysical Journal* **655** (1): 128–134. arXiv:astro-ph/0603588. Bibcode:2007ApJ...655..128G. doi:10.1086/509909.

[71] "Four Pillars". Cambridge Cosmology: Hot Big Bang. Retrieved 4 March 2016.

[72] Sadoulet, B. (2010). "Direct Searches for Dark Matter". *Astro2010: The Astronomy and Astrophysics Decadal Survey*. National Academies Press. Retrieved 12 March 2012.

[73] Cahn, R. (2010). "For a Comprehensive Space-Based Dark Energy Mission". *Astro2010: The Astronomy and Astrophysics Decadal Survey*. National Academies Press. Retrieved 12 March 2012.

[74] Peacock (1999), chapter 3

[75] Srianand, R.; Petitjean, P.; Ledoux, C. (2000). "The microwave background temperature at the redshift of 2.33771". *Nature* **408** (6815): 931–935. arXiv:astro-ph/0012222. Bibcode:2000Natur.408..931S. doi:10.1038/35050020. Lay summary – *European Southern Observatory* (December 2000).

[76] Bennett, C. L.; et al. (2013). "Nine-Year Wilkinson Microwave Anisotropy Probe (WMAP) Observations: Final Maps and Results". arXiv:1212.5225.

[77] Gannon, M. (21 December 2012). "New 'Baby Picture' of Universe Unveiled". Space.com. Retrieved 21 December 2012.

[78] Wright, E. L. (2004). "Theoretical Overview of Cosmic Microwave Background Anisotropy". In W. L. Freedman. *Measuring and Modeling the Universe*. Carnegie Observatories Astrophysics Series. Cambridge University Press. p. 291. arXiv:astro-ph/0305591. ISBN 0-521-75576-X.

[79] White, M. (1999). *Anisotropies in the CMB*. Proceedings of the Los Angeles Meeting, DPF 99 (UCLA). arXiv:astro-ph/9903232. Bibcode:1999dpf..conf.....W.

[80] A. Melchiorri et. al. (1999). "A measurement of Omega from the North American test flight of BOOMERANG". *The Astrophysical Journal* (Institute of Physics) (536). Retrieved 2015-05-15.

[81] P. de Bernardis; et al. (2000). "A Flat Universe from High-Resolution Maps of the Cosmic Microwave Background Radiation". *Nature* (Nature Publishing Group) **404**: 955–959. arXiv:astro-ph/0004404. doi:10.1038/35010035. PMID 10801117.

[82] A. D. Miller; et al. (1999). "A Measurement of the Angular Power Spectrum of the Cosmic Microwave Background from l = 100 to 400". *The Astrophysical Journal Letters* **524** (1): L1–L4. arXiv:astro-ph/9906421. Bibcode:1999ApJ...524L...1M. doi:10.1086/312293.

[83] Steigman, G. (2005). "Primordial Nucleosynthesis: Successes And Challenges". *International Journal of Modern Physics E* **15**: 1–36. arXiv:astro-ph/0511534. Bibcode:2006IJMPE..15....1S. doi:10.1142/S0218301306004028.

[84] Barbara Sue Ryden (2003). *Introduction to cosmology*. Addison-Wesley. ISBN 978-0-8053-8912-8.

[85] Bertschinger, E. (2001). "Cosmological Perturbation Theory and Structure Formation". arXiv:astro-ph/0101009 [astro-ph].

[86] Bertschinger, E. (1998). "Simulations of Structure Formation in the Universe". *Annual Review of Astronomy and Astrophysics* **36** (1): 599–654. Bibcode:1998ARA&A..36..599B. doi:10.1146/annurev.astro.36.1.599.

[87] Fumagalli, M.; O'Meara, J. M.; Prochaska, J. X. (2011). "Detection of Pristine Gas Two Billion Years After the Big Bang". *Science* **334** (6060): 1245–9. arXiv:1111.2334. Bibcode:2011Sci...334.1245F. doi:10.1126/science.1213581. PMID 22075722.

[88] "Astronomers Find Clouds of Primordial Gas from the Early Universe, Just Moments After Big Bang". Science Daily. 10 November 2011. Retrieved 13 November 2011.

[89] Perley, D. (21 February 2005). "Determination of the Universe's Age, t_o". University of California Berkeley, Astronomy Department. Retrieved 27 January 2012.

[90] Srianand, R.; Noterdaeme, P.; Ledoux, C.; Petitjean, P. (2008). "First detection of CO in a high-redshift damped Lyman-α system". *Astronomy and Astrophysics* **482** (3): L39. Bibcode:2008A&A...482L..39S. doi:10.1051/0004-6361:200809727.

[91] Avgoustidis, A.; Luzzi, G.; Martins, C. J. A. P.; Monteiro, A. M. R. V. L. (2011). "Constraints on the CMB temperature-redshift dependence from SZ and distance measurements". arXiv:1112.1862v1 [astro-ph.CO].

[92] Belusevic, R. (2008). *Relativity, Astrophysics and Cosmology*. Wiley-VCH. p. 16. ISBN 3-527-40764-2.

[93] Castelvecchi, Davide; Witze, Witze (February 11, 2016). "Einstein's gravitational waves found at last". *Nature News*. doi:10.1038/nature.2016.19361. Retrieved 2016-02-11.

[94] B. P. Abbott et al. (LIGO Scientific Collaboration and Virgo Collaboration) (2016). "Observation of Gravitational Waves from a Binary Black Hole Merger". *Physical Review Letters* **116** (6). arXiv:1602.03837. Bibcode:2016PhRvL.116f1102A. doi:10.1103/PhysRevLett.116.061102.

[95] "Gravitational waves detected 100 years after Einstein's prediction | NSF - National Science Foundation". *www.nsf.gov*. Retrieved 2016-02-11.

[96] http://www.bbc.com/news/science-environment-35524440 Einstein's gravitational waves 'seen' from black holes

[97] http://www.scientificamerican.com/article/the-future-of-gravitational-wave-astronomy/ The Future of Gravitational Wave Astronomy

[98] Earman, John; Mosterín, Jesús (March 1999). "A Critical Look at Inflationary Cosmology". *Philosophy of Science* **66** (1): 1–49. doi:10.1086/392675. JSTOR 188736.

[99] Penrose, R. (1979). Hawking, S. W.; Israel, W., eds. *Singularities and Time-Asymmetry. General Relativity: An Einstein Centenary Survey* (Cambridge University Press). pp. 581–638.

[100] Penrose, R. (1989). Fergus, E. J., ed. *Difficulties with Inflationary Cosmology. Proceedings of the 14th Texas Symposium on Relativistic Astrophysics* (New York Academy of Sciences). pp. 249–264. doi:10.1111/j.1749-6632.1989.tb50513.x.

[101] Steinhardt, Paul J. (April 2011). "The inflation debate: Is the theory at the heart of modern cosmology deeply flawed?". *Scientific American*: 18–25.

[102] Kolb and Turner, chapter 6

[103] Sakharov, A. D. (1967). "Violation of CP Invariance, C Asymmetry and Baryon Asymmetry of the Universe". *Zhurnal Eksperimental'noi i Teoreticheskoi Fiziki, Pisma* (in Russian) **5**: 32.

(Translated in *Journal of Experimental and Theoretical Physics Letters* **5**, 24 (1967).)

[104] Mortonson, Michael J.; Weinberg, David H.; White, Martin (Dec 2013). "Dark Energy: A Short Review" (PDF). *Particle Data Group 2014 Review of Particle Physics*.

[105] Rugh, S.E.; Zinkernagel, H. (December 2002). "The quantum vacuum and the cosmological constant problem". *Studies in History and Philosophy of Science Part B: Studies in History and Philosophy of Modern Physics* **33** (4): 663–705. doi:10.1016/S1355-2198(02)00033-3.

[106] Keel, B. (October 2009). "Dark Matter". Retrieved 24 July 2013.

[107] Yao, W. M.; et al. (2006). "Review of Particle Physics: Dark Matter" (PDF). *Journal of Physics G* **33** (1): 1–1232. arXiv:astro-ph/0601168. Bibcode:2006JPhG...33....1Y. doi:10.1088/0954-3899/33/1/001.

[108] Bullock, James. "Notes on the Missing Satellites Problem" (PDF). *XX Canary Islands Winter School of Astrophysics on Local Group Cosmology*.

[109] Diemand, Jürg; Zemp, Marcel; Moore, Ben; Stadel, Joachim; Carollo, C. Marcella (December 2005). "Cusps in cold dark matter haloes". *Monthly Notices of the Royal Astronomical Society* **364** (2): 665–673. arXiv:astro-ph/0504215. Bibcode:2005MNRAS.364..665D. doi:10.1111/j.1365-2966.2005.09601.x.

[110] Dodelson, Scott (Dec 2011). "The Real Problem with MOND" (PDF). *Honorable Mention, Gravity Research Foundation 2011 Awards*.

[111] Kolb and Turner (1988), chapter 8

[112] D. N. Spergel; et al. (2007). "Three-Year Wilkinson Microwave Anisotropy Probe (WMAP) Observations: Implications for Cosmology" (PDF). *The Astrophysical Journal Supplement Series* **170**: 377–408. arXiv:astro-ph/0603449. Bibcode:2007ApJS..170..377S. doi:10.1086/513700.

[113] R. Penrose (2007). *The Road to Reality*. Vintage books. ISBN 0-679-77631-1.

[114] Dicke, R. H.; Peebles, P. J. E. Hawking, S. W.; Israel, W., eds. *The big bang cosmology—enigmas and nostrums. General Relativity: an Einstein centenary survey* (Cambridge University Press). pp. 504–517.

[115] Griswold, Britt (2012). "What is the Ultimate Fate of the Universe?". *Universe 101 Big Bang Theory*. NASA.

[116] Fred C. Adams & Gregory Laughlin (1997). "A dying Universe: the long-term fate and evolution of astrophysical objects". *Reviews of Modern Physics* **69** (2): 337–372. arXiv:astro-ph/9701131. Bibcode:1997RvMP...69..337A. doi:10.1103/RevModPhys.69.337..

[117] Caldwell, R. R; Kamionkowski, M.; Weinberg, N. N. (2003). "Phantom Energy and Cosmic Doomsday". *Physical Review Letters* **91** (7): 071301. arXiv:astro-ph/0302506. Bibcode:2003PhRvL..91g1301C. doi:10.1103/PhysRevLett.91.071301. PMID 12935004.

[118] Hawking, S. W.; Ellis, G. F. R. (1973). *The Large Scale Structure of Space-Time*. Cambridge (UK): Cambridge University Press. ISBN 0-521-09906-4.

[119] Hartle, J. H.; Hawking, S. (1983). "Wave Function of the Universe". *Physical Review D* **28** (12): 2960–2975. Bibcode:1983PhRvD..28.2960H. doi:10.1103/PhysRevD.28.2960.

[120] Bird, P. (2011). "Determining the Big Bang State Vector" (PDF).

[121] Langlois, D. (2002). "Brane Cosmology: An Introduction". *Progress of Theoretical Physics Supplement* **148**: 181–212. arXiv:hep-th/0209261. Bibcode:2002PThPS.148..181L. doi:10.1143/PTPS.148.181.

[122] Linde, A. (2002). "Inflationary Theory versus Ekpyrotic/Cyclic Scenario". arXiv:hep-th/0205259 [hep-th].

[123] Than, K. (2006). "Recycled Universe: Theory Could Solve Cosmic Mystery". Space.com. Retrieved 3 July 2007.

[124] Kennedy, B. K. (2007). "What Happened Before the Big Bang?". Archived from the original on 4 July 2007. Retrieved 3 July 2007.

[125] Linde, A. (1986). "Eternal Chaotic Inflation". *Modern Physics Letters A* **1** (2): 81–85. Bibcode:1986MPLA....1...81L. doi:10.1142/S0217732386000129.

[126] Linde, A. (1986). "Eternally Existing Self-Reproducing Chaotic Inflationary Universe". *Physics Letters B* **175** (4): 395–400. Bibcode:1986PhLB..175..395L. doi:10.1016/0370-2693(86)90611-8.

[127] Harris, J. F. (2002). *Analytic philosophy of religion*. Springer. p. 128. ISBN 978-1-4020-0530-5.

[128] Frame, T. (2009). *Losing my religion*. UNSW Press. pp. 137–141. ISBN 978-1-921410-19-2.

[129] Harrison, P. (2010). *The Cambridge Companion to Science and Religion*. Cambridge University Press. p. 9. ISBN 978-0-521-71251-4.

[130] Harris 2002, p. 129

[131] Craig, William Lane (1999). "The ultimate question of origins: God and the beginning of the Universe". *Astrophysics and Space Science*. 269-270 (1–4): 723–740. doi:10.1007/978-94-011-4114-7_85. ISBN 978-94-010-5801-8.

[132] Asad, Muhammad (1984). *The Message of the Qur'ân*. Gibraltar, Spain: Dar al-Andalus Limited. ISBN 1904510000.

[133] Sagan, C. (1988). *introduction to A Brief History of Time by Stephen Hawking*. Bantam Books. pp. X. ISBN 0-553-34614-8. ... a universe with no edge in space, no beginning or end in time, and nothing for a Creator to do.

Books

- Farrell, John (2005). *The Day Without Yesterday: Lemaitre, Einstein, and the Birth of Modern Cosmology*. New York, NY: Thunder's Mouth Press. ISBN 1-56025-660-5.

- Kolb, E.; Turner, M. (1988). *The Early Universe*. Addison–Wesley. ISBN 0-201-11604-9.

- Peacock. J. (1999). *Cosmological Physics*. Cambridge University Press. ISBN 0-521-42270-1.

- Woolfson. M. (2013). *Time, Space, Stars and Man: The Story of Big Bang (2nd edition)*. World Scientific Publishing. ISBN 978-1-84816-933-3.

1.8.13   Further reading

For an annotated list of textbooks and monographs, see Physical cosmology § Textbooks.

- Alpher, R. A.; Herman. R. (1988). "Reflections on Early Work on 'Big Bang' Cosmology". *Physics Today* **8** (8): 24–34. Bibcode:1988PhT....41h..24A. doi:10.1063/1.881126.

- "Cosmic Journey: A History of Scientific Cosmology". American Institute of Physics.

- Barrow, J. D. (1994). *The Origin of the Universe*. Weidenfeld & Nicolson. ISBN 0-297-81497-4.

- Davies. P. C. W. (1992). *The Mind of God: The Scientific Basis for a Rational World*. Simon & Schuster. ISBN 0-671-71069-9.

- Feuerbacher, B.; Scranton. R. (2006). "Evidence for the Big Bang". TalkOrigins.

- Mather, J. C.; Boslough, J. (1996). *The Very First Light: The True Inside Story of the Scientific Journey Back to the Dawn of the Universe*. Basic Books. p. 300. ISBN 0-465-01575-1.

- Riordan, Michael; Zajc, William (May 2006). "The First Few Microseconds" (PDF). Scientific American.

- Singh. S. (2004). Big Bang: The Origins of the Universe. Fourth Estate. ISBN 0-00-716220-0.

- "Misconceptions about the Big Bang" (PDF). Scientific American. March 2005.

- Weinberg. S. (1993). *The First Three Minutes: A Modern View of the Origin of the Universe*. Basic Books. ISBN 0-465-02437-8.

1.8.14   External links

- big-bang model at *Encyclopædia Britannica*

- The Story of the Big Bang - STFC funded project explaining the history of the universe in easy-to-understand language

- Big Bang Cosmology WMAP

- The Big Bang - NASA Science

- Big bang model with animated graphics

- Cosmology at DMOZ

- Evidence for the Big Bang

1.9   Black hole

For other uses, see Black hole (disambiguation).

Simulation of gravitational lensing by a black hole, which distorts the image of a galaxy in the background

A **black hole** is a region of spacetime exhibiting such strong gravitational effects that nothing—including particles and electromagnetic radiation such as light—can escape from inside it.[1] The theory of general relativity predicts that a sufficiently compact mass can deform spacetime to form a black hole.[2][3] The boundary of the region from which no escape is possible is called the event horizon. Although crossing the event horizon has enormous effect on the fate of the object crossing it, it appears to have no locally detectable features. In many ways a black hole acts like an ideal black body, as it reflects no light.[4][5] Moreover, quantum field theory in curved spacetime predicts that event horizons emit Hawking radiation, with the same spectrum as a black body of a temperature inversely proportional to its mass. This temperature is on the order of billionths of a kelvin for black holes of stellar mass, making it essentially impossible to observe.

Objects whose gravitational fields are too strong for light to escape were first considered in the 18th century by John Michell and Pierre-Simon Laplace. The first modern solution of general relativity that would characterize a black hole was found by Karl Schwarzschild in 1916, although its interpretation as a region of space from which nothing can escape was first published by David Finkelstein in 1958. Black holes were long considered a mathematical curiosity; it was during the 1960s that theoretical work showed they were a generic prediction of general relativity. The discovery of neutron stars sparked interest in gravitationally collapsed compact objects as a possible astrophysical reality.

Black holes of stellar mass are expected to form when very massive stars collapse at the end of their life cycle. After a black hole has formed, it can continue to grow by absorbing mass from its surroundings. By absorbing other stars and merging with other black holes, supermassive black holes of millions of solar masses ($M\odot$) may form. There is general consensus that supermassive black holes exist in the centers of most galaxies.

Despite its invisible interior, the presence of a black hole can be inferred through its interaction with other matter and with electromagnetic radiation such as visible light. Matter that falls onto a black hole can form an external accretion disk heated by friction, forming some of the brightest objects in the universe. If there are other stars orbiting a black hole, their orbits can be used to determine the black hole's mass and location. Such observations can be used to exclude possible alternatives such as neutron stars. In this way, astronomers have identified numerous stellar black hole candidates in binary systems, and established that the radio source known as Sagittarius A*, at the core of our own Milky Way galaxy, contains a supermassive black hole of about 4.3 million solar masses.

On 11 February 2016, the LIGO collaboration announced the first observation of gravitational waves; because these waves were generated from a black hole merger it was the first ever direct detection of a binary black hole merger.[6] On 15 June 2016, a second detection of a gravitational wave event from colliding black holes was announced.[7]

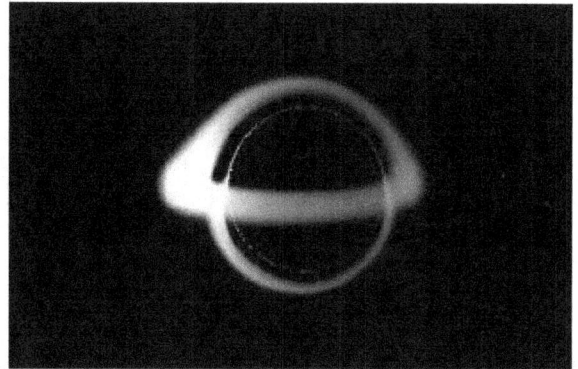

Predicted appearance of non-rotating black hole with toroidal ring of ionised matter, such as has been proposed[8] as a model for Sagittarius A*. The asymmetry is due to the Doppler effect resulting from the enormous orbital speed needed for centrifugal balance of the very strong gravitational attraction of the hole.

Simulated view of a black hole in front of the Large Magellanic Cloud. Note the gravitational lensing effect, which produces two enlarged but highly distorted views of the Cloud. Across the top, the Milky Way disk appears distorted into an arc.

1.9.1 History

The idea of a body so massive that even light could not escape was first put forward by John Michell in a letter written in 1783 to Henry Cavendish of the Royal Society:

> If the semi-diameter of a sphere of the same density as the Sun were to exceed that of the Sun in the proportion of 500 to 1, a body falling from an infinite height towards it would have acquired at its surface greater velocity than that of light, and consequently supposing light to be attracted by the same force in proportion to its vis inertiae, with other bodies, all light emitted from such a body would be made to return towards it by its

own proper gravity.
— John Michell[9]

In 1796, mathematician Pierre-Simon Laplace promoted the same idea in the first and second editions of his book *Exposition du système du Monde* (it was removed from later editions).[10][11] Such "dark stars" were largely ignored in the nineteenth century, since it was not understood how a massless wave such as light could be influenced by gravity.[12]

General relativity

In 1915, Albert Einstein developed his theory of general relativity, having earlier shown that gravity does influence light's motion. Only a few months later, Karl Schwarzschild found a solution to the Einstein field equations, which describes the gravitational field of a point mass and a spherical mass.[13] A few months after Schwarzschild, Johannes Droste, a student of Hendrik Lorentz, independently gave the same solution for the point mass and wrote more extensively about its properties.[14][15] This solution had a peculiar behaviour at what is now called the Schwarzschild radius, where it became singular, meaning that some of the terms in the Einstein equations became infinite. The nature of this surface was not quite understood at the time. In 1924, Arthur Eddington showed that the singularity disappeared after a change of coordinates (see Eddington–Finkelstein coordinates), although it took until 1933 for Georges Lemaître to realize that this meant the singularity at the Schwarzschild radius was an unphysical coordinate singularity.[16] Arthur Eddington did however comment on the possibility of a star with mass compressed to the Schwarzschild radius in a 1926 book, noting that Einstein's theory allows us to rule out overly large densities for visible stars like Betelgeuse because "a star of 250 million km radius could not possibly have so high a density as the sun. Firstly, the force of gravitation would be so great that light would be unable to escape from it, the rays falling back to the star like a stone to the earth. Secondly, the red shift of the spectral lines would be so great that the spectrum would be shifted out of existence. Thirdly, the mass would produce so much curvature of the space-time metric that space would close up around the star, leaving us outside (i.e., nowhere)."[17][18]

In 1931, Subrahmanyan Chandrasekhar calculated, using special relativity, that a non-rotating body of electron-degenerate matter above a certain limiting mass (now called the Chandrasekhar limit at 1.4 $M\odot$) has no stable solutions.[19] His arguments were opposed by many of his contemporaries like Eddington and Lev Landau, who argued that some yet unknown mechanism would stop the collapse.[20] They were partly correct: a white dwarf slightly more massive than the Chandrasekhar limit will collapse into a neutron star,[21] which is itself stable because of the Pauli exclusion principle. But in 1939, Robert Oppenheimer and others predicted that neutron stars above approximately 3 $M\odot$ (the Tolman–Oppenheimer–Volkoff limit) would collapse into black holes for the reasons presented by Chandrasekhar, and concluded that no law of physics was likely to intervene and stop at least some stars from collapsing to black holes.[22]

Oppenheimer and his co-authors interpreted the singularity at the boundary of the Schwarzschild radius as indicating that this was the boundary of a bubble in which time stopped. This is a valid point of view for external observers, but not for infalling observers. Because of this property, the collapsed stars were called "frozen stars",[23] because an outside observer would see the surface of the star frozen in time at the instant where its collapse takes it inside the Schwarzschild radius.

Golden age

See also: History of general relativity

In 1958, David Finkelstein identified the Schwarzschild surface as an event horizon, "a perfect unidirectional membrane: causal influences can cross it in only one direction".[24] This did not strictly contradict Oppenheimer's results, but extended them to include the point of view of infalling observers. Finkelstein's solution extended the Schwarzschild solution for the future of observers falling into a black hole. A complete extension had already been found by Martin Kruskal, who was urged to publish it.[25]

These results came at the beginning of the golden age of general relativity, which was marked by general relativity and black holes becoming mainstream subjects of research. This process was helped by the discovery of pulsars in 1967,[26][27] which, by 1969, were shown to be rapidly rotating neutron stars.[28] Until that time, neutron stars, like black holes, were regarded as just theoretical curiosities; but the discovery of pulsars showed their physical relevance and spurred a further interest in all types of compact objects that might be formed by gravitational collapse.

In this period more general black hole solutions were found. In 1963, Roy Kerr found the exact solution for a rotating black hole. Two years later, Ezra Newman found the axisymmetric solution for a black hole that is both rotating and electrically charged.[29] Through the work of Werner Israel,[30] Brandon Carter,[31][32] and David Robinson[33] the no-hair theorem emerged, stating that a stationary black hole solution is completely described by the three parameters of the Kerr–Newman metric: mass, angular momentum, and electric charge.[34]

At first, it was suspected that the strange features of the black hole solutions were pathological artifacts from the symmetry conditions imposed, and that the singularities would not appear in generic situations. This view was held in particular by Vladimir Belinsky, Isaak Khalatnikov, and Evgeny Lifshitz, who tried to prove that no singularities appear in generic solutions. However, in the late 1960s Roger Penrose[35] and Stephen Hawking used global techniques to prove that singularities appear generically.[36]

Work by James Bardeen, Jacob Bekenstein, Carter, and Hawking in the early 1970s led to the formulation of black hole thermodynamics.[37] These laws describe the behaviour of a black hole in close analogy to the laws of thermodynamics by relating mass to energy, area to entropy, and surface gravity to temperature. The analogy was completed when Hawking, in 1974, showed that quantum field theory predicts that black holes should radiate like a black body with a temperature proportional to the surface gravity of the black hole.[38]

The first use of the term "black hole" in print was by journalist Ann Ewing in her article *"'Black Holes' in Space"*, dated 18 January 1964, which was a report on a meeting of the American Association for the Advancement of Science.[39] John Wheeler used the term "black hole" at a lecture in 1967, leading some to credit him with coining the phrase. After Wheeler's use of the term, it was quickly adopted in general use.

1.9.2 Properties and structure

The no-hair theorem states that, once it achieves a stable condition after formation, a black hole has only three independent physical properties: mass, charge, and angular momentum.[34] Any two black holes that share the same values for these properties, or parameters, are indistinguishable according to classical (i.e. non-quantum) mechanics.

These properties are special because they are visible from outside a black hole. For example, a charged black hole repels other like charges just like any other charged object. Similarly, the total mass inside a sphere containing a black hole can be found by using the gravitational analog of Gauss's law, the ADM mass, far away from the black hole.[40] Likewise, the angular momentum can be measured from far away using frame dragging by the gravitomagnetic field.

When an object falls into a black hole, any information about the shape of the object or distribution of charge on it is evenly distributed along the horizon of the black hole, and is lost to outside observers. The behavior of the horizon in this situation is a dissipative system that is closely analogous to that of a conductive stretchy membrane with friction and electrical resistance—the membrane paradigm.[41] This is different from other field theories such as electromagnetism, which do not have any friction or resistivity at the microscopic level, because they are time-reversible. Because a black hole eventually achieves a stable state with only three parameters, there is no way to avoid losing information about the initial conditions: the gravitational and electric fields of a black hole give very little information about what went in. The information that is lost includes every quantity that cannot be measured far away

from the black hole horizon, including approximately conserved quantum numbers such as the total baryon number and lepton number. This behavior is so puzzling that it has been called the black hole information loss paradox.[42][43]

Physical properties

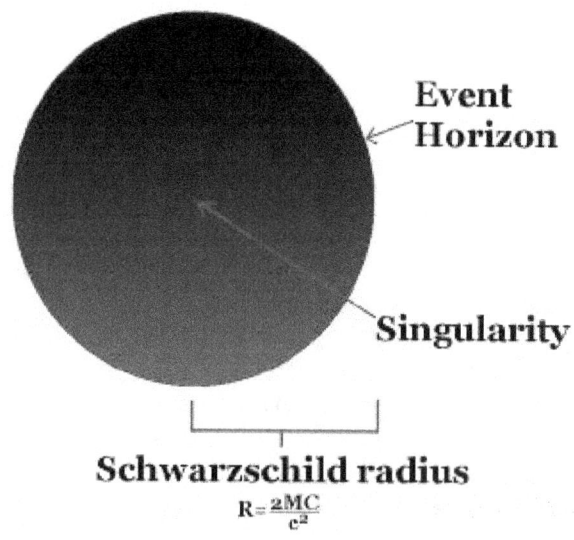

A simple illustration of a non-spinning black hole

The simplest static black holes have mass but neither electric charge nor angular momentum. These black holes are often referred to as Schwarzschild black holes after Karl Schwarzschild who discovered this solution in 1916.[13] According to Birkhoff's theorem, it is the only vacuum solution that is spherically symmetric.[44] This means that there is no observable difference between the gravitational field of such a black hole and that of any other spherical object of the same mass. The popular notion of a black hole "sucking in everything" in its surroundings is therefore only correct near a black hole's horizon; far away, the external gravitational field is identical to that of any other body of the same mass.[45]

Solutions describing more general black holes also exist. Non-rotating charged black holes are described by the Reissner–Nordström metric, while the Kerr metric describes a non-charged rotating black hole. The most general stationary black hole solution known is the Kerr–Newman metric, which describes a black hole with both charge and angular momentum.[46]

While the mass of a black hole can take any positive value, the charge and angular momentum are constrained by the mass. In Planck units, the total electric charge Q and the total angular momentum J are expected to satisfy

$$Q^2 + \left(\tfrac{J}{M}\right)^2 \leq M^2$$

for a black hole of mass M. Black holes satisfying this inequality are called extremal. Solutions of Einstein's equations that violate this inequality exist, but they do not possess an event horizon. These solutions have so-called naked singularities that can be observed from the outside, and hence are deemed *unphysical*. The cosmic censorship hypothesis rules out the formation of such singularities, when they are created through the gravitational collapse of realistic matter.[2] This is supported by numerical simulations.[47]

Due to the relatively large strength of the electromagnetic force, black holes forming from the collapse of stars are expected to retain the nearly neutral charge of the star. Rotation, however, is expected to be a common feature of compact objects. The black-hole candidate binary X-ray source GRS 1915+105[48] appears to have an angular momentum near the maximum allowed value.

Black holes are commonly classified according to their mass, independent of angular momentum J or electric charge Q. The size of a black hole, as determined by the radius of the event horizon, or Schwarzschild radius, is roughly proportional to the mass M through

$$r_{\mathrm{sh}} = \frac{2GM}{c^2} \approx 2.95 \, \frac{M}{M_{\mathrm{Sun}}} \, \mathrm{km},$$

where r_{sh} is the Schwarzschild radius and *MSun* is the mass of the Sun.[49] This relation is exact only for black holes with zero charge and angular momentum; for more general black holes it can differ up to a factor of 2.

Event horizon

Main article: Event horizon

The defining feature of a black hole is the appearance of an event horizon—a boundary in spacetime through which matter and light can only pass inward towards the mass of the black hole. Nothing, not even light, can escape from inside the event horizon. The event horizon is referred to as such because if an event occurs within the boundary, information from that event cannot reach an outside observer, making it impossible to determine if such an event occurred.[51]

As predicted by general relativity, the presence of a mass deforms spacetime in such a way that the paths taken by particles bend towards the mass.[52] At the event horizon of

a black hole, this deformation becomes so strong that there are no paths that lead away from the black hole.

To a distant observer, clocks near a black hole appear to tick more slowly than those further away from the black hole.[53] Due to this effect, known as gravitational time dilation, an object falling into a black hole appears to slow as it approaches the event horizon, taking an infinite time to reach it.[54] At the same time, all processes on this object slow down, from the view point of a fixed outside observer, causing any light emitted by the object to appear redder and dimmer, an effect known as gravitational redshift.[55] Eventually, the falling object becomes so dim that it can no longer be seen.

On the other hand, indestructible observers falling into a black hole do not notice any of these effects as they cross the event horizon. According to their own clocks, which appear to them to tick normally, they cross the event horizon after a finite time without noting any singular behaviour; it is impossible to determine the location of the event horizon from local observations.[56]

The shape of the event horizon of a black hole is always approximately spherical.[Note 2][59] For non-rotating (static) black holes the geometry of the event horizon is precisely spherical, while for rotating black holes the sphere is oblate.

Singularity

Main article: Gravitational singularity

At the center of a black hole, as described by general relativity, lies a gravitational singularity, a region where the spacetime curvature becomes infinite.[60] For a non-rotating black hole, this region takes the shape of a single point and for a rotating black hole, it is smeared out to form a ring singularity that lies in the plane of rotation.[61] In both cases, the singular region has zero volume. It can also be shown that the singular region contains all the mass of the black hole solution.[62] The singular region can thus be thought of as having infinite density.

Observers falling into a Schwarzschild black hole (*i.e.*, non-rotating and not charged) cannot avoid being carried into the singularity, once they cross the event horizon. They can prolong the experience by accelerating away to slow their descent, but only up to a limit; after attaining a certain ideal velocity, it is best to free fall the rest of the way.[63] When they reach the singularity, they are crushed to infinite density and their mass is added to the total of the black hole. Before that happens, they will have been torn apart by the growing tidal forces in a process sometimes referred to as spaghettification or the "noodle effect".[64]

In the case of a charged (Reissner–Nordström) or rotating

(Kerr) black hole, it is possible to avoid the singularity. Extending these solutions as far as possible reveals the hypothetical possibility of exiting the black hole into a different spacetime with the black hole acting as a wormhole.[65] The possibility of traveling to another universe is however only theoretical, since any perturbation would destroy this possibility.[66] It also appears to be possible to follow closed timelike curves (returning to one's own past) around the Kerr singularity, which lead to problems with causality like the grandfather paradox.[67] It is expected that none of these peculiar effects would survive in a proper quantum treatment of rotating and charged black holes.[68]

The appearance of singularities in general relativity is commonly perceived as signaling the breakdown of the theory.[69] This breakdown, however, is expected; it occurs in a situation where quantum effects should describe these actions, due to the extremely high density and therefore particle interactions. To date, it has not been possible to combine quantum and gravitational effects into a single theory, although there exist attempts to formulate such a theory of quantum gravity. It is generally expected that such a theory will not feature any singularities.[70][71]

Photon sphere

Main article: Photon sphere

The photon sphere is a spherical boundary of zero thickness in which photons that move on tangents to that sphere would be trapped in a circular orbit about the black hole. For non-rotating black holes, the photon sphere has a radius 1.5 times the Schwarzschild radius. Their orbits would be dynamically unstable, hence any small perturbation, such as a particle of infalling matter, would cause an instability that would grow over time, either setting the photon on an outward trajectory causing it to escape the black hole, or on an inward spiral where it would eventually cross the event horizon.[72]

While light can still escape from the photon sphere, any light that crosses the photon sphere on an inbound trajectory will be captured by the black hole. Hence any light that reaches an outside observer from the photon sphere must have been emitted by objects between the photon sphere and the event horizon.[72]

Other compact objects, such as neutron stars, can also have photon spheres.[73] This follows from the fact that the gravitational field *external* to a spherically-symmetric object is governed by the Schwarzschild metric, which depends only on the object's mass rather than the radius of the object, hence any object whose radius shrinks to smaller than 1.5 times the Schwarzschild radius will have a photon sphere.

Ergosphere

Main article: Ergosphere
Rotating black holes are surrounded by a region of space-

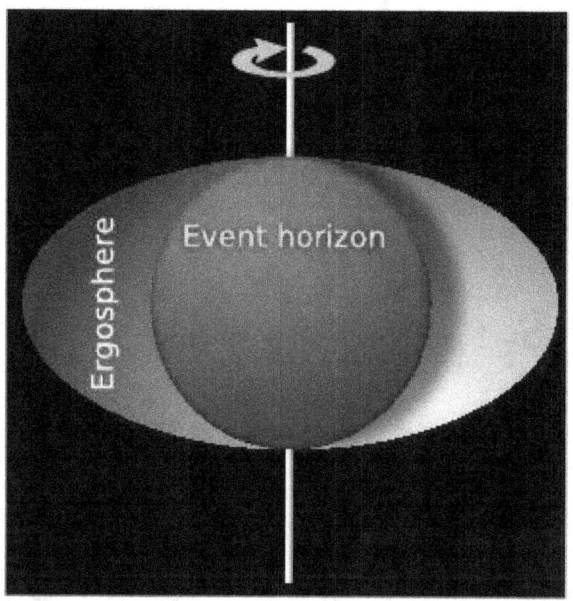

The ergosphere is an oblate spheroid region outside of the event horizon, where objects cannot remain stationary.

time in which it is impossible to stand still, called the ergosphere. This is the result of a process known as frame-dragging; general relativity predicts that any rotating mass will tend to slightly "drag" along the spacetime immediately surrounding it. Any object near the rotating mass will tend to start moving in the direction of rotation. For a rotating black hole, this effect is so strong near the event horizon that an object would have to move faster than the speed of light in the opposite direction to just stand still.[74]

The ergosphere of a black hole is a volume whose inner boundary is the black hole's event horizon and an outer boundary of an oblate spheroid, which coincides with the event horizon at the poles but noticeably wider around the equator. The outer boundary is sometimes called the *ergosurface*.

Objects and radiation can escape normally from the ergosphere. Through the Penrose process, objects can emerge from the ergosphere with more energy than they entered. This energy is taken from the rotational energy of the black hole causing the latter to slow.[75]

Innermost stable circular orbit (ISCO)

Main article: Innermost stable circular orbit

In Newtonian gravity, test particles can stably orbit at arbitrary distances from a central object. In general relativity, however, there exists an innermost stable circular orbit (often called the ISCO), inside of which, any infinitesimal perturbations to a circular orbit will lead to inspiral into the black hole.[76] The location of the ISCO depends on the spin of the black hole, in the case of a Schwarzschild black hole (spin zero) is:

$$r_{isco} = 3\,r_s = \frac{6\,GM}{c^2}.$$

and decreases with increasing spin.

1.9.3   Formation and evolution

Considering the exotic nature of black holes, it may be natural to question if such bizarre objects could exist in nature or to suggest that they are merely pathological solutions to Einstein's equations. Einstein himself wrongly thought that black holes would not form, because he held that the angular momentum of collapsing particles would stabilize their motion at some radius.[77] This led the general relativity community to dismiss all results to the contrary for many years. However, a minority of relativists continued to contend that black holes were physical objects,[78] and by the end of the 1960s, they had persuaded the majority of researchers in the field that there is no obstacle to the formation of an event horizon.

Once an event horizon forms, Penrose proved, general relativity without quantum mechanics requires that a singularity will form within.[35] Shortly afterwards, Hawking showed that many cosmological solutions that describe the Big Bang have singularities without scalar fields or other exotic matter (see "Penrose–Hawking singularity theorems"). The Kerr solution, the no-hair theorem, and the laws of black hole thermodynamics showed that the physical properties of black holes were simple and comprehensible, making them respectable subjects for research.[79] The primary formation process for black holes is expected to be the gravitational collapse of heavy objects such as stars, but there are also more exotic processes that can lead to the production of black holes.

Gravitational collapse

Main article: Gravitational collapse

Gravitational collapse occurs when an object's internal pressure is insufficient to resist the object's own gravity. For stars this usually occurs either because a star has too little "fuel" left to maintain its temperature through stellar

nucleosynthesis, or because a star that would have been stable receives extra matter in a way that does not raise its core temperature. In either case the star's temperature is no longer high enough to prevent it from collapsing under its own weight.[80] The collapse may be stopped by the degeneracy pressure of the star's constituents, allowing the condensation of matter into an exotic denser state. The result is one of the various types of compact star. The type of compact star formed depends on the mass of the remnant of the original star left after the outer layers have been blown away. Such explosions, from a supernova explosion or by pulsations, leads to planetary nebula. Note that this mass can be substantially less than the original star. Remnants exceeding 5 $M\odot$ are produced by stars that were over 20 $M\odot$ before the collapse.[80]

If the mass of the remnant exceeds about 3–4 $M\odot$ (the Tolman–Oppenheimer–Volkoff limit[22]), either because the original star was very heavy or because the remnant collected additional mass through accretion of matter, even the degeneracy pressure of neutrons is insufficient to stop the collapse. No known mechanism (except possibly quark degeneracy pressure, see quark star) is powerful enough to stop the implosion and the object will inevitably collapse to form a black hole.[80]

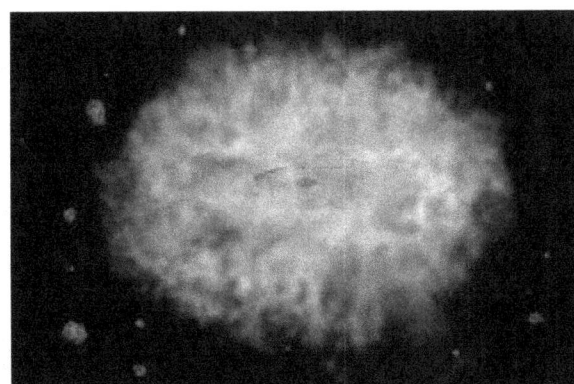

Artist's impression of supermassive black hole seed.[81]

The gravitational collapse of heavy stars is assumed to be responsible for the formation of stellar mass black holes. Star formation in the early universe may have resulted in very massive stars, which upon their collapse would have produced black holes of up to 10^3 $M\odot$. These black holes could be the seeds of the supermassive black holes found in the centers of most galaxies.[82] It has further been suggested that supermassive black holes with typical masses of ~10^5 $M\odot$ could have formed from the direct collapse of gas clouds in the young universe.[83] Some candidates for such objects have been found in observations of the young universe.[83]

While most of the energy released during gravitational collapse is emitted very quickly, an outside observer does not actually see the end of this process. Even though the collapse takes a finite amount of time from the reference frame of infalling matter, a distant observer would see the infalling material slow and halt just above the event horizon, due to gravitational time dilation. Light from the collapsing material takes longer and longer to reach the observer, with the light emitted just before the event horizon forms delayed an infinite amount of time. Thus the external observer never sees the formation of the event horizon; instead, the collapsing material seems to become dimmer and increasingly red-shifted, eventually fading away.[84]

Primordial black holes in the Big Bang Gravitational collapse requires great density. In the current epoch of the universe these high densities are only found in stars, but in the early universe shortly after the big bang densities were much greater, possibly allowing for the creation of black holes. The high density alone is not enough to allow the formation of black holes since a uniform mass distribution will not allow the mass to bunch up. In order for primordial black holes to form in such a dense medium, there must be initial density perturbations that can then grow under their own gravity. Different models for the early universe vary widely in their predictions of the size of these perturbations. Various models predict the creation of black holes, ranging from a Planck mass to hundreds of thousands of solar masses.[85] Primordial black holes could thus account for the creation of any type of black hole.

High-energy collisions

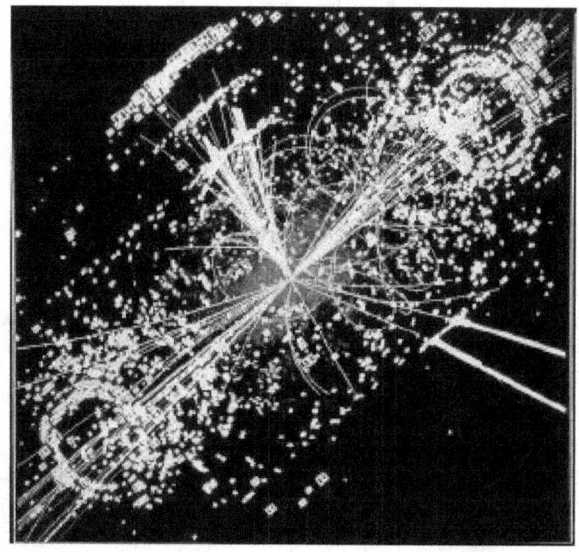

A simulated event in the CMS detector, a collision in which a micro black hole may be created.

Gravitational collapse is not the only process that could cre-

ate black holes. In principle, black holes could be formed in high-energy collisions that achieve sufficient density. As of 2002, no such events have been detected, either directly or indirectly as a deficiency of the mass balance in particle accelerator experiments.[86] This suggests that there must be a lower limit for the mass of black holes. Theoretically, this boundary is expected to lie around the Planck mass ($m_P = \sqrt{\hbar c/G} \approx 1.2 \times 10^{19}$ GeV/$c^2 \approx 2.2 \times 10^{-8}$ kg), where quantum effects are expected to invalidate the predictions of general relativity.[87] This would put the creation of black holes firmly out of reach of any high-energy process occurring on or near the Earth. However, certain developments in quantum gravity suggest that the Planck mass could be much lower: some braneworld scenarios for example put the boundary as low as 1 TeV/c^2.[88] This would make it conceivable for micro black holes to be created in the high-energy collisions that occur when cosmic rays hit the Earth's atmosphere, or possibly in the Large Hadron Collider at CERN. These theories are very speculative, and the creation of black holes in these processes is deemed unlikely by many specialists.[89] Even if micro black holes could be formed, it is expected that they would evaporate in about 10^{-25} seconds, posing no threat to the Earth.[90]

Growth

Once a black hole has formed, it can continue to grow by absorbing additional matter. Any black hole will continually absorb gas and interstellar dust from its surroundings and omnipresent cosmic background radiation. This is the primary process through which supermassive black holes seem to have grown.[82] A similar process has been suggested for the formation of intermediate-mass black holes found in globular clusters.[91]

Another possibility for black hole growth, is for a black hole to merge with other objects such as stars or even other black holes. Although not necessary for growth, this is thought to have been important, especially for the early development of supermassive black holes, which could have formed from the coagulation of many smaller objects.[82] The process has also been proposed as the origin of some intermediate-mass black holes.[92][93]

Evaporation

Main article: Hawking radiation

In 1974, Hawking predicted that black holes are not entirely black but emit small amounts of thermal radiation;[38] this effect has become known as Hawking radiation. By applying quantum field theory to a static black hole background, he determined that a black hole should emit particles that display a perfect black body spectrum. Since Hawking's publication, many others have verified the result through various approaches.[94] If Hawking's theory of black hole radiation is correct, then black holes are expected to shrink and evaporate over time as they lose mass by the emission of photons and other particles.[38] The temperature of this thermal spectrum (Hawking temperature) is proportional to the surface gravity of the black hole, which, for a Schwarzschild black hole, is inversely proportional to the mass. Hence, large black holes emit less radiation than small black holes.[95]

A stellar black hole of 1 $M\odot$ has a Hawking temperature of about 100 nanokelvins. This is far less than the 2.7 K temperature of the cosmic microwave background radiation. Stellar-mass or larger black holes receive more mass from the cosmic microwave background than they emit through Hawking radiation and thus will grow instead of shrink. To have a Hawking temperature larger than 2.7 K (and be able to evaporate), a black hole would need a mass less than the Moon. Such a black hole would have a diameter of less than a tenth of a millimeter.[96]

If a black hole is very small, the radiation effects are expected to become very strong. Even a black hole that is heavy compared to a human would evaporate in an instant. A black hole with the mass of a car would have a diameter of about 10^{-24} m and take a nanosecond to evaporate, during which time it would briefly have a luminosity of more than 200 times that of the Sun. Lower-mass black holes are expected to evaporate even faster; for example, a black hole of mass 1 TeV/c^2 would take less than 10^{-88} seconds to evaporate completely. For such a small black hole, quantum gravitation effects are expected to play an important role and could hypothetically make such a small black hole stable, although current developments in quantum gravity do not indicate so.[97][98]

The Hawking radiation for an astrophysical black hole is predicted to be very weak and would thus be exceedingly difficult to detect from Earth. A possible exception, however, is the burst of gamma rays emitted in the last stage of the evaporation of primordial black holes. Searches for such flashes have proven unsuccessful and provide stringent limits on the possibility of existence of low mass primordial black holes.[99] NASA's Fermi Gamma-ray Space Telescope launched in 2008 will continue the search for these flashes.[100]

1.9.4   Observational evidence

By their very nature, black holes do not directly emit any electromagnetic radiation other than the hypothetical Hawking radiation, so astrophysicists searching for black holes must generally rely on indirect observations. For ex-

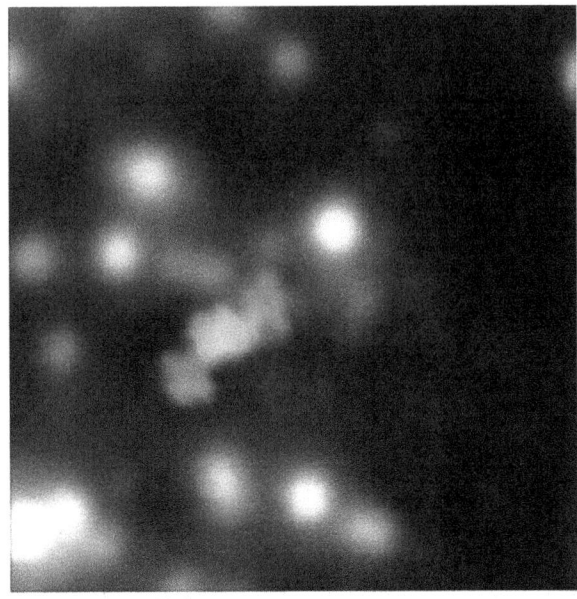

Gas cloud ripped apart by black hole at the centre of the Milky Way.[101]

ample, a black hole's existence can sometimes be inferred by observing its gravitational interactions with its surroundings. However, the Event Horizon Telescope (EHT), run by MIT's Haystack Observatory, is an attempt to directly observe the immediate environment of the event horizon of Sagittarius A*, the black hole at the centre of the Milky Way. The first image of the event horizon may appear as early as 2016.[102] The existence of magnetic fields just outside the event horizon of Sagittarius A*, which were predicted by theoretical studies of black holes, was confirmed by the EHT in 2015.[103][104]

Detection of gravitational waves from merging black holes

On 24 September 2015 the LIGO gravitational wave observatory made the first-ever successful observation of gravitational waves.[6][105] The signal was consistent with theoretical predictions for the gravitational waves produced by the merger of two black holes: one with about 36 solar masses, and the other around 29 solar masses.[6][106] This observation provides the most concrete evidence for the existence of black holes to date. For instance, the gravitational wave signal suggests that the separation of the two object prior to merger was just 350 km (or roughly 4 times the Schwarzschild radius corresponding to the inferred masses). The objects must therefore have been extremely compact, leaving black holes as the most plausible interpretation.[6]

More importantly, the signal observed by LIGO also included the start of the post-merger ringdown, the signal produced as the newly formed compact object settles down to a stationary state. Arguably, the ringdown is the most direct way of observing a black hole.[107] From the LIGO signal it is possible to extract the frequency and damping time of the dominant mode of the ringdown. From these it is possible to infer the mass and angular momentum of the final object, which match independent predictions from numerical simulations of the merger.[108] The frequency and decay time of the dominant mode are determined by the geometry of the photon sphere. Hence, observation of this mode confirms the presence of a photon sphere, however it cannot exclude possible exotic alternatives to black holes that are compact enough to have a photon sphere.[107]

The observation also provides the first observational evidence for the existence of stellar-mass black hole binaries. Furthermore, it is the first observational evidence of stellar-mass black holes weighing 25 solar masses or more.[109]

Proper motions of stars orbiting Sagittarius A*

The proper motions of stars near the center of our own Milky Way provide strong observational evidence that these stars are orbiting a supermassive black hole.[110] Since 1995, astronomers have tracked the motions of 90 stars orbiting an invisible object coincident with the radio source Sagittarius A*. By fitting their motions to Keplerian orbits, the astronomers were able to infer, in 1998, that a 2.6 million $M\odot$ object must be contained in a volume with a radius of 0.02 light-years to cause the motions of those stars.[111] Since then, one of the stars—called S2—has completed a full orbit. From the orbital data, astronomers were able to make refine the calculations of the mass to 4.3 million $M\odot$ and a radius of less than 0.002 lightyears for the object causing the orbital motion of those stars.[110] The upper limit on the object's size is still too large to test whether it is smaller than its Schwarzschild radius; nevertheless, these observations strongly suggest that the central object is a supermassive black hole as there are no other plausible scenarios for confining so much invisible mass into such a small volume.[111] Additionally, there is some observational evidence that this object might possess an event horizon, a feature unique to black holes.[112]

Accretion of matter

See also: Accretion disc

Due to conservation of angular momentum, gas falling into the gravitational well created by a massive object will typically form a disc-like structure around the object. Artists' impressions such as the accompanying representation of a black hole with corona commonly depict the black hole as

Black hole with corona, X-ray source (artist's concept).[113]

if it were a flat-space material body hiding the part of the disc just behind it, but detailed mathematical modelling[114] shows that the image of the disc would actually be distorted by the bending of light that originated behind the black hole in such a way that the upper side of the disc would be entirely visible, while there would be a partially visible secondary image of the underside of the disk.

Predicted view from outside the horizon of a Schwarzschild black hole lit by a thin accretion disc

Within such a disc, friction would cause angular momentum to be transported outward, allowing matter to fall further inward, thus releasing potential energy and increasing the temperature of the gas.[115]

When the accreting object is a neutron star or a black hole, the gas in the inner accretion disc orbits at very high speeds because of its proximity to the compact object. The resulting friction is so significant that it heats the inner disc to temperatures at which it emits vast amounts of electromagnetic radiation (mainly X-rays). These bright X-ray sources may be detected by telescopes. This process of accretion is one of the most efficient energy-producing processes known; up to 40% of the rest mass of the accreted material can be emitted as radiation.[115] (In nuclear fusion only about 0.7% of the rest mass will be emitted as energy.) In many cases, accretion discs are accompanied by relativistic jets that are emitted along the poles, which carry away much of the energy. The mechanism for the creation

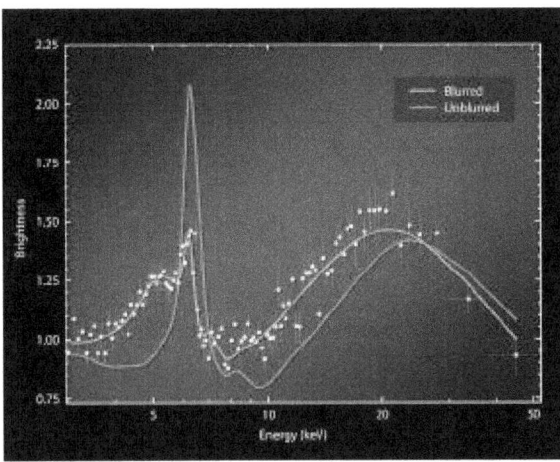

Blurring of X-rays near Black hole (NuSTAR; 12 August 2014).[113]

of these jets is currently not well understood.

As such, many of the universe's more energetic phenomena have been attributed to the accretion of matter on black holes. In particular, active galactic nuclei and quasars are believed to be the accretion discs of supermassive black holes.[116] Similarly, X-ray binaries are generally accepted to be binary star systems in which one of the two stars is a compact object accreting matter from its companion.[116] It has also been suggested that some ultraluminous X-ray sources may be the accretion disks of intermediate-mass black holes.[117]

In November 2011 the first direct observation of a quasar accretion disk around a supermassive black hole was reported.[118][119]

X-ray binaries

See also: X-ray binary

X-ray binaries are binary star systems that emit a majority

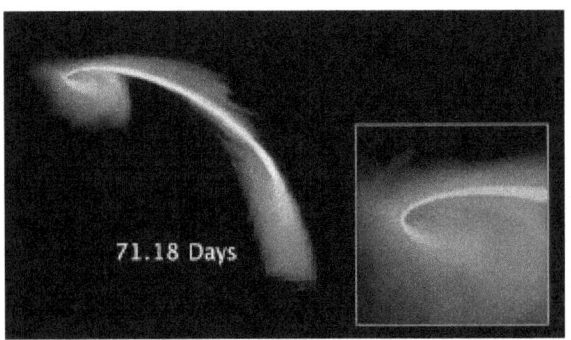

A computer simulation of a star being consumed by a black hole. The blue dot indicates the location of the black hole.

of their radiation in the X-ray part of the spectrum. These

A Chandra X-Ray Observatory image of Cygnus X-1, which was the first strong black hole candidate discovered

X-ray emissions are generally thought to result when one of the stars (compact object) accretes matter from another (regular) star. The presence of an ordinary star in such a system provides a unique opportunity for studying the central object and to determine if it might be a black hole.

This animation compares the X-ray 'heartbeats' of GRS 1915 and IGR J17091, two black holes that ingest gas from companion stars.

If such a system emits signals that can be directly traced back to the compact object, it cannot be a black hole. The absence of such a signal does, however, not exclude the possibility that the compact object is a neutron star. By studying the companion star it is often possible to obtain the orbital parameters of the system and to obtain an estimate for the mass of the compact object. If this is much larger than the Tolman–Oppenheimer–Volkoff limit (that is, the maximum mass a neutron star can have before it collapses) then the object cannot be a neutron star and is generally expected to be a black hole.[116]

The first strong candidate for a black hole, Cygnus X-1, was discovered in this way by Charles Thomas Bolton,[120] Louise Webster and Paul Murdin[121] in 1972.[122][123] Some doubt, however, remained due to the uncertainties that result from the companion star being much heavier than the candidate black hole.[116] Currently, better candidates for black holes are found in a class of X-ray binaries called soft X-ray transients.[116] In this class of system, the companion star is of relatively low mass allowing for more accurate estimates of the black hole mass. Moreover, these systems are actively emit X-rays for only several months once every 10–50 years. During the period of low X-ray emission (called quiescence), the accretion disc is extremely faint allowing detailed observation of the companion star during this period. One of the best such candidates is V404 Cyg.

Quiescence and advection-dominated accretion flow
The faintness of the accretion disc of an X-ray binary during quiescence is suspected to be caused by the flow of mass entering a mode called an advection-dominated accretion flow (ADAF). In this mode, almost all the energy generated by friction in the disc is swept along with the flow instead of radiated away. If this model is correct, then it forms strong qualitative evidence for the presence of an event horizon,[124] since if the object at the center of the disc had a solid surface, it would emit large amounts of radiation as the highly energetic gas hits the surface, an effect that is observed for neutron stars in a similar state.[115]

Quasi-periodic oscillations Main article: Quasi-periodic oscillations

The X-ray emissions from accretion disks sometimes flicker at certain frequencies. These signals are called quasi-periodic oscillations and are thought to be caused by material moving along the inner edge of the accretion disk (the innermost stable circular orbit). As such their frequency is linked to the mass of the compact object. They can thus be used as an alternative way to determine the mass of candidate black holes.[125]

Galactic nuclei

See also: Active galactic nucleus
Astronomers use the term "active galaxy" to describe galaxies with unusual characteristics, such as unusual spectral line emission and very strong radio emission. Theoretical and observational studies have shown that the activity in these active galactic nuclei (AGN) may be explained by the presence of supermassive black holes, which can be millions of times more massive than stellar ones. The models

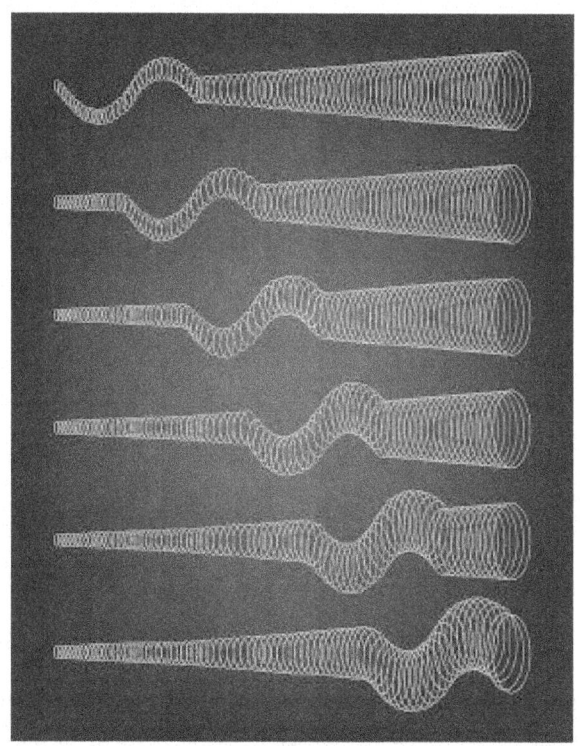

Magnetic waves, called Alfvén S-waves, flow from the base of black hole jets.

of these AGN consist of a central black hole that may be millions or billions of times more massive than the Sun; a disk of gas and dust called an accretion disk; and two jets perpendicular to the accretion disk.[126][127]

Although supermassive black holes are expected to be found in most AGN, only some galaxies' nuclei have been more carefully studied in attempts to both identify and measure the actual masses of the central supermassive black hole candidates. Some of the most notable galaxies with supermassive black hole candidates include the Andromeda Galaxy, M32, M87, NGC 3115, NGC 3377, NGC 4258, NGC 4889, NGC 1277, OJ 287, APM 08279+5255 and the Sombrero Galaxy.[129]

It is now widely accepted that the center of nearly every galaxy, not just active ones, contains a supermassive black hole.[130] The close observational correlation between the mass of this hole and the velocity dispersion of the host galaxy's bulge, known as the M-sigma relation, strongly suggests a connection between the formation of the black hole and the galaxy itself.[131]

Detection of unusually bright X-Ray flare from Sagittarius A, a black hole in the center of the Milky Way galaxy on 5 January 2015.[128]*

Simulation of gas cloud after close approach to the black hole at the centre of the Milky Way.[132]

Microlensing (proposed)

Another way that the black hole nature of an object may be tested in the future is through observation of effects caused by a strong gravitational field in their vicinity. One such effect is gravitational lensing: The deformation of spacetime around a massive object causes light rays to be deflected much as light passing through an optic lens. Observations have been made of weak gravitational lensing, in which light rays are deflected by only a few arcseconds. However, it has never been directly observed for a black hole.[133] One possibility for observing gravitational lensing by a black hole would be to observe stars in orbit around the black hole. There are several candidates for such an observation in orbit around Sagittarius A*.[133]

Alternatives

See also: Exotic star

The evidence for stellar black holes strongly relies on the existence of an upper limit for the mass of a neutron star. The size of this limit heavily depends on the assumptions made about the properties of dense matter. New exotic phases of matter could push up this bound.[116] A phase of free quarks at high density might allow the existence of dense quark stars,[134] and some supersymmetric models predict the existence of Q stars.[135] Some extensions of the standard model posit the existence of preons as fundamental building blocks of quarks and leptons, which could hypothetically form preon stars.[136] These hypothetical models could potentially explain a number of observations of stellar black hole candidates. However, it can be shown from arguments in general relativity that any such object will have a maximum mass.[116]

Since the average density of a black hole inside its Schwarzschild radius is inversely proportional to the square of its mass, supermassive black holes are much less dense than stellar black holes (the average density of a 10^8 $M_\odot$ black hole is comparable to that of water).[116] Consequently, the physics of matter forming a supermassive black hole is much better understood and the possible alternative explanations for supermassive black hole observations are much more mundane. For example, a supermassive black hole could be modelled by a large cluster of very dark objects. However, such alternatives are typically not stable enough to explain the supermassive black hole candidates.[116]

The evidence for the existence of stellar and supermassive black holes implies that in order for black holes to not form, general relativity must fail as a theory of gravity, perhaps due to the onset of quantum mechanical corrections. A much anticipated feature of a theory of quantum gravity is that it will not feature singularities or event horizons and thus black holes would not be real artifacts.[137] In 2002,[138] much attention has been drawn by the fuzzball model in string theory. Based on calculations for specific situations in string theory, the proposal suggests that generically the individual states of a black hole solution do not have an event horizon or singularity, but that for a classical/semi-classical observer the statistical average of such states appears just as an ordinary black hole as deduced from general relativity.[139]

1.9.5   Open questions

Entropy and thermodynamics

Further information: Black hole thermodynamics
In 1971. Hawking showed under general conditions[Note 3]

$$S = \frac{1}{4} \frac{c^3 k}{G\hbar} A$$

The formula for the Bekenstein–Hawking entropy (S) of a black hole, which depends on the area of the black hole (A). The constants are the speed of light (c), the Boltzmann constant (k), Newton's constant (G), and the reduced Planck constant (ħ).

that the total area of the event horizons of any collection of classical black holes can never decrease, even if they collide and merge.[140] This result, now known as the second law of black hole mechanics, is remarkably similar to the second law of thermodynamics, which states that the total entropy of a system can never decrease. As with classical objects at absolute zero temperature, it was assumed that black holes had zero entropy. If this were the case, the second law of thermodynamics would be violated by entropy-laden matter entering a black hole, resulting in a decrease of the total entropy of the universe. Therefore, Bekenstein proposed that a black hole should have an entropy, and that it should be proportional to its horizon area.[141]

The link with the laws of thermodynamics was further strengthened by Hawking's discovery that quantum field theory predicts that a black hole radiates blackbody radiation at a constant temperature. This seemingly causes a violation of the second law of black hole mechanics, since the radiation will carry away energy from the black hole causing it to shrink. The radiation, however also carries away entropy, and it can be proven under general assumptions that the sum of the entropy of the matter surrounding a black hole and one quarter of the area of the horizon as measured in Planck units is in fact always increasing. This allows the formulation of the first law of black hole mechanics as an analogue of the first law of thermodynamics, with the mass acting as energy, the surface gravity as temperature and the area as entropy.[141]

One puzzling feature is that the entropy of a black hole scales with its area rather than with its volume, since entropy is normally an extensive quantity that scales linearly with the volume of the system. This odd property led Gerard 't Hooft and Leonard Susskind to propose the holographic principle, which suggests that anything that happens in a volume of spacetime can be described by data on the bound-

ary of that volume.[142]

Although general relativity can be used to perform a semi-classical calculation of black hole entropy, this situation is theoretically unsatisfying. In statistical mechanics, entropy is understood as counting the number of microscopic configurations of a system that have the same macroscopic qualities (such as mass, charge, pressure, etc.). Without a satisfactory theory of quantum gravity, one cannot perform such a computation for black holes. Some progress has been made in various approaches to quantum gravity. In 1995. Andrew Strominger and Cumrun Vafa showed that counting the microstates of a specific supersymmetric black hole in string theory reproduced the Bekenstein–Hawking entropy.[143] Since then, similar results have been reported for different black holes both in string theory and in other approaches to quantum gravity like loop quantum gravity.[144]

Information loss paradox

Main article: Black hole information paradox

Because a black hole has only a few internal parameters, most of the information about the matter that went into forming the black hole is lost. Regardless of the type of matter which goes into a black hole, it appears that only information concerning the total mass, charge, and angular momentum are conserved. As long as black holes were thought to persist forever this information loss is not that problematic, as the information can be thought of as existing inside the black hole, inaccessible from the outside. However, black holes slowly evaporate by emitting Hawking radiation. This radiation does not appear to carry any additional information about the matter that formed the black hole, meaning that this information appears to be gone forever.[145]

The question whether information is truly lost in black holes (the black hole information paradox) has divided the theoretical physics community (see Thorne–Hawking–Preskill bet). In quantum mechanics, loss of information corresponds to the violation of vital property called unitarity, which has to do with the conservation of probability. It has been argued that loss of unitarity would also imply violation of conservation of energy.[146] Over recent years evidence has been building that indeed information and unitarity are preserved in a full quantum gravitational treatment of the problem.[147]

1.9.6 See also

- List of nearest black holes

- Black brane
- Black hole complementarity
- Black hole starship
- Black holes in fiction
- Black string
- BTZ black hole
- Dumb hole
- General relativity
- Kugelblitz (astrophysics)
- List of black holes
- Susskind-Hawking battle
- Timeline of black hole physics
- White hole
- Wormhole

1.9.7   Notes

[1] The set of possible paths, or more accurately the future light cone containing all possible world lines (in this diagram the light cone is represented by the V-shaped region bounded by arrows representing light ray world lines), is tilted in this way in Eddington–Finkelstein coordinates (the diagram is a "cartoon" version of an Eddington–Finkelstein coordinate diagram), but in other coordinates the light cones are not tilted in this way, for example in Schwarzschild coordinates they simply narrow without tilting as one approaches the event horizon, and in Kruskal–Szekeres coordinates the light cones don't change shape or orientation at all.[59]

[2] This is true only for 4-dimensional spacetimes. In higher dimensions more complicated horizon topologies like a black ring are possible.[57][58]

[3] In particular, he assumed that all matter satisfies the weak energy condition.

1.9.8   References

[1] Wald 1984, pp. 299–300

[2] Wald, R. M. (1997). "Gravitational Collapse and Cosmic Censorship". arXiv:gr-qc/9710068 [gr-qc].

[3] Overbye, Dennis (8 June 2015). "Black Hole Hunters". *NASA*. Retrieved 8 June 2015.

[4] Schutz, Bernard F. (2003). *Gravity from the ground up*. Cambridge University Press. p. 110. ISBN 0-521-45506-5.

[5] Davies, P. C. W. (1978). "Thermodynamics of Black Holes" (PDF). *Reports on Progress in Physics* **41** (8): 1313–1355. Bibcode:1978RPPh...41.1313D. doi:10.1088/0034-4885/41/8/004.

[6] Abbott, B.P.; et al. (2016). "Observation of Gravitational Waves from a Binary Black Hole Merger". *Phys. Rev. Lett.* **116**: 061102. arXiv:1602.03837. Bibcode:2016PhRvL.116f1102A. doi:10.1103/PhysRevLett.116.061102.

[7] Overbye, Dennis (15 June 2016). "Scientists Hear a Second Chirp From Colliding Black Holes". *New York Times*. Retrieved 15 June 2016.

[8] O. Straub, F.H. Vincent, M.A. Abramowicz, E. Gourgoulhon, T. Paumard, "Modelling the black hole silhouette in Sgr A* with ion tori,*Astron. Astroph 543 (2012) A8*

[9] Michell, J. (1784). "On the Means of Discovering the Distance, Magnitude, &c. of the Fixed Stars, in Consequence of the Diminution of the Velocity of Their Light, in Case Such a Diminution Should be Found to Take Place in any of Them, and Such Other Data Should be Procured from Observations, as Would be Farther Necessary for That Purpose". *Philosophical Transactions of the Royal Society* **74** (0): 35–57. Bibcode:1784RSPT...74...35M. doi:10.1098/rstl.1784.0008. JSTOR 106576.

[10] Gillispie, C. C. (2000). *Pierre-Simon Laplace, 1749–1827: a life in exact science*. Princeton paperbacks. Princeton University Press. p. 175. ISBN 0-691-05027-9.

[11] Israel, W. (1989). "Dark stars: the evolution of an idea". In Hawking, S. W.; Israel, W. *300 Years of Gravitation*. Cambridge University Press. ISBN 978-0-521-37976-2.

[12] Thorne 1994, pp. 123–124

[13] Schwarzschild, K. (1916). "Über das Gravitationsfeld eines Massenpunktes nach der Einsteinschen Theorie". *Sitzungsberichte der Königlich Preussischen Akademie der Wissenschaften* **7**: 189–196. and Schwarzschild, K. (1916). "Über das Gravitationsfeld einer Kugel aus inkompressibler Flüssigkeit nach der Einsteinschen Theorie". *Sitzungsberichte der Königlich Preussischen Akademie der Wissenschaften* **18**: 424–434.

[14] Droste, J. (1917). "On the field of a single centre in Einstein's theory of gravitation, and the motion of a particle in that field" (PDF). *Proceedings Royal Academy Amsterdam* **19** (1): 197–215.

[15] Kox, A. J. (1992). "General Relativity in the Netherlands: 1915–1920". In Eisenstaedt, J.; Kox, A. J. *Studies in the history of general relativity*. Birkhäuser. p. 41. ISBN 978-0-8176-3479-7.

[16] 't Hooft, G. (2009). "Introduction to the Theory of Black Holes" (PDF). Institute for Theoretical Physics / Spinoza Institute: 47–48.

[17] Eddington, Arthur (1926). *The Internal Constitution of the Stars*. Cambridge University Press. p. 6.

[18] Kip Thorne comments on this quote on pages 134-135 of his book *Black Holes and Time Warps*, writing that "The first conclusion was the Newtonian version of light not escaping; the second was a semi-accurate, relativistic description; and the third was typical Eddingtonian hyperbole ... when a star is as small as the critical circumference, the curvature is strong but not infinite, and space is definitely not wrapped around the star. Eddington may have known this, but his description made a good story, and it captured in a whimsical way the spirit of Schwarzschild's spacetime curvature."

[19] Venkataraman, G. (1992). *Chandrasekhar and his limit*. Universities Press. p. 89. ISBN 81-7371-035-X.

[20] Detweiler, S. (1981). "Resource letter BH-1: Black holes". *American Journal of Physics* 49 (5): 394–400. Bibcode:1981AmJPh..49..394D. doi:10.1119/1.12686.

[21] Harpaz, A. (1994). *Stellar evolution*. A K Peters. p. 105. ISBN 1-56881-012-1.

[22] Oppenheimer, J. R.; Volkoff, G. M. (1939). "On Massive Neutron Cores". *Physical Review* 55 (4): 374–381. Bibcode:1939PhRv...55..374O. doi:10.1103/PhysRev.55.374.

[23] Ruffini, R.; Wheeler, J. A. (1971). "Introducing the black hole" (PDF). *Physics Today* 24 (1): 30–41. Bibcode:1971PhT....24a..30R. doi:10.1063/1.3022513.

[24] Finkelstein, D. (1958). "Past-Future Asymmetry of the Gravitational Field of a Point Particle". *Physical Review* 110 (4): 965–967. Bibcode:1958PhRv..110..965F. doi:10.1103/PhysRev.110.965.

[25] Kruskal, M. (1960). "Maximal Extension of Schwarzschild Metric". *Physical Review* 119 (5): 1743. Bibcode:1960PhRv..119.1743K. doi:10.1103/PhysRev.119.1743.

[26] Hewish, A.; et al. (1968). "Observation of a Rapidly Pulsating Radio Source". *Nature* 217 (5130): 709–713. Bibcode:1968Natur.217..709H. doi:10.1038/217709a0

[27] Pilkington, J. D. H.; et al. (1968). "Observations of some further Pulsed Radio Sources". *Nature* 218 (5137): 126–129. Bibcode:1968Natur.218..126P. doi:10.1038/218126a0

[28] Hewish, A. (1970). "Pulsars". *Annual Review of Astronomy and Astrophysics* 8 (1): 265–296. Bibcode:1970ARA&A...8..265H. doi:10.1146/annurev.aa.08.090170.001405.

[29] Newman, E. T.; et al. (1965). "Metric of a Rotating, Charged Mass". *Journal of Mathematical Physics* 6 (6): 918. Bibcode:1965JMP.....6..918N. doi:10.1063/1.1704351

[30] Israel, W. (1967). "Event Horizons in Static Vacuum Space-Times". *Physical Review* 164 (5): 1776. Bibcode:1967PhRv..164.1776I. doi:10.1103/PhysRev.164.1776.

[31] Carter, B. (1971). "Axisymmetric Black Hole Has Only Two Degrees of Freedom". *Physical Review Letters* 26 (6): 331. Bibcode:1971PhRvL..26..331C. doi:10.1103/PhysRevLett.26.331.

[32] Carter, B. (1977). "The vacuum black hole uniqueness theorem and its conceivable generalisations". *Proceedings of the 1st Marcel Grossmann meeting on general relativity*. pp. 243–254.

[33] Robinson, D. (1975). "Uniqueness of the Kerr Black Hole". *Physical Review Letters* 34 (14): 905. Bibcode:1975PhRvL..34..905R. doi:10.1103/PhysRevLett.34.905.

[34] Heusler, M. (1998). "Stationary Black Holes: Uniqueness and Beyond". *Living Reviews in Relativity* 1 (6). doi:10.12942/lrr-1998-6. Archived from the original on 1999-02-03. Retrieved 2011-02-08.

[35] Penrose, R. (1965). "Gravitational Collapse and Space-Time Singularities". *Physical Review Letters* 14 (3): 57. Bibcode:1965PhRvL..14...57P. doi:10.1103/PhysRevLett.14.57.

[36] Ford, L. H. (2003). "The Classical Singularity Theorems and Their Quantum Loopholes". *International Journal of Theoretical Physics* 42 (6): 1219. doi:10.1023/A:1025754515197.

[37] Bardeen, J. M.; Carter, B.; Hawking, S. W. (1973). "The four laws of black hole mechanics". *Communications in Mathematical Physics* 31 (2): 161–170. Bibcode:1973CMaPh..31..161B. doi:10.1007/BF01645742. MR MR0334798. Zbl 1125.83309.

[38] Hawking, S. W. (1974). "Black hole explosions?". *Nature* 248 (5443): 30–31. Bibcode:1974Natur.248...30H. doi:10.1038/248030a0.

[39] Quinion, M. (26 April 2008). "Black Hole". *World Wide Words*. Retrieved 2008-06-17.

[40] Carroll 2004, p. 253

[41] Thorne, K. S.; Price, R. H. (1986). *Black holes: the membrane paradigm*. Yale University Press. ISBN 978-0-300-03770-8.

[42] Anderson, Warren G. (1996). "The Black Hole Information Loss Problem". *Usenet Physics FAQ*. Retrieved 2009-03-24.

[43] Preskill, J. (1994-10-21). *Black holes and information: A crisis in quantum physics* (PDF). Caltech Theory Seminar.

[44] Hawking & Ellis 1973, Appendix B

[45] Seeds, Michael A.; Backman, Dana E. (2007). *Perspectives on Astronomy*. Cengage Learning. p. 167. ISBN 0-495-11352-2

[46] Shapiro, S. L.; Teukolsky, S. A. (1983). *Black holes, white dwarfs, and neutron stars: the physics of compact objects*. John Wiley and Sons. p. 357. ISBN 0-471-87316-0.

[47] Berger, B. K. (2002). "Numerical Approaches to Spacetime Singularities". *Living Reviews in Relativity* **5**: 1. arXiv:gr-qc/0201056. Bibcode:2002LRR.....5....1B. doi:10.12942/lrr-2002-1. Retrieved 2007-08-04.

[48] McClintock, J. E.; Shafee, R.; Narayan, R.; Remillard, R. A.; Davis, S. W.; Li, L.-X. (2006). "The Spin of the Near-Extreme Kerr Black Hole GRS 1915+105". *Astrophysical Journal* **652** (1): 518–539. arXiv:astro-ph/0606076. Bibcode:2006ApJ...652..518M. doi:10.1086/508457.

[49] Wald 1984, pp. 124–125

[50] Thorne, Misner & Wheeler 1973, p. 848

[51] Wheeler 2007, p. 179

[52] Carroll 2004, Ch. 5.4 and 7.3

[53] Carroll 2004, p. 217

[54] Carroll 2004, p. 218

[55] "Inside a black hole". *Knowing the universe and its secrets*. Retrieved 2009-03-26.

[56] Carroll 2004, p. 222

[57] Emparan, R.; Reall, H. S. (2008). "Black Holes in Higher Dimensions". *Living Reviews in Relativity* **11** (6). arXiv:0801.3471. Bibcode:2008LRR....11....6E. doi:10.12942/lrr-2008-6. Retrieved 2011-02-10.

[58] Obers, N. A. (2009). Papantonopoulos, Eleftherios, ed. "Black Holes in Higher-Dimensional Gravity". *Lecture Notes in Physics*. Lecture Notes in Physics **769**: 211–258. arXiv:0802.0519. doi:10.1007/978-3-540-88460-6. ISBN 978-3-540-88459-0.

[59] hawking & ellis 1973, Ch. 9.3

[60] Carroll 2004, p. 205

[61] Carroll 2004, pp. 264–265

[62] Carroll 2004, p. 252

[63] Lewis, G. F.; Kwan, J. (2007). "No Way Back: Maximizing Survival Time Below the Schwarzschild Event Horizon". *Publications of the Astronomical Society of Australia* **24** (2): 46–52. arXiv:0705.1029. Bibcode:2007PASA...24...46L. doi:10.1071/AS07012.

[64] Wheeler 2007, p. 182

[65] Carroll 2004, pp. 257–259 and 265–266

[66] Droz, S.; Israel, W.; Morsink, S. M. (1996). "Black holes: the inside story". *Physics World* **9** (1): 34–37. Bibcode:1996PhyW....9...34D.

[67] Carroll 2004, p. 266

[68] Poisson, E.; Israel, W. (1990). "Internal structure of black holes". *Physical Review D* **41** (6): 1796. Bibcode:1990PhRvD..41.1796P. doi:10.1103/PhysRevD.41.1796.

[69] Wald 1984, p. 212

[70] Hamade, R. (1996). "Black Holes and Quantum Gravity". *Cambridge Relativity and Cosmology*. University of Cambridge. Retrieved 2009-03-26.

[71] Palmer, D. "Ask an Astrophysicist: Quantum Gravity and Black Holes". NASA. Retrieved 2009-03-26.

[72] Nitta, Daisuke; Chiba, Takeshi; Sugiyama, Naoshi (September 2011). "Shadows of colliding black holes". *Physical Review D* **84** (6). arXiv:1106.2425. Bibcode:2011PhRvD..84f3008N. doi:10.1103/PhysRevD.84.063008

[73] Nemiroff, R. J. (1993). "Visual distortions near a neutron star and black hole". *American Journal of Physics* **61** (7): 619. arXiv:astro-ph/9312003. Bibcode:1993AmJPh..61..619N. doi:10.1119/1.17224.

[74] Carroll 2004, Ch. 6.6

[75] Carroll 2004, Ch. 6.7

[76] Thorne, Misner & Wheeler 1973

[77] Einstein, A. (1939). "On A Stationary System With Spherical Symmetry Consisting of Many Gravitating Masses". *Annals of Mathematics* **40** (4): 922–936. doi:10.2307/1968902. JSTOR 1968902.

[78] Kerr, R. P. (2009). "The Kerr and Kerr-Schild metrics". In Wiltshire, D. L.; Visser, M.; Scott, S. M. *The Kerr Spacetime*. Cambridge University Press. arXiv:0706.1109. ISBN 978-0-521-88512-6.

[79] Hawking, S. W.; Penrose, R. (January 1970). "The Singularities of Gravitational Collapse and Cosmology". *Proceedings of the Royal Society A* **314** (1519): 529–548. Bibcode:1970RSPSA.314..529H. doi:10.1098/rspa.1970.0021. JSTOR 2416467.

[80] Carroll 2004, Section 5.8

[81] "Artist's impression of supermassive black hole seed". Retrieved 27 May 2016.

[82] Rees, M. J.; Volonteri, M. (2007). "Massive black holes: formation and evolution". In Karas, V.; Matt, G. *Black Holes from Stars to Galaxies—Across the Range of Masses*. Cambridge University Press. pp. 51–58. arXiv:astro-ph/0701512. ISBN 978-0-521-86347-6.

[83] Pacucci, F.; Ferrara, A.; Grazian, A.; Fiore, F.; Giallongo, E. (2016). "First Identification of Direct Collapse Black Hole Candidates in the Early Universe in CANDELS/GOODS-S". *Mon. Not. Roy. Astron. Soc.* **459**: 1432. arXiv:1603.08522. doi:10.1093/mnras/stw725.

[84] Penrose, R. (2002). "Gravitational Collapse: The Role of General Relativity" (PDF). *General Relativity and Gravitation* **34** (7): 1141. Bibcode:2002GReGr..34.1141P. doi:10.1023/A:1016578408204.

[85] Carr, B. J. (2005). "Primordial Black Holes: Do They Exist and Are They Useful?". In Suzuki, H.; Yokoyama, J.; Suto, Y.; Sato, K. *Inflating Horizon of Particle Astrophysics and Cosmology*. Universal Academy Press. arXiv:astro-ph/0511743. ISBN 4-946443-94-0.

[86] Giddings, S. B.; Thomas, S. (2002). "High energy colliders as black hole factories: The end of short distance physics". *Physical Review D* **65** (5): 056010. arXiv:hep-ph/0106219. Bibcode:2002PhRvD..65e6010G. doi:10.1103/PhysRevD.65.056010.

[87] Harada, T. (2006). "Is there a black hole minimum mass?". *Physical Review D* **74** (8): 084004. arXiv:gr-qc/0609055. Bibcode:2006PhRvD..74h4004H. doi:10.1103/PhysRevD.74.084004.

[88] Arkani-Hamed, N.; Dimopoulos, S.; Dvali, G. (1998). "The hierarchy problem and new dimensions at a millimeter". *Physics Letters B* **429** (3–4): 263. arXiv:hep-ph/9803315. Bibcode:1998PhLB..429..263A. doi:10.1016/S0370-2693(98)00466-3.

[89] LHC Safety Assessment Group. "Review of the Safety of LHC Collisions" (PDF). CERN.

[90] Cavaglià, M. (2010). "Particle accelerators as black hole factories?". *Einstein-Online* (Max Planck Institute for Gravitational Physics (Albert Einstein Institute)) **4**: 1010.

[91] Vesperini, E.; McMillan, S. L. W.; d'Ercole, A.; et al. (2010). "Intermediate-Mass Black Holes in Early Globular Clusters". *The Astrophysical Journal Letters* **713** (1): L41–L44. arXiv:1003.3470. Bibcode:2010ApJ...713L..41V. doi:10.1088/2041-8205/713/1/L41.

[92] Zwart, S. F. P.; Baumgardt, H.; Hut, P.; et al. (2004). "Formation of massive black holes through runaway collisions in dense young star clusters". *Nature* **428** (6984): 724–6. arXiv:astro-ph/0402622. Bibcode:2004Natur.428..724P. doi:10.1038/nature02448. PMID 15085124.

[93] O'Leary, R. M.; Rasio, F. A.; Fregeau, J. M.; et al. (2006). "Binary Mergers and Growth of Black Holes in Dense Star Clusters". *The Astrophysical Journal* **637** (2): 937. arXiv:astro-ph/0508224. Bibcode:2006ApJ...637..937O. doi:10.1086/498446.

[94] Page, D. N. (2005). "Hawking radiation and black hole thermodynamics". *New Journal of Physics* **7**: 203. arXiv:hep-th/0409024. Bibcode:2005NJPh....7..203P. doi:10.1088/1367-2630/7/1/203.

[95] Carroll 2004, Ch. 9.6

[96] "Evaporating black holes?". *Einstein online*. Max Planck Institute for Gravitational Physics. 2010. Retrieved 2010-12-12.

[97] Giddings, S. B.; Mangano, M. L. (2008). "Astrophysical implications of hypothetical stable TeV-scale black holes". *Physical Review D* **78** (3): 035009. arXiv:0806.3381. Bibcode:2008PhRvD..78c5009G. doi:10.1103/PhysRevD.78.035009.

[98] Peskin, M. E. (2008). "The end of the world at the Large Hadron Collider?". *Physics* **1**: 14. Bibcode:2008PhyOJ...1...14P. doi:10.1103/Physics.1.14.

[99] Fichtel, C. E.; Bertsch, D. L.; Dingus, B. L.; et al. (1994). "Search of the energetic gamma-ray experiment telescope (EGRET) data for high-energy gamma-ray microsecond bursts". *Astrophysical Journal* **434** (2): 557–559. Bibcode:1994ApJ...434..557F. doi:10.1086/174758.

[100] Naeye, R. "Testing Fundamental Physics". NASA. Retrieved 2008-09-16.

[101] "Ripped Apart by a Black Hole". *ESO Press Release*. Retrieved 19 July 2013.

[102] O'Neill, Ian (2 July 2015). "Event Horizon Telescope Will Probe Spacetime's Mysteries". *Discovery News*. Retrieved 6 December 2015.

[103] Johnson, M. D.; Fish, V. L.; Doeleman, S. S.; Marrone, D. P.; Plambeck, R. L.; Wardle, J. F. C.; Akiyama, K.; Asada, K.; Beaudoin, C. (2015-12-04). "Resolved magnetic-field structure and variability near the event horizon of Sagittarius A*". *Science* **350** (6265): 1242–1245. arXiv:1512.01220. Bibcode:2015Sci...350.1242J. doi:10.1126/science.aac7087. ISSN 0036-8075.

[104] "Event Horizon Telescope Reveals Magnetic Fields at Milky Way's Central Black Hole". *cfa.harvard.edu*. 3 December 2015. Retrieved 12 January 2016.

[105] Overbye, Dennis (11 February 2016). "Physicists Detect Gravitational Waves, Proving Einstein Right". *New York Times*. Retrieved 11 February 2016.

[106] Abbott, Benjamin P.; et al. (LIGO Scientific Collaboration and Virgo Collaboration) (11 February 2016). "Properties of the binary black hole merger GW150914". arXiv:1602.03840.

[107] Cardoso, V., Franzin, E., and Pani, P. (2016). "Is the gravitational-wave ringdown a probe of the event horizon?". *Physical Review Letters*. arXiv:1602.07309. Bibcode:2016PhRvL.116f1102A. doi:10.1103/PhysRevLett.116.061102.

[108] Abbott, Benjamin P.; et al. (LIGO Scientific Collaboration and Virgo Collaboration) (11 February 2016). "Tests of general relativity with GW150914". LIGO. Retrieved 12 February 2016.

[109] "Astrophysical Implications of the Binary Black Hole Merger GW150914". *Astrophys. J. Lett.* **818** (2): L22. arXiv:1602.03846. Bibcode:2016ApJ...818L..22A. doi:10.3847/2041-8205/818/2/L22.

[110] Gillessen, S.; Eisenhauer, F.; Trippe, S.; et al. (2009). "Monitoring Stellar Orbits around the Massive Black Hole in the Galactic Center". *The Astrophysical Journal* **692** (2): 1075. arXiv:0810.4674. Bibcode:2009ApJ...692.1075G. doi:10.1088/0004-637X/692/2/1075.

[111] Ghez, A. M.; Klein, B. L.; Morris, M.; et al. (1998). "High Proper-Motion Stars in the Vicinity of Sagittarius A*: Evidence for a Supermassive Black Hole at the Center of Our Galaxy". *The Astrophysical Journal* **509** (2): 678. arXiv:astro-ph/9807210. Bibcode:1998ApJ...509..678G. doi:10.1086/306528.

[112] Broderick, Avery; Loeb, Abraham; Narayan, Ramesh (August 2009). "The Event Horizon of Sagittarius A*". *The Astrophysical Journal* **701**: 1357. arXiv:0903.1105. Bibcode:2009ApJ...701.1357B. doi:10.1088/0004-637X/701/2/1357.

[113] "NASA's NuSTAR Sees Rare Blurring of Black Hole Light". *NASA*. 12 August 2014. Retrieved 12 August 2014.

[114] "Short-cut method of solution of geodesic equations for Schwarzchild black hole", J.A. Marck, Class.Quant. Grav. 13 (1996) 393-402.

[115] McClintock, J. E.; Remillard, R. A. (2006). "Black Hole Binaries". In Lewin, W.; van der Klis, M. *Compact Stellar X-ray Sources*. Cambridge University Press. arXiv:astro-ph/0306213. ISBN 0-521-82659-4. section 4.1.5.

[116] Celotti, A.; Miller, J. C.; Sciama, D. W. (1999). "Astrophysical evidence for the existence of black holes". *Classical and Quantum Gravity* **16** (12A): A3–A21. arXiv:astro-ph/9912186. doi:10.1088/0264-9381/16/12A/301.

[117] Winter, L. M.; Mushotzky, R. F.; Reynolds, C. S. (2006). "XMM-Newton Archival Study of the Ultraluminous X-Ray Population in Nearby Galaxies". *The Astrophysical Journal* **649** (2): 730. arXiv:astro-ph/0512480. Bibcode:2006ApJ...649..730W. doi:10.1086/506579.

[118] information@eso.org. "Hubble directly observes the disc around a black hole". *www.spacetelescope.org*. Retrieved 2016-03-07.

[119] Muñoz, José A.; Mediavilla, Evencio; Kochanek, Christopher S.; Falco, Emilio; Mosquera, Ana María (2011-12-01). "A Study of Gravitational Lens Chromaticity with the Hubble Space Telescope". *The Astrophysical Journal* **742** (2): 67. arXiv:1107.5932. Bibcode:2011ApJ...742...67M. doi:10.1088/0004-637X/742/2/67. ISSN 0004-637X.

[120] Bolton, C. T. (1972). "Identification of Cygnus X-1 with HDE 226868". *Nature* **235** (5336): 271–273. Bibcode:1972Natur.235..271B. doi:10.1038/235271b0.

[121] Webster, B. L.; Murdin, P. (1972). "Cygnus X-1—a Spectroscopic Binary with a Heavy Companion ?". *Nature* **235** (5332): 37–38. Bibcode:1972Natur.235..37W. doi:10.1038/235037a0.

[122] Rolston, B. (10 November 1997). "The First Black Hole". *The bulletin*. University of Toronto. Archived from the original on 2008-05-02. Retrieved 2008-03-11.

[123] Shipman, H. L.; Yu, Z; Du, Y.W (1 January 1975). "The implausible history of triple star models for Cygnus X-1 Evidence for a black hole". *Astrophysical Letters* **16** (1): 9–12. Bibcode:1975ApL....16....9S. doi:10.1016/S0304-8853(99)00384-4.

[124] Narayan, R.; McClintock, J. (2008). "Advection-dominated accretion and the black hole event horizon". *New Astronomy Reviews* **51** (10–12): 733. arXiv:0803.0322. Bibcode:2008NewAR..51..733N. doi:10.1016/j.newar.2008.03.002.

[125] "NASA scientists identify smallest known black hole" (Press release). Goddard Space Flight Center. 2008-04-01. Retrieved 2009-03-14.

[126] Krolik, J. H. (1999). *Active Galactic Nuclei*. Princeton University Press. Ch. 1.2. ISBN 0-691-01151-6.

[127] Sparke, L. S.; Gallagher, J. S. (2000). *Galaxies in the Universe: An Introduction*. Cambridge University Press. Ch. 9.1. ISBN 0-521-59740-4.

[128] Chou, Felicia; Anderson, Janet; Watzke, Megan (5 January 2015). "RELEASE 15-001 - NASA's Chandra Detects Record-Breaking Outburst from Milky Way's Black Hole". *NASA*. Retrieved 6 January 2015.

[129] Kormendy, J.; Richstone, D. (1995). "Inward Bound—The Search For Supermassive Black Holes In Galactic Nuclei". *Annual Review of Astronomy and Astrophysics* **33** (1): 581–624. Bibcode:1995ARA&A..33..581K. doi:10.1146/annurev.aa.33.090195.003053.

[130] King, A. (2003). "Black Holes, Galaxy Formation, and the MBH-σ Relation". *The Astrophysical Journal Letters* **596** (1): 27–29. arXiv:astro-ph/0308342. Bibcode:2003ApJ...596L..27K. doi:10.1086/379143.

[131] Ferrarese, L.; Merritt, D. (2000). "A Fundamental Relation Between Supermassive Black Holes and their Host Galaxies". *The Astrophysical Journal Letters* **539** (1): 9–12. arXiv:astro-ph/0006053. Bibcode:2000ApJ...539L...9F. doi:10.1086/312838.

[132] "A Black Hole's Dinner is Fast Approaching". *ESO Press Release*. Retrieved 6 February 2012.

[133] Bozza, V. (2010). "Gravitational Lensing by Black Holes". *General Relativity and Gravitation* **42** (42): 2269–2300. arXiv:0911.2187. Bibcode:2010GReGr..42.2269B. doi:10.1007/s10714-010-0988-2.

[134] Kovacs, Z.; Cheng, K. S.; Harko, T. (2009). "Can stellar mass black holes be quark stars?". *Monthly Notices of the Royal Astronomical Society* **400** (3): 1632–1642. arXiv:0908.2672. Bibcode:2009MNRAS.400.1632K. doi:10.1111/j.1365-2966.2009.15571.x.

[135] Kusenko, A. (2006). "Properties and signatures of supersymmetric Q-balls". arXiv:hep-ph/0612159.

[136] Hansson, J.; Sandin, F. (2005). "Preon stars: a new class of cosmic compact objects". *Physics Letters B* **616** (1–2): 1. arXiv:astro-ph/0410417. Bibcode:2005PhLB..616....1H. doi:10.1016/j.physletb.2005.04.034.

[137] Kiefer, C. (2006). "Quantum gravity: general introduction and recent developments". *Annalen der Physik* **15** (1–2): 129. arXiv:gr-qc/0508120. Bibcode:2006AnP...518..129K. doi:10.1002/andp.200510175.

[138] "[NKS, Mathur states, 't Hooft-Polyakov monopoles, and Ward-Takahashi identities] - A New Kind of Science: The NKS Forum". *wolframscience.com*. Retrieved 12 April 2015.

[139] Skenderis, K.; Taylor, M. (2008). "The fuzzball proposal for black holes". *Physics Reports* **467** (4–5): 117. arXiv:0804.0552. Bibcode:2008PhR...467..117S. doi:10.1016/j.physrep.2008.08.001.

[140] Hawking, S. W. (1971). "Gravitational Radiation from Colliding Black Holes". *Physical Review Letters* **26** (21): 1344–1346. Bibcode:1971PhRvL..26.1344H. doi:10.1103/PhysRevLett.26.1344.

[141] Wald, R. M. (2001). "The Thermodynamics of Black Holes". *Living Reviews in Relativity* **4**: 6. arXiv:gr-qc/9912119. Bibcode:2001LRR.....4....6W. doi:10.12942/lrr-2001-6. Retrieved 2011-02-10.

[142] 't Hooft, G. (2001). "The Holographic Principle". In Zichichi, A. *Basics and highlights in fundamental physics*. Subnuclear series **37**. World Scientific. arXiv:hep-th/0003004. ISBN 978-981-02-4536-8.

[143] Strominger, A.; Vafa, C. (1996). "Microscopic origin of the Bekenstein-Hawking entropy". *Physics Letters B* **379** (1–4): 99. arXiv:hep-th/9601029. Bibcode:1996PhLB..379...99S. doi:10.1016/0370-2693(96)00345-0.

[144] Carlip, S. (2009). "Black Hole Thermodynamics and Statistical Mechanics". *Lecture Notes in Physics*. Lecture Notes in Physics **769**: 89. arXiv:0807.4520. doi:10.1007/978-3-540-88460-6_3. ISBN 978-3-540-88459-0.

[145] Hawking, S. W. "Does God Play Dice?". *www.hawking.org.uk*. Retrieved 2009-03-14.

[146] Giddings, S. B. (1995). "The black hole information paradox". *Particles, Strings and Cosmology*. Johns Hopkins Workshop on Current Problems in Particle Theory 19 and the PASCOS Interdisciplinary Symposium 5. arXiv:hep-th/9508151.

[147] Mathur, S. D. (2011). *The information paradox: conflicts and resolutions*. XXV International Symposium on Lepton Photon Interactions at High Energies. arXiv:1201.2079.

1.9.9 Further reading

Popular reading

- Ferguson, Kitty (1991). *Black Holes in Space-Time*. Watts Franklin. ISBN 0-531-12524-6.

- Hawking, Stephen (1988). *A Brief History of Time*. Bantam Books, Inc. ISBN 0-553-38016-8.

- Hawking, Stephen; Penrose, Roger (1996). *The Nature of Space and Time*. Princeton University Press. ISBN 0-691-03791-4.

- Melia, Fulvio (2003). *The Black Hole at the Center of Our Galaxy*. Princeton U Press. ISBN 978-0-691-09505-9.

- Melia, Fulvio (2003). *The Edge of Infinity. Supermassive Black Holes in the Universe*. Cambridge U Press. ISBN 978-0-521-81405-8.

- Pickover, Clifford (1998). *Black Holes: A Traveler's Guide*. Wiley, John & Sons, Inc. ISBN 0-471-19704-1.

- Thorne, Kip S. (1994). *Black Holes and Time Warps*. Norton, W. W. & Company, Inc. ISBN 0-393-31276-3.

- Wheeler, J. Craig (2007). *Cosmic Catastrophes* (2nd ed.). Cambridge University Press. ISBN 0-521-85714-7.

University textbooks and monographs

- Carroll, Sean M. (2004). *Spacetime and Geometry*. Addison Wesley. ISBN 0-8053-8732-3., the lecture notes on which the book was based are available for free from Sean Carroll's website.

- Carter, B. (1973). "Black hole equilibrium states". In DeWitt, B. S.; DeWitt, C. *Black Holes*.

- Chandrasekhar, Subrahmanyan (1999). *Mathematical Theory of Black Holes*. Oxford University Press. ISBN 0-19-850370-9.

- Frolov, V. P.; Novikov, I. D. (1998). "Black hole physics".

- Frolov, Valeri P.; Zelnikov, Andrei (2011). *Introduction to Black Hole Physics*. Oxford: Oxford University Press. ISBN 978-0-19-969229-3. Zbl 1234.83001.

- Hawking, S. W.; Ellis, G. F. R. (1973). *Large Scale Structure of space time*. Cambridge University Press. ISBN 0-521-09906-4.

- Melia, Fulvio (2007). *The Galactic Supermassive Black Hole*. Princeton U Press. ISBN 978-0-691-13129-0.

- Misner, Charles; Thorne, Kip S.; Wheeler, John (1973). *Gravitation*. W. H. Freeman and Company. ISBN 0-7167-0344-0.

- Taylor, Edwin F.; Wheeler, John Archibald (2000). *Exploring Black Holes*. Addison Wesley Longman. ISBN 0-201-38423-X.

- Wald, Robert M. (1984). *General Relativity*. University of Chicago Press. ISBN 978-0-226-87033-5.

- Wald, Robert M. (1992). *Space, Time, and Gravity: The Theory of the Big Bang and Black Holes*. University of Chicago Press. ISBN 0-226-87029-4.

- Black holes Teviet Creighton, Richard H. Price Scholarpedia 3(1):4277. doi:10.4249/scholarpedia.4277

Review papers

- Gallo, Elena; Marolf, Donald (2009). "Resource Letter BH-2: Black Holes". *American Journal of Physics* 77 (4): 294. arXiv:0806.2316. Bibcode:2009AmJPh..77..294G. doi:10.1119/1.3056569.

- Hughes, Scott A. (2005). "Trust but verify: The case for astrophysical black holes". arXiv:hep-ph/0511217. Lecture notes from 2005 SLAC Summer Institute.

1.9.10   External links

-

- Black Holes on *In Our Time* at the BBC. (listen now)

- Stanford Encyclopedia of Philosophy: "Singularities and Black Holes" by Erik Curiel and Peter Bokulich.

- Black Holes: Gravity's Relentless Pull—Interactive multimedia Web site about the physics and astronomy of black holes from the Space Telescope Science Institute

- Frequently Asked Questions (FAQs) on Black Holes

- "Schwarzschild Geometry"

- Advanced Mathematics of Black Hole Evaporation

- Hubble site

Videos

- 16-year-long study tracks stars orbiting Milky Way black hole

- Movie of Black Hole Candidate from Max Planek Institute

- Nature.com 2015-04-20 3D simulations of colliding black holes

- Computer visualisation of the signal detected by LIGO

- Two Black Holes Merge into One (based upon the signal GW150914

1.10   Hierarchy problem

In theoretical physics, the **hierarchy problem** is the large discrepancy between aspects of the weak force and gravity.[1] There is no scientific consensus on why, for example, the weak force is 10^{32} times stronger than gravity.

1.10.1   Technical definition

A hierarchy problem occurs when the fundamental value of some physical parameter, such as a coupling constant or a mass, in some Lagrangian is vastly different from its effective value, which is the value that gets measured in an experiment. This happens because the effective value is related to the fundamental value by a prescription known as renormalization, which applies corrections to it. Typically the renormalized value of parameters are close to their fundamental values, but in some cases, it appears that there has been a delicate cancellation between the fundamental quantity and the quantum corrections. Hierarchy problems are related to fine-tuning problems and problems of naturalness.

Studying renormalization in hierarchy problems is difficult, because such quantum corrections are usually power-law divergent, which means that the shortest-distance physics are most important. Because we do not know the precise details of the shortest-distance theory of physics, we cannot even address how this delicate cancellation between two large terms occurs. Therefore, researchers are led to postulate new physical phenomena that resolve hierarchy problems without fine tuning.

1.10.2 The Higgs mass

In particle physics, the most important **hierarchy problem** is the question that asks why the weak force is 10^{32} times stronger than gravity. Both of these forces involve constants of nature, Fermi's constant for the weak force and Newton's constant for gravity. Furthermore, if the Standard Model is used to calculate the quantum corrections to Fermi's constant, it appears that Fermi's constant is surprisingly large and is expected to be closer to Newton's constant, unless there is a delicate cancellation between the bare value of Fermi's constant and the quantum corrections to it.

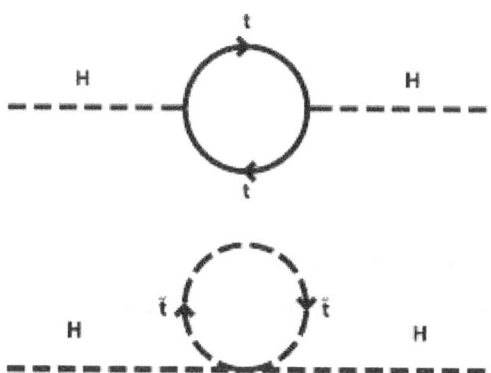

Cancellation of the Higgs boson quadratic mass renormalization between fermionic top quark loop and scalar stop squark tadpole Feynman diagrams in a supersymmetric extension of the Standard Model

More technically, the question is why the Higgs boson is so much lighter than the Planck mass (or the grand unification energy, or a heavy neutrino mass scale): one would expect that the large quantum contributions to the square of the Higgs boson mass would inevitably make the mass huge, comparable to the scale at which new physics appears, unless there is an incredible fine-tuning cancellation between the quadratic radiative corrections and the bare mass.

It should be remarked that the problem cannot even be formulated in the strict context of the Standard Model, for the Higgs mass cannot be calculated. In a sense, the problem amounts to the worry that a future theory of fundamental particles, in which the Higgs boson mass will be calculable, should not have excessive fine-tunings.

One proposed solution, popular amongst many physicists, is that one may solve the hierarchy problem via supersymmetry. Supersymmetry can explain how a tiny Higgs mass can be protected from quantum corrections. Supersymmetry removes the power-law divergences of the radiative corrections to the Higgs mass and solves the hierarchy problem as long as the supersymmetric particles are light enough to satisfy the Barbieri–Giudice criterion.[2] This still leaves open the mu problem, however. Currently the tenets of supersymmetry are being tested at the LHC, although no evidence has been found so far for supersymmetry.

1.10.3 Theoretical solutions

Supersymmetric solution

Each particle that couples to the Higgs field has a Yukawa coupling λ_f. The coupling with the Higgs field for fermions gives an interaction term $\mathcal{L}_{\text{Yukawa}} = -\lambda_f \bar{\psi} H \psi$, with ψ being the Dirac Field and H the Higgs Field. Also, the mass of a fermion is proportional to its Yukawa coupling, meaning that the Higgs boson will couple most to the most massive particle. This means that the most significant corrections to the Higgs mass will originate from the heaviest particles, most prominently the top quark. By applying the Feynman rules, one gets the quantum corrections to the Higgs mass squared from a fermion to be:

$$\Delta m_H^2 = -\frac{|\lambda_f|^2}{8\pi^2}[\Lambda_{\text{UV}}^2 + ...].$$

The Λ_{UV} is called the ultraviolet cutoff and is the scale up to which the Standard Model is valid. If we take this scale to be the Planck scale, then we have the quadratically diverging Lagrangian. However, suppose there existed two complex scalars (taken to be spin 0) such that:

$\lambda_S = |\lambda_f|^2$ (the couplings to the Higgs are exactly the same).

Then by the Feynman rules, the correction (from both scalars) is:

$$\Delta m_H^2 = 2 \times \frac{\lambda_S}{16\pi^2}[\Lambda_{\text{UV}}^2 + ...].$$

(Note that the contribution here is positive. This is because of the spin-statistics theorem, which means that fermions will have a negative contribution and bosons a positive contribution. This fact is exploited.)

This gives a total contribution to the Higgs mass to be zero if we include both the fermionic and bosonic particles. Supersymmetry is an extension of this that creates 'superpartners' for all Standard Model particles.

This section adapted from Stephen P. Martin's "A Supersymmetry Primer" on arXiv.[3]

Conformal solution

Without supersymmetry, a solution to the hierarchy problem has been proposed using just the Standard Model. The idea can be traced back to the fact that the term in the Higgs field that produces the uncontrolled quadratic correction upon renormalization is the quadratic one. If the Higgs field had no mass term, then no hierarchy problem arises. But by missing a quadratic term in the Higgs field, one must find a way to recover the breaking of electroweak symmetry through a non-null vacuum expectation value. This can be obtained using the Weinberg–Coleman mechanism with terms in the Higgs potential arising from quantum corrections. Mass obtained in this way is far too small with respect to what is seen in accelerator facilities and so a conformal Standard Model needs more than one Higgs particle. This proposal has been put forward in 2006 by Krzysztof Meissner and Hermann Nicolai[4] and is currently under scrutiny. But if no further excitation is observed beyond the one seen so far at LHC, this model would have to be abandoned.

Solution via extra dimensions

If we live in a 3+1 dimensional world, then we calculate the Gravitational Force via Gauss' law for gravity:

$$\mathbf{g}(\mathbf{r}) = -Gm\frac{\mathbf{e_r}}{r^2}$$

which is simply Newton's law of gravitation. Note that Newton's constant G can be rewritten in terms of the Planck mass.

$$\frac{1}{M_{\text{Pl}}^2}$$

If we extend this idea to δ extra dimensions, then we get:

$$\mathbf{g}(\mathbf{r}) = -m\frac{\mathbf{e_r}}{M_{\text{Pl}_{3+1+\delta}}^{2+\delta}\,r^{2+\delta}}$$

where $M_{\text{Pl}_{3+1+\delta}}$ is the $3+1+\delta$ dimensional Planck mass. However, we are assuming that these extra dimensions are the same size as the normal 3+1 dimensions. Let us say that the extra dimensions are of size $n \lll$ than normal dimensions. If we let $r \ll n$, then we get (2). However, if we let $r \gg n$, then we get our usual Newton's law. However, when $r \gg n$, the flux in the extra dimensions becomes a constant, because there is no extra room for gravitational flux to flow through. Thus the flux will be proportional to n^δ because this is the flux in the extra dimensions. The formula is:

$$\mathbf{g}(\mathbf{r}) = -m\frac{\mathbf{e_r}}{M_{\text{Pl}_{3+1+\delta}}^{2+\delta}\,r^2 n^\delta}$$

$$-m\frac{\mathbf{e_r}}{M_{\text{Pl}}^2 r^2} = -m\frac{\mathbf{e_r}}{M_{\text{Pl}_{3+1+\delta}}^{2+\delta}\,r^2 n^\delta}$$

which gives:

$$\frac{1}{M_{\text{Pl}}^2 r^2} = \frac{1}{M_{\text{Pl}_{3+1+\delta}}^{2+\delta}\,r^2 n^\delta} \Rightarrow$$

$$M_{\text{Pl}}^2 = M_{\text{Pl}_{3+1+\delta}}^{2+\delta}\,n^\delta.$$

Thus the fundamental Planck mass (the extra dimensional one) could actually be small, meaning that gravity is actually strong, but this must be compensated by the number of the extra dimensions and their size. Physically, this means that gravity is weak because there is a loss of flux to the extra dimensions.

This section adapted from "Quantum Field Theory in a Nutshell" by A. Zee.[5]

Braneworld models   Main article: Brane cosmology

In 1998 Nima Arkani-Hamed, Savas Dimopoulos, and Gia Dvali proposed the **ADD model**, also known as the model with large extra dimensions, an alternative scenario to explain the weakness of gravity relative to the other forces.[6][7] This theory requires that the fields of the Standard Model are confined to a four-dimensional membrane, while gravity propagates in several additional spatial dimensions that are large compared to the Planck scale.[8]

In 1998/99 Merab Gogberashvili published on the arXiv (and subsequently in peer-reviewed journals) a number of articles where he showed that if the Universe is considered as a thin shell (a mathematical synonym for "brane") expanding in 5-dimensional space then it is possible to obtain one scale for particle theory corresponding to the 5-dimensional cosmological constant and Universe thickness, and thus to solve the hierarchy problem.[9][10][11] It was also shown that four-dimensionality of the Universe is the result of stability requirement since the extra component of the Einstein field equations giving the localized solution for matter fields coincides with one of the conditions of stability.

Subsequently, there were proposed the closely related Randall–Sundrum scenarios which offered their solution to the hierarchy problem.

Finite Groups It has also been noted that the group order of the Baby Monster group is of the right order of magnitude, 4×10^{33}. It is known that the Monster Group is related to the symmetries of a particular bosonic string theory on the Leech lattice. However, there's no physical reason for why the size of the Monster Group or its subgroups should appear in the Lagrangian. Most physicists think this is merely a coincidence. Another coincidence is that in *reduced* Planck units, the Higgs mass is approximately $48.|M|^{-1/3} = 125.5$ GeV where $|M|$ is the order of the Monster group. This suggests that the smallness of the Higgs mass may be due to a redundancy caused by a symmetry of the extra dimensions, which must be divided out. There are other groups that are also of the right order of magnitude for example $Weyl(E_8 \times E_8)$.

Extra dimensions Until now, no experimental or observational evidence of extra dimensions has been officially reported. Analyses of results from the Large Hadron Collider severely constrain theories with large extra dimensions.[12] However, extra dimensions could explain why the gravity force is so weak, and why the expansion of the universe is faster than expected.[13]

1.10.4 The cosmological constant

In physical cosmology, current observations in favor of an accelerating universe imply the existence of a tiny, but nonzero cosmological constant. This is a hierarchy problem very similar to that of the Higgs boson mass problem, since the cosmological constant is also very sensitive to quantum corrections. It is complicated, however, by the necessary involvement of general relativity in the problem and may be a clue that we do not understand gravity on long distance scales (such as the size of the universe today). While quintessence has been proposed as an explanation of the acceleration of the Universe, it does not actually address the cosmological constant hierarchy problem in the technical sense of addressing the large quantum corrections. Supersymmetry does not address the cosmological constant problem, since supersymmetry cancels the $M^4 P_{lanck}$ contribution, but not the $M^2 P_{lanck}$ one (quadratically diverging).

1.10.5 See also

- CP violation

- Little hierarchy problem

- Quantum triviality

1.10.6 References

[1] "The Hierarchy Problem | Of Particular Significance". *Profmattstrassler.com*. Retrieved 2015-12-13.

[2] R. Barbieri; G. F. Giudice (1988). "Upper Bounds on Supersymmetric Particle Masses". *Nucl. Phys.* B **306**: 63. Bibcode:1988NuPhB.306...63B. doi:10.1016/0550-3213(88)90171-X.

[3] Stephen P. Martin, A Supersymmetry Primer

[4] K. Meissner; H. Nicolai (2006). "Conformal Symmetry and the Standard Model". *Physics Letters* **B648**: 312–317. arXiv:hep-th/0612165. Bibcode:2007PhLB..648..312M. doi:10.1016/j.physletb.2007.03.023.

[5] Zee, A. (2003). "Quantum field theory in a nutshell". Princeton University Press. Bibcode:2003qftn.book.....Z.

[6] N. Arkani-Hamed; S. Dimopoulos; G. Dvali (1998). "The Hierarchy problem and new dimensions at a millimeter". *Physics Letters* **B429**: 263–272. arXiv:hep-ph/9803315. Bibcode:1998PhLB..429..263A. doi:10.1016/S0370-2693(98)00466-3.

[7] N. Arkani-Hamed; S. Dimopoulos; G. Dvali (1999). "Phenomenology, astrophysics and cosmology of theories with submillimeter dimensions and TeV scale quantum gravity". *Physical Review* **D59**: 086004. arXiv:hep-ph/9807344. Bibcode:1999PhRvD..59h6004A. doi:10.1103/PhysRevD.59.086004.

[8] For a pedagogical introduction, see M. Shifman (2009). *Large Extra Dimensions: Becoming acquainted with an alternative paradigm*. Crossing the boundaries: Gauge dynamics at strong coupling. Singapore: World Scientific. arXiv:0907.3074.

[9] M. Gogberashvili, *Hierarchy problem in the shell universe model*, Arxiv:hep-ph/9812296.

[10] M. Gogberashvili, *Our world as an expanding shell*, Arxiv: hep-ph/9812365.

[11] M. Gogberashvili, *Four dimensionality in noncompact Kaluza-Klein model*. Arxiv:hep-ph/9904383.

[12] "[1311.2006] Search for Quantum Black-Hole Production in High-Invariant-Mass Lepton+Jet Final States Using Proton-Proton Collisions at sqrt(s) = 8 TeV and the ATLAS Detector". *Arxiv.org*. doi:10.1103/PhysRevLett.112.091804. Retrieved 2015-12-13.

[13] "Extra dimensions, gravitons, and tiny black holes". *Home.web.cern.ch.* 20 January 2012. Retrieved 2015-12-13.

1.11   Quantum triviality

In a quantum field theory, charge screening can restrict the value of the observable "renormalized" charge of a classical theory. If the only allowed value of the renormalized charge is zero, the theory is said to be "trivial" or noninteracting. Thus, surprisingly, a classical theory that appears to describe interacting particles can, when realized as a quantum field theory, become a "trivial" theory of noninteracting free particles. This phenomenon is referred to as **quantum triviality**. Strong evidence supports the idea that a field theory involving only a scalar Higgs boson is trivial in four spacetime dimensions,[1][2] but the situation for realistic models including other particles in addition to the Higgs boson is not known in general. Nevertheless, because the Higgs boson plays a central role in the Standard Model of particle physics, the question of triviality in Higgs models is of great importance.

This Higgs triviality is similar to the Landau pole problem in quantum electrodynamics, where this quantum theory may be inconsistent at very high momentum scales unless the renormalized charge is set to zero, i.e., unless the field theory has no interactions. The Landau pole question is generally considered to be of minor academic interest for quantum electrodynamics because of the inaccessibly large momentum scale at which the inconsistency appears. This is not however the case in theories that involve the elementary scalar Higgs boson, as the momentum scale at which a "trivial" theory exhibits inconsistencies may be accessible to present experimental efforts such as at the LHC. In these Higgs theories, the interactions of the Higgs particle with itself are posited to generate the masses of the W and Z bosons, as well as lepton masses like those of the electron and muon. If realistic models of particle physics such as the Standard Model suffer from triviality issues, the idea of an elementary scalar Higgs particle may have to be modified or abandoned.

The situation becomes more complex in theories that involve other particles however. In fact, the addition of other particles can turn a trivial theory into a nontrivial one, at the cost of introducing constraints. Depending on the details of the theory, the Higgs mass can be bounded or even predictable.[2] These quantum triviality constraints are in sharp contrast to the picture one derives at the classical level, where the Higgs mass is a free parameter.

1.11.1   Triviality and the renormalization group

Modern considerations of triviality are usually formulated in terms of the real-space renormalization group, largely developed by Kenneth Wilson and others. Investiga-tions of triviality are usually performed in the context of lattice gauge theory. A deeper understanding of the physical meaning and generalization of the renormalization process, which goes beyond the dilatation group of conventional *renormalizable* theories, came from condensed matter physics. Leo P. Kadanoff's paper in 1966 proposed the "block-spin" renormalization group.[3] The *blocking idea* is a way to define the components of the theory at large distances as aggregates of components at shorter distances.

This approach covered the conceptual point and was given full computational substance[4] in the extensive important contributions of Kenneth Wilson. The power of Wilson's ideas was demonstrated by a constructive iterative renormalization solution of a long-standing problem, the Kondo problem, in 1974, as well as the preceding seminal developments of his new method in the theory of second-order phase transitions and critical phenomena in 1971. He was awarded the Nobel prize for these decisive contributions in 1982.

In more technical terms, let us assume that we have a theory described by a certain function Z of the state variables $\{s_i\}$ and a certain set of coupling constants $\{J_k\}$. This function may be a partition function, an action, a Hamiltonian, etc. It must contain the whole description of the physics of the system.

Now we consider a certain blocking transformation of the state variables $\{s_i\} \rightarrow \{\tilde{s}_i\}$, the number of $\tilde{s}_i$ must be lower than the number of s_i. Now let us try to rewrite the Z function *only* in terms of the $\tilde{s}_i$. If this is achievable by a certain change in the parameters, $\{J_k\} \rightarrow \{\tilde{J}_k\}$, then the theory is said to be **renormalizable**. The most important information in the RG flow are its **fixed points**. The possible macroscopic states of the system, at a large scale, are given by this set of fixed points. If these fixed points correspond to a free field theory, the theory is said to be **trivial**.

1.11.2   Historical background

The first evidence of possible triviality of quantum field theories was obtained by Landau, Abrikosov, and Khalatnikov[5][6][7] who obtained the following relation of the observable charge g_{obs} with the "bare" charge g_0.

where m is the mass of the particle, and Λ is the momentum cut-off. If g_0 is finite, then g_{obs} tends to zero in the limit of infinite cut-off Λ.

In fact, the proper interpretation of Eq.1 consists in its inversion, so that g_0 (related to the length scale $1/\Lambda$) is chosen to give a correct value of g_{obs}.

The growth of g_0 with Λ invalidates Eqs. (1) and (2) in the region $g_0 \approx 1$ (since they were obtained for $g_0 \ll 1$) and the existence of the "Landau pole" in Eq.2 has no physical meaning.

The actual behavior of the charge $g(\mu)$ as a function of the momentum scale μ is determined by the full Gell-Mann–Low equation

which gives Eqs.(1),(2) if it is integrated under conditions $g(\mu) = g_{obs}$ for $\mu = m$ and $g(\mu) = g_0$ for $\mu = \Lambda$, when only the term with β_2 is retained in the right hand side.

The general behavior of $g(\mu)$ relies on the appearance of the function $\beta(g)$. According to the classification by Bogoliubov and Shirkov,[8] there are three qualitatively different situations:

1. if $\beta(g)$ has a zero at the finite value g*, then growth of g is saturated, i.e. $g(\mu) \to g*$ for $\mu \to \infty$;

2. if $\beta(g)$ is non-alternating and behaves as $\beta(g) \propto g^\alpha$ with $\alpha \leq 1$ for large g , then the growth of $g(\mu)$ continues to infinity;

3. if $\beta(g) \propto g^\alpha$ with $\alpha > 1$ for large g , then $g(\mu)$ is divergent at finite value μ_0 and the real Landau pole arises: the theory is internally inconsistent due to indeterminacy of $g(\mu)$ for $\mu > \mu_0$.

The latter case corresponds to the quantum triviality in the full theory (beyond its perturbation context), as can be seen by reductio ad absurdum. Indeed, if g_{obs} is finite, the theory is internally inconsistent. The only way to avoid it, is to tend μ_0 to infinity, which is possible only for $g_{obs} \to 0$.

1.11.3 Conclusions

As a result, the question of whether the Standard Model of particle physics is nontrivial (and whether elementary scalar Higgs particles can exist) remains a serious unresolved question. Theoretical proofs evidencing triviality of the scalar field theory have appeared, but the nature of the quantum field theories interacting with it remains an open question[9][10][11] implications for the Standard Model and the resulting Higgs Boson mass bounds have also been discussed.[12][13][14]

1.11.4 See also

- Hierarchy problem

1.11.5 References

[1] R. Fernandez, J. Froehlich, A. D. Sokal (1992). Random Walks, Critical Phenomena, and Triviality in Quantum Field Theory. Springer. ISBN 0-387-54358-9.

[2] D. J. E. Callaway (1988). "Triviality Pursuit: Can Elementary Scalar Particles Exist?". Physics Reports 167 (5): 241–320. Bibcode:1988PhR...167..241C. doi:10.1016/0370-1573(88)90008-7.

[3] L.P. Kadanoff (1966): "Scaling laws for Ising models near T_c". Physics (Long Island City, N.Y.) 2, 263.

[4] K.G. Wilson(1975): The renormalization group: critical phenomena and the Kondo problem. Rev. Mod. Phys. 47, 4, 773.

[5] L. D. Landau, A. A. Abrikosov, and I. M. Khalatnikov (1954). Doklady Akademii Nauk SSSR 95: 497. Missing or empty |title= (help)

[6] L. D. Landau; A. A. Abrikosov & I. M. Khalatnikov (1954). Doklady Akademii Nauk SSSR 95: 773. Missing or empty |title= (help)

[7] L. D. Landau; A. A. Abrikosov & I. M. Khalatnikov (1954). Doklady Akademii Nauk SSSR 95: 1177. Missing or empty |title= (help)

[8] N. N. Bogoliubov; D. V. Shirkov (1980). Introduction to the Theory of Quantized Fields (3rd ed.). John Wiley & Sons. ISBN 978-0-471-04223-5.

[9] Callaway, D.; Petronzio, R. (1987). "Is the standard model Higgs mass predictable?". Nuclear Physics B 292: 497. Bibcode:1987NuPhB.292..497C. doi:10.1016/0550-3213(87)90657-2.

[10] I. M. Suslov (2010). "Asymptotic Behavior of the β Function in the φ^4 Theory: A Scheme Without Complex Parameters". Journal of Experimental and Theoretical Physics 111 (3): 450. arXiv:1010.4317. Bibcode:2010JETP..111..450S. doi:10.1134/S1063776110090153.

[11] Frasca, Marco (2011). Mapping theorem and Green functions in Yang-Mills theory (PDF). The many faces of QCD. Trieste: Proceedings of Science. p. 039. arXiv:1011.3643. Retrieved 2011-08-27.

[12] Callaway, D. J. E. (1984). "Non-triviality of gauge theories with elementary scalars and upper bounds on Higgs masses". Nuclear Physics B 233 (2): 189. Bibcode:1984NuPhB.233..189C. doi:10.1016/0550-3213(84)90410-3.

[13] Lindner, M. (1986). "Implications of triviality for the standard model". Zeitschrift für Physik C 31: 295. Bibcode:1986ZPhyC..31..295L. doi:10.1007/BF01479540.

[14] Urs Heller, Markus Klomfass, Herbert Neuberger, and Pavlos Vranas, (1993). "Numerical analysis of the Higgs mass triviality bound". Nucl. Phys., B405: 555-573.

1.12   CP violation

In particle physics, **CP violation** (CP standing for **charge parity**) is a violation of the postulated **CP-symmetry** (or **charge conjugation parity symmetry**): the combination of C-symmetry (charge conjugation symmetry) and P-symmetry (parity symmetry). CP-symmetry states that the laws of physics should be the same if a particle is interchanged with its antiparticle (C symmetry), and when its spatial coordinates are inverted ("mirror" or P symmetry). The discovery of CP violation in 1964 in the decays of neutral kaons resulted in the Nobel Prize in Physics in 1980 for its discoverers James Cronin and Val Fitch.

It plays an important role both in the attempts of cosmology to explain the dominance of matter over antimatter in the present Universe, and in the study of weak interactions in particle physics.

1.12.1   CP-symmetry

CP-symmetry, often called just *CP*, is the product of two symmetries: C for charge conjugation, which transforms a particle into its antiparticle, and P for parity, which creates the mirror image of a physical system. The strong interaction and electromagnetic interaction seem to be invariant under the combined CP transformation operation, but this symmetry is slightly violated during certain types of weak decay. Historically, CP-symmetry was proposed to restore order after the discovery of parity violation in the 1950s.

The idea behind parity symmetry is that the equations of particle physics are invariant under mirror inversion. This leads to the prediction that the mirror image of a reaction (such as a chemical reaction or radioactive decay) occurs at the same rate as the original reaction. Parity symmetry appears to be valid for all reactions involving electromagnetism and strong interactions. Until 1956, parity conservation was believed to be one of the fundamental geometric conservation laws (along with conservation of energy and conservation of momentum). However, in 1956 a careful critical review of the existing experimental data by theoretical physicists Tsung-Dao Lee and Chen Ning Yang revealed that while parity conservation had been verified in decays by the strong or electromagnetic interactions, it was untested in the weak interaction. They proposed several possible direct experimental tests. The first test based on beta decay of cobalt-60 nuclei was carried out in 1956 by a group led by Chien-Shiung Wu, and demonstrated conclusively that weak interactions violate the P symmetry or, as the analogy goes, some reactions did not occur as often as their mirror image.

Overall, the symmetry of a quantum mechanical system can be restored if another symmetry S can be found such that the combined symmetry PS remains unbroken. This rather subtle point about the structure of Hilbert space was realized shortly after the discovery of P violation, and it was proposed that charge conjugation was the desired symmetry to restore order.

Simply speaking, charge conjugation is a symmetry between particles and antiparticles, and so CP-symmetry was proposed in 1957 by Lev Landau as the true symmetry between matter and antimatter. In other words, a process in which all particles are exchanged with their antiparticles was assumed to be equivalent to the mirror image of the original process.

CP violation in the Standard Model

"Direct" CP violation is allowed in the Standard Model if a complex phase appears in the CKM matrix describing quark mixing, or the PMNS matrix describing neutrino mixing. A necessary condition for the appearance of the complex phase is the presence of at least three generations of quarks (if fewer generations are present, the complex phase parameter can be absorbed into redefinitions of the quark fields).

The reason why such a complex phase causes CP violation is not immediately obvious, but can be seen as follows. Consider any given particles (or sets of particles) a and b, and their antiparticles $\bar{a}$ and $\bar{b}$. Now consider the processes $a \to b$ and the corresponding antiparticle process $\bar{a} \to \bar{b}$, and denote their amplitudes M and $\bar{M}$ respectively. Before CP violation, these terms must be the *same* complex number. We can separate the magnitude and phase by writing $M = |M|e^{i\theta}$. If a phase term is introduced from (e.g.) the CKM matrix, denote it $e^{i\phi}$. Note that $\bar{M}$ contains the conjugate matrix to M, so it picks up a phase term $e^{-i\phi}$. Now we have:

$$M = |M|e^{i\theta}e^{i\phi}$$

$$\bar{M} = |M|e^{i\theta}e^{-i\phi}$$

Physically measurable reaction rates are proportional to $|M|^2$, thus so far nothing is different. However, consider that there are *two different routes* (e.g. intermediate states) for $a \to b$. Now we have:

$$M = |M_1|e^{i\theta_1}e^{i\phi_1} + |M_2|e^{i\theta_2}e^{i\phi_2}$$

$$\bar{M} = |M_1|e^{i\theta_1}e^{-i\phi_1} + |M_2|e^{i\theta_2}e^{-i\phi_2}$$

Some further calculation gives:

$$|M|^2 - |\bar{M}|^2 = 4|M_1||M_2|\sin(\theta_1 - \theta_2)\sin(\phi_1 - \phi_2)$$

Thus, we see that a complex phase gives rise to processes that proceed at different rates for particles and antiparticles, and CP is violated.

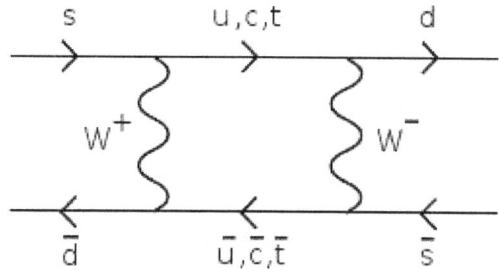

The two box diagrams above are the Feynman diagrams providing the leading contributions to the amplitude of K0-K0 oscillation

1.12.2 Experimental status

Indirect CP violation

In 1964, James Cronin, Val Fitch and coworkers provided clear evidence (which was first announced at the 12th ICHEP conference in Dubna) that CP-symmetry could be broken. This work[1] won them the 1980 Nobel Prize. This discovery showed that weak interactions violate not only the charge-conjugation symmetry C between particles and antiparticles and the P or parity, but also their combination. The discovery shocked particle physics and opened the door to questions still at the core of particle physics and of cosmology today. The lack of an exact CP-symmetry, but also the fact that it is so nearly a symmetry, created a great puzzle.

Only a weaker version of the symmetry could be preserved by physical phenomena, which was CPT symmetry. Besides C and P, there is a third operation, time reversal (T), which corresponds to reversal of motion. Invariance under time reversal implies that whenever a motion is allowed by the laws of physics, the reversed motion is also an allowed one. The combination of CPT is thought to constitute an exact symmetry of all types of fundamental interactions. Because of the CPT symmetry, a violation of the CP-symmetry is equivalent to a violation of the T symmetry. CP violation implied nonconservation of T, provided that the long-held CPT theorem was valid. In this theorem, regarded as one of the basic principles of quantum field theory, charge conjugation, parity, and time reversal are applied together.

Direct CP violation

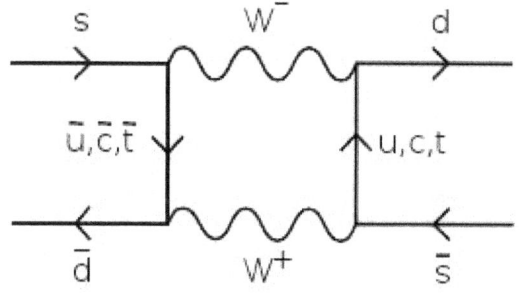

Kaon oscillation box diagram

The kind of CP violation discovered in 1964 was linked to the fact that neutral kaons can transform into their antiparticles (in which each quark is replaced with the other's antiquark) and vice versa, but such transformation does not occur with exactly the same probability in both directions; this is called *indirect* CP violation. Despite many searches, no other manifestation of CP violation was discovered until the 1990s, when the NA31 experiment at CERN suggested evidence for CP violation in the decay process of the very same neutral kaons (*direct* CP violation). The observation was somewhat controversial, and final proof for it came in 1999 from the KTeV experiment at Fermilab[2] and the NA48 experiment at CERN.[3]

In 2001, a new generation of experiments, including the BaBar Experiment at the Stanford Linear Accelerator Center (SLAC)[4] and the Belle Experiment at the High Energy Accelerator Research Organisation (KEK)[5] in Japan, observed direct CP violation in a different system, namely in decays of the B mesons.[6] A large number of CP violation processes in B meson decays have now been discovered. Before these "B-factory" experiments, there was a logical possibility that all CP violation was confined to kaon physics. However, this raised the question of why it's *not* extended to the strong force, and furthermore, why this is not predicted in the unextended Standard Model, despite the model being undeniably accurate with "normal" phenomena.

In 2011, a hint of CP violation in decays of neutral D mesons was reported by the LHCb experiment at CERN using 0.6 fb⁻¹ of Run 1 data.[7] However, the same measurement using the full 3.0 fb⁻¹ Run 1 sample was consistent with CP symmetry.[8]

In 2013 LHCb announced discovery of CP violation in strange B meson decays,[9] so did BaBar and Belle scientists in 2015.[10]

1.12.3   Strong CP problem

There is no experimentally known violation of the CP-symmetry in quantum chromodynamics. As there is no known reason for it to be conserved in QCD specifically, this is a "fine tuning" problem known as the strong CP problem.

QCD does not violate the CP-symmetry as easily as the electroweak theory; unlike the electroweak theory in which the gauge fields couple to chiral currents constructed from the fermionic fields, the gluons couple to vector currents. Experiments do not indicate any CP violation in the QCD sector. For example, a generic CP violation in the strongly interacting sector would create the electric dipole moment of the neutron which would be comparable to 10^{-18} e·m while the experimental upper bound is roughly one trillionth that size.

This is a problem because at the end, there are natural terms in the QCD Lagrangian that are able to break the CP-symmetry.

$$\mathcal{L} = -\frac{1}{4}F_{\mu\nu}F^{\mu\nu} - \frac{n_f g^2 \theta}{32\pi^2}F_{\mu\nu}\tilde{F}^{\mu\nu} + \bar{\psi}(i\gamma^\mu D_\mu - m e^{i\theta'\gamma_5})\psi$$

For a nonzero choice of the θ angle and the chiral phase of the quark mass θ′ one expects the CP-symmetry to be violated. One usually assumes that the chiral quark mass phase can be converted to a contribution to the total effective $\bar{\theta}$ angle, but it remains to be explained why this angle is extremely small instead of being of order one; the particular value of the θ angle that must be very close to zero (in this case) is an example of a fine-tuning problem in physics, and is typically solved by physics beyond the Standard Model.

There are several proposed solutions to solve the strong CP problem. The most well-known is Peccei–Quinn theory, involving new scalar particles called axions. A newer, more radical approach not requiring the axion is a theory involving two time dimensions first proposed in 1998 by Bars, Deliduman, and Andreev.[11]

Little CP problem

The little CP problem is a term coined by Lisa Randall. It refers to an issue related to the enhanced new physics contributions to the electric dipole moment (EDM) of the neutron in flavor anarchic models.[12]

1.12.4   CP violation and the matter–antimatter imbalance

Main article: Baryogenesis

The universe is made chiefly of matter, rather than consisting of equal parts of matter and antimatter as might be expected. It can be demonstrated that, to create an imbalance in matter and antimatter from an initial condition of balance, the Sakharov conditions must be satisfied, one of which is the existence of CP violation during the extreme conditions of the first seconds after the Big Bang. Explanations which do not involve CP violation are less plausible, since they rely on the assumption that the matter–antimatter imbalance was present at the beginning, or on other admittedly exotic assumptions.

The Big Bang should have produced equal amounts of matter and antimatter if CP-symmetry was preserved; as such, there should have been total cancellation of both— protons should have cancelled with antiprotons, electrons with positrons, neutrons with antineutrons, and so on. This would have resulted in a sea of radiation in the universe with no matter. Since this is not the case, after the Big Bang, physical laws must have acted differently for matter and antimatter, i.e. violating CP-symmetry.

The Standard Model contains at least three sources of CP violation. The first of these, involving the Cabibbo–Kobayashi–Maskawa matrix in the quark sector, has been observed experimentally and can only account for a small portion of the CP violation required to explain the matter-antimatter asymmetry. The strong interaction should also violate CP, in principle, but the failure to observe the electric dipole moment of the neutron in experiments suggests that any CP violation in the strong sector is also too small to account for the necessary CP violation in the early universe. The third source of CP violation is the Pontecorvo–Maki–Nakagawa–Sakata matrix in the lepton sector. Current neutrino experiments are not yet sensitive enough to allow experimental observation of CP violation in the lepton sector, but the NOvA experiment currently under construction could observe some small fraction of possible CP violating phases and proposed neutrino experiments Hyper-Kamiokande and LBNE will be sensitive to a relatively large fraction of CP violating phases. Further into the future, a neutrino factory could be sensitive to nearly all possible CP violating phases. If neutrinos are Majorana fermions, the PMNS matrix could have two independent CP violating phases leading to a fourth source of CP violation within the Standard Model. The experimental evidence for Majorana neutrinos would be the observation of neutrinoless double-beta decay. As of September 2013, the best limits come from the GERDA experiment. CP violation in the lepton sector generates a matter-antimatter asymmetry through a process called leptogenesis. This could become the preferred explanation in the Standard Model for the matter-antimatter asymmetry of the universe once CP violation is experimentally confirmed in the lepton sector.

If CP violation in the lepton sector is experimentally determined to be too small to account for matter-antimatter asymmetry, some new physics beyond the Standard Model would be required to explain additional sources of CP violation. Fortunately, it is generally the case that adding new particles and/or interactions to the Standard Model introduces new sources of CP violation since CP is not a symmetry of nature.

1.12.5 See also

- B-factory

- LHCb

- BTeV experiment

- Cabibbo–Kobayashi–Maskawa matrix

- Penguin diagram

- Neutral particle oscillation

1.12.6 References

[1] "Evidence for the 2π Decay of the K0 2 Meson System". *Physical Review Letters* **13**: 138. 1964. Bibcode:1964PhRvL..13..138C. doi:10.1103/PhysRevLett.13.138.

[2] "Observation of Direct CP Violation in KS,L→ππ Decays". *Physical Review Letters* **83**: 22. 1999. arXiv:hep-ex/9905060. Bibcode:1999PhRvL..83...22A. doi:10.1103/PhysRevLett.83.22.

[3] NA48 Collaboration, V. Fanti, A. Lai, D. Marras, L. Musa; et al. (1999). "A new measurement of direct CP violation in two pion decays of the neutral kaon". *Physics Letters B* **465** (1–4): 335–348. arXiv:hep-ex/9909022. Bibcode:1999PhLB..465..335F. doi:10.1016/S0370-2693(99)01030-8.

[4] "Measurement of CP-Violating Asymmetries in B⁰ Decays to CP Eigenstates". *Physical Review Letters* **86**: 2515. 2001. arXiv:hep-ex/0102030. Bibcode:2001PhRvL..86.2515A. doi:10.1103/PhysRevLett.86.2515.

[5] "Observation of Large CP Violation in the Neutral B Meson System". *Physical Review Letters* **87**. 2001. arXiv:hep-ex/0107061. Bibcode:2001PhRvL..87i1802A. doi:10.1103/PhysRevLett.87.091802.

[6] Rodgers, Peter (August 2001). "Where did all the antimatter go?". *Physics World*. p. 11.

[7] Carbone, A. (2012). "A search for time-integrated CP violation in D⁰→h⁻h⁺ decays". arXiv:1210.8257.

[8] LHCb Collaboration (2014). "Measurement of CP asymmetry in D⁰→K⁺K⁻ and D⁰→π⁺π⁻ decays". *JHEP* **7**: 41. arXiv:1405.2797. Bibcode:2014JHEP...07..041A. doi:10.1007/JHEP07(2014)041.

[9] http://journals.aps.org/prl/pdf/10.1103/PhysRevLett.110.221601

[10] http://authors.library.caltech.edu/61145/2/1505.04147v2.pdf

[11] I. Bars; C. Deliduman; O. Andreev (1998). "Gauged Duality, Conformal Symmetry, and Spacetime with Two Times". *Physical Review D* **58** (6): 066004. arXiv:hep-th/9803188. Bibcode:1998PhRvD..58f6004B. doi:10.1103/PhysRevD.58.066004.

[12] Kadosh, Avihay; Pallante, Elisabetta (2011). "CP violation and FCNC in a warped A_4 flavor model". *Journal of High Energy Physics* **2011** (6). arXiv:1101.5420. Bibcode:2011JHEP...06..121K. doi:10.1007/JHEP06(2011)121.

1.12.7 Further reading

- Sozzi, M.S. (2008). *Discrete symmetries and CP violation*. Oxford University Press. ISBN 978-0-19-929666-8.

- G. C. Branco; L. Lavoura; J. P. Silva (1999). *CP violation*. Clarendon Press. ISBN 0-19-850399-7.

- I. Bigi; A. Sanda (1999). *CP violation*. Cambridge University Press. ISBN 0-521-44349-0.

- Michael Beyer, ed. (2002). *CP Violation in Particle, Nuclear and Astrophysics*. Springer. ISBN 3-540-43705-3. (*A collection of essays introducing the subject, with an emphasis on experimental results.*)

- L. Wolfenstein (1989). *CP violation*. North–Holland Publishing. ISBN 0-444-88081-X. (*A compilation of reprints of numerous important papers on the topic, including papers by T.D. Lee, Cronin, Fitch, Kobayashi and Maskawa, and many others.*)

- David J. Griffiths (1987). *Introduction to Elementary Particles*. John Wiley & Sons. ISBN 0-471-60386-4.

- Bigi, I. (1997). "CP Violation – An Essential Mystery in Nature's Grand Design". *Surveys of High Energy Physics* **12**: 269–336. arXiv:hep-ph/9712475. Bibcode:1997hep.ph..12475B. doi:10.1080/01422419808228861.

- Mark Trodden (1998). "Electroweak Baryogenesis". *Reviews of Modern Physics* **71** (5): 1463. arXiv:hep-ph/9803479.   Bibcode:1999RvMP...71.1463T. doi:10.1103/RevModPhys.71.1463.

- Davide Castelvecchi. "What is direct CP-violation?". SLAC. Retrieved 2009-07-01.

1.12.8   External links

- Cern Courier article

1.13   Cosmological constant

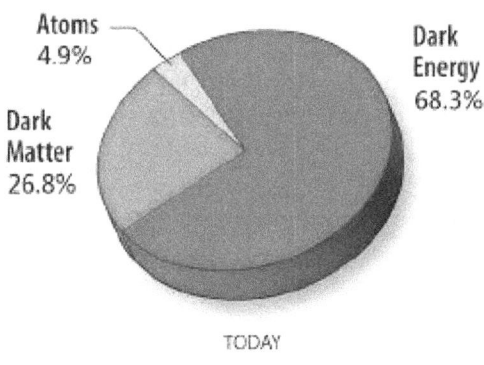

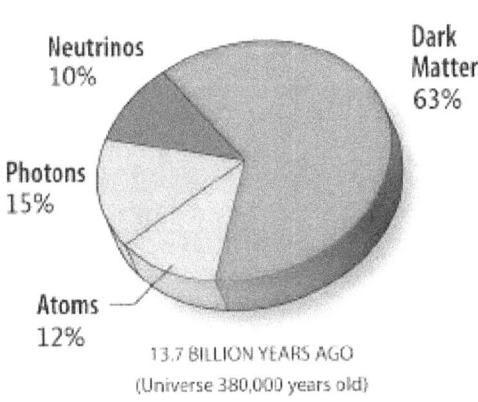

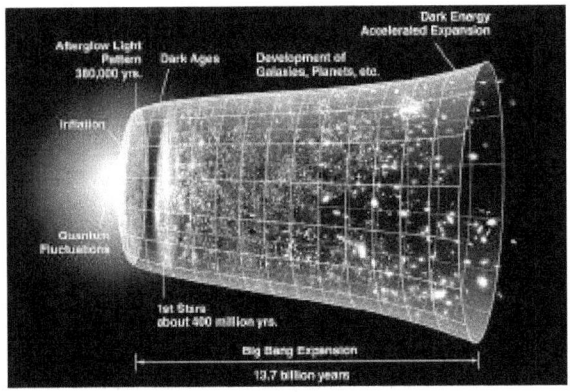

Sketch of the timeline of the universe in the ΛCDM model. The accelerated expansion in the last third of the timeline represents the dark-energy dominated era.

Estimated ratios of dark matter and dark energy (which may be the cosmological constant) in the universe. According to current theories of physics, dark energy now dominates as the largest source of energy of the universe, in contrast to earlier epochs when it was insignificant.

In cosmology, the **cosmological constant** (usually denoted by the Greek capital letter lambda: Λ) is the value of the energy density of the vacuum of space. It was originally introduced by Albert Einstein in 1917[1] as an addition to his theory of general relativity to "hold back gravity" and achieve a static universe, which was the accepted view at the time. Einstein abandoned the concept after Hubble's 1929 discovery that all galaxies outside the Local Group (the group that contains the Milky Way Galaxy) are moving away from each other, implying an overall expanding universe. From 1929 until the early 1990s, most cosmology researchers assumed the cosmological constant to be zero.

Since the 1990s, several developments in observational cosmology, especially the discovery of the accelerating universe from distant supernovae in 1998 (in addition to independent evidence from the cosmic microwave background and large galaxy redshift surveys), have shown that around 68% of the mass–energy density of the universe can be attributed to dark energy.[2] While dark energy is poorly understood at a fundamental level, the main required proper-ties of dark energy are that it functions as a type of anti-gravity, it dilutes much more slowly than matter as the universe expands, and it clusters much more weakly than matter, or perhaps not at all. The cosmological constant is the simplest possible form of dark energy since it is constant in both space and time, and this leads to the current standard model of cosmology known as the Lambda-CDM model, which provides a good fit to many cosmological observations as of 2016.

1.13.1   Equation

The cosmological constant Λ appears in Einstein's field equation in the form of

$$R_{\mu\nu} - \frac{1}{2}R\,g_{\mu\nu} + \Lambda\,g_{\mu\nu} = \frac{8\pi G}{c^4}T_{\mu\nu},$$

where R and g describe the structure of spacetime, T pertains to matter and energy affecting that structure, and G

and c are conversion factors that arise from using traditional units of measurement. When Λ is zero, this reduces to the original field equation of general relativity. When T is zero, the field equation describes empty space (the vacuum).

The cosmological constant has the same effect as an intrinsic energy density of the vacuum, ϱ_{vac} (and an associated pressure). In this context, it is commonly moved onto the right-hand side of the equation, and defined with a proportionality factor of 8π: $\Lambda = 8\pi\varrho_{vac}$, where unit conventions of general relativity are used (otherwise factors of G and c would also appear, i.e. $\Lambda = 8\pi \, (G/c^2)\varrho_{vac} = \kappa \, \varrho_{vac}$, where κ is Einstein's constant). It is common to quote values of energy density directly, though still using the name "cosmological constant", with convention $8\pi \, G = 1$. (In fact, the true dimension of Λ is a length^{-2} and it has the value of ~$1 \, 10^{-52}$ m^{-2} or in reduced Planck units : ~$3 \, 10^{-122}$, calculated with the best present (2015) values of $\Omega\Lambda = 0.6911 \pm 0.0062$ and $H_0 = 67.74 \pm 0.46$ km/s / Mpc $= 2.195 \pm 0.015 \, 10^{-18}$ s^{-1}).

A positive vacuum energy density resulting from a cosmological constant implies a negative pressure, and vice versa. If the energy density is positive, the associated negative pressure will drive an accelerated expansion of the universe, as observed. (See dark energy and cosmic inflation for details.)

$\Omega\Lambda$ (Omega Lambda)

Instead of the cosmological constant itself, cosmologists often refer to the ratio between the energy density due to the cosmological constant and the critical density of the universe, the tipping point for a sufficient density to stop the universe from expanding forever. This ratio is usually denoted $\Omega\Lambda$, and is estimated to be 0.6911 ± 0.0062, according to results published by the Planck Collaboration in 2015.[3]

In a flat universe $\Omega\Lambda$ is the fraction of the energy of the universe due to the cosmological constant, i.e., what we would intuitively call the fraction of the universe that is made up of dark energy. Note that this value changes over time: the critical density changes with cosmological time, but the energy density due to the cosmological constant remains unchanged throughout the history of the universe: the amount of dark energy increases as the universe grows, while the amount of matter does not.

Equation of state

Another ratio that is used by scientists is the equation of state, usually denoted w, which is the ratio of pressure that dark energy puts on the universe to the energy per unit volume.[4] This ratio is $w = -1$ for a true cosmological constant, and is generally different for alternative time-varying forms of vacuum energy such as quintessence.

1.13.2 History

Einstein included the cosmological constant as a term in his field equations for general relativity because he was dissatisfied that otherwise his equations did not allow, apparently, for a static universe: gravity would cause a universe that was initially at dynamic equilibrium to contract. To counteract this possibility, Einstein added the cosmological constant.[5] However, soon after Einstein developed his static theory, observations by Edwin Hubble indicated that the universe appears to be expanding; this was consistent with a cosmological solution to the *original* general relativity equations that had been found by the mathematician Friedmann, working on the Einstein equations of general relativity. Einstein later reputedly referred to his failure to accept the validation of his equations—when they had predicted the expansion of the universe in theory, before it was demonstrated in observation of the cosmological red shift—as the "biggest blunder" of his life.[6][7]

In fact, adding the cosmological constant to Einstein's equations does not lead to a static universe at equilibrium because the equilibrium is unstable: if the universe expands slightly, then the expansion releases vacuum energy, which causes yet more expansion. Likewise, a universe that contracts slightly will continue contracting.[8]:59

However, the cosmological constant remained a subject of theoretical and empirical interest. Empirically, the onslaught of cosmological data in the past decades strongly suggests that our universe has a positive cosmological constant.[5] The explanation of this small but positive value is an outstanding theoretical challenge (*see the section below*).

Finally, it should be noted that some early generalizations of Einstein's gravitational theory, known as classical unified field theories, either introduced a cosmological constant on theoretical grounds or found that it arose naturally from the mathematics. For example, Sir Arthur Stanley Eddington claimed that the cosmological constant version of the vacuum field equation expressed the "epistemological" property that the universe is "self-gauging", and Erwin Schrödinger's pure-affine theory using a simple variational principle produced the field equation with a cosmological term.

1.13.3   Positive value

Observations announced in 1998 of distance–redshift relation for Type Ia supernovae[9][10] indicated that the expansion of the universe is accelerating. When combined with measurements of the cosmic microwave background radiation these implied a value of $\Omega\Lambda \approx 0.7$,[11] a result which has been supported and refined by more recent measurements. There are other possible causes of an accelerating universe, such as quintessence, but the cosmological constant is in most respects the simplest solution. Thus, the current standard model of cosmology, the Lambda-CDM model, includes the cosmological constant, which is measured to be on the order of 10^{-52} m^{-2}, in metric units. Multiplied by other constants that appear in the equations, it is often expressed as 10^{-52} m^{-2}, 10^{-35} s^{-2}, 10^{-47} GeV^4, 10^{-29} g/cm^3.[12] In terms of Planck units, and as a natural dimensionless value, the cosmological constant, Λ, is on the order of 10^{-122}.[13]

As was only recently seen, by works of 't Hooft, Susskind[14] and others, a positive cosmological constant has surprising consequences, such as a finite maximum entropy of the observable universe (see the holographic principle).

1.13.4   Predictions

Quantum field theory

See also: Cosmological constant problem

A major outstanding problem is that most quantum field theories predict a huge value for the quantum vacuum. A common assumption is that the quantum vacuum is equivalent to the cosmological constant. Although no theory exists that supports this assumption, arguments can be made in its favor.[15]

Such arguments are usually based on dimensional analysis and effective field theory. If the universe is described by an effective local quantum field theory down to the Planck scale, then we would expect a cosmological constant of the order of M_{pl}^4. As noted above, the measured cosmological constant is smaller than this by a factor of 10^{-120}. This discrepancy has been called "the worst theoretical prediction in the history of physics!".[16]

Some supersymmetric theories require a cosmological constant that is exactly zero, which further complicates things. This is the *cosmological constant problem*, the worst problem of fine-tuning in physics: there is no known natural way to derive the tiny cosmological constant used in cosmology from particle physics.

Anthropic principle

One possible explanation for the small but non-zero value was noted by Steven Weinberg in 1987 following the anthropic principle.[17] Weinberg explains that if the vacuum energy took different values in different domains of the universe, then observers would necessarily measure values similar to that which is observed: the formation of life-supporting structures would be suppressed in domains where the vacuum energy is much larger. Specifically, if the vacuum energy is negative and its absolute value is substantially larger than it appears to be in the observed universe (say, a factor of 10 larger), holding all other variables (e.g. matter density) constant, that would mean that the universe is closed; furthermore, its lifetime would be shorter than the age of our universe, possibly too short for intelligent life to form. On the other hand, a universe with a large positive cosmological constant would expand too fast, preventing galaxy formation. According to Weinberg, domains where the vacuum energy is compatible with life would be comparatively rare. Using this argument, Weinberg predicted that the cosmological constant would have a value of less than a hundred times the currently accepted value.[18] In 1992, Weinberg refined this prediction of the cosmological constant to 5 to 10 times the matter density.[19]

This argument depends on a lack of a variation of the distribution (spatial or otherwise) in the vacuum energy density, as would be expected if dark energy were the cosmological constant. There is no evidence that the vacuum energy does vary, but it may be the case if, for example, the vacuum energy is (even in part) the potential of a scalar field such as the residual inflaton (also see quintessence). Another theoretical approach that deals with the issue is that of multiverse theories, which predict a large number of "parallel" universes with different laws of physics and/or values of fundamental constants. Again, the anthropic principle states that we can only live in one of the universes that is compatible with some form of intelligent life. Critics claim that these theories, when used as an explanation for fine-tuning, commit the inverse gambler's fallacy.

In 1995, Weinberg's argument was refined by Alexander Vilenkin to predict a value for the cosmological constant that was only ten times the matter density,[20] i.e. about three times the current value since determined.

Cyclic model

More recent work has suggested the problem may be indirect evidence of a cyclic universe possibly as allowed by string theory. With every cycle of the universe (Big Bang then eventually a Big Crunch) taking about a trillion (10^{12}) years, "the amount of matter and radiation in the universe is reset, but the cosmological constant is not. Instead, the cosmological constant gradually diminishes over many cycles to the small value observed today."[21] Critics respond that, as the authors acknowledge in their paper, the model "entails ... the same degree of tuning required in any cosmological model".[22]

1.13.5 See also

- Higgs mechanism

- Lambdavacuum solution

- Naturalness (physics)

- Quantum electrodynamics

- de Sitter relativity

- Unruh effect

1.13.6 References

[1] Einstein, A (1917). "Kosmologische Betrachtungen zur allgemeinen Relativitaetstheorie". *Sitzungsberichte der Königlich Preussischen Akademie der Wissenschaften Berlin*. part 1: 142–152.

[2] What is Dark Energy?, Space.com, 1 May 2013

[3] Collaboration, Planck, PAR Ade, N Aghanim, C Armitage-Caplan, M Arnaud, et al., Planck 2015 results. XIII. Cosmological parameters. arXiv preprint 1502.1589v2 , 6 Feb 2015.

[4] Hogan, Jenny (2007). "Welcome to the Dark Side". *Nature* **448** (7151): 240–245. Bibcode:2007Natur.448..240H. doi:10.1038/448240a. PMID 17637630.

[5] Urry, Meg (2008). *The Mysteries of Dark Energy*. Yale Science. Yale University.

[6] Gamov, George (1970). *My World Line*. Viking Press. p. 44. ISBN 978-0670503766

[7] Rosen, Rebecca J. "Einstein Likely Never Said One of His Most Oft-Quoted Phrases". *The Atlantic*. The Atlantic Media Company. Retrieved 10 August 2013.

[8] Barbara Sue Ryden (2003). *Introduction to cosmology*. Addison-Wesley. ISBN 978-0-8053-8912-8.

[9] Riess, A.; et al. (September 1998). "Observational Evidence from Supernovae for an Accelerating Universe and a Cosmological Constant". *The Astronomical Journal* **116** (3): 1009–1038. arXiv:astro-ph/9805201. Bibcode:1998AJ....116.1009R. doi:10.1086/300499.

[10] Perlmutter, S.; et al. (June 1999). "Measurements of Omega and Lambda from 42 High-Redshift Supernovae". *The Astrophysical Journal* **517** (2): 565–586. arXiv:astro-ph/9812133. Bibcode:1999ApJ...517..565P. doi:10.1086/307221.

[11] See e.g. Baker, Joanne C.; et al. (1999). "Detection of cosmic microwave background structure in a second field with the Cosmic Anisotropy Telescope". *Monthly Notices of the Royal Astronomical Society* **308** (4): 1173–1178. arXiv:astro-ph/9904415. Bibcode:1999MNRAS.308.1173B. doi:10.1046/j.1365-8711.1999.02829.x.

[12] Tegmark, Max; et al. (2004). "Cosmological parameters from SDSS and WMAP". *Physical Review D* **69** (103501): 103501. arXiv:astro-ph/0310723. Bibcode:2004PhRvD..69j3501T. doi:10.1103/PhysRevD.69.103501.

[13] John D. Barrow The Value of the Cosmological Constant

[14] Lisa Dyson, Matthew Kleban, Leonard Susskind: "Disturbing Implications of a Cosmological Constant"

[15] Rugh, S; Zinkernagel, H. (2001). "The Quantum Vacuum and the Cosmological Constant Problem". *Studies in History and Philosophy of Modern Physics* **33** (4): 663–705. doi:10.1016/S1355-2198(02)00033-3.

[16] MP Hobson; GP Efstathiou; AN Lasenby (2006). *General Relativity: An introduction for physicists* (Reprinted with corrections 2007 ed.). Cambridge University Press. p. 187. ISBN 978-0-521-82951-9.

[17] Weinberg, S (1987). "Anthropic Bound on the Cosmological Constant". *Phys. Rev. Lett.* **59** (22): 2607–2610. Bibcode:1987PhRvL.59.2607W. doi:10.1103/PhysRevLett.59.2607. PMID 10035596.

[18] Alexander Vilenkin, *Many Worlds in One: The Search for Other Universes*. ISBN 978-0-8090-9523-0, pp. 138–9

[19] Weinberg, Steven (1993). *Dreams of a Final Theory: the search for the fundamental laws of nature*. Vintage Press. p. 182. ISBN 0-09-922391-0.

[20] Alexander Vilenkin, *Many Worlds in One: The Search for Other universes*, ISBN 978-0-8090-9523-0, p. 146, which references Vilenkin' *Predictions from quantum cosmology*, Physical Review Letters, vol 74, p. 846 (1995)

[21] 'Cyclic universe' can explain cosmological constant, NewScientistSpace, 4 May 2006

[22] Steinhardt, P. J.; Turok, N. (2002-04-25). "A Cyclic Model of the Universe". *Science* **296** (5572): 1436–1439. arXiv:hep-th/0111030v2. Bibcode:2002Sci...296.1436S. doi:10.1126/science.1070462. PMID 11976408. Retrieved 2012-04-29.

- Michael, E., University of Colorado, Department of Astrophysical and Planetary Sciences, "The Cosmological Constant"

- Ferguson, Kitty (1991). *Stephen Hawking: Quest For A Theory of Everything*, Franklin Watts. ISBN 0-553-29895-X.

- John D. Barrow; John K. Webb (June 2005). "Inconstant Constants". *Scientific American*.

- *Beyond the Cosmological Standard Model*[1] (2014)

1.13.7   External links

- Cosmological constant (astronomy) at *Encyclopædia Britannica*

- Carroll, Sean M., *"The Cosmological Constant"* (short), *"The Cosmological Constant"* (extended).

- News story: More evidence for dark energy being the cosmological constant

- Cosmological constant article from Scholarpedia

- Copeland, Ed; Merrifield, Mike. "Λ – Cosmological Constant". *Sixty Symbols*. Brady Haran for the University of Nottingham.

[1] Austin Joyce; Bhuvnesh Jain; Justin Khoury; Mark Trodden (2014). "Beyond the Cosmological Standard Model".

Chapter 2

Quantum Gravity

2.1 Loop quantum gravity

Loop quantum gravity (**LQG**) is a theory that attempts to describe the quantum properties of the universe and gravity. It is also a theory of quantum spacetime because, according to general relativity, gravity is a manifestation of the geometry of spacetime. LQG is an attempt to merge quantum mechanics and general relativity.

From the point of view of Einstein's theory, it comes as no surprise that all attempts to treat gravity simply like one more quantum force (on par with electromagnetism and the nuclear forces) have failed. According to Einstein, gravity is not a force – it is a property of space-time itself. Loop quantum gravity is an attempt to develop a quantum theory of gravity based directly on Einstein's geometrical formulation. The main output of the theory is a physical picture of space where space is granular. The granularity is a direct consequence of the quantization. It has the same nature as the granularity of the photons in the quantum theory of electromagnetism and the discrete levels of the energy of the atoms. Here, it is space itself that is discrete. In other words, there is a minimum distance possible to travel through it.

More precisely, space can be viewed as an extremely fine fabric or network "woven" of finite loops. These networks of loops are called spin networks. The evolution of a spin network over time is called a spin foam. The predicted size of this structure is the Planck length, which is approximately 10^{-35} meters. According to the theory, there is no meaning to distance at scales smaller than the Planck scale. Therefore, LQG predicts that not just matter, but space itself, has an atomic structure.

Today LQG is a vast area of research, developing in several directions, which involves about 30 research groups worldwide.[1] They all share the basic physical assumptions and the mathematical description of quantum space. The full development of the theory is being pursued in two directions: the more traditional canonical loop quantum gravity, and the newer covariant loop quantum gravity, more commonly called spin foam theory.

Research into the physical consequences of the theory is proceeding in several directions. Among these, the most well-developed is the application of LQG to cosmology, called loop quantum cosmology (LQC). LQC applies LQG ideas to the study of the early universe and the physics of the Big Bang. Its most spectacular consequence is that the evolution of the universe can be continued beyond the Big Bang. The Big Bang appears thus to be replaced by a sort of cosmic Big Bounce.

2.1.1 History

Main article: History of loop quantum gravity

In 1986, Abhay Ashtekar reformulated Einstein's general relativity in a language closer to that of the rest of fundamental physics. Shortly after, Ted Jacobson and Lee Smolin realized that the formal equation of quantum gravity, called the Wheeler–DeWitt equation, admitted solutions labelled by loops when rewritten in the new Ashtekar variables. Carlo Rovelli and Lee Smolin defined a nonperturbative and background-independent quantum theory of gravity in terms of these loop solutions. Jorge Pullin and Jerzy Lewandowski understood that the intersections of the loops are essential for the consistency of the theory, and the theory should be formulated in terms of intersecting loops, or graphs.

In 1994, Rovelli and Smolin showed that the quantum operators of the theory associated to area and volume have a discrete spectrum. That is, geometry is quantized. This result defines an explicit basis of states of quantum geometry, which turned out to be labelled by Roger Penrose's spin networks, which are graphs labelled by spins.

The canonical version of the dynamics was put on firm ground by Thomas Thiemann, who defined an anomaly-free Hamiltonian operator, showing the existence of a mathematically consistent background-independent theory. The covariant or spinfoam version of the dynamics developed during several decades, and crystallized in 2008, from the joint work of research groups in France, Canada, UK, Poland, and Germany, lead to the definition of a family of transition amplitudes, which in the classical limit can be shown to be related to a family of truncations of general relativity.[2] The finiteness of these amplitudes was proven in 2011.[3][4] It requires the existence of a positive cosmological constant, and this is consistent with observed acceleration in the expansion of the Universe.

2.1.2 General covariance and background independence

Main articles: General covariance, background-independent and diffeomorphism

In theoretical physics, general covariance is the invariance of the form of physical laws under arbitrary differentiable coordinate transformations. The essential idea is that co-ordinates are only artifices used in describing nature, and hence should play no role in the formulation of fundamental physical laws. A more significant requirement is the principle of general relativity that states that the laws of physics take the same form in all reference systems. This is a generalization of the principle of special relativity which states that the laws of physics take the same form in all inertial frames.

In mathematics, a diffeomorphism is an isomorphism in the category of smooth manifolds. It is an invertible function that maps one differentiable manifold to another, such that both the function and its inverse are smooth. These are the defining symmetry transformations of General Relativity since the theory is formulated only in terms of a differentiable manifold.

In general relativity, general covariance is intimately related to "diffeomorphism invariance". This symmetry is one of the defining features of the theory. However, it is a common misunderstanding that "diffeomorphism invariance" refers to the invariance of the physical predictions of a theory under arbitrary coordinate transformations; this is untrue and in fact every physical theory is invariant under coordinate transformations this way. Diffeomorphisms, as mathematicians define them, correspond to something much more radical; intuitively a way they can be envisaged is as simultaneously dragging all the physical fields (including the gravitational field) over the bare differentiable manifold while staying in the same coordinate system. Diffeomorphisms are the true symmetry transformations of general relativity, and come about from the assertion that the formulation of the theory is based on a bare differentiable manifold, but not on any prior geometry — the theory is background-independent (this is a profound shift, as all physical theories before general relativity had as part of their formulation a prior geometry). What is preserved under such transformations are the coincidences between the values the gravitational field take at such and such a "place" and the values the matter fields take there. From these relationships one can form a notion of matter being located with respect to the gravitational field, or vice versa. This is what Einstein discovered: that physical entities are located with respect to one another only and not with respect to the spacetime manifold. As Carlo Rovelli puts it: "No more fields on spacetime: just fields on fields".[5] This is the true meaning of the saying "The stage disappears and becomes one of the actors"; space-time as a "container" over which physics takes place has no objective physical meaning and instead the gravitational interaction is represented as just one of the fields forming the world. This is known as the relationalist interpretation of space-time. The realization by Einstein that general relativity should be interpreted this way is the origin of his remark "Beyond my wildest expectations".

In LQG this aspect of general relativity is taken seriously and this symmetry is preserved by requiring that the physical states remain invariant under the generators of diffeomorphisms. The interpretation of this condition is well understood for purely spatial diffeomorphisms. However, the understanding of diffeomorphisms involving time (the Hamiltonian constraint) is more subtle because it is related to dynamics and the so-called "problem of time" in general relativity.[6] A generally accepted calculational framework to account for this constraint has yet to be found.[7][8] A plausible candidate for the quantum hamiltonian constraint is the operator introduced by Thiemann.[9]

LQG is formally background independent. The equations of LQG are not embedded in, or dependent on, space and time (except for its invariant topology). Instead, they are expected to give rise to space and time at distances which are large compared to the Planck length. The issue of background independence in LQG still has some unresolved subtleties. For example, some derivations require a fixed choice of the topology, while any consistent quantum theory of gravity should include topology change as a dynamical process.

2.1.3 Constraints and their Poisson bracket algebra

Main articles: Poisson bracket and Hamiltonian constraint

The constraints of classical canonical general relativity

Main article: Lie derivative

In the Hamiltonian formulation of ordinary classical mechanics the Poisson bracket is an important concept. A "canonical coordinate system" consists of canonical position and momentum variables that satisfy canonical Poisson-bracket relations,

$$\{q_i, p_j\} = \delta_{ij}$$

where the Poisson bracket is given by

$$\{f, g\} = \sum_{i=1}^{N} \left(\frac{\partial f}{\partial q_i} \frac{\partial g}{\partial p_i} - \frac{\partial f}{\partial p_i} \frac{\partial g}{\partial q_i} \right) .$$

for arbitrary phase space functions $f(q_i, p_j)$ and $g(q_i, p_j)$. With the use of Poisson brackets, the Hamilton's equations can be rewritten as,

$$\dot{q}_i = \{q_i, H\} ,$$
$$\dot{p}_i = \{p_i, H\} .$$

These equations describe a "flow" or orbit in phase space generated by the Hamiltonian H. Given any phase space function $F(q, p)$, we have

$$\frac{d}{dt} F(q_i, p_i) = \{F, H\}.$$

Let us consider constrained systems, of which General relativity is an example. In a similar way the Poisson bracket between a constraint and the phase space variables generates a flow along an orbit in (the unconstrained) phase space generated by the constraint. There are three types of constraints in Ashtekar's reformulation of classical general relativity:

$SU(2)$ Gauss gauge constraints   The Gauss constraints

$$G_j(x) = 0 .$$

This represents an infinite number of constraints one for each value of x. These come about from re-expressing General relativity as an $SU(2)$ Yang–Mills type gauge theory (Yang–Mills is a generalization of Maxwell's theory where the gauge field transforms as a vector under Gauss transformations, that is, the Gauge field is of the form $A_a^i(x)$ where i is an internal index. See Ashtekar variables). These infinite number of Gauss gauge constraints can be smeared with test fields with internal indices, $\lambda^j(x)$,

$$G(\lambda) = \int d^3x \, G_j(x) \lambda^j(x) .$$

which we demand vanish for any such function. These smeared constraints defined with respect to a suitable space of smearing functions give an equivalent description to the original constraints.

In fact Ashtekar's formulation may be thought of as ordinary $SU(2)$ Yang–Mills theory together with the following special constraints, resulting from diffeomorphism invariance, and a Hamiltonian that vanishes. The dynamics of such a theory are thus very different from that of ordinary Yang–Mills theory.

Spatial diffeomorphisms constraints   The spatial diffeomorphism constraints

$$C_a(x) = 0$$

can be smeared by the so-called shift functions $\vec{N}(x)$ to give an equivalent set of smeared spatial diffeomorphism constraints,

$$C(\vec{N}) = \int d^3x \, C_a(x) N^a(x) .$$

These generate spatial diffeomorphisms along orbits defined by the shift function $N^a(x)$.

Hamiltonian constraints   The Hamiltonian

$$H(x) = 0$$

can be smeared by the so-called lapse functions $N(x)$ to give an equivalent set of smeared Hamiltonian constraints,

$$H(N) = \int d^3x \, H(x) N(x) .$$

These generate time diffeomorphisms along orbits defined by the lapse function $N(x)$.

In Ashtekar formulation the gauge field $A_a^i(x)$ is the configuration variable (the configuration variable being analogous to q in ordinary mechanics) and its conjugate momentum is the (densitized) triad (electrical field) $\tilde{E}_i^a(x)$. The constraints are certain functions of these phase space variables.

We consider the action of the constraints on arbitrary phase space functions. An important notion here is the Lie derivative, $\mathcal{L}_V$, which is basically a derivative operation that infinitesimally "shifts" functions along some orbit with tangent vector V.

The Poisson bracket algebra

Of particular importance is the Poisson bracket algebra formed between the (smeared) constraints themselves as it completely determines the theory. In terms of the above smeared constraints the constraint algebra amongst the Gauss' law reads,

$$\{G(\lambda), G(\mu)\} = G([\lambda, \mu])$$

where $[\lambda, \mu]^k = \lambda_i \mu_j \epsilon^{ijk}$. And so we see that the Poisson bracket of two Gauss' law is equivalent to a single Gauss' law evaluated on the commutator of the smearings. The Poisson bracket amongst spatial diffeomorphisms constraints reads

$$\{C(\vec{N}), C(\vec{M})\} = C(\mathcal{L}_{\vec{N}}\vec{M})$$

and we see that its effect is to "shift the smearing". The reason for this is that the smearing functions are not functions of the canonical variables and so the spatial diffeomorphism does not generate diffeomorphims on them. They do however generate diffeomorphims on everything else. This is equivalent to leaving everything else fixed while shifting the smearing .The action of the spatial diffeomorphism on the Gauss law is

$$\{C(\vec{N}), G(\lambda)\} = G(\mathcal{L}_{\vec{N}}\lambda) .$$

again, it shifts the test field λ . The Gauss law has vanishing Poisson bracket with the Hamiltonian constraint. The spatial diffeomorphism constraint with a Hamiltonian gives a Hamiltonian with its smearing shifted.

$$\{C(\vec{N}), H(M)\} = H(\mathcal{L}_{\vec{N}}M) .$$

Finally, the poisson bracket of two Hamiltonians is a spatial diffeomorphism.

$$\{H(N), H(M)\} = C(K)$$

where K is some phase space function. That is, it is a sum over infinitesimal spatial diffeomorphisms constraints where the coefficients of proportionality are not constants but have non-trivial phase space dependence.

A (Poisson bracket) Lie algebra, with constraints C_I , is of the form

$$\{C_I, C_J\} = f_{IJ}^K C_K$$

where f_{IJ}^K are constants (the so-called structure constants). The above Poisson bracket algebra for General relativity does not form a true Lie algebra as we have structure functions rather than structure constants for the Poisson bracket between two Hamiltonians. This leads to difficulties.

Dirac observables

The constraints define a constraint surface in the original phase space. The gauge motions of the constraints apply to all phase space but have the feature that they leave the constraint surface where it is, and thus the orbit of a point in the hypersurface under gauge transformations will be an orbit entirely within it. Dirac observables are defined as phase space functions, O , that Poisson commute with all the constraints when the constraint equations are imposed,

$$\{G_j, O\}_{G_j=C_a=H=0} = \{C_a, O\}_{G_j=C_a=H=0} = \{H, O\}_{G_j=C_a=H=0} = 0 .$$

that is, they are quantities defined on the constraint surface that are invariant under the gauge transformations of the theory.

Then, solving only the constraint $G_j = 0$ and determining the Dirac observables with respect to it leads us back to the

ADM phase space with constraints H, C_a . The dynamics of general relativity is generated by the constraints, it can be shown that six Einstein equations describing time evolution (really a gauge transformation) can be obtained by calculating the Poisson brackets of the three-metric and its conjugate momentum with a linear combination of the spatial diffeomorphism and Hamiltonian constraint. The vanishing of the constraints, giving the physical phase space, are the four other Einstein equations.[10]

2.1.4 Quantization of the constraints – the equations of quantum general relativity

Pre-history and Ashtekar new variables

Main articles: Frame fields in general relativity, Ashtekar variables and Self-dual Palatini action

Many of the technical problems in canonical quantum gravity revolve around the constraints. Canonical general relativity was originally formulated in terms of metric variables, but there seemed to be insurmountable mathematical difficulties in promoting the constraints to quantum operators because of their highly non-linear dependence on the canonical variables. The equations were much simplified with the introduction of Ashtekars new variables. Ashtekar variables describe canonical general relativity in terms of a new pair canonical variables closer to that of gauge theories. The first step consists of using densitized triads $\tilde{E}_i^a$ (a triad E_i^a is simply three orthogonal vector fields labeled by $i = 1, 2, 3$ and the densitized triad is defined by $\tilde{E}_i^a = \sqrt{\det(q)}E_i^a$) to encode information about the spatial metric,

$$\det(q)q^{ab} = \tilde{E}_i^a \tilde{E}_j^b \delta^{ij} .$$

(where δ^{ij} is the flat space metric, and the above equation expresses that q^{ab} , when written in terms of the basis E_i^a , is locally flat). (Formulating general relativity with triads instead of metrics was not new.) The densitized triads are not unique, and in fact one can perform a local in space rotation with respect to the internal indices i . The canonically conjugate variable is related to the extrinsic curvature by $K_a^i = K_{ab}\tilde{E}^{ai}/\sqrt{\det(q)}$. But problems similar to using the metric formulation arise when one tries to quantize the theory. Ashtekar's new insight was to introduce a new configuration variable,

$$A_a^i = \Gamma_a^i - iK_a^i$$

that behaves as a complex SU(2) connection where Γ_a^i is related to the so-called spin connection via $\Gamma_a^i = \Gamma_{ajk}e^{jki}$. Here A_a^i is called the chiral spin connection. It defines a covariant derivative $\mathcal{D}_a$. It turns out that $\tilde{E}_i^a$ is the conjugate momentum of A_a^i , and together these form Ashtekar's new variables.

The expressions for the constraints in Ashtekar variables; the Gauss's law, the spatial diffeomorphism constraint and the (densitized) Hamiltonian constraint then read:

$$G^i = \mathcal{D}_a \tilde{E}^a_i = 0$$

$$C_a = \tilde{E}^b_i F^i_{ab} - A^i_a(\mathcal{D}_b \tilde{E}^b_i) = V_a - A^i_a G^i = 0,$$

$$\tilde{H} = \epsilon_{ijk} \tilde{E}^a_i \tilde{E}^b_j F^k_{ab} = 0$$

respectively. where F^i_{ab} is the field strength tensor of the connection A^i_a and where V_a is referred to as the vector constraint. The above-mentioned local in space rotational invariance is the original of the $SU(2)$ gauge invariance here expressed by the Gauss law. Note that these constraints are polynomial in the fundamental variables. unlike as with the constraints in the metric formulation. This dramatic simplification seemed to open up the way to quantizing the constraints. (See the article Self-dual Palatini action for a derivation of Ashtekar's formulism).

With Ashtekar's new variables, given the configuration variable A^i_a, it is natural to consider wavefunctions $\Psi(A^i_a)$. This is the connection representation. It is analogous to ordinary quantum mechanics with configuration variable q and wavefunctions $\psi(q)$. The configuration variable gets promoted to a quantum operator via:

$$\hat{A}^i_a \Psi(A) = A^i_a \Psi(A).$$

(analogous to $\hat{q}\psi(q) = q\psi(q)$) and the triads are (functional) derivatives,

$$\hat{\tilde{E}}^a_i \Psi(A) = -i \frac{\delta \Psi(A)}{\delta A^i_a}.$$

(analogous to $\hat{p}\psi(q) = -i\hbar d\psi(q)/dq$). In passing over to the quantum theory the constraints become operators on a kinematic Hilbert space (the unconstrained $SU(2)$ Yang–Mills Hilbert space). Note that different ordering of the A 's and $\tilde{E}$'s when replacing the $\tilde{E}$'s with derivatives give rise to different operators - the choice made is called the factor ordering and should be chosen via physical reasoning. Formally they read

$$\hat{G}_j|\psi\rangle = 0$$

$$\hat{C}_a|\psi\rangle = 0$$

$$\hat{\tilde{H}}|\psi\rangle = 0.$$

There are still problems in properly defining all these equations and solving them. For example, the Hamiltonian constraint Ashtekar worked with was the densitized version instead of the original Hamiltonian, that is, he worked with $\tilde{H} = \sqrt{\det(q)}H$. There were serious difficulties in promoting this quantity to a quantum operator. Moreover, although Ashtekar variables had the virtue of simplifying the Hamiltonian, they are complex. When one quantizes the theory, it is difficult to ensure that one recovers real general

relativity as opposed to complex general relativity.

Quantum constraints as the equations of quantum general relativity

We now move on to demonstrate an important aspect of the quantum constraints. We consider Gauss' law only. First we state the classical result that the Poisson bracket of the smeared Gauss' law $G(\lambda) = \int d^3x \lambda^j (D_a E^a)^j$ with the connections is

$$\{G(\lambda), A^i_a\} = \partial_a \lambda^i + g\epsilon^{ijk} A^j_a \lambda^k = (D_a \lambda)^i.$$

The quantum Gauss' law reads

$$\hat{G}_j \Psi(A) = -iD_a \frac{\delta \lambda \Psi[A]}{\delta A^j_a} = 0.$$

If one smears the quantum Gauss' law and study its action on the quantum state one finds that the action of the constraint on the quantum state is equivalent to shifting the argument of Ψ by an infinitesimal (in the sense of the parameter λ small) gauge transformation,

$$\left[1 + \int d^3x \lambda^i(x)\hat{G}_j\right]\Psi(A) = \Psi[A + D\lambda] = \Psi[A],$$

and the last identity comes from the fact that the constraint annihilates the state. So the constraint, as a quantum operator, is imposing the same symmetry that its vanishing imposed classically: it is telling us that the functions $\Psi[A]$ have to be gauge invariant functions of the connection. The same idea is true for the other constraints.

Therefore, the two step process in the classical theory of solving the constraints $C_I = 0$ (equivalent to solving the admissibility conditions for the initial data) and looking for the gauge orbits (solving the 'evolution' equations) is replaced by a one step process in the quantum theory, namely looking for solutions Ψ of the quantum equations $\hat{C}_I \Psi = 0$. This is because it obviously solves the constraint at the quantum level and it simultaneously looks for states that are gauge invariant because $\hat{C}_I$ is the quantum generator of gauge transformations (gauge invariant functions are constant along the gauge orbits and thus characterize them).[11] Recall that, at the classical level, solving the admissibility conditions and evolution equations was equivalent to solving all of Einstein's field equations, this underlines the central role of the quantum constraint equations in canonical quantum gravity.

Introduction of the loop representation

Main articles: Holonomy. Wilson loop and Knot invariant

It was in particular the inability to have good control over the space of solutions to the Gauss' law and spatial diffeomorphism constraints that led Rovelli and Smolin to consider a new representation - the loop representation in gauge theories and quantum gravity.[12]

We need the notion of a holonomy. A holonomy is a measure of how much the initial and final values of a spinor or vector differ after parallel transport around a closed loop; it is denoted

$$h_\gamma[A] \, .$$

Knowledge of the holonomies is equivalent to knowledge of the connection, up to gauge equivalence. Holonomies can also be associated with an edge; under a Gauss Law these transform as

$$(h'_e)_{\alpha\beta} = U^{-1}_{\alpha\gamma}(x)(h_e)_{\gamma\sigma}U_{\sigma\beta}(y) \, .$$

For a closed loop $x = y$ if we take the trace of this, that is, putting $\alpha = \beta$ and summing we obtain

$$(h'_e)_{\alpha\alpha} = U^{-1}_{\alpha\gamma}(x)(h_e)_{\gamma\sigma}U_{\sigma\alpha}(x) = [U_{\sigma\alpha}(x)U^{-1}_{\alpha\gamma}(x)](h_e)_{\gamma\sigma} = \delta_{\sigma\gamma}(h_e)_{\gamma\sigma} = (h_e)_{\gamma\gamma}$$

or

$$\text{Tr}\, h'_\gamma = \text{Tr}\, h_\gamma \, .$$

The trace of an holonomy around a closed loop is written

$$W_\gamma[A]$$

and is called a Wilson loop. Thus Wilson loops are gauge invariant. The explicit form of the Holonomy is

$$h_\gamma[A] = \mathcal{P}\exp\left\{ -\int_{\gamma_0}^{\gamma_1} ds\, \dot\gamma^a A^i_a(\gamma(s))T_i \right\}$$

where γ is the curve along which the holonomy is evaluated, and s is a parameter along the curve, $\mathcal{P}$ denotes path ordering meaning factors for smaller values of s appear to the left, and T_i are matrices that satisfy the SU(2) algebra

$$[T^i, T^j] = 2i\epsilon^{ijk}T^k \, .$$

The Pauli matrices satisfy the above relation. It turns out that there are infinitely many more examples of sets of matrices that satisfy these relations, where each set comprises $(N+1) \times (N+1)$ matrices with $N = 1, 2, 3, \ldots$, and where none of these can be thought to 'decompose' into two or more examples of lower dimension. They are called different irreducible representations of the SU(2) algebra. The most fundamental representation being the Pauli matrices. The holonomy is labelled by a half integer $N/2$ according to the irreducible representation used.

The use of Wilson loops explicitly solves the Gauss gauge constraint. To handle the spatial diffeomorphism constraint we need to go over to the loop representation. As Wilson loops form a basis we can formally expand any Gauss gauge invariant function as,

$$\Psi[A] = \sum_\gamma \Psi[\gamma]W_\gamma[A] \, .$$

This is called the loop transform. We can see the analogy with going to the momentum representation in quan-

tum mechanics(see Position and momentum space). There one has a basis of states $\exp(ikx)$ labelled by a number k and one expands

$$\psi[x] = \int dk\,\psi(k)\exp(ikx) \, .$$

and works with the coefficients of the expansion $\psi(k)$.

The inverse loop transform is defined by

$$\Psi[\gamma] = \int[dA]\Psi[A]W_\gamma[A] \, .$$

This defines the loop representation. Given an operator $\hat O$ in the connection representation,

$$\Phi[A] = \hat O\Psi[A] \qquad Eq\ 1 \, .$$

one should define the corresponding operator $\hat O'$ on $\Psi[\gamma]$ in the loop representation via,

$$\Phi[\gamma] = \hat O'\Psi[\gamma] \qquad Eq\ 2 \, .$$

where $\Phi[\gamma]$ is defined by the usual inverse loop transform,

$$\Phi[\gamma] = \int[dA]\Phi[A]W_\gamma[A] \qquad Eq\ 3. \, .$$

A transformation formula giving the action of the operator $\hat O'$ on $\Psi[\gamma]$ in terms of the action of the operator $\hat O$ on $\Psi[A]$ is then obtained by equating the R.H.S. of $Eq\ 2$ with the R.H.S. of $Eq\ 3$ with $Eq\ 1$ substituted into $Eq\ 3$, namely

$$\hat O'\Psi[\gamma] = \int[dA]W_\gamma[A]\hat O\Psi[A] \, .$$

or

$$\hat O'\Psi[\gamma] = \int[dA](\hat O^\dagger W_\gamma[A])\Psi[A] \, .$$

where by $\hat O^\dagger$ we mean the operator $\hat O$ but with the reverse factor ordering (remember from simple quantum mechanics where the product of operators is reversed under conjugation). We evaluate the action of this operator on the Wilson loop as a calculation in the connection representation and rearranging the result as a manipulation purely in terms of loops (one should remember that when considering the action on the Wilson loop one should choose the operator one wishes to transform with the opposite factor ordering to the one chosen for its action on wavefunctions $\Psi[A]$). This gives the physical meaning of the operator $\hat O'$. For example, if $\hat O^\dagger$ corresponded to a spatial diffeomorphism, then this can be thought of as keeping the connection field A of $W_\gamma[A]$ where it is while performing a spatial diffeomorphism on γ instead. Therefore, the meaning of $\hat O'$ is a spatial diffeomorphism on γ , the argument of $\Psi[\gamma]$.

In the loop representation we can then solve the spatial diffeomorphism constraint by considering functions of loops $\Psi[\gamma]$ that are invariant under spatial diffeomorphisms of the loop γ . That is, we construct what mathematicians call knot invariants. This opened up an unexpected connection between knot theory and quantum gravity.

What about the Hamiltonian constraint? Let us go back to the connection representation. Any collection of non-intersecting Wilson loops satisfy Ashtekar's quantum Hamiltonian constraint. This can be seen from the following. With a particular ordering of terms and replacing $\tilde{E}_i^a$ by a derivative, the action of the quantum Hamiltonian constraint on a Wilson loop is

$$\hat{\tilde{H}}^\dagger W_\gamma[A] = -\epsilon_{ijk}\hat{F}_{ab}^k \frac{\delta}{\delta A_a^i} \frac{\delta}{\delta A_b^j} W_\gamma[A] .$$

When a derivative is taken it brings down the tangent vector, $\dot{\gamma}^a$, of the loop, γ. So we have something like

$$\hat{F}_{ab}^i \dot{\gamma}^a \dot{\gamma}^b .$$

However, as F_{ab}^i is anti-symmetric in the indices a and b this vanishes (this assumes that γ is not discontinuous anywhere and so the tangent vector is unique). Now let us go back to the loop representation.

We consider wavefunctions $\Psi[\gamma]$ that vanish if the loop has discontinuities and that are knot invariants. Such functions solve the Gauss law, the spatial diffeomorphism constraint and (formally) the Hamiltonian constraint. Thus we have identified an infinite set of exact (if only formal) solutions to all the equations of quantum general relativity![12] This generated a lot of interest in the approach and eventually led to LQG.

Geometric operators, the need for intersecting Wilson loops and spin network states

The easiest geometric quantity is the area. Let us choose coordinates so that the surface Σ is characterized by $x^3 = 0$. The area of small parallelogram of the surface Σ is the product of length of each side times $\sin\theta$ where θ is the angle between the sides. Say one edge is given by the vector $\vec{u}$ and the other by $\vec{v}$ then,

$$A = \|\vec{u}\|\|\vec{v}\|\sin\theta = \sqrt{\|\vec{u}\|^2\|\vec{v}\|^2(1-\cos^2\theta)} = \sqrt{\|\vec{u}\|^2\|\vec{v}\|^2 - (\vec{u}\cdot\vec{v})^2}$$

In the space spanned by x^1 and x^2 we have an infinitesimal parallelogram described by $\vec{u} = \vec{e}_1 dx^1$ and $\vec{v} = \vec{e}_2 dx^2$. Using $q_{AB}^{(2)} = \vec{e}_A \cdot \vec{e}_B$ (where the indices A and B run from 1 to 2), we get the area of the surface Σ to be given by

$$A_\Sigma = \int_\Sigma dx^1 dx^2 \sqrt{\det(q^{(2)})}$$

where $\det(q^{(2)}) = q_{11}q_{22} - q_{12}^2$ and is the determinant of the metric induced on Σ. The latter can be rewritten $\det(q^{(2)}) = \epsilon^{AB}\epsilon^{CD} q_{AC}q_{BD}/2$ where the indices $A \ldots D$ go from 1 to 2. This can be further rewritten as

$$\det(q^{(2)}) = \frac{\epsilon^{3ab}\epsilon^{3cd} q_{ac}q_{bc}}{2} .$$

The standard formula for an inverse matrix is

$$q^{ab} = \frac{\epsilon^{acd}\epsilon^{bef} q_{ce}q_{df}}{3!\det(q)}$$

Note the similarity between this and the expression for $\det(q^{(2)})$. But in Ashtekar variables we have $\tilde{E}_i^a \tilde{E}^{bi} =$ $\det(q)q^{ab}$. Therefore,

$$A_\Sigma = \int_\Sigma dx^1 dx^2 \sqrt{\tilde{E}_i^3 \tilde{E}^{3i}} .$$

According to the rules of canonical quantization we should promote the triads $\tilde{E}_i^3$ to quantum operators,

$$\hat{\tilde{E}}_i^3 \sim \frac{\delta}{\delta A_3^i} .$$

It turns out that the area A_Σ can be promoted to a well defined quantum operator despite the fact that we are dealing with product of two functional derivatives and worse we have a square-root to contend with as well.[13] Putting $N = 2J$, we talk of being in the J-th representation. We note that $\sum_i T^i T^i = J(J+1)\mathbb{1}$. This quantity is important in the final formula for the area spectrum. We simply state the result below,

$$\hat{A}_\Sigma W_\gamma[A] = 8\pi\ell_{\text{Planck}}^2 \beta \sum_I \sqrt{j_I(j_I+1)} W_\gamma[A]$$

where the sum is over all edges I of the Wilson loop that pierce the surface Σ.

The formula for the volume of a region R is given by

$$V = \int_R d^3x \sqrt{\det(q)} = \frac{1}{6}\int_R dx^3 \sqrt{\epsilon_{abc}\epsilon^{ijk}\tilde{E}_i^a \tilde{E}_j^b \tilde{E}_k^c} .$$

The quantization of the volume proceeds the same way as with the area. As we take the derivative, and each time we do so we bring down the tangent vector $\dot{\gamma}^a$, when the volume operator acts on non-intersecting Wilson loops the result vanishes. Quantum states with non-zero volume must therefore involve intersections. Given that the anti-symmetric summation is taken over in the formula for the volume we would need at least intersections with three non-coplanar lines. Actually it turns out that one needs at least four-valent vertices for the volume operator to be non-vanishing.

We now consider Wilson loops with intersections. We assume the real representation where the gauge group is $SU(2)$. Wilson loops are an over complete basis as there are identities relating different Wilson loops. These come about from the fact that Wilson loops are based on matrices (the holonomy) and these matrices satisfy identities. Given any two $SU(2)$ matrices $\mathbb{A}$ and $\mathbb{B}$ it is easy to check that,

$$\text{Tr}(\mathbb{A})\text{Tr}(\mathbb{B}) = \text{Tr}(\mathbb{A}\mathbb{B}) + \text{Tr}(\mathbb{A}\mathbb{B}^{-1}) .$$

This implies that given two loops γ and η that intersect, we will have,

$$W_\gamma[A]W_\eta[A] = W_{\gamma\circ\eta}[A] + W_{\gamma\circ\eta^{-1}}[A]$$

where by η^{-1} we mean the loop η traversed in the opposite direction and $\gamma\circ\eta$ means the loop obtained by going around the loop γ and then along η. See figure below. Given that the matrices are unitary one has that $W_\gamma[A] = W_{\gamma^{-1}}[A]$. Also given the cyclic property of the matrix traces (i.e. $Tr(\mathbb{A}\mathbb{B}) = Tr(\mathbb{B}\mathbb{A})$) one has that $W_{\gamma\circ\eta}[A] = W_{\eta\circ\gamma}[A]$.

These identities can be combined with each other into further identities of increasing complexity adding more loops.
These identities are the so-called Mandelstam identities.
Spin networks certain are linear combinations of intersecting Wilson loops designed to address the over completeness
introduced by the Mandelstam identities (for trivalent intersections they eliminate the over-completeness entirely) and
actually constitute a basis for all gauge invariant functions.

*Graphical representation of the simplest non-trivial Mandelstam
identity relating different Wilson loops.*

As mentioned above the holonomy tells one how to propagate test spin half particles. A spin network state assigns an
amplitude to a set of spin half particles tracing out a path in
space, merging and splitting. These are described by spin
networks γ : the edges are labelled by spins together with
'intertwiners' at the vertices which are prescription for how
to sum over different ways the spins are rerouted. The sum
over rerouting are chosen as such to make the form of the
intertwiner invariant under Gauss gauge transformations.

Real variables, modern analysis and LQG

Main article: Hamiltonian constraint of LQG

Let us go into more detail about the technical difficulties
associated with using Ashtekar's variables:

With Ashtekar's variables one uses a complex connection
and so the relevant gauge group as actually $SL(2, \mathbb{C})$ and
not $SU(2)$. As $SL(2, \mathbb{C})$ is non-compact it creates serious problems for the rigorous construction of the necessary
mathematical machinery. The group $SU(2)$, on the other
hand, is compact and the needed constructions have been
developed.

As mentioned above, because Ashtekar's variables are complex the resulting general relativity is complex. To recover the real theory, one has to impose what are known
as the "reality conditions." These require that the densitized
triad be real and that the real part of the Ashtekar connection equals the compatible spin connection (the compatibil-
ity condition being $\nabla_a e^l_b = 0$) determined by the desitized triad. The expression for compatible connection Γ^i_a
is rather complicated and as such non-polynomial formula
enters through the back door.

Before we state the next difficulty we should give a definition; a tensor density of weight W transforms like an ordinary tensor, except that in addition the W th power of the

Jacobian,

$$ J = \left| \frac{\partial x^a}{\partial x^\delta} \right| $$

appears as a factor, i.e.

$$ T'^{a\cdots}_{b\cdots} = J^W \frac{\partial x'^a}{\partial x^c} \cdots \frac{\partial x^d}{\partial x'^b} T^{c\cdots}_{d\cdots} . $$

It turns out that it is impossible, on general grounds, to
construct a UV-finite, diffeomorphism non-violating operator corresponding to $\sqrt{\det(q)}H$. The reason is that the
rescaled Hamiltonian constraint is a scalar density of weight
two while it can be shown that only scalar densities of weight
one have a chance to result in a well defined operator. Thus,
one is forced to work with the original unrescaled, density one-valued, Hamiltonian constraint. However, this is
non-polynomial and the whole virtue of the complex variables is questioned. In fact, all the solutions constructed for
Ashtekar's Hamiltonian constraint only vanished for finite
regularization (physics), however, this violates spatial diffeomorphism invariance.

Without the implementation and solution of the Hamiltonian constraint no progress can be made and no reliable
predictions are possible.

To overcome the first problem one works with the configuration variable

$$ A^i_a = \Gamma^i_a + \beta K^i_a $$

where β is real (as pointed out by Barbero, who introduced
real variables some time after Ashtekar's variables[14][15]).
The Guass law and the spatial diffeomorphism constraints
are the same. In real Ashtekar variables the Hamiltonian is

$$ H = \frac{\epsilon_{ijk}F^k_{ab}\tilde{E}^a_i\tilde{E}^b_j}{\sqrt{\det(q)}} + 2\frac{\beta^2+1}{\beta^2} \frac{(\tilde{E}^a_i\tilde{E}^b_j - \tilde{E}^a_j\tilde{E}^b_i)}{\sqrt{\det(q)}}(A^i_a - \Gamma^i_a)(A^j_b - \Gamma^j_b) = H_E + H' . $$

The complicated relationship between Γ^i_a and the desitized
triads causes serious problems upon quantization. It is with
the choice $\beta = \pm i$ that the second more complicated term
is made to vanish. However, as mentioned above Γ^i_a reap-
pears in the reality conditions. Also we still have the problem of the $1/\sqrt{\det(q)}$ factor.

Thiemann was able to make it work for real β . First he
could simplify the troublesome $1/\sqrt{\det(q)}$ by using the
identity

$$ \{A^k_c, V\} = \frac{\epsilon_{abc}\epsilon^{ijk}\tilde{E}^a_i\tilde{E}^b_j}{\sqrt{\det(q)}} $$

where V is the volume. The A^k_c and V can be promoted
to well defined operators in the loop representation and the
Poisson bracket is replaced by a commutator upon quantization; this takes care of the first term. It turns out that
a similar trick can be used to treat the second term. One
introduces the quantity

$$ K = \int d^3x K^i_a \tilde{E}^a_i $$

and notes that

$$K_a^i = \{A_a^i, K\} .$$

We are then able to write

$$A_a^i - \Gamma_a^i = \beta K_a^i = \beta\{A_a^i, K\} .$$

The reason the quantity K is easier to work with at the time of quantization is that it can be written as

$$K = -\{V, \int d^3x H_E\}$$

where we have used that the integrated densitized trace of the extrinsic curvature, K, is the "time derivative of the volume".

In the long history of canonical quantum gravity formulating the Hamiltonian constraint as a quantum operator (Wheeler–DeWitt equation) in a mathematically rigorous manner has been a formidable problem. It was in the loop representation that a mathematically well defined Hamiltonian constraint was finally formulated in 1996.[9] We leave more details of its construction to the article Hamiltonian constraint of LQG. This together with the quantum versions of the Gauss law and spatial diffeomorphism constrains written in the loop representation are the central equations of LQG (modern canonical quantum General relativity).

Finding the states that are annihilated by these constraints (the physical states), and finding the corresponding physical inner product, and observables is the main goal of the technical side of LQG.

A very important aspect of the Hamiltonian operator is that it only acts at vertices (a consequence of this is that Thiemann's Hamiltonian operator, like Ashtekar's operator, annihilates non-intersecting loops except now it is not just formal and has rigorous mathematical meaning). More precisely, its action is non-zero on at least vertices of valence three and greater and results in a linear combination of new spin networks where the original graph has been modified by the addition of lines at each vertex together and a change in the labels of the adjacent links of the vertex.

Implementation and solution the quantum constraints

Main articles: spectrum, dual space, Rigged Hilbert space and quantum configuration space

We solve, at least approximately, all the quantum constraint equations and for the physical inner product to make physical predictions.

Before we move on to the constraints of LQG, lets us consider certain cases. We start with a kinematic Hilbert space $\mathcal{H}_{Kin}$ as so is equipped with an inner product—the kinematic inner product $\langle \phi, \psi \rangle_{Kin}$.

i) Say we have constraints $\hat{C}_I$ whose zero eigenvalues lie in their discrete spectrum. Solutions of the first constraint, $\hat{C}_1$, correspond to a subspace of the kinematic Hilbert space, $\mathcal{H}_1 \subset \mathcal{H}_{Kin}$. There will be a projection operator P_1 mapping $\mathcal{H}_{Kin}$ onto $\mathcal{H}_1$. The kinematic inner product structure is easily employed to provide the inner product structure after solving this first constraint; the new inner product $\langle \phi, \psi \rangle_1$ is simply

$$\langle \phi, \psi \rangle_1 = \langle P\phi, P\psi \rangle_{Kin}$$

They are based on the same inner product and are states normalizable with respect to it.

ii) The zero point is not contained in the point spectrum of all the $\hat{C}_I$, there is then no non-trivial solution $\Psi \in \mathcal{H}_{Kin}$ to the system of quantum constraint equations $\hat{C}_I \Psi = 0$ for all I .

For example, the zero eigenvalue of the operator

$$\hat{C} = \left(i\frac{d}{dx} - k \right)$$

on $L_2(\mathbb{R}, dx)$ lies in the continuous spectrum $\mathbb{R}$ but the formal "eigenstate" $\exp(-ikx)$ is not normalizable in the kinematic inner product,

$$\int_{-\infty}^{\infty} dx\psi^*(x)\psi(x) = \int_{-\infty}^{\infty} dx e^{ikx} e^{-ikx} = \int_{-\infty}^{\infty} dx = \infty$$

and so does not belong to the kinematic Hilbert space $\mathcal{H}_{Kin}$. In these cases we take a dense subset $\mathcal{S}$ of $\mathcal{H}_{Kin}$ (intuitively this means either any point in $\mathcal{S}$ is either in $\mathcal{H}_{Kin}$ or arbitrarily close to a point in $\mathcal{H}_{Kin}$) with very good convergence properties and consider its dual space $\mathcal{S}'$ (intuitively these map elements of $\mathcal{S}$ onto finite complex numbers in a linear manner), then $\mathcal{S} \subset \mathcal{H}_{Kin} \subset \mathcal{S}'$ (as $\mathcal{S}'$ contains distributional functions). The constraint operator is then implemented on this larger dual space, which contains distributional functions, under the adjoint action on the operator. One looks for solutions on this larger space. This comes at the price that the solutions must be given a new Hilbert space inner product with respect to which they are normalizable (see article on rigged Hilbert space). In this case we have a generalized projection operator on the new space of states. We cannot use the above formula for the new inner product as it diverges, instead the new inner product is given by the simply modification of the above,

$$\langle \phi, \psi \rangle_1 = \langle P\phi, \psi \rangle_{Kin}.$$

The generalized projector P is known as a rigging map.

Implementation and solution the quantum constraints of LQG.

Let us move to LQG, additional complications will arise from that one cannot define an operator for the quantum spatial diffeomorphism constraint as the infinitesimal generator of finite diffeomorphism transformations and the fact the constraint algebra is not a Lie algebra due to the bracket between two Hamiltonian constraints.

Implementation and solution the Gauss constraint:

One does not actually need to promote the Gauss constraint to an operator since we can work directly with Gauss-gauge-invariant functions (that is, one solves the constraint classically and quantizes only the phase space reduced with respect to the Gauss constraint). The Gauss law is solved by the use of spin network states. They provide a basis for the Kinematic Hilbert space $\mathcal{H}_{\text{Kin}}$.

Implementation of the quantum spatial diffeomorphism constraint:

It turns out that one cannot define an operator for the quantum spatial diffeomorphism constraint as the infinitesimal generator of finite diffeomorphism transformations, represented on $\mathcal{H}_{\text{Kin}}$. The representation of finite diffeomorphisms is a family of unitary operators $\hat{U}_\varphi$ acting on a spin-network state ψ_γ by

$$\hat{U}_\varphi \psi_\gamma := \psi_{\varphi \circ \gamma}.$$

for any spatial diffeomorphism φ on Σ. To understand why one cannot define an operator for the quantum spatial diffeomorphism constraint consider what is called a 1-parameter subgroup φ_t in the group of spatial diffeomorphisms, this is then represented as a 1-parameter unitary group U_{φ_t} on $\mathcal{H}_{\text{Kin}}$. However, $\hat{U}_{\varphi_t}$ is not weakly continuous since the subspace $\psi_{\varphi_t \circ \gamma}$ belongs to and the subspace ψ_γ belongs to are orthogonal to each other no matter how small the parameter t is. So one always has

$$| < \psi_\gamma | \hat{U}_{\varphi_t} | \psi_\gamma >_{Kin} - < \psi_\gamma | \psi_\gamma >_{Kin} | = < \psi_\gamma | \psi_\gamma >_{Kin} \neq 0.$$

even in the limit when t goes to zero. Therefore, the infinitesimal generator of $\hat{U}_{\varphi_t}$ does not exist.

Solution of the spatial diffeomorphism constraint.

The spatial diffeomorphism constraint has been solved. The induced inner product $< \cdot, \cdot >_{Diff}$ on $\mathcal{H}_{\text{Diff}}$ (we do not pursue the details) has a very simple description in terms of spin network states: given two spin networks s and s', with associated spin network states ψ_s and $\psi_{s'}$, the inner product is 1 if s and s' are related to each other by a spatial diffeomorphism and zero otherwise.

We have provided a description of the implemented and complete solution of the kinematic constraints, the Gauss and spatial diffeomorphisms constraints which will be the same for any background-independent gauge field theory. The feature that distinguishes such different theories is the Hamiltonian constraint which is the only one that depends on the Lagrangian of the classical theory.

Problem arising from the Hamiltonian constraint.

Details of the implementation the quantum Hamiltonian constraint and solutions are treated in a different article Hamiltonian constraint of LQG. However, in this article we introduce an approximation scheme for the formal solution of the Hamiltonian constraint operator given in the section below on spinfoams. Here we just mention issues that arises with the Hamiltonian constraint.

The Hamiltonian constraint maps diffeomorphism invariant states onto non-diffeomorphism invariant states as so does not preserve the diffeomorphism Hilbert space $\mathcal{H}_{\text{Diff}}$. This is an unavoidable consequence of the operator algebra, in particular the commutator:

$$[\vec{C}(\vec{N}), \hat{H}(M)] \propto \hat{H}(\mathcal{L}_{\vec{N}} M)$$

as can be seen by applying this to $\psi_s \in \mathcal{H}_{Diff}$.

$$(\vec{C}(\vec{N})\hat{H}(M) - \hat{H}(M)\vec{C}(\vec{N}))\psi_s \propto \hat{H}(\mathcal{L}_{\vec{N}} M)\psi_s$$

and using $\vec{C}(\vec{N})\psi_s = 0$ to obtain

$$\vec{C}(\vec{N})[\hat{H}(M)\psi_s] \propto \hat{H}(\mathcal{L}_{\vec{N}} M)\psi_s \neq 0$$

and so $\hat{H}(M)\psi_s$ is not in $\mathcal{H}_{Diff}$.

This means that one cannot just solve the spatial diffeomorphism constraint and then the Hamiltonian constraint. This problem can be circumvented by the introduction of the master constraint, with its trivial operator algebra, one is then able in principle to construct the physical inner product from $\mathcal{H}_{\text{Diff}}$.

2.1.5 Spin foams

Main articles: spin network, spin foam, BF model and Barrett–Crane model

In loop quantum gravity (LQG), a spin network represents a "quantum state" of the gravitational field on a 3-dimensional hypersurface. The set of all possible spin networks (or, more accurately, "s-knots" - that is, equivalence classes of spin networks under diffeomorphisms) is countable; it constitutes a basis of LQG Hilbert space.

In physics, a spin foam is a topological structure made out of two-dimensional faces that represents one of the configurations that must be summed to obtain a Feynman's path integral (functional integration) description of quantum gravity. It is closely related to loop quantum gravity.

Spin foam derived from the Hamiltonian constraint operator

The Hamiltonian constraint generates 'time' evolution. Solving the Hamiltonian constraint should tell us how quantum states evolve in 'time' from an initial spin network state to a final spin network state. One approach to solving the Hamiltonian constraint starts with what is called the Dirac delta function. This is a rather singular function of the real line, denoted $\delta(x)$, that is zero everywhere except at $x = 0$ but whose integral is finite and nonzero. It can be represented as a Fourier integral.

$$\delta(x) = \int e^{ikx} dk .$$

One can employ the idea of the delta function to impose the condition that the Hamiltonian constraint should vanish. It is obvious that

$$\prod_{x \in \Sigma} \delta(\hat{H}(x))$$

is non-zero only when $\hat{H}(x) = 0$ for all x in Σ. Using this we can 'project' out solutions to the Hamiltonian constraint. With analogy to the Fourier integral given above, this (generalized) projector can formally be written as

$$\int [dN] e^{i \int d^3x N(x)\hat{H}(x)} .$$

Interestingly, this is formally spatially diffeomorphism-invariant. As such it can be applied at the spatially diffeomorphism-invariant level. Using this the physical inner product is formally given by

$$\left\langle \int [dN] e^{i \int d^3x N(x)\hat{H}(x)} s_{\text{int}} s_{\text{fin}} \right\rangle_{\text{Diff}}$$

where s_{int} are the initial spin network and s_{fin} is the final spin network.

The exponential can be expanded

$$\left\langle \int [dN](1 \quad + \quad i \int d^3x N(x)\hat{H}(x) \quad + \right.$$
$$\frac{i^2}{2!} [\int d^3x N(x)\hat{H}(x)][\int d^3x' N(x')\hat{H}(x')] \quad +$$
$$\left. \ldots) s_{\text{int}}, s_{\text{fin}} \right\rangle_{\text{Diff}}$$

and each time a Hamiltonian operator acts it does so by adding a new edge at the vertex. The summation over different sequences of actions of $\hat{H}$ can be visualized as a summation over different histories of 'interaction vertices' in the 'time' evolution sending the initial spin network to the final spin network. This then naturally gives rise to the two-complex (a combinatorial set of faces that join along edges, which in turn join on vertices) underlying the spin foam description; we evolve forward an initial spin network sweeping out a surface, the action of the Hamiltonian constraint operator is to produce a new planar surface starting at the vertex. We are able to use the action of the Hamiltonian constraint on the vertex of a spin network state to associate an amplitude to each "interaction" (in analogy to Feynman diagrams). See figure below. This opens up a way of trying to directly link canonical LQG to a path integral description. Now just as a spin networks describe quantum space, each configuration contributing to these path integrals, or sums over history, describe 'quantum space-time'. Because

of their resemblance to soap foams and the way they are labeled John Baez gave these 'quantum space-times' the name 'spin foams'.

The action of the Hamiltonian constraint translated to the path integral or so-called spin foam description. A single node splits into three nodes, creating a spin foam vertex. $N(x_n)$ is the value of N at the vertex and H_{nop} are the matrix elements of the Hamiltonian constraint $\hat{H}$.

There are however severe difficulties with this particular approach, for example the Hamiltonian operator is not self-adjoint, in fact it is not even a normal operator (i.e. the operator does not commute with its adjoint) and so the spectral theorem cannot be used to define the exponential in general. The most serious problem is that the $\hat{H}(x)$'s are not mutually commuting, it can then be shown the formal quantity $\int [dN] e^{i \int d^3x N(x)\hat{H}(x)}$ cannot even define a (generalized) projector. The master constraint (see below) does not suffer from these problems and as such offers a way of connecting the canonical theory to the path integral formulation.

Spin foams from BF theory

It turns out there are alternative routes to formulating the path integral, however their connection to the Hamiltonian formalism is less clear. One way is to start with the BF theory. This is a simpler theory to general relativity. It has no local degrees of freedom and as such depends only on topological aspects of the fields. BF theory is what is known as a topological field theory. Surprisingly, it turns out that general relativity can be obtained from BF theory by imposing a constraint.[16] BF theory involves a field B_{ab}^{IJ} and if one chooses the field B to be the (anti-symmetric) product of two tetrads

$$B_{ab}^{IJ} = \frac{1}{2}(E_a^I E_b^J - E_b^I E_a^J)$$

(tetrads are like triads but in four spacetime dimensions), one recovers general relativity. The condition that the B field be given by the product of two tetrads is called the simplicity constraint. The spin foam dynamics of the topological field theory is well understood. Given the spin foam 'interaction' amplitudes for this simple theory, one then tries to implement the simplicity conditions to obtain a path integral for general relativity. The non-trivial task of constructing a spin foam model is then reduced to the question of how this simplicity constraint should be imposed in the quantum theory. The first attempt at this was the famous Barrett–Crane model.[17] However this model was shown to be problematic, for example there did not seem to be enough degrees of freedom to ensure the correct classical limit.[18] It has been argued that the simplicity constraint was imposed too strongly at the quantum level and should only be imposed in the sense of expectation values just as with the Lorenz gauge condition $\partial_\mu \hat{A}^\mu$ in the Gupta–Bleuler formalism of quantum electrodynamics. New models have now been put forward, sometimes motivated by imposing the simplicity conditions in a weaker sense.

Another difficulty here is that spin foams are defined on a discretization of spacetime. While this presents no problems for a topological field theory as it has no local degrees of freedom, it presents problems for GR. This is known as the problem triangularization dependence.

Modern formulation of spin foams

Just as imposing the classical simplicity constraint recovers general relativity from BF theory, one expects an appropriate quantum simplicity constraint will recover quantum gravity from quantum BF theory.

Much progress has been made with regard to this issue by Engle, Pereira, and Rovelli[19] and Freidel and Krasnov[20] in defining spin foam interaction amplitudes with much better behaviour.

An attempt to make contact between EPRL-FK spin foam and the canonical formulation of LQG has been made.[21]

Spin foam derived from the master constraint operator

See below.

2.1.6 The semiclassical limit

What is the semiclassical limit?

Main articles: Correspondence principle and classical limit

The **classical limit** or **correspondence limit** is the ability of a physical theory to approximate or "recover" classical mechanics when considered over special values of its parameters.[22] The classical limit is used with physical theories that predict non-classical behavior.

In physics, the **correspondence principle** states that the behavior of systems described by the theory of quantum mechanics (or by the old quantum theory) reproduces classical physics in the limit of large quantum numbers. In other words, it says that for large orbits and for large energies, quantum calculations must agree with classical calculations.[23]

The principle was formulated by Niels Bohr in 1920,[24] though he had previously made use of it as early as 1913 in developing his model of the atom.[25]

There are two basic requirements in establishing the semiclassical limit of any quantum theory:

i) reproduction of the Poisson brackets (of the diffeomorphism constraints in the case of general relativity). This is extremely important because, as noted above, the Poisson bracket algebra formed between the (smeared) constraints themselves completely determines the classical theory. This is analogous to establishing Ehrenfest's theorem:

ii) the specification of a complete set of classical observables whose corresponding operators (see complete set of commuting observables for the quantum mechanical definition of a complete set of observables) when acted on by appropriate semiclassical states reproduce the same classical variables with small quantum corrections (a subtle point is that states that are semiclassical for one class of observables may not be semiclassical for a different class of observables[26]).

This may be easily done, for example, in ordinary quantum mechanics for a particle but in general relativity this becomes a highly non-trivial problem as we will see below.

Why might LQG not have general relativity as its semiclassical limit?

Any candidate theory of quantum gravity must be able to reproduce Einstein's theory of general relativity as a classical limit of a quantum theory. This is not guaranteed because of a feature of quantum field theories which is that they have different sectors, these are analogous to the different phases that come about in the thermodynamical limit of statistical systems. Just as different phases are physically different, so are different sectors of a quantum field theory. It may turn out that LQG belongs to an unphysical sector - one in which one does not recover general relativity in the semiclassical limit (in fact there might not be any physical sector at all).

Moreover, the physical Hilbert space H_{phys} must contain enough semiclassical states to guarantee that the quantum theory one obtains can return to the classical theory when $h \to 0$. In order to guarantee this one must avoid quantum anomalies at all cost, because if we do not there will be restrictions on the physical Hilbert space that have no counterpart in the classical theory, implying that the quantum theory has less degrees of freedom than the classical theory.

Theorems establishing the uniqueness of the loop representation as defined by Ashtekar et al. (i.e. a certain concrete realization of a Hilbert space and associated operators reproducing the correct loop algebra - the realization that everybody was using) have been given by two groups (Lewandowski, Okolow, Sahlmann and Thiemann;[27] and Christian Fleischhack[28]). Before this result was established it was not known whether there could be other examples of Hilbert spaces with operators invoking the same loop algebra, other realizations, not equivalent to the one that had been used so far. These uniqueness theorems imply no others exist and so if LQG does not have the correct semiclassical limit then this would mean the end of the loop representation of quantum gravity altogether.

Difficulties checking the semiclassical limit of LQG

There are difficulties in trying to establish LQG gives Einstein's theory of general relativity in the semiclassical limit. There are a number of particular difficulties in establishing the semiclassical limit:

1. There is no operator corresponding to infinitesimal spatial diffeomorphisms (it is not surprising that the theory has no generator of infinitesimal spatial 'translations' as it predicts spatial geometry has a discrete nature, compare to the situation in condensed matter). Instead it must be approximated by finite spatial diffeomorphisms and so the Poisson bracket structure of the classical theory is not exactly reproduced. This problem can be circumvented with the introduction of the so-called master constraint (see below)[29]

2. There is the problem of reconciling the discrete combinatorial nature of the quantum states with the continuous nature of the fields of the classical theory.

3. There are serious difficulties arising from the structure of the Poisson brackets involving the spatial diffeomorphism and Hamiltonian constraints. In particular, the algebra of (smeared) Hamiltonian constraints does not close, it is proportional to a sum over infinitesimal spatial diffeomorphisms (which, as we have just noted, does not exist in the quantum theory) where the coefficients of proportionality are not constants but have non-trivial phase space dependence – as such it does not form a Lie algebra. However, the situation is much improved by the introduction of the master constraint.[29]

4. The semiclassical machinery developed so far is only appropriate to non-graph-changing operators, however, Thiemann's Hamiltonian constraint is a graph-changing operator – the new graph it generates has degrees of freedom upon which the coherent state does

not depend and so their quantum fluctuations are not suppressed. There is also the restriction, so far, that these coherent states are only defined at the Kinematic level, and now one has to lift them to the level of $\mathcal{H}_{Diff}$ and $\mathcal{H}_{Phys}$. It can be shown that Thiemann's Hamiltonian constraint is required to be graph changing in order to resolve problem 3 in some sense. The master constraint algebra however is trivial and so the requirement that it be graph changing can be lifted and indeed non-graph changing master constraint operators have been defined.

5. Formulating observables for classical general relativity is a formidable problem by itself because of its non-linear nature and space-time diffeomorphism invariance. In fact a systematic approximation scheme to calculate observables has only been recently developed.[30][31]

Difficulties in trying to examine the semiclassical limit of the theory should not be confused with it having the wrong semiclassical limit.

Progress in demonstrating LQG has the correct semiclassical limit

Much details here to be written up...

Concerning issue number 2 above one can consider so-called weave states. Ordinary measurements of geometric quantities are macroscopic, and planckian discreteness is smoothed out. The fabric of a T-shirt is analogous. At a distance it is a smooth curved two-dimensional surface. But a closer inspection we see that it is actually composed of thousands of one-dimensional linked threads. The image of space given in LQG is similar, consider a very large spin network formed by a very large number of nodes and links, each of Planck scale. But probed at a macroscopic scale, it appears as a three-dimensional continuous metric geometry.

As far as the editor knows problem 4 of having semiclassical machinery for non-graph changing operators is as the moment still out of reach.

To make contact with familiar low energy physics it is mandatory to have to develop approximation schemes both for the physical inner product and for Dirac observables.

The spin foam models have been intensively studied can be viewed as avenues toward approximation schemes for the physical inner product.

Markopoulou et al. adopted the idea of noiseless subsystems in an attempt to solve the problem of the low energy limit in background independent quantum gravity theories[32][33][34] The idea has even led to the intriguing possibility of matter of the standard model being identified with emergent degrees of freedom from some versions of LQG (see section below: *LQG and related research programs*).

As Wightman emphasized in the 1950s, in Minkowski QFTs the $n-$ point functions§

$$W(x_1, \ldots, x_n) = \langle 0| \phi(x_n) \ldots \phi(x_1)|0 \rangle .$$

completely determine the theory. In particular, one can calculate the scattering amplitudes from these quantities. As explained below in the section on the *Background independent scattering amplitudes*, in the background-independent context, the $n-$ point functions refer to a state and in gravity that state can naturally encode information about a specific geometry which can then appear in the expressions of these quantities. To leading order LQG calculations have been shown to agree in an appropriate sense with the $n-$ point functions calculated in the effective low energy quantum general relativity.

2.1.7 Improved dynamics and the master constraint

Main articles: Hamiltonian (quantum mechanics), Hamiltonian constraint of LQG and Friedrichs extension

The master constraint

Thiemann's master constraint should not be confused with the master equation which has to do with random processes. The Master Constraint Programme for Loop Quantum Gravity (LQG) was proposed as a classically equivalent way to impose the infinite number of Hamiltonian constraint equations

$$H(x) = 0$$

(x being a continuous index) in terms of a single master constraint,

$$M = \int d^3x \frac{|H(x)|^2}{\sqrt{\det(q(x))}} .$$

which involves the square of the constraints in question. Note that $H(x)$ were infinitely many whereas the master constraint is only one. It is clear that if M vanishes then so do the infinitely many $H(x)$'s. Conversely, if all the $H(x)$'s vanish then so does M , therefore they are equivalent. The master constraint M involves an appropriate averaging over all space and so is invariant under spatial diffeomorphisms (it is invariant under spatial "shifts" as it is a summation over all such spatial "shifts" of a quantity that transforms as a scalar). Hence its Poisson bracket with the (smeared) spatial diffeomorphism constraint, $C(\vec{N})$, is simple:

$$\{M, C(\vec{N})\} = 0 .$$

(it is $su(2)$ invariant as well). Also, obviously as any quantity Poisson commutes with itself, and the master constraint being a single constraint, it satisfies

$$\{M, M\} = 0 .$$

We also have the usual algebra between spatial diffeomorphisms. This represents a dramatic simplification of the Poisson bracket structure, and raises new hope in understanding the dynamics and establishing the semiclassical limit.[135]

An initial objection to the use of the master constraint was that on first sight it did not seem to encode information about the observables; because the Mater constraint is quadratic in the constraint, when one computes its Poisson bracket with any quantity, the result is proportional to the constraint, therefore it always vanishes when the constraints are imposed and as such does not select out particular phase space functions. However, it was realized that the condition

$$\{\{M, O\}, O\}_{M=0} = 0$$

is equivalent to O being a Dirac observable. So the master constraint does capture information about the observables. Because of its significance this is known as the master equation.[135]

That the master constraint Poisson algebra is an honest Lie algebra opens up the possibility of using a certain method, known as group averaging, in order to construct solutions of the infinite number of Hamiltonian constraints, a physical inner product thereon and Dirac observables via what is known as refined algebraic quantization RAQ[136]

The quantum master constraint

Define the quantum master constraint (regularisation issues aside) as

$$\hat{M} := \int d^3x \left(\widehat{\frac{H}{\det(q(x))^{1/4}}} \right)^\dagger (x) \left(\widehat{\frac{H}{\det(q(x))^{1/4}}} \right) (x) .$$

Obviously,

$$\left(\widehat{\frac{H}{\det(q(x))^{1/4}}} \right) (x)\Psi = 0$$

for all x implies $\hat{M}\Psi = 0$. Conversely, if $\hat{M}\Psi = 0$ then

$$0 = \langle \Psi, \hat{M}\Psi \rangle = \int d^3x \left\| \left(\widehat{\frac{H}{\det(q(x))^{1/4}}} \right) (x)\Psi \right\|^2 \qquad Eq\ 4$$

implies

$$\left(\widehat{\frac{H}{\det(q(x))^{1/4}}} \right) (x)\Psi = 0 .$$

What is done first is, we are able to compute the matrix elements of the would-be operator $\hat{M}$, that is, we compute the quadratic form Q_M. It turns out that as Q_M is a graph changing, diffeomorphism invariant quadratic form it cannot exist on the kinematic Hilbert space H_{Kin}, and must be defined on H_{Diff}. The fact that the master constraint operator $\hat{M}$ is densely defined on H_{Diff}, it is obvious that $\hat{M}$ is a positive and symmetric operator in H_{Diff}. Therefore, the quadratic form Q_M associated with $\hat{M}$ is closable. The closure of Q_M is the quadratic form of a unique self-adjoint operator $\overline{M}$, called the Friedrichs extension of $\hat{M}$. We relabel $\overline{M}$ as $\hat{M}$ for simplicity. (Note that the presence of an inner product, viz Eq 4, means there are no superfluous solutions i.e. there are no Ψ such that $\left(\widehat{\frac{H}{\det(q(x))^{3/4}}} \right)(x)\Psi \neq 0$ but for which $\hat{M}\Psi = 0$).

It is also possible to construct a quadratic form Q_{M_E} for what is called the extended master constraint (discussed below) on H_{Kin} which also involves the weighted integral of the square of the spatial diffeomorphism constraint (this is possible because Q_{M_E} is not graph changing).

The spectrum of the master constraint may not contain zero due to normal or factor ordering effects which are finite but similar in nature to the infinite vacuum energies of background-dependent quantum field theories. In this case it turns out to be physically correct to replace $\hat{M}$ with $\hat{M}' := \hat{M} - min(spec(\hat{M}))\mathbb{1}$ provided that the "normal ordering constant" vanishes in the classical limit, that is, $\lim_{\hbar\to 0} min(spec(\hat{M})) = 0$, so that $\hat{M}'$ is a valid quantisation of M.

Testing the master constraint

The constraints in their primitive form are rather singular, this was the reason for integrating them over test functions to obtain smeared constraints. However, it would appear that the equation for the master constraint, given above, is even more singular involving the product of two primitive constraints (although integrated over space). Squaring the constraint is dangerous as it could lead to worsened ultra-violent behaviour of the corresponding operator and hence the master constraint programme must be approached with due care.

In doing so the master constraint programme has been satisfactorily tested in a number of model systems with non-trivial constraint algebras, free and interacting field theories.[37][38][39][40][41] The master constraint for LQG was established as a genuine positive self-adjoint operator and the physical Hilbert space of LQG was shown to be non-empty,[42] an obvious consistency test LQG must pass to be a viable theory of quantum General relativity.

Applications of the master constraint

The master constraint has been employed in attempts to approximate the physical inner product and define more rigorous path integrals.[43][44][45][46]

The Consistent Discretizations approach to LQG,[47][48] is an application of the master constraint program to construct the physical Hilbert space of the canonical theory.

Spin foam from the master constraint

It turns out that the master constraint is easily generalized to incorporate the other constraints. It is then referred to as the extended master constraint, denoted M_E. We can define the extended master constraint which imposes both the Hamiltonian constraint and spatial diffeomorphism constraint as a single operator,

$$M_E = \int_\Sigma d^3x \frac{H(x)^2 - q^{ab}V_a(x)V_b(x)}{\sqrt{\det(q)}}.$$

Setting this single constraint to zero is equivalent to $H(x) = 0$ and $V_a(x) = 0$ for all x in Σ. This constraint implements the spatial diffeomorphism and Hamiltonian constraint at the same time on the Kinematic Hilbert space. The physical inner product is then defined as

$$\langle \phi, \psi \rangle_{\text{Phys}} = \lim_{T\to\infty}\left\langle \phi, \int_{-T}^{T} dt e^{it\hat{M}_E}\psi \right\rangle$$

(as $\delta(\hat{M}_E) = \lim_{T\to\infty}\int_{-T}^{T} dt e^{it\hat{M}_E}$). A spin foam representation of this expression is obtained by splitting the t-parameter in discrete steps and writing

$$e^{it\hat{M}_E} = \lim_{n\to\infty}[e^{it\hat{M}_E/n}]^n = \lim_{n\to\infty}[1 + it\hat{M}_E/n]^n.$$

The spin foam description then follows from the application of $[1 + it\hat{M}_E/n]$ on a spin network resulting in a linear combination of new spin networks whose graph and labels have been modified. Obviously an approximation is made by truncating the value of n to some finite integer. An advantage of the extended master constraint is that we are working at the kinematic level and so far it is only here we have access semiclassical coherent states. Moreover, one can find none graph changing versions of this master constraint operator, which are the only type of operators appropriate for these coherent states.

Algebraic quantum gravity

The master constraint programme has evolved into a fully combinatorial treatment of gravity known as Algebraic Quantum Gravity (AQG).[49] The non-graph changing master constraint operator is adapted in the framework of algebraic quantum gravity. While AQG is inspired by LQG, it differs drastically from it because in AQG there is fundamentally no topology or differential structure - it is background independent in a more generalized sense and could possibly have something to say about topology change. In this new formulation of quantum gravity AQG semiclassical states always control the fluctuations of all present degrees of freedom. This makes the AQG semiclassical analysis superior over that of LQG, and progress has been made in establishing it has the correct semiclassical limit and providing contact with familiar low energy physics.[50][51] See Thiemann's book for details.

2.1.8   Physical applications of LQG

Black hole entropy

Main articles: Black hole thermodynamics, Isolated horizon and Immirzi parameter

The Immirzi parameter (also known as the Barbero-Immirzi parameter) is a numerical coefficient appearing in loop quantum gravity. It may take real or imaginary values.

An artist depiction of two black holes merging, a process in which the laws of thermodynamics are upheld.

Black hole thermodynamics is the area of study that seeks to reconcile the laws of thermodynamics with the existence of black hole event horizons. The no hair conjecture of general relativity states that a black hole is characterized only by its mass, its charge, and its angular momentum; hence, it has no entropy. It appears, then, that one can violate the second law of thermodynamics by dropping an object with nonzero entropy into a black hole.[52] Work by Stephen Hawking and Jacob Bekenstein showed that one can preserve the second law of thermodynamics by assigning to each black hole a *black-hole entropy*

$$S_{\mathrm{BH}} = \frac{k_{\mathrm{B}} A}{4 \ell_{\mathrm{P}}^2},$$

where A is the area of the hole's event horizon, k_{B} is the Boltzmann constant, and $\ell_{\mathrm{P}} = \sqrt{G\hbar/c^3}$ is the Planck length.[53] The fact that the black hole entropy is also the maximal entropy that can be obtained by the Bekenstein bound (wherein the Bekenstein bound becomes an equality) was the main observation that led to the holographic principle.[52]

An oversight in the application of the no-hair theorem is the assumption that the relevant degrees of freedom accounting for the entropy of the black hole must be classical in nature; what if they were purely quantum mechanical instead and had non-zero entropy? Actually, this is what is realized in the LQG derivation of black hole entropy, and can be seen as a consequence of its background-independence – the classical black hole spacetime comes about from the semiclassical limit of the quantum state of the gravitational field, but there are many quantum states that have the same semiclassical limit. Specifically, in LQG[54] it is possible to associate a quantum geometrical interpretation to the microstates: These are the quantum geometries of the horizon which are consistent with the area, A, of the black hole and the topology of the horizon (i.e. spherical). LQG offers a geometric explanation of the finiteness of the entropy and of the proportionality of the area of the horizon.[55][56] These calculations have been generalized to rotating black holes.[57]

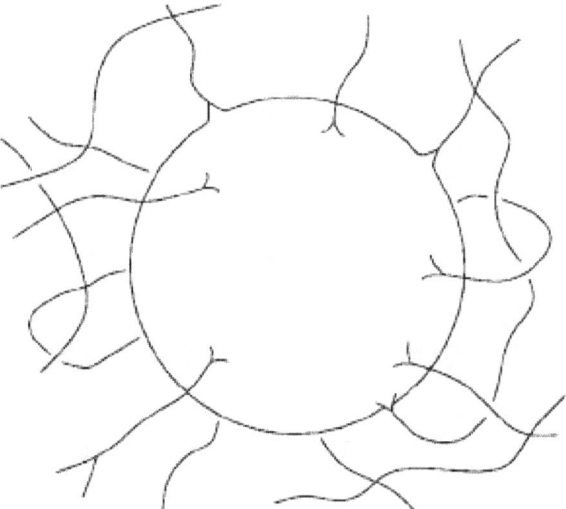

Representation of quantum geometries of the horizon. Polymer excitations in the bulk puncture the horizon, endowing it with quantized area. Intrinsically the horizon is flat except at punctures where it acquires a quantized deficit angle or quantized amount of curvature. These deficit angles add up to 4π.

It is possible to derive, from the covariant formulation of

full quantum theory (Spinfoam) the correct relation between energy and area (1st law), the Unruh temperature and the distribution that yields Hawking entropy.[58] The calculation makes use of the notion of dynamical horizon and is done for non-extremal black holes.

A recent success of the theory in this direction is the computation of the entropy of all non singular black holes directly from theory and independent of Immirzi parameter.[59] The result is the expected formula $S = A/4$, where S is the entropy and A the area of the black hole, derived by Bekenstein and Hawking on heuristic grounds. This is the only known derivation of this formula from a fundamental theory, for the case of generic non singular black holes. Older attempts at this calculation had difficulties. The problem was that although Loop quantum gravity predicted that the entropy of a black hole is proportional to the area of the event horizon, the result depended on a crucial free parameter in the theory, the above-mentioned Immirzi parameter. However, there is no known computation of the Immirzi parameter, so it had to be fixed by demanding agreement with Bekenstein and Hawking's calculation of the black hole entropy.

Loop quantum cosmology

Main articles: loop quantum cosmology, Big bounce and inflation (cosmology)

The popular and technical literature makes extensive references to LQG-related topic of loop quantum cosmology. LQC was mainly developed by Martin Bojowald, it was popularized Loop quantum cosmology in *Scientific American* for predicting a Big Bounce prior to the Big Bang. Loop quantum cosmology (LQC) is a symmetry-reduced model of classical general relativity quantized using methods that mimic those of loop quantum gravity (LQG) that predicts a "quantum bridge" between contracting and expanding cosmological branches.

Achievements of LQC have been the resolution of the big bang singularity, the prediction of a Big Bounce, and a natural mechanism for inflation (cosmology).

LQC models share features of LQG and so is a useful toy model. However, the results obtained are subject to the usual restriction that a truncated classical theory, then quantized, might not display the true behaviour of the full theory due to artificial suppression of degrees of freedom that might have large quantum fluctuations in the full theory. It has been argued that singularity avoidance in LQC are by mechanisms only available in these restrictive models and that singularity avoidance in the full theory can still be obtained but by a more subtle feature of LQG.[60][61]

Loop quantum gravity phenomenology

Quantum gravity effects are notoriously difficult to measure because the Planck length is so incredibly small. However recently physicists have started to consider the possibility of measuring quantum gravity effects mostly from astrophysical observations and gravitational wave detectors. The energy of those fluctuations at scales this small cause space-perturbations which are visible at higher scales.

Background independent scattering amplitudes

Loop quantum gravity is formulated in a background-independent language. No spacetime is assumed a priori, but rather it is built up by the states of theory themselves - however scattering amplitudes are derived from n-point functions (Correlation function (quantum field theory)) and these, formulated in conventional quantum field theory, are functions of points of a background space-time. The relation between the background-independent formalism and the conventional formalism of quantum field theory on a given spacetime is far from obvious, and it is far from obvious how to recover low-energy quantities from the full background-independent theory. One would like to derive the n-point functions of the theory from the background-independent formalism, in order to compare them with the standard perturbative expansion of quantum general relativity and therefore check that loop quantum gravity yields the correct low-energy limit.

A strategy for addressing this problem has been suggested;[62] the idea is to study the boundary amplitude, namely a path integral over a finite space-time region, seen as a function of the boundary value of the field.[63] In conventional quantum field theory, this boundary amplitude is well–defined[64][65] and codes the physical information of the theory; it does so in quantum gravity as well, but in a fully background–independent manner.[66] A generally covariant definition of n-point functions can then be based on the idea that the distance between physical points –arguments of the n-point function is determined by the state of the gravitational field on the boundary of the spacetime region considered.

Progress has been made in calculating background independent scattering amplitudes this way with the use of spin foams. This is a way to extract physical information from the theory. Claims to have reproduced the correct behaviour for graviton scattering amplitudes and to have recovered classical gravity have been made. "We have calculated Newton's law starting from a world with no space and no time." - Carlo Rovelli.

2.1.9   Gravitons, string theory, supersymmetry, extra dimensions in LQG

Main articles: graviton, string theory, supersymmetry, Kaluza–Klein theory and supergravity

Some quantum theories of gravity posit a spin-2 quantum field that is quantized, giving rise to gravitons. In string theory one generally starts with quantized excitations on top of a classically fixed background. This theory is thus described as background dependent. Particles like photons as well as changes in the spacetime geometry (gravitons) are both described as excitations on the string worldsheet. The background dependence of string theory can have important physical consequences, such as determining the number of quark generations. In contrast, loop quantum gravity, like general relativity, is manifestly background independent, eliminating the background required in string theory. Loop quantum gravity, like string theory, also aims to overcome the nonrenormalizable divergences of quantum field theories.

LQG never introduces a background and excitations living on this background, so LQG does not use gravitons as building blocks. Instead one expects that one may recover a kind of semiclassical limit or weak field limit where something like "gravitons" will show up again. In contrast, gravitons play a key role in string theory where they are among the first (massless) level of excitations of a superstring.

LQG differs from string theory in that it is formulated in 3 and 4 dimensions and without supersymmetry or Kaluza-Klein extra dimensions, while the latter requires both to be true. There is no experimental evidence to date that confirms string theory's predictions of supersymmetry and Kaluza–Klein extra dimensions. In a 2003 paper A dialog on quantum gravity,[67] Carlo Rovelli regards the fact LQG is formulated in 4 dimensions and without supersymmetry as a strength of the theory as it represents the most parsimonious explanation, consistent with current experimental results, over its rival string/M-theory. Proponents of string theory will often point to the fact that, among other things, it demonstrably reproduces the established theories of general relativity and quantum field theory in the appropriate limits, which Loop Quantum Gravity has struggled to do. In that sense string theory's connection to established physics may be considered more reliable and less speculative, at the mathematical level. Loop Quantum Gravity has nothing to say about the matter(fermions) in the universe.

Since LQG has been formulated in 4 dimensions (with and without supersymmetry), and M-theory requires supersymmetry and 11 dimensions, a direct comparison between the two has not been possible. It is possible to extend mainstream LQG formalism to higher-dimensional super-

gravity, general relativity with supersymmetry and Kaluza–Klein extra dimensions should experimental evidence establish their existence. It would therefore be desirable to have higher-dimensional Supergravity loop quantizations at one's disposal in order to compare these approaches. In fact a series of recent papers have been published attempting just this.[68][69][70][71][72][73][74][75] Most recently, Thiemann (and alumni) have made progress toward calculating black hole entropy for supergravity in higher dimensions. It will be interesting to compare these results to the corresponding super string calculations.[76][77]

Loop Quantum gravity like other theories of gravity remains unfalsifiable since spin foams exist at super-planck scales , it is impossible for a collider to probe those lengths in the foreseeable future.

2.1.10   LQG and related research programs

Main articles: noncommutative geometry, twistor theory, entropic gravity, Sundance Bilson-Thompson, Asymptotic safety in quantum gravity, Causal dynamical triangulation, group field theory and consistent discretizations

Several research groups have attempted to combine LQG with other research programs: Johannes Aastrup, Jesper M. Grimstrup et al. research combines noncommutative geometry with loop quantum gravity,[78] Laurent Freidel, Simone Speziale, et al., spinors and twistor theory with loop quantum gravity,[79] and Lee Smolin et al. with Verlinde entropic gravity and loop gravity.[80] Stephon Alexander, Antonino Marciano and Lee Smolin have attempted to explain the origins of weak force chirality in terms of Ashketar's variables, which describe gravity as chiral,[81] and LQG with Yang–Mills theory fields[82] in four dimensions. Sundance Bilson-Thompson, Hackett et al.,[83][84] has attempted to introduce standard model via LQG"s degrees of freedom as an emergent property (by employing the idea noiseless subsystems a useful notion introduced in more general situation for constrained systems by Fotini Markopoulou-Kalamara et al.[85]) LQG has also drawn philosophical comparisons with causal dynamical triangulation[86] and asymptotically safe gravity,[87] and the spinfoam with group field theory and AdS/CFT correspondence.[88] Smolin and Wen have suggested combining LQG with String-net liquid, tensors, and Smolin and Fotini Markopoulou-Kalamara Quantum Graphity. There is the consistent discretizations approach. Also, Pullin and Gambini provide a framework to connect the path integral and canonical approaches to quantum gravity. They may help reconcile the spin foam and canonical loop representation approaches. Recent research by Chris Duston and Matilde Marcolli introduces topology change via topspin networks.[89]

2.1.11 Problems and comparisons with alternative approaches

Main article: List of unsolved problems in physics

Some of the major unsolved problems in physics are theoretical, meaning that existing theories seem incapable of explaining a certain observed phenomenon or experimental result. The others are experimental, meaning that there is a difficulty in creating an experiment to test a proposed theory or investigate a phenomenon in greater detail.

Can quantum mechanics and general relativity be realized as a fully consistent theory (perhaps as a quantum field theory)? Is spacetime fundamentally continuous or discrete? Would a consistent theory involve a force mediated by a hypothetical graviton, or be a product of a discrete structure of spacetime itself (as in loop quantum gravity)? Are there deviations from the predictions of general relativity at very small or very large scales or in other extreme circumstances that flow from a quantum gravity theory?

The theory of LQG is one possible solution to the problem of quantum gravity, as is string theory. There are substantial differences however. For example, string theory also addresses unification, the understanding of all known forces and particles as manifestations of a single entity, by postulating extra dimensions and so-far unobserved additional particles and symmetries. Contrary to this, LQG is based only on quantum theory and general relativity and its scope is limited to understanding the quantum aspects of the gravitational interaction. On the other hand, the consequences of LQG are radical, because they fundamentally change the nature of space and time and provide a tentative but detailed physical and mathematical picture of quantum spacetime.

Presently, no semiclassical limit recovering general relativity has been shown to exist. This means it remains unproven that LQG's description of spacetime at the Planck scale has the right continuum limit (described by general relativity with possible quantum corrections). Specifically, the dynamics of the theory is encoded in the Hamiltonian constraint, but there is no candidate Hamiltonian.[90] Other technical problems include finding off-shell closure of the constraint algebra and physical inner product vector space, coupling to matter fields of Quantum field theory, fate of the renormalization of the graviton in perturbation theory that lead to ultraviolet divergence beyond 2-loops (see One-loop Feynman diagram in Feynman diagram).[90]

While there has been a recent proposal relating to observation of naked singularities,[91] and doubly special relativity as a part of a program called loop quantum cosmology,

there is no experimental observation for which loop quantum gravity makes a prediction not made by the Standard Model or general relativity (a problem that plagues all current theories of quantum gravity). Because of the above-mentioned lack of a semiclassical limit, LQG has not yet even reproduced the predictions made by general relativity.

An alternative criticism is that general relativity may be an effective field theory, and therefore quantization ignores the fundamental degrees of freedom.

2.1.12 See also

2.1.13 Notes

[1] Rovelli, Carlo (August 2008). "Loop Quantum Gravity" (PDF). *CERN*. Retrieved 14 September 2014.

[2] Rovelli, C. (2011). "Zakopane lectures on loop gravity". arXiv:1102.3660 [gr-qc].

[3] Muxin, H. (2011). "Cosmological constant in loop quantum gravity vertex amplitude". *Physical Review D* **84** (6): 064010. arXiv:1105.2212. Bibcode:2011PhRvD..84f4010H. doi:10.1103/PhysRevD.84.064010.

[4] Fairbairn, W. J.; Meusburger, C. (2011). "q-Deformation of Lorentzian spin foam models". arXiv:1112.2511 [gr-qc].

[5] Rovelli, C. (2004). *Quantum Gravity*. Cambridge Monographs on Mathematical Physics. p. 71. ISBN 978-0-521-83733-0.

[6] Kauffman, S.; Smolin, L. (7 April 1997). "A Possible Solution For The Problem Of Time In Quantum Cosmology". *Edge.org*. Retrieved 2014-08-20.

[7] Smolin, L. (2006). "The Case for Background Independence". In Rickles, D.; French, S.; Saatsi, J. T. *The Structural Foundations of Quantum Gravity*. Clarendon Press. pp. 196ff. arXiv:hep-th/0507235. ISBN 978-0-19-926969-3.

[8] Rovelli, C. (2004). *Quantum Gravity*. Cambridge Monographs on Mathematical Physics. p. 13ff. ISBN 978-0-521-83733-0.

[9] Thiemann, T. (1996). "Anomaly-free formulation of nonperturbative, four-dimensional Lorentzian quantum gravity". *Physics Letters B* **380**: 257–264. arXiv:gr-qc/9606088. Bibcode:1996PhLB..380..257T. doi:10.1016/0370-2693(96)00532-1.

[10] Baez, J.; de Muniain, J. P. (1994). *Gauge Fields, Knots and Quantum Gravity*. Series on Knots and Everything. Vol. 4. World Scientific. Part III, chapter 4. ISBN 978-981-02-1729-7.

[11] Thiemann, T. (2003). "Lectures on Loop Quantum Gravity". *Lecture Notes in Physics* **631**: 41–135. arXiv:gr-qc/0210094. Bibcode:2003LNP...631...41T. doi:10.1007/978-3-540-45230-0_3.

[12] Rovelli, C.; Smolin, L. (1988). "Knot Theory and Quantum Gravity". *Physical Review Letters* **61** (10): 1155–1958. Bibcode:1988PhRvL..61.1155R. doi:10.1103/PhysRevLett.61.1155.

[13] Gambini, R.; Pullin, J. (2011). *A First Course in Loop Quantum Gravity*. Oxford University Press. Section 8.2. ISBN 978-0-19-959075-9.

[14] Fernando, J.; Barbero, G. (1995). "Reality Conditions and Ashtekar Variables: A Different Perspective". *Physical Review D* **51**: 5498–5506. arXiv:gr-qc/9410013. Bibcode:1995PhRvD..51.5498B. doi:10.1103/PhysRevD.51.5498.

[15] Fernando, J.; Barbero, G. (1995). "Real Ashtekar Variables for Lorentzian Signature Space-times". *Physical Review D* **51**: 5507–5520. arXiv:gr-qc/9410014. Bibcode:1995PhRvD..51.5507B. doi:10.1103/PhysRevD.51.5507.

[16] Bojowald, M.; Alejandro, P. "Spin Foam Quantization and Anomalies". arXiv:gr-qc/0303026 [gr-qc].

[17] Barrett, J.; Crane, L. (2000). "A Lorentzian signature model for quantum general relativity". *Classical and Quantum Gravity* **17**: 3101–3118. arXiv:gr-qc/9904025. Bibcode:2000CQGra..17.3101B. doi:10.1088/0264-9381/17/16/302.

[18] Rovelli, C.; Alesci, E. (2007). "The complete LQG propagator I. Difficulties with the Barrett–Crane vertex". *Physical Review D* **76**: 104012. arXiv:hep-th/0703074. Bibcode:2007PhRvD..76b4012B. doi:10.1103/PhysRevD.76.024012.

[19] Engle, J.; Pereira, R.; Rovelli, C. (2009). "Loop-Quantum-Gravity Vertex Amplitude". *Physical Review Letters* **99**: 161301. arXiv:0705.2388. Bibcode:2007PhRvL..99p1301E. doi:10.1103/physrevlett.99.161301.

[20] Freidel, L.; Krasnov, K. (2008). "A new spin foam model for 4D gravity". *Classical and Quantum Gravity* **25**: 125018. arXiv:0708.1595. Bibcode:2008CQGra..25l5018F. doi:10.1088/0264-9381/25/12/125018.

[21] Alesci, E.; Thiemann, T.; Zipfel, A. (2011). "Linking covariant and canonical LQG: new solutions to the Euclidean Scalar Constraint". arXiv:1109.1290.

[22] Bohm, D. (1989). *Quantum Theory*. Dover Publications. ISBN 978-0-486-65969-5.

[23] Tipler, P.; Llewellyn, R. (2008). *Modern Physics* (5th ed.). W. H. Freeman and Co. pp. 160–161. ISBN 978-0-7167-7550-8.

[24] Bohr, N. (1920). "Über die Serienspektra der Element". *Zeitschrift für Physik* **2** (5): 423–478. Bibcode:1920ZPhy....2..423B. doi:10.1007/BF01329978. (English translation in Bohr 1976, pp. 241–282)

[25] Jammer, M. (1989). *The Conceptual Development of Quantum Mechanics* (2nd ed.). Tomash Publishers. Section 3.2. ISBN 978-0-88318-617-6.

[26] Ashtekar, A.; Bombelli, L.; Corichi, A. (2005). "Semiclassical States for Constrained Systems". *Physical Review D* **72**: 025008. arXiv:hep-ph/0504114. Bibcode:2005PhRvD..72a5008C. doi:10.1103/PhysRevD.72.015008.

[27] Lewandowski, J.; Okołów, A.; Sahlmann, H.; Thiemann, T. (2005). "Uniqueness of Diffeomorphism Invariant States on Holonomy-Flux Algebras". *Communications in Mathematical Physics* **267**: 703–733. arXiv:gr-qc/0504147. Bibcode:2006CMaPh.267..703L. doi:10.1007/s00220-006-0100-7.

[28] Fleischhack, C. (2006). "Irreducibility of the Weyl algebra in loop quantum gravity". *Physical Review Letters* **97**: 061302. Bibcode:2006PhRvL..97f1302F. doi:10.1103/physrevlett.97.061302.

[29] Thiemann, T. (2008). *Modern Canonical General Relativity*. Cambridge Monographs on Mathematical Physics. Cambridge University Press. Section 10.6. ISBN 978-0-521-74187-3.

[30] "Partial and Complete Observables for Hamiltonian Constrained Systems". *General Relativity and Gravitation* **39**: 1891–1927. 2007. arXiv:gr-qc/0411013. Bibcode:2007GReGr..39.1891D. doi:10.1007/s10714-007-0495-2.

[31] "Partial and Complete Observables for Canonical General Relativity". *Classical and Quantum Gravity* **23**: 6155–6184. arXiv:gr-qc/0507106. Bibcode:2006CQGra..23.6155D. doi:10.1088/0264-9381/23/22/006.

[32] Dreyer, O.; Markopoulou, f.; Smolin, L. (2006). "Symmetry and entropy of black hole horizons". *Nuclear Physics B* **774**: 1–13. arXiv:hep-th/0409056. Bibcode:2006NuPhB.744....1D. doi:10.1016/j.nuclphysb.2006.02.045.

[33] Kribs, D. W.; Markopoulou, F. "Geometry from quantum particles". arXiv:gr-qc/0510052.

[34] Markopoulou, F.; Poulin, D. "Noiseless subsystems and the low energy limit of spin foam models" (unpublished).

[35] *The Phoenix Project: Master Constraint Programme for Loop Quantum Gravity*. Class.Quant.Grav.23:2211-2248.2006 or http://fr.arxiv.org/pdf/gr-qc/0305080

[36] *Modern Canonical Quantum General Relativity* by Thomas Thiemann

[37] *Testing the Master Constraint Programme for Loop Quantum Gravity I. General Framework*, Bianca Dittrich, Thomas Thiemann, Class.Quant.Grav. 23 (2006) 1025-1066.

[38] *Testing the Master Constraint Programme for Loop Quantum Gravity II. Finite Dimensional Systems*, Bianca Dittrich, Thomas Thiemann, Class.Quant.Grav. 23 (2006) 1067-1088.

[39] *Testing the Master Constraint Programme for Loop Quantum Gravity III. SL(2,R) Models*, Bianca Dittrich, Thomas Thiemann, Class.Quant.Grav. 23 (2006) 1089-1120.

[40] *Testing the Master Constraint Programme for Loop Quantum Gravity IV. Free Field Theories*, Bianca Dittrich, Thomas Thiemann, Class.Quant.Grav. 23 (2006) 1121-1142.

[41] *Testing the Master Constraint Programme for Loop Quantum Gravity V. Interacting Field Theories*, Bianca Dittrich, Thomas Thiemann, Class.Quant.Grav. 23 (2006) 1143-1162.

[42] *Quantum Spin Dynamics VIII. The Master Constraint*, Thomas Thiemann, Class.Quant.Grav. 23 (2006) 2249-2266.

[43] *Approximating the physical inner product of Loop Quantum Cosmology*, Benjamin Bahr, Thomas Thiemann, Class.Quant.Grav.24:2109-2138,2007.

[44] *On the Relation between Operator Constraint --, Master Constraint --, Reduced Phase Space --, and Path Integral Quantisation*, Muxin Han, Thomas Thiemann, Class.Quant.Grav.27:225019,2010.

[45] *On the Relation between Rigging Inner Product and Master Constraint Direct Integral Decomposition*, Muxin Han, Thomas Thiemann, J.Math.Phys.51:092501,2010.

[46] *A Path-integral for the Master Constraint of Loop Quantum Gravity*, Muxin Han, Class.Quant.Grav.27:215009,2010

[47] *Emergent diffeomorphism invariance in a discrete loop quantum gravity model*, Rodolfo Gambini, Jorge Pullin, Class.Quant.Grav.26:035002,2009

[48] Section 10.2.2 *A First Course in Loop quantum Gravity*, Rodolfo Gambinni, Jorge Pullin, Oxford University Press, first published 2011.

[49] *Algebraic Quantum Gravity (AQG) I. Conceptual Setup*, K. Giesel, T. Thiemann, Class.Quant.Grav.24:2465-2498,2007.

[50] *Algebraic Quantum Gravity (AQG) II. Semiclassical Analysis*, K. Giesel, T. Thiemann, Class.Quant.Grav.24:2499-2564,2007.

[51] *Algebraic Quantum Gravity (AQG) III. Semiclassical Perturbation Theory*, K. Giesel, T. Thiemann, Class.Quant.Grav.24:2565-2588,2007.

[52] Bousso, Raphael (2002). "The Holographic Principle". *Reviews of Modern Physics* **74** (3): 825–874. arXiv:hep-th/0203101. Bibcode:2002RvMP...74..825B. doi:10.1103/RevModPhys.74.825.

[53] Majumdar, Parthasarathi (1998). "Black Hole Entropy and Quantum Gravity" **73**: 147. arXiv:gr-qc/9807045. Bibcode:1999InJPB..73..147M.

[54] See List of loop quantum gravity researchers

[55] Rovelli, Carlo (1996). "Black Hole Entropy from Loop Quantum Gravity". *Physical Review Letters* **77** (16): 3288–3291. arXiv:gr-qc/9603063. Bibcode:1996PhRvL..77.3288R. doi:10.1103/PhysRevLett.77.3288.

[56] Ashtekar, Abhay; Baez, John; Corichi, Alejandro; Krasnov, Kirill (1998). "Quantum Geometry and Black Hole Entropy". *Physical Review Letters* **80** (5): 904–907. arXiv:gr-qc/9710007. Bibcode:1998PhRvL..80..904A. doi:10.1103/PhysRevLett.80.904.

[57] *Quantum horizons and black hole entropy: Inclusion of distortion and rotation*, Abhay Ashtekar, Jonathan Engle, Chris Van Den Broeck, Class.Quant.Grav.22:L27-L34, 2005.

[58] Bianchi, Eugenio (2012). "Entropy of Non-Extremal Black Holes from Loop Gravity". arXiv:1204.5122.

[59] http://inspirehep.net/record/940357?ln=en. http://inspirehep.net/record/1111991.

[60] *On (Cosmological) Singularity Avoidance in Loop Quantum Gravity*, Johannes Brunnemann, Thomas Thiemann, Class.Quant.Grav. 23 (2006) 1395-1428.

[61] *Unboundedness of Triad -- Like Operators in Loop Quantum Gravity*, Johannes Brunnemann, Thomas Thiemann, Class.Quant.Grav. 23 (2006) 1429-1484.

[62] L. Modesto, C. Rovelli:*Particle scattering in loop quantum gravity*, Phys Rev Lett 95 (2005) 191301

[63] R Oeckl, *A 'general boundary' formulation for quantum mechanics and quantum gravity*, Phys Lett B575 (2003) 318-324 ; *Schrodinger's cat and the clock: lessons for quantum gravity*, Class Quant Grav 20 (2003) 5371-5380l

[64] F. Conrady, C. Rovelli *Generalized Schrodinger equation in Euclidean field theory*", Int J Mod Phys A 19, (2004) 1-32.

[65] L Doplicher, *Generalized Tomonaga-Schwinger equation from the Hadamard formula*, Phys Rev D70 (2004) 064037

[66] F. Conrady, L. Doplicher, R. Oeckl, C. Rovelli, M. Testa, *Minkowski vacuum in background independent quantum gravity*, Phys Rev D69 (2004) 064019.

[67] http://arxiv.org/abs/arXiv:hep-th/0310077

[68] *New Variables for Classical and Quantum Gravity in all Dimensions I. Hamiltonian Analysis*, Norbert Bodendorfer, Thomas Thiemann, Andreas Thurn, Class. Quantum Grav. 30 (2013) 045001

[69] *New Variables for Classical and Quantum Gravity in all Dimensions II. Lagrangian Analysis*, Norbert Bodendorfer, Thomas Thiemann, Andreas Thurn, Quantum Grav. 30 (2013) 045002

[70] *New Variables for Classical and Quantum Gravity in all Dimensions III. Quantum Theory*, Norbert Bodendorfer, Thomas Thiemann, Andreas Thurn, Class. Quantum Grav. 30 (2013) 045003

[71] *New Variables for Classical and Quantum Gravity in all Dimensions IV. Matter Coupling*, Norbert Bodendorfer, Thomas Thiemann, Andreas Thurn, Class. Quantum Grav. 30 (2013) 045004

[72] *On the Implementation of the Canonical Quantum Simplicity Constraint*, Norbert Bodendorfer, Thomas Thiemann, Andreas Thurn, Class. Quantum Grav. 30 (2013) 045005

[73] *Towards Loop Quantum Supergravity (LQSG) I. Rarita-Schwinger Sector*, Norbert Bodendorfer, Thomas Thiemann, Andreas Thurn, Class. Quantum Grav. 30 (2013) 045006

[74] *Towards Loop Quantum Supergravity (LQSG) II. p-Form Sector*, Norbert Bodendorfer, Thomas Thiemann, Andreas Thurn, Class. Quantum Grav. 30 (2013) 045007

[75] *Towards Loop Quantum Supergravity (LQSG)*, Norbert Bodendorfer, Thomas Thiemann, Andreas Thurn, Phys. Lett. B 711: 205-211 (2012)

[76] *New Variables for Classical and Quantum Gravity in all Dimensions V. Isolated Horizon Boundary Degrees of Freedom*, Norbert Bodendorfer, Thomas Thiemann, Andreas Thurn, http://uk.arxiv.org/pdf/1304.2679.

[77] *Black hole entropy from loop quantum gravity in higher dimensions*, Norbert Bodendorfer http://uk.arxiv.org/pdf/1307.5029

[78] http://arxiv.org/abs/1203.6164

[79] http://arxiv.org/abs/1006.0199

[80] http://arxiv.org/abs/1001.3668

[81] http://arxiv.org/abs/1212.5246

[82] http://arxiv.org/abs/1105.3480

[83] *Quantum gravity and the standard model*, Sundance O. Bilson-Thompson, Fotini Markopoulou, Lee Smolin, Class.Quant.Grav.24:3975-3994,2007.

[84] For a precise review and outlook of this research see: *Emergent Braided Matter of Quantum Geometry*, Sundance Bilson-Thompson, Jonathan Hackett, Louis Kauffman, Yidun Wan, SIGMA 8 (2012), 014, 43 pages.

[85] *Constrained Mechanics and Noiseless Subsystems*, Tomasz Konopka, Fotini Markopoulou, arXiv:gr-qc/0601028.

[86] http://www.perimeterinstitute.ca/people/renate-loll

[87] wwnpqft.inln.cnrs.fr/pdf/Bianchi.pdf

[88] http://arxiv.org/abs/0804.0632

[89] http://arxiv.org/abs/1308.2934

[90] Nicolai, Hermann; Peeters, Kasper; Zamaklar, Marija (2005). "Loop quantum gravity: an outside view". *Classical and Quantum Gravity* **22** (19): R193–R247. arXiv:hep-th/0501114. Bibcode:2005CQGra..22R.193N. doi:10.1088/0264-9381/22/19/R01.

[91] Goswami; Joshi, Pankaj S.; Singh, Parampreet; et al. (2006). "Quantum evaporation of a naked singularity". *Physical Review Letters* **96** (3): 31302. arXiv:gr-qc/0506129. Bibcode:2006PhRvL..96c1302G. doi:10.1103/PhysRevLett.96.031302.

2.1.14   References

• Topical Reviews

  • Rovelli, Carlo (2011). "Zakopane lectures on loop gravity". arXiv:1102.3660.

  • Rovelli, Carlo (1998). "Loop Quantum Gravity". *Living Reviews in Relativity* **1**. Retrieved 2008-03-13.

  • Thiemann, Thomas (2003). "Lectures on Loop Quantum Gravity". *Lectures Notes in Physics*. Lecture Notes in Physics **631**: 41–135. arXiv:gr-qc/0210094. Bibcode:2003LNP...631...41T. doi:10.1007/978-3-540-45230-0_3. ISBN 978-3-540-40810-9.

  • Ashtekar, Abhay; Lewandowski, Jerzy (2004). "Background Independent Quantum Gravity: A Status Report". *Classical and Quantum Gravity* **21** (15): R53–R152. arXiv:gr-qc/0404018. Bibcode:2004CQGra..21R..53A. doi:10.1088/0264-9381/21/15/R01.

  • Carlo Rovelli and Marcus Gaul, *Loop Quantum Gravity and the Meaning of Diffeomorphism Invariance*, e-print available as gr-qc/9910079.

  • Lee Smolin, *The case for background independence*, e-print available as hep-th/0507235.

  • Alejandro Corichi, *Loop Quantum Geometry: A primer*, e-print available as .

  • Alejandro Perez, *Introduction to loop quantum gravity and spin foams*, e-print available as .

  • Hermann Nicolai and Kasper Peeters *Loop and spin foam quantum gravity: A Brief guide for beginners.*, e-print available as .

• Popular books:

- Carlo Rovelli, "Reality is not what it seems", Penguin, 2016.

- Martin Bojowald, *Once Before Time: A Whole Story of the Universe* 2010.

- Carlo Rovelli, *What is Time? What is space?*, Di Renzo Editore, Roma, 2006.

- Lee Smolin, *Three Roads to Quantum Gravity*, 2001

- Magazine articles:

  - Lee Smolin, "Atoms of Space and Time", *Scientific American*, January 2004

  - Martin Bojowald, "Following the Bouncing Universe", *Scientific American*, October 2008

- Easier introductory, expository or critical works:

  - Abhay Ashtekar, *Gravity and the quantum*, e-print available as gr-qc/0410054 (2004)

  - John C. Baez and Javier Perez de Muniain, *Gauge Fields, Knots and Quantum Gravity*, World Scientific (1994)

  - Carlo Rovelli, *A Dialog on Quantum Gravity*, e-print available as hep-th/0310077 (2003)

  - Rodolfo Gambini and Jorge Pullin, *A First Course in Loop Quantum Gravity*, Oxford (2011)

  - Carlo Rovelli and Francesca Vidotto, *Covariant Loop Quantum Gravity*, Cambridge (2014); draft available online

- More advanced introductory/expository works:

  - Carlo Rovelli, *Quantum Gravity*, Cambridge University Press (2004); draft available online

  - Thomas Thiemann, *Introduction to modern canonical quantum general relativity*, e-print available as gr-qc/0110034

  - Thomas Thiemann, *Introduction to Modern Canonical Quantum General Relativity*, Cambridge University Press (2007)

  - Abhay Ashtekar, *New Perspectives in Canonical Gravity*, Bibliopolis (1988).

  - Abhay Ashtekar, *Lectures on Non-Perturbative Canonical Gravity*, World Scientific (1991)

  - Rodolfo Gambini and Jorge Pullin, *Loops, Knots, Gauge Theories and Quantum Gravity*, Cambridge University Press (1996)

  - Hermann Nicolai, Kasper Peeters, Marija Zamaklar, *Loop quantum gravity: an outside view*, e-print available as hep-th/0501114

  - H. Nicolai and K. Peeters, *Loop and Spin Foam Quantum Gravity: A Brief Guide for Beginners*, e-print available as hep-th/0601129

  - T. Thiemann The LQG – String: Loop Quantum Gravity Quantization of String Theory (2004)

- Conference proceedings:

  - John C. Baez (ed.), *Knots and Quantum Gravity*

- Fundamental research papers:

  - Ashtekar, Abhay (1986). "New variables for classical and quantum gravity". *Physical Review Letters* **57** (18): 2244–2247. Bibcode:1986PhRvL..57.2244A. doi:10.1103/PhysRevLett.57.2244. PMID 10033673

  - Ashtekar, Abhay (1987). "New Hamiltonian formulation of general relativity". *Physical Review D* **36** (6): 1587–1602. Bibcode:1987PhRvD..36.1587A. doi:10.1103/PhysRevD.36.1587

  - Roger Penrose, *Angular momentum: an approach to combinatorial space-time* in *Quantum Theory and Beyond*, ed. Ted Bastin, Cambridge University Press, 1971

  - Rovelli, Carlo; Smolin, Lee (1988). "Knot theory and quantum gravity". *Physical Review Letters* **61** (10): 1155–1158. Bibcode:1988PhRvL..61.1155R. doi:10.1103/PhysRevLett.61.1155.

  - Rovelli, Carlo; Smolin, Lee (1990). "Loop space representation of quantum general relativity". *Nuclear Physics* **B331**: 80–152.

  - Carlo Rovelli and Lee Smolin, *Discreteness of area and volume in quantum gravity*, Nucl. Phys., **B442** (1995) 593-622, e-print available as gr-qc/9411005

  - Kuchař, Karel (1973). "Canonical Quantization of Gravity". In Israel, Werner. *Relativity, Astrophysics and Cosmology*. D. Reidel. pp. 237–288. ISBN 90-277-0369-8.

  - Thiemann, Thomas (2006). "Loop Quantum Gravity: An Inside View". *Approaches to Fundamental Physics*. Lecture Notes in Physics **721**: 185–263. arXiv:hep-th/0608210. Bibcode:2007LNP...721..185T. doi:10.1007/978-3-540-71117-9_10. ISBN 978-3-540-71115-5.

2.1.15   External links

- "Loop Quantum Gravity" by Carlo Rovelli Physics World, November 2003

- Quantum Foam and Loop Quantum Gravity

- Abhay Ashtekar: Semi-Popular Articles . Some excellent popular articles suitable for beginners about space, time, GR, and LQG.

- Loop Quantum Gravity: Lee Smolin.

- Loop Quantum Gravity on arxiv.org

- A list of LQG references catered to fresh graduates

- Loop Quantum Gravity Lectures Online by Lee Smolin

- Spin networks, spin foams and loop quantum gravity

- Wired magazine, News: *Moving Beyond String Theory*

- April 2006 Scientific American Special Issue, *A Matter of Time*, has Lee Smolin LQG Article *Atoms of Space and Time*

- September 2006, The Economist, article *Looping the loop*

- Gamma-ray Large Area Space Telescope: http://glast.gsfc.nasa.gov/

- Zeno meets modern science. Article from Acta Physica Polonica B by Z.K. Silagadze.

- Did pre-big bang universe leave its mark on the sky? - According to a model based on "loop quantum gravity" theory, a parent universe that existed before ours may have left an imprint (*New Scientist*, 10 April 2008)

2.2   Causal dynamical triangulation

Causal dynamical triangulation (abbreviated as **CDT**) invented by Renate Loll, Jan Ambjørn and Jerzy Jurkiewicz, and popularized by Fotini Markopoulou and Lee Smolin, is an approach to quantum gravity that like loop quantum gravity is background independent.

This means that it does not assume any pre-existing arena (dimensional space), but rather attempts to show how the spacetime fabric itself evolves.

The Loops '05 conference, hosted by many loop quantum gravity theorists, included several presentations which discussed CDT in great depth, and revealed it to be a pivotal insight for theorists. It has sparked considerable interest

as it appears to have a good semi-classical description. At large scales, it re-creates the familiar 4-dimensional spacetime, but it shows spacetime to be 2-d near the Planck scale, and reveals a fractal structure on slices of constant time. These interesting results agree with the findings of Lauscher and Reuter, who use an approach called Quantum Einstein Gravity, and with other recent theoretical work. A brief article appeared in the February 2007 issue of *Scientific American*, which gives an overview of the theory, explained why some physicists are excited about it, and put it in historical perspective. The same publication gives CDT, and its primary authors, a feature article in its July 2008 issue.

2.2.1   Introduction

Near the Planck scale, the structure of spacetime itself is supposed to be constantly changing due to quantum fluctuations. CDT theory uses a triangulation process which varies dynamically and follows deterministic rules, to map out how this can evolve into dimensional spaces similar to that of our universe.

The results of researchers suggest that this is a good way to model the early universe, and describe its evolution. Using a structure called a simplex, it divides spacetime into tiny triangular sections. A simplex is the multidimensional analogue of a triangle; a 3-simplex is usually called a tetrahedron, while the 4-simplex, which is the basic building block in this theory, is also known as the pentachoron. Each simplex is geometrically flat, but simplices can be "glued" together in a variety of ways to create curved spacetimes, where previous attempts at triangulation of quantum spaces have produced jumbled universes with far too many dimensions, or minimal universes with too few.

CDT avoids this problem by allowing only those configurations in which the timelines of all joined edges of simplices agree.

2.2.2   Derivation

CDT is a modification of quantum Regge calculus where spacetime is discretized by approximating it with a piecewise linear manifold in a process called triangulation. In this process, a d-dimensional spacetime is considered as formed by space slices that are labeled by a discrete time variable t. Each space slice is approximated by a simplicial manifold composed by regular $(d-1)$-dimensional simplices and the connection between these slices is made by a piecewise linear manifold of d-simplices. In place of a smooth manifold there is a network of triangulation nodes, where space is locally flat (within each simplex) but globally curved, as with the individual faces and the overall surface of a geodesic dome. The line segments which make up each triangle can represent either a space-like or time-like extent, depending on whether they lie on a given time slice, or connect a vertex at time *t* with one at time *t* + 1. The crucial development is that the network of simplices is constrained to evolve in a way that preserves causality. This allows a path integral to be calculated non-perturbatively, by summation of all possible (allowed) configurations of the simplices, and correspondingly, of all possible spatial geometries.

Simply put, each individual simplex is like a building block of spacetime, but the edges that have a time arrow must agree in direction, wherever the edges are joined. This rule preserves causality, a feature missing from previous "triangulation" theories. When simplexes are joined in this way, the complex evolves in an orderly fashion, and eventually creates the observed framework of dimensions. CDT builds upon the earlier work of Barrett and Crane, and Baez and Barret, but by introducing the causality constraint as a fundamental rule (influencing the process from the very start), Loll, Ambjørn, and Jurkiewicz created something different.

2.2.3   Advantages and disadvantages

CDT derives the observed nature and properties of spacetime from a small set of assumptions, without adjusting factors. The idea of deriving what is observed from first principles is very attractive to physicists. CDT models the character of spacetime both in the ultra-microscopic realm near the Planck scale, and at the scale of the cosmos, so CDT may provide insights into the nature of reality.

Evaluation of the observable implications of CDT relies heavily on Monte Carlo simulation by computer. Some feel that this makes CDT an inelegant quantum gravity theory. Also, it has been argued that discrete time-slicing may not accurately reproduce all possible modes of a dynamical system. However, research by Markopoulou and Smolin demonstrates that the cause for those concerns may be limited. Therefore, many physicists still regard this line of reasoning as promising.

2.2.4   Related theories

CDT has some similarities with loop quantum gravity, especially with its spin foam formulations. For example, the Lorentzian Barrett–Crane model is essentially a nonperturbative prescription for computing path integrals, just like CDT. There are important differences, however. Spin foam formulations of quantum gravity use different degrees of freedom and different Lagrangians. For example, in CDT, the distance, or "the interval", between any two points in a given triangulation can be calculated exactly (triangulations are eigenstates of the distance operator). This is not

true for spin foams or loop quantum gravity in general.

Another approach to quantum gravity that is closely related to causal dynamical triangulation is called causal sets. Both CDT and causal sets attempt to model the spacetime with a discrete causal structure. The main difference between the two is that the causal set approach is relatively general, whereas CDT assumes a more specific relationship between the lattice of spacetime events and geometry. Consequently, the Lagrangian of CDT is constrained by the initial assumptions to the extent that it can be written down explicitly and analyzed (see, for example, hep-th/0505154, page 5), whereas there is more freedom in how one might write down an action for causal-set theory.

2.2.5   See also

- Asymptotic safety in quantum gravity
- Causal sets
- Fractal cosmology
- Loop quantum gravity
- 5-cell
- Planck scale
- Quantum gravity
- Regge calculus
- Simplex
- Simplicial manifold
- Spin foam

2.2.6   References

- Quantum gravity: progress from an unexpected direction
- Jan Ambjørn, Jerzy Jurkiewicz, and Renate Loll - "The Self-Organizing Quantum Universe", Scientific American, July 2008
- Alpert, Mark "The Triangular Universe" Scientific American page 24, February 2007
- Ambjørn, J.; Jurkiewicz, J.; Loll, R. - Quantum Gravity or the Art of Building Spacetime
- Loll, R.; Ambjørn, J.; Jurkiewicz, J. - The Universe from Scratch - a less technical recent overview
- Loll, R.; Ambjørn, J.; Jurkiewicz, J. - Reconstructing the Universe - a technically detailed overview

- Markopoulou, Fotini; Smolin, Lee - Gauge Fixing in Causal Dynamical Triangulations - shows that varying the time-slice gives similar results

Early papers on the subject:

- R. Loll, *Discrete Lorentzian Quantum Gravity*, arXiv: hep-th/0011194v1 21 Nov 2000

- J Ambjørn, A. Dasgupta, J. Jurkiewicz, and R. Loll, *A Lorentzian cure for Euclidean troubles*, arXiv:hep-th/0201104 v1 14 Jan 2002

- Causal dynamical triangulation on arxiv.org

2.2.7   External links

- Renate Loll's talk at Loops '05

- John Baez' talk at Loops '05

- Pentatope: from MathWorld

- Simplex: from MathWorld

- Tetrahedron: from MathWorld

- (Re-)Constructing the Universe from Renate Loll's homepage

- Renate Loll on the Quantum Origins of Space and Time as broadcast by TVO

2.3   Canonical quantum gravity

In physics, **canonical quantum gravity** is an attempt to quantize the canonical formulation of general relativity (or **canonical gravity**). It is a Hamiltonian formulation of Einstein's general theory of relativity. The basic theory was outlined by Bryce DeWitt in a seminal 1967 paper, and based on earlier work by Peter G. Bergmann using the so-called canonical quantization techniques for constrained Hamiltonian systems invented by Paul Dirac. Dirac's approach allows the quantization of systems that include gauge symmetries using Hamiltonian techniques in a fixed gauge choice. Newer approaches based in part on the work of DeWitt and Dirac include the Hartle–Hawking state, Regge calculus, the Wheeler–DeWitt equation and loop quantum gravity.

2.3.1   Canonical quantization

Main articles: Phase space, Poisson brackets, Hilbert space, canonical commutation relation and Schrödinger equation

In the Hamiltonian formulation of ordinary classical mechanics the Poisson bracket is an important concept. A "canonical coordinate system" consists of canonical position and momentum variables that satisfy canonical Poisson-bracket relations,

$$\{q_i, p_j\} = \delta_{ij}$$

where the Poisson bracket is given by

$$\{f, g\} = \sum_{i=1}^{N} \left(\frac{\partial f}{\partial q_i} \frac{\partial g}{\partial p_i} - \frac{\partial f}{\partial p_i} \frac{\partial g}{\partial q_i} \right).$$

for arbitrary phase space functions $f(q_i, p_j)$ and $g(q_i, p_j)$. With the use of Poisson brackets, the Hamilton's equations can be rewritten as,

$$\dot{q}_i = \{q_i, H\},$$
$$\dot{p}_i = \{p_i, H\}.$$

These equations describe a "flow" or orbit in phase space generated by the Hamiltonian H. Given any phase space function $F(q, p)$, we have

$$\frac{d}{dt} F(q_i, p_i) = \{F, H\}.$$

In canonical quantization the phase space variables are promoted to quantum operators on a Hilbert space and the Poisson bracket between phase space variables is replaced by the canonical commutation relation:

$$[\hat{q}, \hat{p}] = i\hbar.$$

In the so-called position representation this commutation relation is realized by the choice:

$$\hat{q}\psi(q) = q\psi(q) \text{ and } \hat{p}\psi(q) = -i\hbar \frac{d}{dq}\psi(q)$$

The dynamics are described by Schrödinger equation:

$$i\hbar \frac{\partial}{\partial t} \psi = \hat{H}\psi$$

where $\hat{H}$ is the operator formed from the Hamiltonian $H(q, p)$ with the replacement $q \mapsto q$ and $p \mapsto -i\hbar\frac{d}{dq}$.

2.3.2   Canonical quantization with constraints

Main articles: Gauge symmetry, Hole argument and Diffeomorphism

Canonical classical general relativity is an example of a fully constrained theory. In constrained theories there are different kinds of phase space: the unrestricted (also called kinematic) phase space on which constraint functions are defined and the reduced phase space on which the constraints have already been solved. For canonical quantization in general terms, phase space is replaced by an appropriate Hilbert space and phase space variables are to be promoted to quantum operators.

In Dirac's approach to quantization the unrestricted phase space is replaced by the so-called kinematic Hilbert space and the constraint functions replaced by constraint operators implemented on the kinematic Hilbert space, solutions are then searched for. These quantum constraint equations are the central equations of canonical quantum general relativity, at least in the Dirac approach which is the approach usually taken.

In theories with constraints there is also the reduced phase space quantization where the constraints are solved at the classical level and the phase space variables of the reduced phase space are then promoted to quantum operators, however this approache was thought to be impossible in General relativity as it seemed to be equivalent to finding a general solution to the classical field equations. However, with the fairly recent development of a systematic approximation scheme for calculating observables of General relativity (for the first time) by Bianca Dittrich, based on ideas introduced by Carlo Rovelli, a viable scheme for a reduced phase space quantization of Gravity has been developed by Thomas Thiemann. However it is not fully equivalent to the Dirac quantization as the `clock-variables' must be taken to be classical in the reduced phase space quantization, as apposed to the case in the Dirac quantization.

A common misunderstanding is that coordinate transformations are the gauge symmetries of general relativity, when actually the true gauge symmetries are diffeomorphisms as defined by a mathematician (see the Hole argument) – which are much more radical. The first class constraints of general relativity are the spatial diffeomorphism constraint and the Hamiltonian constraint (also known as the Wheeler-De Witt equation) and imprint the spatial and temporal diffeomorphism invariance of the theory respectively. Imposing these constraints classically are basically admissibility conditions on the initial data, also they generate the `evolution' equations (really gauge transformations) via the Poisson bracket. Importantly the Poisson bracket algebra between the constraints fully determines the classical theory – this is something that must in some way be reproduced in the semi-classical limit of canonical quantum gravity for it to be a viable theory of quantum gravity.

In Dirac's approach it turns out that the first class quantum constraints imposed on a wavefunction also generate gauge transformations. Thus the two step process in the classical theory of solving the constraints $C_I = 0$ (equivalent to solving the admissibility conditions for the initial data) and looking for the gauge orbits (solving the `evolution' equations) is replaced by a one step process in the quantum theory, namely looking for solutions Ψ of the quantum equations $\hat{C}_I \Psi = 0$. This is because it obviously solves the constraint at the quantum level and it simultaneously looks for states that are gauge invariant because $\hat{C}_I$ is the quantum generator of gauge transformations. At the classical level, solving the admissibility conditions and evolution equations are equivalent to solving all of Einstein's field equations, this underlines the central role of the quantum constraint equations in Dirac's approach to canonical quantum gravity.

2.3.3 Canonical quantization, Diffeomorphism invariance and Manifest Finiteness

Main articles: Hole argument, Diffeomorphism and Renormalization

A diffeomorphism can be thought of as simultaneously `dragging' the metric (gravitational field) and matter fields over the bare manifold while staying in the same coordinate system, and so are more radical than invariance under a mere coordinate transformation. This symmetry arises from the subtle requirement that the laws of general relativity cannot depend on any a-priori given space-time geometry.

This diffeomorphism invariance has an important implication: canonical quantum gravity will be manifestly finite as the ability to `drag' the metric function over the bare manifold means that small and large `distances' between abstractly defined coordinate points are gauge-equivalent! A more rigorous argument has been provided by Lee Smolin:

"A background independent operator must always be finite. This is because the regulator scale and the background metric are always introduced together in the regularization procedure. This is necessary, because the scale that the regularization parameter refers to must be described in terms of a background metric or coordinate chart introduced in the construction of the regulated operator. Because of this the dependence of the regulated operator on the cutoff, or regulator parameter, is related to its dependence on the background metric. When one takes the limit of the regulator parameter going to zero one isolates the non-vanishing terms. If these have any dependence on the regulator parameter (which would be the case if the term is blowing up) then it must also have dependence on the background metric. Conversely, if the terms that are nonvanishing in the limit the regulator is removed have no dependence on the background metric, it must be finite."

In fact, as mentioned below, Thomas Thiemann has explicitly demonstrated that loop quantum gravity(a well developed version of canonical quantum gravity) is manifestly finite even in the presence of all forms of matter! So there is no need for renormalization and the elimination of infinities.

In perturbative quantum gravity (from which the non-renormalization arguments originate), as with any perturbative scheme, one makes the assumption that the unperturbed starting point is qualitatively the same as the true quantum state – so perturbative quantum gravity makes the physically unwarranted assumption that the true structure of quantum space-time can be approximated by a smooth classical (usually Minkowski) spacetime. Canonical quantum gravity on the other hand makes no such assumption and instead allows the theory itself tell you, in principle, what the true structure of quantum space-time is. A long held expectation is that in a theory of quantum geometry such as canonical quantum gravity that geometric quantities such as area and volume become quantum observables and take non-zero discrete values, providing a natural regulator which eliminates infinities from the theory including those coming from matter contributions. This 'quantization' of geometric observables is in fact realized in loop quantum gravity (LQG).

2.3.4   Canonical quantization in metric variables

Main article: Diffeomorphism

The quantization is based on decomposing the metric tensor as follows.

$$g_{\mu\nu}dx^{\mu}\,dx^{\nu} = (-N^2+\beta_k\beta^k)dt^2+2\beta_k\,dx^k\,dt+\gamma_{ij}\,dx^i\,dx^j$$

where the summation over repeated indices is implied, the index 0 denotes time $\tau = x^0$. Greek indices run over all values $0, \ldots, 3$ and Latin indices run over spatial values $1, \ldots, 3$. The function N is called the **lapse function** and the functions β_k are called the **shift functions**. The spatial indices are raised and lowered using the spatial metric γ_{ij} and its inverse γ^{ij} : $\gamma_{ij}\gamma^{jk} = \delta_i{}^k$ and $\beta^i = \gamma^{ij}\beta_j$, $\gamma = \det\gamma_{ij}$, where δ is the Kronecker delta. Under this decomposition the Einstein–Hilbert Lagrangian becomes, up to total derivatives,

$$L = \int d^3x\,N\gamma^{1/2}(K_{ij}K^{ij} - K^2 + {}^{(3)}R)$$

where ${}^{(3)}R$ is the spatial scalar curvature computed with respect to the Riemannian metric γ_{ij} and K_{ij} is the extrinsic curvature,

$$K_{ij} = -\frac{1}{2}(\mathcal{L}_n\gamma)_{ij} = \frac{1}{2}N^{-1}\left(\nabla_j\beta_i + \nabla_i\beta_j - \frac{\partial\gamma_{ij}}{\partial t}\right),$$

where $\mathcal{L}$ denotes Lie-differentiation, n is the unit normal to surfaces of constant t and ∇_i denotes covariant differentiation with respect to the metric γ_{ij}. Note that $\gamma_{\mu\nu} = g_{\mu\nu} + n_\mu n_\nu$. DeWitt writes that the Lagrangian "has the classic form 'kinetic energy minus potential energy,' with the extrinsic curvature playing the role of kinetic energy and the negative of the intrinsic curvature that of potential energy." While this form of the Lagrangian is manifestly invariant under redefinition of the spatial coordinates, it makes general covariance opaque.

Since the lapse function and shift functions may be eliminated by a gauge transformation, they do not represent physical degrees of freedom. This is indicated in moving to the Hamiltonian formalism by the fact that their conjugate momenta, respectively π and π^i, vanish identically (on shell and off shell). These are called *primary constraints* by Dirac. A popular choice of gauge, called synchronous gauge, is $N = 1$ and $\beta_i = 0$, although they can, in principle, be chosen to be any function of the coordinates. In this case, the Hamiltonian takes the form

$$H = \int d^3x\mathcal{H}.$$

where

$$\mathcal{H} = \frac{1}{2}\gamma^{-1/2}(\gamma_{ik}\gamma_{jl} + \gamma_{il}\gamma_{jk} - \gamma_{ij}\gamma_{kl})\pi^{ij}\pi^{kl} - \gamma^{1/2\,(3)}R$$

and π^{ij} is the momentum conjugate to γ_{ij}. Einstein's equations may be recovered by taking Poisson brackets with the Hamiltonian. Additional on-shell constraints, called *secondary constraints* by Dirac, arise from the consistency of the Poisson bracket algebra. These are $\mathcal{H} = 0$ and $\nabla_j\pi^{ij} = 0$. This is the theory which is being quantized in approaches to canonical quantum gravity.

It can be shown that six Einstein equations describing time evolution (really a gauge transformation) can be obtained by calculating the Poisson brackets of the three-metric and its conjugate momentum with a linear combination of the spatial diffeomorphism and Hamiltonian constraint. The vanishing of the constraints, giving the physical phase space, are the four other Einstein equations. That is, we have:

Spatial diffeomorphisms constraints

$C_a(x) = 0$

of which there are an infinite number – one for value of x . can be smeared by the so-called shift functions $\vec{N}(x)$ to give an equivalent set of smeared spatial diffeomorphism constraints,

$$C(\vec{N}) = \int d^3x C_a(x) N^a(x) .$$

These generate spatial diffeomorphisms along orbits defined by the shift function $N^a(x)$.

Hamiltonian constraints

$H(x) = 0$

of which there are an infinite number, can be smeared by the so-called lapse functions $N(x)$ to give an equivalent set of smeared Hamiltonian constraints,

$$H(N) = \int d^3x H(x) N(x) .$$

as mentioned above, the Poission bracket structure between the (smeared) constraints is important because they fully determine the classical theory, and must be reproduced in the semi-classical limit of any theory of quantum gravity.

2.3.5 The Wheeler-De-Witt equation

Hamiltonian constraint of LQG

The Wheeler-De-Witt equation (sometimes called the Hamiltonian constraint, sometimes the Einstein-Schrödinger equation) is rather central as it encodes the dynamics at the quantum level. It is analogous to Schrödinger's equation, except as the time coordinate, t , is unphysical, a physical wavefunction can't depend on t and hence 'Schrödinger's equation' reduces to a constraint:

$\hat{H}\Psi = 0$.

Using metric variables lead to seemingly un-summountable mathematical difficulties when trying to promote the classical expression to a well-defined quantum operator, and as such decades went by without making progress via this approach. This problem was circumvented and the formulation of a well-defined Wheeler-De-Witt equation was first accomplished with the introduction of Ashtekar-Barbero variables and the loop representation, this well defined operator formulated by Thomas Thiemann.

Before this development the Wheeler-De-Witt equation had only been formulated in symmetry-reduced models, such as quantum cosmology.

2.3.6 Canonical quantization in Ashtekar-Barbero variables and LQG

Main articles: Ashtekar variables, holonomy, Wilson loop and Loop quantum gravity

Many of the technical problems in canonical quantum gravity revolve around the constraints. Canonical general relativity was originally formulated in terms of metric variables, but there seemed to be insurmountable mathematical difficulties in promoting the constraints to quantum operators because of their highly non-linear dependence on the canonical variables. The equations were much simplified with the introduction of Ashtekars new variables. Ashtekar variables describe canonical general relativity in terms of a new pair canonical variables closer to that of gauge theories. In doing so it introduced an additional constraint, on top of the spatial diffeomorphism and Hamiltonian constraint, the Gauss gauge constraint.

The loop representation is a quantum hamiltonian representation of gauge theories in terms of loops. The aim of the loop representation, in the context of Yang-Mills theories is to avoid the redundancy introduced by Gauss gauge symmetries allowing to work directly in the space of Gauss gauge invariant states. The use of this representation arose naturally from the Ashtekar-Barbero representation as it provides an exact non-perturbative description and also because the spatial diffeomorphism constraint is easily dealt with within this representation.

Within the loop representation Thiemann has provided a well defined canonical theory in the presence of all forms of matter and explicitly demonstrated it to be manifestly finite! So there is no need for renormalization. However, as LQG approach is well suited to describe physics at the Planck scale, there are difficulties in making contact with familiar low energy physics and establishing it has the correct semi-classical limit.

2.3.7 The problem of time

All canonical theories of general relativity have to deal with the problem of time. In quantum gravity, the problem of time is a conceptual conflict between general relativity and quantum mechanics. In canonical general relativity, time is just another coordinate as a result of general covariance. In quantum field theories, especially in the Hamiltonian formulation, the formulation is split between three dimensions of space, and one dimension of time. Roughly speaking, the problem of time is that there is none in general relativity. This is because in general relativity the Hamiltonian is a constraint that must vanish. However, in any canonical theory, the Hamiltonian generates time translations.

Therefore, we arrive at the conclusion that "nothing moves" ("there is no time") in general relativity. Since "there is no time", the usual interpretation of quantum mechanics measurements at given moments of time breaks down. This problem of time is the broad banner for all interpretational problems of the formalism.

2.3.8 The problem of quantum cosmology

The problem of quantum cosmology is that the physical states that solve the constraints of canonical quantum gravity represent quantum states of the entire universe and as such exclude an outside observer, however an outside observer is a crucial element in most interpretations of quantum mechanics.

2.3.9 See also

- ADM formalism

- Ashtekar variables

- Canonical quantization

- Diffeomorphism

- Hole argument

- Regge Calculus

- Loop quantum gravity is one of this family of theories.

- Loop quantum cosmology (LQC) is a finite, symmetry reduced model of loop quantum gravity.

- Problem of time

2.3.10 Sources and notes

1. Arnowitt, R.; Deser, S.; Misner, C. W. (2008). "The Dynamics of General Relativity". *General Relativity and Gravitation* **40** (9): 1997–2027. arXiv:gr-qc/0405109. Bibcode:2008GReGr..40.1997A. doi:10.1007/s10714-008-0661-1.

   - Originally from Witten, L. (1962). *Gravitation: An Introduction to Current Research*. John Wiley & Sons. pp. 227–265.

2. ^ Bergmann, P. (1966). "Hamilton–Jacobi and Schrödinger Theory in Theories with First-Class Hamiltonian Constraints". *Physical Review* **144** (4): 1078–1080. Bibcode:1966PhRv..144.1078B. doi:10.1103/PhysRev.144.1078.

3. ^ Dewitt, B. (1967). "Quantum Theory of Gravity. I. The Canonical Theory". *Physical Review* **160** (5): 1113–1148. Bibcode:1967PhRv..160.1113D. doi:10.1103/PhysRev.160.1113.

4. ^ Dirac, P. A. M. (1958). "Generalized Hamiltonian Dynamics". *Proceedings of the Royal Society of London A* **246** (1246): 326–332. Bibcode:1958RSPSA.246..326D. doi:10.1098/rspa.1958.0141. JSTOR 100496.

5. ^ Thiemann, T. (1996). "Anomaly-free formulation of non-perturbative, four-dimensional Lorentzian quantum gravity". *Phys.Lett.* **B380**: 257–264.

6. Dirac, P. A. M. (1958). "The Theory of Gravitation in Hamiltonian Form". *Proceedings of the Royal Society of London A* **246** (1246): 333–343. Bibcode:1958RSPSA.246..333D. doi:10.1098/rspa.1958.0142. JSTOR 100497.

7. Dirac, P. A. M. (1959). "Fixation of Coordinates in the Hamiltonian Theory of Gravitation". *Physical Review* **114** (3): 924–930. Bibcode:1959PhRv..114..924D. doi:10.1103/PhysRev.114.924.

8. Dirac, P. A. M. (1964). *Lectures on quantum mechanics*. Yeshiva University. ISBN 0-387-51916-5.

2.3.11 Notes

2.4 Superfluid vacuum theory

Superfluid vacuum theory (**SVT**), sometimes known as the **BEC vacuum theory**, is an approach in theoretical physics and quantum mechanics where the fundamental physical vacuum (non-removable background) is viewed as superfluid or as a Bose–Einstein condensate (BEC).

The microscopic structure of this physical vacuum is currently unknown and is a subject of intensive studies in SVT. An ultimate goal of this approach is to develop scientific models that unify quantum mechanics (describing three of the four known fundamental interactions) with gravity, making SVT a candidate for the theory of quantum gravity and describing all known interactions in the Universe, at both microscopic and astronomic scales, as different manifestations of the same entity, superfluid vacuum.

2.4.1 History

The concept of a luminiferous aether as a medium sustaining electromagnetic waves was discarded after the advent of the special theory of relativity. The aether, as conceived in classical physics leads to several contradictions; in particular, aether having a definite velocity at each space-time point will exhibit a preferred direction. This conflicts with the relativistic requirement that all directions within a light cone are equivalent. However, as early as in 1951 P.A.M. Dirac published two papers where he pointed out that we should take into account quantum fluctuations in the flow of the aether.[1][2] His arguments involve the application of the uncertainty principle to the velocity of aether at any space-time point, implying that the velocity will not be a well-defined quantity. In fact, it will be distributed over various possible values. At best, one could represent the aether by a wave function representing the perfect vacuum state for which all aether velocities are equally probable. These works can be regarded as the birth point of the theory.

Inspired by the Dirac ideas, K. P. Sinha, C. Sivaram and E. C. G. Sudarshan published in 1975 a series of papers that suggested a new model for the aether according to which it is a superfluid state of fermion and anti-fermion pairs, describable by a macroscopic wave function.[3][4][5] They noted that particle-like small fluctuations of superfluid background obey the Lorentz symmetry, even if the superfluid itself is non-relativistic. Nevertheless, they decided to treat the superfluid as the relativistic matter - by putting it into the stress–energy tensor of the Einstein field equations. This did not allow them to describe the relativistic gravity as a small fluctuation of the superfluid vacuum, as subsequent authors have noted.

Since then, several theories have been proposed within the SVT framework. They differ in how the structure and properties of the background superfluid must look like. In absence of observational data which would rule out some of them, these theories are being pursued independently.

2.4.2 Relation to other concepts and theories

Lorentz and Galilean symmetries

According to the approach, the background superfluid is assumed to be essentially non-relativistic whereas the Lorentz symmetry is not an exact symmetry of Nature but rather the approximate description valid only for small fluctuations. An observer who resides inside such vacuum and is capable of creating or measuring the small fluctuations would observe them as relativistic objects - unless their energy and momentum are sufficiently high to make the Lorentz-breaking corrections detectable.[6] If the energies and momenta are below the excitation threshold then the superfluid background behaves like the ideal fluid, therefore, the Michelson–Morley-type experiments would observe no drag force from such aether.[1][2]

Further, in the theory of relativity the Galilean symmetry (pertinent to our macroscopic non-relativistic world) arises as the approximate one - when particles' velocities are small compared to speed of light in vacuum. In SVT one does not need to go through Lorentz symmetry to obtain the Galilean one - the dispersion relations of most non-relativistic superfluids are known to obey the non-relativistic behavior at large momenta.[7][8][9]

To summarize, the fluctuations of vacuum superfluid behave like relativistic objects at "small"[nb 1] momenta (a.k.a. the "phononic limit")

$$E^2 \propto |\vec{p}|^2$$

and like non-relativistic ones

$$E \propto |\vec{p}|^2$$

at large momenta. The yet unknown nontrivial physics is believed to be located somewhere between these two regimes.

Relativistic quantum field theory

In the relativistic quantum field theory the physical vacuum is also assumed to be some sort of non-trivial medium to which one can associate certain energy. This is because the concept of absolutely empty space (or "mathematical vacuum") contradicts to the postulates of quantum mechanics. According to QFT, even in absence of real particles the background is always filled by pairs of creating and annihilating virtual particles. However, a direct attempt to describe such medium leads to the so-called ultraviolet divergences. In some QFT models, such as quantum electrodynamics, these problems can be "solved" using the renormalization technique, namely, replacing the diverging physical values by their experimentally measured values. In other theories, such as the quantum general relativity, this trick does not work, and reliable perturbation theory cannot be constructed.

According to SVT, this is because in the high-energy ("ultraviolet") regime the Lorentz symmetry starts failing so dependent theories cannot be regarded valid for all scales of energies and momenta. Correspondingly, while the Lorentz-symmetric quantum field models are obviously a good approximation below the vacuum-energy threshold, in its close vicinity the relativistic description becomes more and more "effective" and less and less natural since one will need to adjust the expressions for the covariant field-theoretical actions by hand.

Curved space-time

According to general relativity, gravitational interaction is described in terms of space-time curvature using the mathematical formalism of Riemannian geometry. This was supported by numerous experiments and observations in the regime of low energies. However, the attempts to quantize general relativity led to various severe problems, therefore, the microscopic structure of gravity is still ill-defined. There may be a fundamental reason for this—the degrees of freedom of general relativity are based on may be only approximate and effective. The question of whether general relativity is an effective theory has been raised for a long time.[10]

According to SVT, the curved space-time arises as the small-amplitude collective excitation mode of the non-relativistic background condensate.[6][11] The mathematical description of this is similar to fluid-gravity analogy which is being used also in the analog gravity models.[12] Thus, relativistic gravity is essentially a long-wavelength theory of the collective modes whose amplitude is small compared to the background one. Outside this requirement the curved-space description of gravity in terms of the Riemannian geometry becomes incomplete or ill-defined.

Cosmological constant

The notion of the cosmological constant makes sense in a relativistic theory only, therefore, within the SVT framework this constant can refer at most to the energy of small fluctuations of the vacuum above a background value but not to the energy of vacuum itself.[13] Thus, in SVT this constant does not have any fundamental physical meaning and the related problems, such as the vacuum catastrophe, simply do not occur in first place.

Gravitational waves and gravitons

According to general relativity, the conventional gravitational wave is:

1. the small fluctuation of curved spacetime which

2. has been separated from its source and propagates independently.

Superfluid vacuum theory brings into question the possibility that a relativistic object possessing both of these properties exists in nature.[11] Indeed, according to the approach, the curved spacetime itself is the small collective excitation of the superfluid background, therefore, the property (1) means that the graviton would be in fact the "small fluctuation of the small fluctuation", which does not look like

a physically robust concept (as if somebody tried to introduce small fluctuations inside a phonon, for instance). As a result, it may be not just a coincidence that in general relativity the gravitational field alone has no well-defined stress–energy tensor, only the pseudotensor one.[14] Therefore, the property (2) cannot be completely justified in a theory with exact Lorentz symmetry which the general relativity is. Though, SVT does not *a priori* forbid an existence of the non-localized wave-like excitations of the superfluid background which might be responsible for the astrophysical phenomena which are currently being attributed to gravitational waves, such as the Hulse–Taylor binary. However, such excitations cannot be correctly described within the framework of a fully relativistic theory.

Mass generation and Higgs boson

The Higgs boson is the spin-0 particle that has been introduced in electroweak theory to give mass to the weak bosons. The origin of mass of the Higgs boson itself is not explained by electroweak theory. Instead, this mass is introduced as a free parameter by means of the Higgs potential, which thus makes it yet another free parameter of the Standard Model.[15] Within the framework of the Standard Model (or its extensions) the theoretical estimates of this parameter's value are possible only indirectly and results differ from each other significantly.[16] Thus, the usage of the Higgs boson (or any other elementary particle with predefined mass) alone is not the most fundamental solution of the mass generation problem but only its reformulation *ad infinitum*. Another known issue of the Glashow–Weinberg–Salam model is the wrong sign of mass term in the (unbroken) Higgs sector for energies above the symmetry-breaking scale.[16 2]

While SVT does not explicitly forbid the existence of the electroweak Higgs particle, it has its own idea of the fundamental mass generation mechanism - elementary particles acquire mass due to the interaction with the vacuum condensate, similarly to the gap generation mechanism in superconductors or superfluids.[11][17] Although this idea is not entirely new, one could recall the relativistic Coleman-Weinberg approach.[18] SVT gives the meaning to the symmetry-breaking relativistic scalar field as describing small fluctuations of background superfluid which can be interpreted as an elementary particle only under certain conditions.[19] In general, one allows two scenarios to happen:

- Higgs boson exists: in this case SVT provides the mass generation mechanism which underlies the electroweak one and explains the origin of mass of the Higgs boson itself:

- Higgs boson does not exist: then the weak bosons acquire mass by directly interacting with the vacuum condensate.

Thus, the Higgs boson, even if it exists, would be a by-product of the fundamental mass generation phenomenon rather than its cause.[19]

Also, some versions of SVT favor a wave equation based on the logarithmic potential rather than on the quartic one. The former potential has not only the Mexican-hat shape, necessary for the spontaneous symmetry breaking, but also some other features which make it more suitable for the vacuum's description.

2.4.3 Logarithmic BEC vacuum theory

In this model the physical vacuum is conjectured to be strongly-correlated quantum Bose liquid whose ground-state wavefunction is described by the logarithmic Schrödinger equation. It was shown that the relativistic gravitational interaction arises as the small-amplitude collective excitation mode whereas relativistic elementary particles can be described by the particle-like modes in the limit of low energies and momenta.[17] The essential difference of this theory from others is that in the logarithmic superfluid the maximal velocity of fluctuations is constant in the leading (classical) order. This allows to fully recover the relativity postulates in the "phononic" (linearized) limit.[11]

The proposed theory has many observational consequences. They are based on the fact that at high energies and momenta the behavior of the particle-like modes eventually becomes distinct from the relativistic one - they can reach the speed of light limit at finite energy.[20] Among other predicted effects is the superluminal propagation and vacuum Cherenkov radiation.[21]

Theory advocates the mass generation mechanism which is supposed to replace or alter the electroweak Higgs one. It was shown that masses of elementary particles can arise as a result of interaction with the superfluid vacuum, similarly to the gap generation mechanism in superconductors.[11][17] For instance, the photon propagating in the average interstellar vacuum acquires a tiny mass which is estimated to be about 10^{-35} electronvolt. One can also derive an effective potential for the Higgs sector which is different from the one used in the Glashow–Weinberg–Salam model, yet it yields the mass generation and it is free of the imaginary-mass problem[nb 2] appearing in the conventional Higgs potential.[19]

2.4.4 See also

- Analog gravity
- Acoustic metric
- Bose–Einstein condensate
- Casimir vacuum
- Hawking radiation
- Induced gravity
- Planck scale
- Planck units
- Hořava–Lifshitz gravity
- Quantum gravity
- Quantum realm
- Sonic black hole
- Vacuum energy

2.4.5 Notes

[1] The term "small" refers here to the linearized limit, in practice the values of these momenta may not be small at all.

[2] If one expands the Higgs potential then the coefficient at the quadratic term appears to be negative. This coefficient has a physical meaning of squared mass of a scalar particle.

2.4.6 References

[1] Dirac. P. A. M. (24 November 1951). "Is there an Æther?". Letters to Nature (Nature) 168 (4282): 906–907. Bibcode:1951Natur.168..906D. doi:10.1038/168906a0. Retrieved 16 October 2012.

[2] Dirac. P. A. M. (26 April 1952). "Is there an Æther?". Nature 169 (4304): 702–702. Bibcode:1952Natur.169..702D. doi:10.1038/169702b0.

[3] K. P. Sinha. C. Sivaram, E. C. G. Sudarshan, Found. Phys. 6, 65 (1976).

[4] K. P. Sinha. C. Sivaram, E. C. G. Sudarshan, Found. Phys. 6, 717 (1976).

[5] K. P. Sinha and E. C. G. Sudarshan. Found. Phys. 8, 823 (1978).

[6] G. E. Volovik, The Universe in a helium droplet, Int. Ser. Monogr. Phys. 117 (2003) 1-507.

[7] N. N. Bogoliubov, Izv. Acad. Nauk USSR 11, 77 (1947).

[8] N.N. Bogoliubov, J. Phys. 11, 23 (1947)

[9] V. L. Ginzburg, L. D. Landau, Zh. Eksp. Teor. Fiz. 20, 1064 (1950).

[10] A. D. Sakharov, Sov. Phys. Dokl. 12, 1040 (1968). This paper was reprinted in Gen. Rel. Grav. 32, 365 (2000) and commented in: M. Visser, Mod. Phys. Lett. A 17, 977 (2002).

[11] K. G. Zloshchastiev, *Spontaneous symmetry breaking and mass generation as built-in phenomena in logarithmic nonlinear quantum theory*, Acta Phys. Polon. B **42** (2011) 261-292 ArXiv:0912.4139.

[12] M. Novello, M. Visser, G. Volovik, *Artificial Black Holes*, World Scientific, River Edge, USA, 2002, p391.

[13] G.E. Volovik, Int. J. Mod. Phys. D15, 1987 (2006) ArXiv: gr-qc/0604062.

[14] L.D. Landau and E.M. Lifshitz, *The Classical Theory of Fields*, (1951), Pergamon Press, chapter 11.96.

[15] V. A. Bednyakov, N. D. Giokaris and A. V. Bednyakov, Phys. Part. Nucl. **39** (2008) 13-36 ArXiv:hep-ph/0703280.

[16] B. Schrempp and M. Wimmer, Prog. Part. Nucl. Phys. **37** (1996) 1-90 ArXiv:hep-ph/9606386.

[17] A. V. Avdeenkov and K. G. Zloshchastiev, *Quantum Bose liquids with logarithmic nonlinearity: Self-sustainability and emergence of spatial extent*, J. Phys. B: At. Mol. Opt. Phys. **44** (2011) 195303. ArXiv:1108.0847.

[18] S. R. Coleman and E. J. Weinberg, Phys. Rev. D7, 1888 (1973).

[19] V. Dzhunushaliev and K.G. Zloshchastiev (2013). "Singularity-free model of electric charge in physical vacuum: Non-zero spatial extent and mass generation". *Cent. Eur. J. Phys.* **11** (3): 325–335. arXiv:1204.6380. Bibcode:2013CEJPh..11..325D. doi:10.2478/s11534-012-0159-z.

[20] K. G. Zloshchastiev, *Logarithmic nonlinearity in theories of quantum gravity: Origin of time and observational consequences*, Grav. Cosmol. **16** (2010) 288-297 ArXiv:0906.4282.

[21] K. G. Zloshchastiev, *Vacuum Cherenkov effect in logarithmic nonlinear quantum theory*, Phys. Lett. A **375** (2011) 2305–2308 ArXiv:1003.0657.

2.5   Twistor theory

In theoretical and mathematical physics, **twistor theory** maps the geometric objects of conventional 3+1 space-time (Minkowski space) into geometric objects in a 4-dimensional space with a Hermitian form of signature (2,2). This space is called twistor space, and its complex valued coordinates are called "twistors."

Twistor theory was first proposed by Roger Penrose in 1967,[1] as a possible path to a theory of quantum gravity. The twistor approach is especially natural for solving the equations of motion of massless fields of arbitrary spin.

In 2003, Edward Witten[2] proposed uniting twistor and string theory by embedding the topological B model of string theory in twistor space. His objective was to model certain Yang–Mills amplitudes. The resulting model has come to be known as twistor string theory (read below). Simone Speziale and collaborators have also applied it to loop quantum gravity.[3]

2.5.1   Details

Twistor theory is unique to 4D Minkowski space and the (2,2) signature, and does not generalize to other dimensions or signatures. At the heart of twistor theory lies the isomorphism between the conformal group Spin(4,2) and SU(2,2), which is the group of unitary transformations of determinant 1 over a four-dimensional complex vector space that leave invariant a Hermitian form of signature (2,2), see classical group.

- $\mathbb{R}^6$ is the real 6D vector space corresponding to the vector representation of Spin(4,2).

- $\mathbf{R}\mathbb{P}^5$ is the real 5D projective representation corresponding to the equivalence class of nonzero points in $\mathbb{R}^6$ under scalar multiplication.

- $\mathbb{M}^c$ corresponds to the subspace of $\mathbf{R}\mathbb{P}^5$ corresponding to vectors of zero norm. This is conformally compactified Minkowski space.

- $\mathbb{T}$ is the 4D complex Weyl spinor representation, called twistor space. It has an invariant Hermitian sesquilinear norm of signature (2,2).

- $\mathbb{P}\mathbb{T}$ is a 3D complex manifold corresponding to projective twistor space.

- $\mathbb{P}\mathbb{T}^+$ is the subspace of $\mathbb{P}\mathbb{T}$ corresponding to projective twistors with positive norm (the sign of the norm, but not its absolute value is projectively invariant). This is a 3D complex manifold.

- $\mathbb{P}\mathbb{N}$ is the subspace of $\mathbb{P}\mathbb{T}$ consisting of null projective twistors (zero norm). This is a real-complex manifold (i.e., it has 5 real dimensions, with four of the real dimensions having a complex structure making them two complex dimensions).

- $\mathbb{P}\mathbb{T}^-$ is the subspace of $\mathbb{P}\mathbb{T}$ of projective twistors with negative norm.

$\mathbb{M}^c$, $\mathbb{PT}^+$, $\mathbb{PN}$ and $\mathbb{PT}^-$ are all homogeneous spaces of the conformal group.

$\mathbb{M}^c$ admits a conformal metric (i.e., an equivalence class of metric tensors under Weyl rescalings) with signature (+++−). Straight null rays map to straight null rays under a conformal transformation and there is a unique canonical isomorphism between null rays in $\mathbb{M}^c$ and points in $\mathbb{PN}$ respecting the conformal group.

In $\mathbb{M}^c$, it is the case that positive and negative frequency solutions cannot be locally separated. However, this is possible in twistor space.

$$\mathbb{PT}^+ \simeq SU(2,2)/[SU(2,1) \times U(1)]$$

2.5.2 Twistor string theory

Main article: Twistor string theory

For many years after Penrose's foundational 1967 paper, twistor theory progressed slowly, in part because of mathematical challenges. Twistor theory also seemed unrelated to ideas in mainstream physics. While twistor theory appeared to say something about quantum gravity, its potential contributions to understanding the other fundamental interactions and particle physics were less obvious.

Witten (2003) proposed a connection between string theory and twistor geometry, called *twistor string theory*. Witten (2004)[2] built on this insight to propose a way to do string theory in twistor space, whose dimensionality is necessarily the same as that of 3+1 Minkowski spacetime. Although Witten has said that "I think twistor string theory is something that only partly works," his work has given new life to the twistor research program. For example, twistor string theory may simplify calculating scattering amplitudes from Feynman diagrams by using a geometric structure called an amplituhedron.

2.5.3 Supertwistors

Witten's twistor string theory is defined on the supertwistor space $\mathbb{CP}^{3|4}$. Supertwistors are a supersymmetric extension of twistors introduced by Alan Ferber in 1978.[4] Along with the standard twistor degrees of freedom, a supertwistor contains N fermionic scalars, where N is the number of supersymmetries. The superconformal algebra can be realized on supertwistor space.

2.5.4 See also

- Penrose transform

- Twistor space
- Invariance mechanics

2.5.5 Notes

[1] Penrose, R. (1967). "Twistor Algebra". *Journal of Mathematical Physics* **8** (2): 345. Bibcode:1967JMP.....8..345P. doi:10.1063/1.1705200.

[2] Witten, Edward (7 October 2004). "Perturbative Gauge Theory as a String Theory in Twistor Space". *Communications in Mathematical Physics* **252** (1-3): 189–258. arXiv:hep-th/0312171. Bibcode:2004CMaPh.252..189W. doi:10.1007/s00220-004-1187-3.

[3] Freidel, Laurent; Speziale, Simone (25 October 2010). "Twistors to twisted geometries". *Physical Review D* **82** (8). arXiv:1006.0199. Bibcode:2010PhRvD..82h4041F. doi:10.1103/PhysRevD.82.084041.

[4] Ferber, A (1978). "Supertwistors and conformal supersymmetry". *Nuclear Physics B* **132**: 55–64. Bibcode:1978NuPhB.132...55F. doi:10.1016/0550-3213(78)90257-2.

2.5.6 Further reading

- Baird, Paul "An Introduction To Twistors"

- Penrose, Roger (1967). "Twistor algebra", *Journal of Mathematical Physics* **8** (2): 345–366, Bibcode:1967JMP.....8..345P, doi:10.1063/1.1705200, MR 0216828

- Penrose, Roger (1968). "Twistor quantisation and curved space-time", *International Journal of Theoretical Physics* (Springer Netherlands) **1**: 61–99, Bibcode:1968IJTP....1...61P, doi:10.1007/BF00668831

- Penrose, Roger (1969). "Solutions of the Zero-Rest-Mass Equations", *Journal of Mathematical Physics* **10** (1): 38–39, Bibcode:1969JMP....10...38P, doi:10.1063/1.1664756

- Penrose, Roger (1977). "The twistor programme", *Reports on Mathematical Physics* **12** (1): 65–76, Bibcode:1977RpMP...12...65P, doi:10.1016/0034-4877(77)90047-7, MR 0465032

- Penrose, Roger (1987) "On the Origins of Twistor Theory" in *Gravitation and Geometry*, a volume in honour of I. Robinson. Naples: Bibliopolis.

- Penrose, Roger (1999) "The Central Programme of Twistor Theory," *Chaos, Solitons and Fractals* 10: 581-611.

- Arkani-Hamed, Nima; Cachazo, Freddy; Cheung, Clifford; Kaplan, Jared (2009) "The S-Matrix in Twistor Space."

- Witten, Edward (2003), "Perturbative Gauge Theory As A String Theory In Twistor Space", *Communications in Mathematical Physics* **252**: 189–258, arXiv:hep-th/0312171, Bibcode:2004CMaPh.252..189W, doi:10.1007/s00220-004-1187-3

2.5.7   External links

- Penrose, Roger (1999) "Einstein's Equation and Twistor Theory: Recent Developments"

- Penrose, Roger; Hadrovich, Fedja. "Twistor Theory."

- Dunajski, Maciej. "Twistor Theory and Differential Equations."

- Hadrovich, Fedja. "Twistor primer."

- Andrew Hodges, "Twistor Theory and the Twistor Programme." Includes many links.

- Huggett, Stephen (2005) "The Elements of Twistor Theory."

- Richard Jozsa (1976) "Applications of Sheaf Cohomology in Twistor Theory."

- Mason, L. J., "The twistor programme and twistor strings:From twistor strings to quantum gravity?"

- Sämann, Christian (2006) "Aspects of Twistor Geometry and Supersymmetric Field Theories within Superstring Theory."

- Sparling, George (1999) "On Time Asymmetry."

- Spradlin, Marcus (2006), "Progress And Prospects In Twistor String Theory."

- MathWorld - Twistors.

- Universe Review "Twistor Theory."

- Twistor newsletter archives

Chapter 3

Theories Beyond The Standard Model

3.1 Supersymmetry

"SUSY" redirects here. For other uses, see Susy (disambiguation).

For the episode of the American TV series *Angel*, see Supersymmetry (Angel).

In particle physics, **Supersymmetry** (**SUSY**) is a proposed type of spacetime symmetry that relates two basic classes of elementary particles: bosons, which have an integer-valued spin, and fermions, which have a half-integer spin.[1] Each particle from one group is associated with a particle from the other, known as its superpartner, the spin of which differs by a half-integer. In a theory with perfectly "unbroken" supersymmetry, each pair of superpartners would share the same mass and internal quantum numbers besides spin. For example, there would be a "selectron" (superpartner electron), a bosonic version of the electron with the same mass as the electron, that would be easy to find in a laboratory. Thus, since no superpartners have been observed, if supersymmetry exists it must be a spontaneously broken symmetry so that superpartners may differ in mass.[2][3] Spontaneously-broken supersymmetry could solve many mysterious problems in particle physics including the hierarchy problem. The simplest realization of spontaneously-broken supersymmetry, the so-called Minimal Supersymmetric Standard Model, is one of the best studied candidates for physics beyond the Standard Model.

There is only indirect evidence and motivation for the existence of supersymmetry. Direct confirmation would entail production of superpartners in collider experiments, such as the Large Hadron Collider (LHC). The first run of the LHC found no evidence for supersymmetry (all results were consistent with the Standard Model), and thus set limits on superpartner masses in supersymmetric theories. While some remain enthusiastic about supersymmetry,[4] this first run at the LHC led some physicists to explore other ideas.[5] The LHC resumed its search for supersymmetry and other new

physics in its second run.

3.1.1 Motivations

There are numerous phenomenological motivations for supersymmetry close to the electroweak scale, as well as technical motivations for supersymmetry at any scale.

The hierarchy problem

Supersymmetry close to the electroweak scale ameliorates the hierarchy problem that afflicts the Standard Model. In the Standard Model, the electroweak scale receives enormous Planck-scale quantum corrections. The observed hierarchy between the electroweak scale and the Planck scale must be achieved with extraordinary fine tuning. In a supersymmetric theory, on the other hand, Planck-scale quantum corrections cancel between partners and super-partners (owing to a minus sign associated with fermionic loops). The hierarchy between the electroweak scale and the Planck scale is achieved in a natural manner, without miraculous fine-tuning.

Gauge coupling unification

The idea that the gauge symmetry groups unify at high-energy is called Grand unification theory. In the Standard Model, however, the weak, strong and electromagnetic couplings fail to unify at high energy. In a supersymmetry theory, the running of the gauge couplings are modified, and precise high-energy unification of the gauge couplings is achieved. The modified running also provides a natural mechanism for radiative electroweak symmetry breaking.

Dark matter

TeV-scale supersymmetry (augmented with a discrete symmetry) typically provides a candidate dark matter parti-

171

cle at a mass scale consistent with thermal relic abundance calculations.[6][7]

Other technical motivations

Supersymmetry is also motivated by solutions to several theoretical problems, for generally providing many desirable mathematical properties, and for ensuring sensible behavior at high energies. Supersymmetric quantum field theory is often much easier to analyze, as many more problems become exactly solvable. When supersymmetry is imposed as a *local* symmetry, Einstein's theory of general relativity is included automatically, and the result is said to be a theory of supergravity. It is also a necessary feature of the most popular candidate for a theory of everything, superstring theory.

Another theoretically appealing property of supersymmetry is that it offers the only "loophole" to the Coleman–Mandula theorem, which prohibits spacetime and internal symmetries from being combined in any nontrivial way, for quantum field theories like the Standard Model with very general assumptions. The Haag-Lopuszanski-Sohnius theorem demonstrates that supersymmetry is the only way spacetime and internal symmetries can be combined consistently.[8]

3.1.2 History

A supersymmetry relating mesons and baryons was first proposed, in the context of hadronic physics, by Hironari Miyazawa during 1966. This supersymmetry did not involve spacetime, that is, it concerned internal symmetry, and was broken badly. Miyazawa's work was largely ignored at the time.[9][10][11][12]

J. L. Gervais and B. Sakita (during 1971),[13] Yu. A. Golfand and E. P. Likhtman (also during 1971), and D.V. Volkov and V.P. Akulov (1972),[14] independently rediscovered supersymmetry in the context of quantum field theory, a radically new type of symmetry of spacetime and fundamental fields, which establishes a relationship between elementary particles of different quantum nature, bosons and fermions, and unifies spacetime and internal symmetries of microscopic phenomena. Supersymmetry with a consistent Lie-algebraic graded structure on which the Gervais–Sakita rediscovery was based directly first arose during 1971[15] in the context of an early version of string theory by Pierre Ramond, John H. Schwarz and André Neveu.

Finally, Julius Wess and Bruno Zumino (during 1974)[16] identified the characteristic renormalization features of four-dimensional supersymmetric field theories, which identified them as remarkable QFTs, and they and Abdus Salam and their fellow researchers introduced early particle physics applications. The mathematical structure of supersymmetry (Graded Lie superalgebras) has subsequently been applied successfully to other topics of physics, ranging from nuclear physics,[17][18] critical phenomena,[19] quantum mechanics to statistical physics. It remains a vital part of many proposed theories of physics.

The first realistic supersymmetric version of the Standard Model was proposed during 1977 by Pierre Fayet and is known as the Minimal Supersymmetric Standard Model or MSSM for short. It was proposed to solve, amongst other things, the hierarchy problem.

3.1.3 Applications

Extension of possible symmetry groups

One reason that physicists explored supersymmetry is because it offers an extension to the more familiar symmetries of quantum field theory. These symmetries are grouped into the Poincaré group and internal symmetries and the Coleman–Mandula theorem showed that under certain assumptions, the symmetries of the S-matrix must be a direct product of the Poincaré group with a compact internal symmetry group or if there is not any mass gap, the conformal group with a compact internal symmetry group. During 1971 Golfand and Likhtman were the first to show that the Poincaré algebra can be extended through introduction of four anticommuting spinor generators (in four dimensions), which later became known as supercharges. During 1975 the Haag-Lopuszanski-Sohnius theorem analyzed all possible superalgebras in the general form, including those with an extended number of the supergenerators and central charges. This extended super-Poincaré algebra paved the way for obtaining a very large and important class of supersymmetric field theories.

The supersymmetry algebra Main article: Supersymmetry algebra

Traditional symmetries of physics are generated by objects that transform by the tensor representations of the Poincaré group and internal symmetries. Supersymmetries, however, are generated by objects that transform by the spinor representations. According to the spin-statistics theorem, bosonic fields commute while fermionic fields anticommute. Combining the two kinds of fields into a single algebra requires the introduction of a $\mathbf{Z}_2$-grading under which the bosons are the even elements and the fermions are the odd elements. Such an algebra is called a Lie superalgebra.

The simplest supersymmetric extension of the Poincaré algebra is the Super-Poincaré algebra. Expressed in terms of two Weyl spinors, has the following anti-commutation relation:

$$\{Q_\alpha, \bar{Q}\beta\} = 2(\sigma^\mu)_{\alpha\beta} P_\mu$$

and all other anti-commutation relations between the Qs and commutation relations between the Qs and Ps vanish. In the above expression $P_\mu = -i\partial_\mu$ are the generators of translation and σ^μ are the Pauli matrices.

There are representations of a Lie superalgebra that are analogous to representations of a Lie algebra. Each Lie algebra has an associated Lie group and a Lie superalgebra can sometimes be extended into representations of a Lie supergroup.

The Supersymmetric Standard Model

Main article: Minimal Supersymmetric Standard Model

Incorporating supersymmetry into the Standard Model requires doubling the number of particles since there is no way that any of the particles in the Standard Model can be superpartners of each other. With the addition of new particles, there are many possible new interactions. The simplest possible supersymmetric model consistent with the Standard Model is the Minimal Supersymmetric Standard Model (MSSM) which can include the necessary additional new particles that are able to be superpartners of those in the Standard Model.

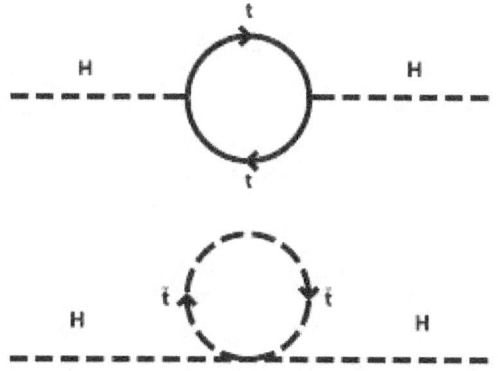

Cancellation of the Higgs boson quadratic mass renormalization between fermionic top quark loop and scalar stop squark tadpole Feynman diagrams in a supersymmetric extension of the Standard Model

One of the main motivations for SUSY comes from the quadratically divergent contributions to the Higgs mass squared. The quantum mechanical interactions of the Higgs boson causes a large renormalization of the Higgs mass and unless there is an accidental cancellation, the natural size of the Higgs mass is the greatest scale possible. This problem is known as the hierarchy problem. Supersymmetry reduces the size of the quantum corrections by having automatic cancellations between fermionic and bosonic Higgs interactions. If supersymmetry is restored at the weak scale, then the Higgs mass is related to supersymmetry breaking which can be induced from small non-perturbative effects explaining the vastly different scales in the weak interactions and gravitational interactions.

In many supersymmetric Standard Models there is a heavy stable particle (such as neutralino) which could serve as a weakly interacting massive particle (WIMP) dark matter candidate. The existence of a supersymmetric dark matter candidate is related closely to R-parity.

The standard paradigm for incorporating supersymmetry into a realistic theory is to have the underlying dynamics of the theory be supersymmetric, but the ground state of the theory does not respect the symmetry and supersymmetry is broken spontaneously. The supersymmetry break can not be done permanently by the particles of the MSSM as they currently appear. This means that there is a new sector of the theory that is responsible for the breaking. The only constraint on this new sector is that it must break supersymmetry permanently and must give superparticles TeV scale masses. There are many models that can do this and most of their details do not matter. In order to parameterize the relevant features of supersymmetry breaking, arbitrary soft SUSY breaking terms are added to the theory which temporarily break SUSY explicitly but could never arise from a complete theory of supersymmetry breaking.

Gauge-coupling unification   Main article: Minimal Supersymmetric Standard Model § Gauge-coupling unification

One piece of evidence for supersymmetry existing is gauge coupling unification. The renormalization group evolution of the three gauge coupling constants of the Standard Model is somewhat sensitive to the present particle content of the theory. These coupling constants do not quite meet together at a common energy scale if we run the renormalization group using the Standard Model.[20] With the addition of minimal SUSY joint convergence of the coupling constants is projected at approximately 10^{16} GeV.[20]

Supersymmetric quantum mechanics

Main article: Supersymmetric quantum mechanics

Supersymmetric quantum mechanics adds the SUSY super-algebra to quantum mechanics as opposed to quantum field theory. Supersymmetric quantum mechanics often becomes relevant when studying the dynamics of supersymmetric solitons, and due to the simplified nature of having fields which are only functions of time (rather than space-time), a great deal of progress has been made in this subject and it is now studied in its own right.

SUSY quantum mechanics involves pairs of Hamiltonians which share a particular mathematical relationship, which are called *partner Hamiltonians*. (The potential energy terms which occur in the Hamiltonians are then known as *partner potentials*.) An introductory theorem shows that for every eigenstate of one Hamiltonian, its partner Hamiltonian has a corresponding eigenstate with the same energy. This fact can be exploited to deduce many properties of the eigenstate spectrum. It is analogous to the original description of SUSY, which referred to bosons and fermions. We can imagine a "bosonic Hamiltonian", whose eigenstates are the various bosons of our theory. The SUSY partner of this Hamiltonian would be "fermionic", and its eigenstates would be the theory's fermions. Each boson would have a fermionic partner of equal energy.

Supersymmetry: Applications to condensed matter physics

SUSY concepts have provided useful extensions to the WKB approximation. Additionally, SUSY has been applied to disorder averaged systems both quantum and non-quantum (through statistical mechanics), the Fokker-Planck equation being an example of a non-quantum theory. The 'supersymmetry' in all these systems arises from the fact that one is modelling one particle and as such the 'statistics' don't matter. The use of the supersymmetry method provides a mathematical rigorous alternative to the replica trick, but only in non-interacting systems, which attempts to address the so-called 'problem of the denominator' under disorder averaging. For more on the applications of supersymmetry in condensed matter physics see the book[21]

Supersymmetry in optics

Integrated optics was recently found[22] to provide a fertile ground on which certain ramifications of SUSY can be explored in readily-accessible laboratory settings. Making use of the analogous mathematical structure of the quantum-mechanical Schrödinger equation and the wave equation governing the evolution of light in one-dimensional settings, one may interpret the refractive index distribution of a structure as a potential landscape in which optical wave packets propagate. In this manner, a new class of functional optical structures with possible applications in phase matching, mode conversion[23] and space-division multiplexing becomes possible. SUSY transformations have been also proposed as a way to address inverse scattering problems in optics and as a one-dimensional transformation optics.[24]

Mathematics

SUSY is also sometimes studied mathematically for its intrinsic properties. This is because it describes complex fields satisfying a property known as holomorphy, which allows holomorphic quantities to be exactly computed. This makes supersymmetric models useful "toy models" of more realistic theories. A prime example of this has been the demonstration of S-duality in four-dimensional gauge theories[25] that interchanges particles and monopoles.

The proof of the Atiyah-Singer index theorem is much simplified by the use of supersymmetric quantum mechanics.

3.1.4 General supersymmetry

Supersymmetry appears in many related contexts of theoretical physics. It is possible to have multiple supersymmetries and also have supersymmetric extra dimensions.

Extended supersymmetry

Main article: Extended supersymmetry

It is possible to have more than one kind of supersymmetry transformation. Theories with more than one supersymmetry transformation are known as extended supersymmetric theories. The more supersymmetry a theory has, the more constrained are the field content and interactions. Typically the number of copies of a supersymmetry is a power of 2, i.e. 1, 2, 4, 8. In four dimensions, a spinor has four degrees of freedom and thus the minimal number of supersymmetry generators is four in four dimensions and having eight copies of supersymmetry means that there are 32 supersymmetry generators.

The maximal number of supersymmetry generators possible is 32. Theories with more than 32 supersymmetry generators automatically have massless fields with spin greater than 2. It is not known how to make massless fields with spin greater than two interact, so the maximal number of supersymmetry generators considered is 32. This is due to the Weinberg-Witten theorem. This corresponds to an $N = 8$ supersymmetry theory. Theories with 32 supersymmetries automatically have a graviton.

For four dimensions there are the following theories, with the corresponding multiplets[126](CPT adds a copy, whenever they are not invariant under such symmetry)

- $N = 1$

Chiral multiplet: $(0,\frac{1}{2})$ Vector multiplet: $(\frac{1}{2},1)$ Gravitino multiplet: $(1,\frac{3}{2})$ Graviton multiplet: $(\frac{3}{2},2)$

- $N = 2$

hypermultiplet: $(-\frac{1}{2},0^2,\frac{1}{2})$ vector multiplet: $(0,\frac{1}{2}^2,1)$ supergravity multiplet: $(1,\frac{3}{2}^2,2)$

- $N = 4$

Vector multiplet: $(-1,-\frac{1}{2}^4,0^6,\frac{1}{2}^4,1)$ Supergravity multiplet: $(0,\frac{1}{2}^4,1^6,\frac{3}{2}^4,2)$

- $N = 8$

Supergravity multiplet: $(-2,-\frac{3}{2}^8,-1^{28},-\frac{1}{2}^{56},0^{70},\frac{1}{2}^{56},1^{28},\frac{3}{2}^8,2)$

Supersymmetry in alternate numbers of dimensions

It is possible to have supersymmetry in dimensions other than four. Because the properties of spinors change drastically between different dimensions, each dimension has its characteristic. In d dimensions, the size of spinors is approximately $2^{d/2}$ or $2^{(d-1)/2}$. Since the maximum number of supersymmetries is 32, the greatest number of dimensions in which a supersymmetric theory can exist is eleven.

3.1.5 Supersymmetry in quantum gravity

Supersymmetry is part of a larger enterprise of theoretical physics to unify everything we know about the universe into a single consistent set of physical principles, known as the quest for a Theory of Everything (TOE). A significant part of this larger enterprise is the quest for a theory of quantum gravity, which would unify the classical theory of general relativity and the Standard Model, which explains the other three basic forces in physics (electromagnetism, the strong interaction, and the weak interaction), and provides a palette of fundamental particles upon which all four forces act. Two of the most active methods of forming a theory of quantum gravity are string theory and loop

quantum gravity (LQG), although in theory, supersymmetry could be a component of other theories as well.

For string theory to be consistent, supersymmetry seems to be required at some level (although it may be a strongly broken symmetry). In particle theory, supersymmetry is recognized as a way to stabilize the hierarchy between the unification scale and the electroweak scale (or the Higgs boson mass), and can also provide a natural dark matter candidate. String theory also requires extra spatial dimensions which have to be compactified as in Kaluza–Klein theory.

Loop quantum gravity (LQG) predicts no additional spatial dimensions, nor anything else about particle physics. These theories can be formulated in three spatial dimensions and one dimension of time, although in some LQG theories dimensionality is an emergent property of the theory, rather than a fundamental assumption of the theory. Also, LQG is a theory of quantum gravity which does not require supersymmetry. Lee Smolin, one of the originators of LQG, has proposed that a loop quantum gravity theory incorporating either supersymmetry or extra dimensions, or both, be called "loop quantum gravity II".

If experimental evidence confirms supersymmetry in the form of supersymmetric particles such as the neutralino that is often believed to be the lightest superpartner, some people believe this would be a major boost to string theory. Since supersymmetry is a required component of string theory, any discovered supersymmetry would be consistent with string theory. If the Large Hadron Collider and other major particle physics experiments fail to detect supersymmetric partners or evidence of extra dimensions, many versions of string theory which had predicted certain low mass superpartners to existing particles may need to be significantly revised. The failure of experiments to discover either supersymmetric partners or extra spatial dimensions, as of 2013, has encouraged loop quantum gravity researchers.

3.1.6 Current status

Supersymmetric models are constrained by a variety of experiments, including measurements of low-energy observables – for example, the anomalous magnetic moment of the muon at Brookhaven; the WMAP dark matter density measurement and direct detection experiments – for example, XENON–100 and LUX; and by particle collider experiments, including B-physics, Higgs phenomenology and direct searches for superpartners (sparticles), at the Large Electron–Positron Collider, Tevatron and the LHC.

Historically, the tightest limits were from direct production at colliders. The first mass limits for squarks and gluinos were made at CERN by the UA1 experiment and the UA2 experiment at the Super Proton Synchrotron. LEP later set very strong limits.,[27] which in 2006 were extended by the D0 experiment at the Tevatron.[28][29] From 2003, WMAP's and Planck's dark matter density measurements have strongly constrained supersymmetry models, which, if they explain dark matter, have to be tuned to invoke a particular mechanism to sufficiently reduce the neutralino density.

Prior to the beginning of the LHC, in 2009 fits of available data to CMSSM and NUHM1 indicated that squarks and gluinos were most likely to have masses in the 500 to 800 GeV range, though values as high as 2.5 TeV were allowed with low probabilities. Neutralinos and sleptons were expected to be quite light, with the lightest neutralino and the lightest stau most likely to be found between 100 and 150 GeV.[30]

The first run of the LHC found no evidence for supersymmetry, and, as a result, surpassed existing experimental limits from the Large Electron–Positron Collider and Tevatron and partially excluded the aforementioned expected ranges.[31]

During 2011 and 2012, the LHC discovered a Higgs boson with a mass of about 125 GeV, and with couplings to fermions and bosons which are consistent with the Standard Model. The MSSM predicts that the mass of the lightest Higgs boson should not be much higher than the mass of the Z boson, and, in the absence of fine tuning (with the supersymmetry breaking scale on the order of 1 TeV), should not exceed 130 GeV. Furthermore, for values of the MSSM parameter $tan\ \beta \leq 3$, it predicts a Higgs mass below 114 GeV over most of the parameter space.[32] This region of Higgs mass was excluded by LEP by 2000. The LHC result is somewhat problematic for the minimal supersymmetric model, as the value of 125 GeV is relatively large for the model and can only be achieved with large radiative loop corrections from top squarks, which many theorists consider to be "unnatural" (see naturalness and fine tuning).[33] On the other hand, the lightest Higgs boson in the MSSM is Standard Model-like, which is consistent with measurements of the Higgs boson couplings at the LHC.

3.1.7   See also

- Supersymmetric gauge theory

- Wess–Zumino model

- Minimal Supersymmetric Standard Model

- Supersymmetry as a quantum group

- Quantum group

- Supercharge

- Superfield

- Supergeometry

- Supergravity

- Supergroup

- Superspace

- Superpartner

3.1.8   References

[1] Haber, Howie. "SUPERSYMMETRY, PART I (THEORY)" (PDF). *Reviews, Tables and Plots.* Particle Data Group (PDG). Retrieved 8 July 2015.

[2] Martin, Stephen P. (1997). "A Supersymmetry Primer". arXiv:hep-ph/9709356.

[3] Dine, Michael (2007). *Supersymmetry and String Theory: Beyond the Standard Model.* p. 169.

[4] Ellis, John. "The Physics Landscape after the Higgs Discovery at the LHC". *arXiv.* Invited plenary talk at SILAFAE 2014. Retrieved 8 July 2015.

[5] Wolchover, Natalie (November 20, 2012). "Supersymmetry Fails Test, Forcing Physics to Seek New Ideas". *Quanta Magazine.*

[6] Jonathan Feng: Supersymmetric Dark Matter *(pdf)*, University of California, Irvine, 11 May 2007

[7] Torsten Bringmann: The WIMP "Miracle" *(pdf)* University of Hamburg

[8] R. Haag, J. T. Lopuszanski and M. Sohnius, "All Possible Generators Of Supersymmetries Of The S Matrix", Nucl. Phys. B 88 (1975) 257

[9] H. Miyazawa (1966). "Baryon Number Changing Currents". *Prog. Theor. Phys.* **36** (6): 1266–1276. Bibcode:1966PThPh..36.1266M. doi:10.1143/PTP.36.1266.

[10] H. Miyazawa (1968). "Spinor Currents and Symmetries of Baryons and Mesons". *Phys. Rev.* **170** (5): 1586–1590. Bibcode:1968PhRv..170.1586M. doi:10.1103/PhysRev.170.1586.

[11] Michio Kaku, *Quantum Field Theory*, ISBN 0-19-509158-2, pg 663.

[12] Peter Freund, *Introduction to Supersymmetry*, ISBN 0-521-35675-X, pages 26-27, 138.

[13] Gervais, J. -L.; Sakita, B. (1971). "Field theory interpretation of supergauges in dual models". *Nuclear Physics B* **34** (2): 632–639. Bibcode:1971NuPhB..34..632G. doi:10.1016/0550-3213(71)90351-8.

[14] D.V. Volkov, V.P. Akulov, Pisma Zh.Eksp.Teor.Fiz. 16 (1972) 621; Phys.Lett. B46 (1973) 109; V.P. Akulov, D.V. Volkov, Teor.Mat.Fiz. 18 (1974) 39

[15] Ramond, P. (1971). "Dual Theory for Free Fermions". *Physical Review D* **3** (10): 2415–2418. Bibcode:1971PhRvD...3.2415R. doi:10.1103/PhysRevD.3.2415.

[16] Wess, J.; Zumino, B. (1974). "Supergauge transformations in four dimensions". *Nuclear Physics B* **70**: 39–50. Bibcode:1974NuPhB..70...39W. doi:10.1016/0550-3213(74)90355-1.

[17] http://users.physik.tu-berlin.de/~{}kleinert/kleinert/?p= supersym suggested here

[18] Iachello, F. (1980). "Dynamical Supersymmetries in Nuclei". *Physical Review Letters* **44** (12): 772–775. Bibcode:1980PhRvL..44..772I. doi:10.1103/PhysRevLett.44.772.

[19] Friedan, D.; Qiu, Z.; Shenker, S. (1984). "Conformal Invariance, Unitarity, and Critical Exponents in Two Dimensions". *Physical Review Letters* **52** (18): 1575–1578. Bibcode:1984PhRvL..52.1575F. doi:10.1103/PhysRevLett.52.1575.

[20] Gordon L. Kane, *The Dawn of Physics Beyond the Standard Model*. Scientific American, June 2003, page 60 and *The frontiers of physics*, special edition, Vol 15, #3, page 8

[21] *Supersymmetry in Disorder and Chaos*, Konstantin Efetov, Cambridge university press, 1997.

[22] Miri, M.-A.; Heinrich, M.; El-Ganainy, R.; Christodoulides, D. N. (2013). "Superymmetric optical structures". *Physical Review Letters* (APS) **110** (23): 233902. arXiv:1304.6646. Bibcode:2013PhRvL.110w3902M. doi:10.1103/PhysRevLett.110.233902. PMID 25167493. Retrieved April 2014.

[23] Heinrich, M.; Miri, M.-A.; Stützer, S.; El-Ganainy, R.; Nolte, S.; Szameit, A.; Christodoulides, D. N. (2014). "Superymmetric mode converters". *Nature Communications* (NPG) **5**: 3698. arXiv:1401.5734. Bibcode:2014NatCo...5E3698H. doi:10.1038/ncomms4698. PMID 24739256. Retrieved April 2014.

[24] Miri, M.-A.; Heinrich, Matthias; Christodoulides, D. N. (2014). "SUSY-inspired one-dimensional transformation optics". *Optica* (OSA) **1** (2): 89. arXiv:1408.0832. doi:10.1364/OPTICA.1.000089. Retrieved August 2014.

[25] Krasnitz, Michael (2002). *Correlation functions in supersymmetric gauge theories from supergravity fluctuafluctuations hHKtions* (PDF). Princeton University Department of Physics: Princeton University Department of Physics. p. 91.

[26] Polchinski,J. *String theory. Vol. 2: Superstring theory and beyond*, Appendix B

[27] LEPSUSYWG, ALEPH, DELPHI, L3 and OPAL experiments, charginos, large m0 LEPSUSYWG/01-03.1

[28] The D0-Collaboration (2009). "Search for associated production of charginos and neutralinos in the trilepton final state using 2.3 fb^{-1} of data". arXiv:0901.0646. Bibcode:2009PhLB..680...34D. doi:10.1016/j.physletb.2009.08.011.

[29] The D0 Collaboration (2006). "Search for squarks and gluinos in events with jets and missing transverse energy using 2.1 fb-1 of pp̄ collision data at s=1.96 TeV". arXiv:0712.3805. Bibcode:2008PhLB..660..449D. doi:10.1016/j.physletb.2008.01.042.

[30] O. Buchmueller; et al. (2009). "Likelihood Functions for Supersymmetric Observables in Frequentist Analyses of the CMSSM and NUHM1". *The European Physical Journal C* **64** (3): 391–415. arXiv:0907.5568. Bibcode:2009EPJC...64..391B. doi:10.1140/epjc/s10052-009-1159-z.

[31] Roszkowski, Leszek; Sessolo, Enrico Maria; Williams, Andrew J. (11 August 2014). "What next for the CMSSM and the NUHM: improved prospects for superpartner and dark matter detection". *Journal of High Energy Physics* **2014** (8). arXiv:1405.4289. Bibcode:2014JHEP...08..067R. doi:10.1007/JHEP08(2014)067.

[32] Marcela Carena and Howard E. Haber; Haber (1970). "Higgs Boson Theory and Phenomenology". *Progress in Particle and Nuclear Physics* **50**: 63–152. arXiv:hep-ph/0208209v3. Bibcode:2003PrPNP..50...63C. doi:10.1016/S0146-6410(02)00177-1.

[33] Patrick Draper; et al. (December 2011). "Implications of a 125 GeV Higgs for the MSSM and Low-Scale SUSY Breaking". *Physical Review D* **85** (9): 095007. arXiv:1112.3068. Bibcode:2012PhRvD..85i5007D. doi:10.1103/PhysRevD.85.095007.

3.1.9   Further reading

- Supersymmetry and Supergravity page in String Theory Wiki lists more books and reviews.

Theoretical introductions, free and online

- S. Martin (2011). "A Supersymmetry Primer". arXiv:hep-ph/9709356.

- Joseph D. Lykken (1996). "Introduction to Supersymmetry". arXiv:hep-th/9612114.

- Manuel Drees (1996). "An Introduction to Supersymmetry". arXiv:hep-ph/9611409.

- Adel Bilal (2001). "Introduction to Supersymmetry". arXiv:hep-th/0101055.

- An Introduction to Global Supersymmetry by Philip Arygres, 2001

Monographs

- Weak Scale Supersymmetry by Howard Baer and Xerxes Tata, 2006.

- Cooper, F.; Khare, A.; Sukhatme, U. (1995). "Supersymmetry and quantum mechanics". *Physics Reports* **251** (5–6): 267–385. arXiv:hep-th/9405029. Bibcode:1995PhR...251..267C. doi:10.1016/0370-1573(94)00080-M. (arXiv:hep-th/9405029).

- Junker, G. (1996). "Supersymmetric Methods in Quantum and Statistical Physics". doi:10.1007/978-3-642-61194-0. ISBN 978-3-540-61591-0..

- Kane, Gordon L., *Supersymmetry: Unveiling the Ultimate Laws of Nature*, Basic Books, New York (2001). ISBN 0-7382-0489-7.

- Kane, Gordon L., and Shifman, M., eds. *The Supersymmetric World: The Beginnings of the Theory*, World Scientific, Singapore (2000). ISBN 981-02-4522-X.

- Müller-Kirsten, Harald J. W., and Wiedemann, Armin, *Introduction to Supersymmetry*, 2nd ed., World Scientific, Singapore (2010). ISBN 978-981-4293-41-9.

- Weinberg, Steven, *The Quantum Theory of Fields, Volume 3: Supersymmetry*, Cambridge University Press, Cambridge, (1999). ISBN 0-521-66000-9.

- Wess, Julius, and Jonathan Bagger, *Supersymmetry and Supergravity*, Princeton University Press, Princeton, (1992). ISBN 0-691-02530-4.

- "Concise Encyclopedia of Supersymmetry". 2003. doi:10.1007/1-4020-4522-0. ISBN 978-1-4020-1338-6.

On experiments

- Bennett GW; Muon (g−2) Collaboration; Bousquet; Brown; Bunce; Carey; Cushman; Danby; Debevec; Deile; Deng; Dhawan; Druzhinin; Duong; Farley; Fedotovich; Gray; Grigoriev; Grosse-Perdekamp; Grossmann; Hare; Hertzog; Huang; Hughes; Iwasaki; Jungmann; Kawall; Khazin; Krienen; Kronkvist; et al. (2004). "Measurement of the negative muon anomalous magnetic moment to 0.7 ppm". *Physical Review Letters* **92** (16): 161802. arXiv:hep-ex/0401008. Bibcode:2004PhRvL..92p1802B.

doi:10.1103/PhysRevLett.92.161802. PMID 15169217.

- Brookhaven National Laboratory (Jan. 8, 2004). *New g−2 measurement deviates further from Standard Model.* Press Release.

- Fermi National Accelerator Laboratory (Sept 25, 2006). *Fermilab's CDF scientists have discovered the quick-change behavior of the B-sub-s meson.* Press Release.

3.1.10 External links

- Supersymmetry (physics) at *Encyclopædia Britannica*

- What do current LHC results (mid-August 2011) imply about supersymmetry? Matt Strassler

- ATLAS Experiment Supersymmetry search documents

- CMS Experiment Supersymmetry search documents

- "Particle wobble shakes up supersymmetry", *Cosmos* magazine, September 2006

- LHC results put supersymmetry theory 'on the spot' BBC news 27/8/2011

- SUSY running out of hiding places BBC news 12/11/2012

- Supersymmetry in optics? "Skulls in the Stars" blog 22/08/2013

3.2 String theory

For a more accessible and less technical introduction to this topic, see Introduction to M-theory.

In physics, **string theory** is a theoretical framework in which the point-like particles of particle physics are replaced by one-dimensional objects called strings. It describes how these strings propagate through space and interact with each other. On distance scales larger than the string scale, a string looks just like an ordinary particle, with its mass, charge, and other properties determined by the vibrational state of the string. In string theory, one of the many vibrational states of the string corresponds to the graviton, a quantum mechanical particle that carries gravitational force. Thus string theory is a theory of quantum gravity.

String theory is a broad and varied subject that attempts to address a number of deep questions of fundamental physics.

String theory has been applied to a variety of problems in black hole physics, early universe cosmology, nuclear physics, and condensed matter physics, and it has stimulated a number of major developments in pure mathematics. Because string theory potentially provides a unified description of gravity and particle physics, it is a candidate for a theory of everything, a self-contained mathematical model that describes all fundamental forces and forms of matter. Despite much work on these problems, it is not known to what extent string theory describes the real world or how much freedom the theory allows to choose the details.

String theory was first studied in the late 1960s as a theory of the strong nuclear force, before being abandoned in favor of quantum chromodynamics. Subsequently, it was realized that the very properties that made string theory unsuitable as a theory of nuclear physics made it a promising candidate for a quantum theory of gravity. The earliest version of string theory, bosonic string theory, incorporated only the class of particles known as bosons. It later developed into superstring theory, which posits a connection called supersymmetry between bosons and the class of particles called fermions. Five consistent versions of superstring theory were developed before it was conjectured in the mid-1990s that they were all different limiting cases of a single theory in eleven dimensions known as M-theory. In late 1997, theorists discovered an important relationship called the AdS/CFT correspondence, which relates string theory to another type of physical theory called a quantum field theory.

One of the challenges of string theory is that the full theory does not have a satisfactory definition in all circumstances. Another issue is that the theory is thought to describe an enormous landscape of possible universes, and this has complicated efforts to develop theories of particle physics based on string theory. These issues have led some in the community to criticize these approaches to physics and question the value of continued research on string theory unification.

3.2.1   Fundamentals

In the twentieth century, two theoretical frameworks emerged for formulating the laws of physics. One of these frameworks was Albert Einstein's general theory of relativity, a theory that explains the force of gravity and the structure of space and time. The other was quantum mechanics, a radically different formalism for describing physical phenomena using probability. By the late 1970s, these two frameworks had proven to be sufficient to explain most of the observed features of the universe, from elementary particles to atoms to the evolution of stars and the universe as a whole.[1]

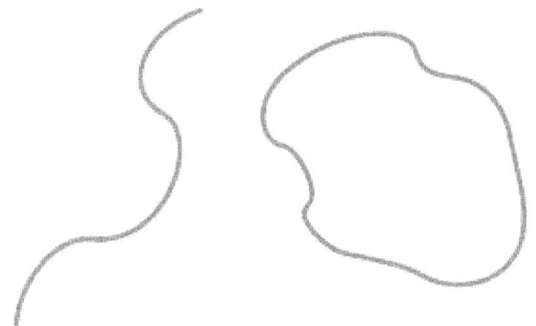

The fundamental objects of string theory are open and closed strings.

In spite of these successes, there are still many problems that remain to be solved. One of the deepest problems in modern physics is the problem of quantum gravity.[1] The general theory of relativity is formulated within the framework of classical physics, whereas the other fundamental forces are described within the framework of quantum mechanics. A quantum theory of gravity is needed in order to reconcile general relativity with the principles of quantum mechanics, but difficulties arise when one attempts to apply the usual prescriptions of quantum theory to the force of gravity.[2] In addition to the problem of developing a consistent theory of quantum gravity, there are many other fundamental problems in the physics of atomic nuclei, black holes, and the early universe.[lower-alpha 1]

String theory is a theoretical framework that attempts to address these questions and many others. The starting point for string theory is the idea that the point-like particles of particle physics can also be modeled as one-dimensional objects called strings. String theory describes how strings propagate through space and interact with each other. In a given version of string theory, there is only one kind of string, which may look like a small loop or segment of ordinary string, and it can vibrate in different ways. On distance scales larger than the string scale, a string will look just like an ordinary particle, with its mass, charge, and other properties determined by the vibrational state of the string. In this way, all of the different elementary particles may be viewed as vibrating strings. In string theory, one of the vibrational states of the string gives rise to the graviton, a quantum mechanical particle that carries gravitational force. Thus string theory is a theory of quantum gravity.[3]

One of the main developments of the past several decades in string theory was the discovery of certain "dualities", mathematical transformations that identify one physical theory with another. Physicists studying string theory have discovered a number of these dualities between different versions of string theory, and this has led to the conjecture that all consistent versions of string theory are subsumed in a single framework known as M-theory.[4]

Studies of string theory have also yielded a number of results on the nature of black holes and the gravitational interaction. There are certain paradoxes that arise when one attempts to understand the quantum aspects of black holes, and work on string theory has attempted to clarify these issues. In late 1997 this line of work culminated in the discovery of the anti-de Sitter/conformal field theory correspondence or AdS/CFT.[5] This is a theoretical result which relates string theory to other physical theories which are better understood theoretically. The AdS/CFT correspondence has implications for the study of black holes and quantum gravity, and it has been applied to other subjects, including nuclear[6] and condensed matter physics.[7][8]

Since string theory incorporates all of the fundamental interactions, including gravity, many physicists hope that it fully describes our universe, making it a theory of everything. One of the goals of current research in string theory is to find a solution of the theory that reproduces the observed spectrum of elementary particles, with a small cosmological constant, containing dark matter and a plausible mechanism for cosmic inflation. While there has been progress toward these goals, it is not known to what extent string theory describes the real world or how much freedom the theory allows to choose the details.[9]

One of the challenges of string theory is that the full theory does not have a satisfactory definition in all circumstances. The scattering of strings is most straightforwardly defined using the techniques of perturbation theory, but it is not known in general how to define string theory nonperturbatively.[10] It is also not clear whether there is any principle by which string theory selects its vacuum state, the physical state that determines the properties of our universe.[11] These problems have led some in the community to criticize these approaches to the unification of physics and question the value of continued research on these problems.[12]

Strings

Main article: String (physics)

The application of quantum mechanics to physical objects such as the electromagnetic field, which are extended in space and time, is known as quantum field theory. In particle physics, quantum field theories form the basis for our understanding of elementary particles, which are modeled as excitations in the fundamental fields.[13]

In quantum field theory, one typically computes the probabilities of various physical events using the techniques of perturbation theory. Developed by Richard Feynman and others in the first half of the twentieth century, pertur-

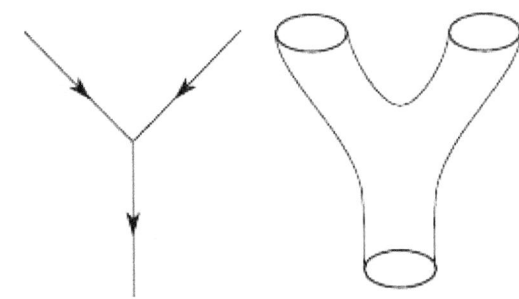

Interaction in the quantum world: worldlines of point-like particles or a worldsheet swept up by closed strings in string theory.

bative quantum field theory uses special diagrams called Feynman diagrams to organize computations. One imagines that these diagrams depict the paths of point-like particles and their interactions.[13]

The starting point for string theory is the idea that the point-like particles of quantum field theory can also be modeled as one-dimensional objects called strings.[14] The interaction of strings is most straightforwardly defined by generalizing the perturbation theory used in ordinary quantum field theory. At the level of Feynman diagrams, this means replacing the one-dimensional diagram representing the path of a point particle by a two-dimensional surface representing the motion of a string.[15] Unlike in quantum field theory, string theory does not have a full non-perturbative definition, so many of the theoretical questions that physicists would like to answer remain out of reach.[16]

In theories of particle physics based on string theory, the characteristic length scale of strings is assumed to be on the order of the Planck length, or 10^{-35} meters, the scale at which the effects of quantum gravity are believed to become significant.[15] On much larger length scales, such as the scales visible in physics laboratories, such objects would be indistinguishable from zero-dimensional point particles, and the vibrational state of the string would determine the type of particle. One of the vibrational states of a string corresponds to the graviton, a quantum mechanical particle that carries the gravitational force.[3]

The original version of string theory was bosonic string theory, but this version described only bosons, a class of particles which transmit forces between the matter particles, or fermions. Bosonic string theory was eventually superseded by theories called superstring theories. These theories describe both bosons and fermions, and they incorporate a theoretical idea called supersymmetry. This is a mathematical relation that exists in certain physical theories between the bosons and fermions. In theories with supersymmetry, each boson has a counterpart which is a fermion, and vice versa.[17]

There are several versions of superstring theory: type I, type IIA, type IIB, and two flavors of heterotic string theory ($SO(32)$ and $E_8 \times E_8$). The different theories allow different types of strings, and the particles that arise at low energies exhibit different symmetries. For example, the type I theory includes both open strings (which are segments with endpoints) and closed strings (which form closed loops), while types IIA, IIB and heterotic include only closed strings.[18]

Extra dimensions

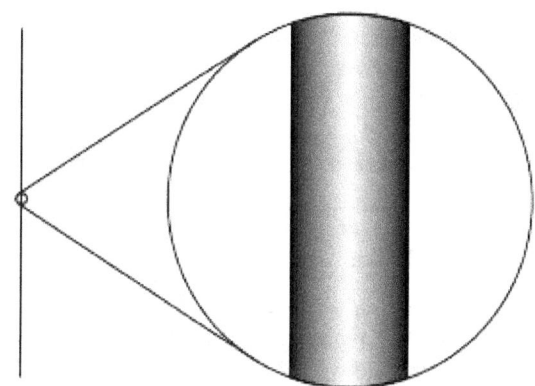

An example of compactification: At large distances, a two dimensional surface with one circular dimension looks one-dimensional.

In everyday life, there are three familiar dimensions of space: height, width and length. Einstein's general theory of relativity treats time as a dimension on par with the three spatial dimensions; in general relativity, space and time are not modeled as separate entities but are instead unified to a four-dimensional spacetime. In this framework, the phenomenon of gravity is viewed as a consequence of the geometry of spacetime.[19]

In spite of the fact that the universe is well described by four-dimensional spacetime, there are several reasons why physicists consider theories in other dimensions. In some cases, by modeling spacetime in a different number of dimensions, a theory becomes more mathematically tractable, and one can perform calculations and gain general insights more easily.[lower-alpha 2] There are also situations where theories in two or three spacetime dimensions are useful for describing phenomena in condensed matter physics.[20] Finally, there exist scenarios in which there could actually be more than four dimensions of spacetime which have nonetheless managed to escape detection.[21]

One notable feature of string theories is that these theories require extra dimensions of spacetime for their mathematical consistency. In bosonic string theory, spacetime is 26-dimensional, while in superstring theory it is ten-dimensional. In order to describe real physical phenomena using string theory, one must therefore imagine scenarios in which these extra dimensions would not be observed in experiments.[22]

A cross section of a quintic Calabi–Yau manifold

Compactification is one way of modifying the number of dimensions in a physical theory. In compactification, some of the extra dimensions are assumed to "close up" on themselves to form circles.[23] In the limit where these curled up dimensions become very small, one obtains a theory in which spacetime has effectively a lower number of dimensions. A standard analogy for this is to consider a multidimensional object such as a garden hose. If the hose is viewed from a sufficient distance, it appears to have only one dimension, its length. However, as one approaches the hose, one discovers that it contains a second dimension, its circumference. Thus, an ant crawling on the surface of the hose would move in two dimensions.[24]

Compactification can be used to construct models in which spacetime is effectively four-dimensional. However, not every way of compactifying the extra dimensions produces a model with the right properties to describe nature. In a viable model of particle physics, the compact extra dimensions must be shaped like a Calabi–Yau manifold.[23] A Calabi–Yau manifold is a special space which is typically taken to be six-dimensional in applications to string theory. It is named after mathematicians Eugenio Calabi and Shing-Tung Yau.[25]

Another approach to reducing the number of dimensions

is the so-called brane-world scenario. In this approach, physicists assume that the observable universe is a four-dimensional subspace of a higher dimensional space. In such models, the force-carrying bosons of particle physics arise from open strings with endpoints attached to the four-dimensional subspace, while gravity arises from closed strings propagating through the larger ambient space. This idea plays an important role in attempts to develop models of real world physics based on string theory, and it provides a natural explanation for the weakness of gravity compared to the other fundamental forces.[26]

Dualities

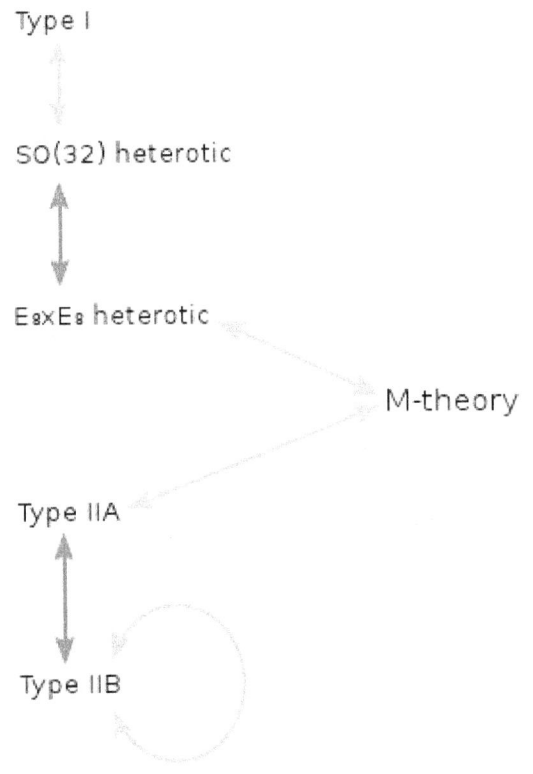

A diagram of string theory dualities. Yellow arrows indicate S-duality. Blue arrows indicate T-duality.

Main articles: S-duality and T-duality

One notable fact about string theory is that the different versions of the theory all turn out to be related in highly nontrivial ways. One of the relationships that can exist between different string theories is called S-duality. This is a relationship which says that a collection of strongly interacting particles in one theory can, in some cases, be viewed as a collection of weakly interacting particles in a completely different theory. Roughly speaking, a collection of particles is said to be strongly interacting if they combine and decay often and weakly interacting if they do so infrequently. Type I string theory turns out to be equivalent by S-duality to the $SO(32)$ heterotic string theory. Similarly, type IIB string theory is related to itself in a nontrivial way by S-duality.[27]

Another relationship between different string theories is T-duality. Here one considers strings propagating around a circular extra dimension. T-duality states that a string propagating around a circle of radius R is equivalent to a string propagating around a circle of radius $1/R$ in the sense that all observable quantities in one description are identified with quantities in the dual description. For example, a string has momentum as it propagates around a circle, and it can also wind around the circle one or more times. The number of times the string winds around a circle is called the winding number. If a string has momentum p and winding number n in one description, it will have momentum n and winding number p in the dual description. For example, type IIA string theory is equivalent to type IIB string theory via T-duality, and the two versions of heterotic string theory are also related by T-duality.[27]

In general, the term *duality* refers to a situation where two seemingly different physical systems turn out to be equivalent in a nontrivial way. Two theories related by a duality need not be string theories. For example, Montonen–Olive duality is example of an S-duality relationship between quantum field theories. The AdS/CFT correspondence is example of a duality which relates string theory to a quantum field theory. If two theories are related by a duality, it means that one theory can be transformed in some way so that it ends up looking just like the other theory. The two theories are then said to be *dual* to one another under the transformation. Put differently, the two theories are mathematically different descriptions of the same phenomena.[28]

Branes

Main article: Brane

In string theory and related theories, a brane is a physical object that generalizes the notion of a point particle to higher dimensions. For example, a point particle can be viewed as a brane of dimension zero, while a string can be viewed as a brane of dimension one. It is also possible to consider higher-dimensional branes. In dimension p, these are called p-branes. The word brane comes from the word "membrane" which refers to a two-dimensional brane.[29]

Branes are dynamical objects which can propagate through spacetime according to the rules of quantum mechanics. They have mass and can have other attributes such as charge. A p-brane sweeps out a $(p+1)$-dimensional volume

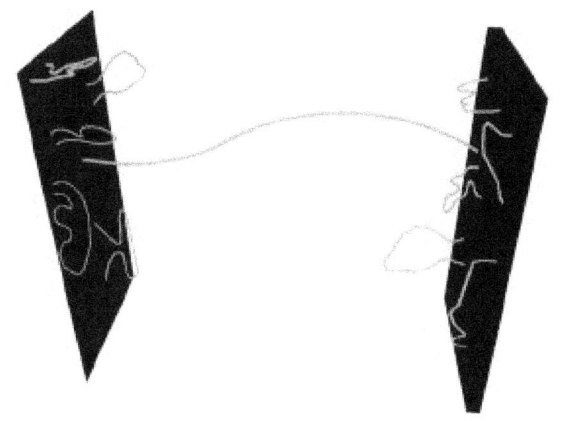

Open strings attached to a pair of D-branes

in spacetime called its *worldvolume*. Physicists often study fields analogous to the electromagnetic field which live on the worldvolume of a brane.[29]

In string theory, D-branes are an important class of branes that arise when one considers open strings. As an open string propagates through spacetime, its endpoints are required to lie on a D-brane. The letter "D" in D-brane refers to a certain mathematical condition on the system known as the Dirichlet boundary condition. The study of D-branes in string theory has led to important results such as the AdS/CFT correspondence, which has shed light on many problems in quantum field theory.[30]

Branes are also frequently studied from a purely mathematical point of view. Mathematically, branes can be described as objects of certain categories, such as the derived category of coherent sheaves on a complex algebraic variety, or the Fukaya category of a symplectic manifold.[31] The connection between the physical notion of a brane and the mathematical notion of a category has led to important mathematical insights in the fields of algebraic and symplectic geometry[32] and representation theory.[33]

3.2.2   M-theory

Main article: M-theory

Prior to 1995, theorists believed that there were five consistent versions of superstring theory (type I, type IIA, type IIB, and two versions of heterotic string theory). This understanding changed in 1995 when Edward Witten suggested that the five theories were just special limiting cases of an eleven-dimensional theory called M-theory. Witten's conjecture was based on the work of a number of other physicists, including Ashoke Sen, Chris Hull, Paul Townsend, and Michael Duff. His announcement led to a

flurry of research activity now known as the second superstring revolution.[34]

Unification of superstring theories

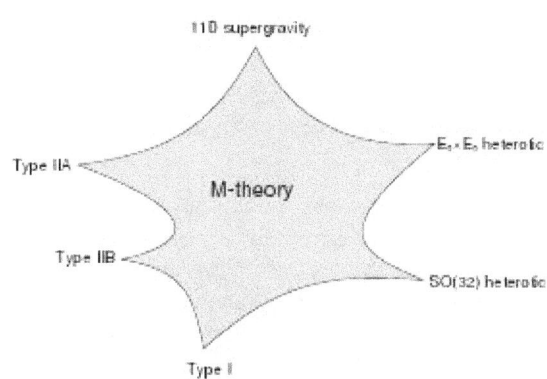

A schematic illustration of the relationship between M-theory, the five superstring theories, and eleven-dimensional supergravity. The shaded region represents a family of different physical scenarios that are possible in M-theory. In certain limiting cases corresponding to the cusps, it is natural to describe the physics using one of the six theories labeled there.

In the 1970s, many physicists became interested in supergravity theories, which combine general relativity with supersymmetry. Whereas general relativity makes sense in any number of dimensions, supergravity places an upper limit on the number of dimensions.[35] In 1978, work by Werner Nahm showed that the maximum spacetime dimension in which one can formulate a consistent supersymmetric theory is eleven.[36] In the same year, Eugene Cremmer, Bernard Julia, and Joel Scherk of the École Normale Supérieure showed that supergravity not only permits up to eleven dimensions but is in fact most elegant in this maximal number of dimensions.[37][38]

Initially, many physicists hoped that by compactifying eleven-dimensional supergravity, it might be possible to construct realistic models of our four-dimensional world. The hope was that such models would provide a unified description of the four fundamental forces of nature: electromagnetism, the strong and weak nuclear forces, and gravity. Interest in eleven-dimensional supergravity soon waned as various flaws in this scheme were discovered. One of the problems was that the laws of physics appear to distinguish between clockwise and counterclockwise, a phenomenon known as chirality. Edward Witten and others observed this chirality property cannot be readily derived by compactifying from eleven dimensions.[38]

In the first superstring revolution in 1984, many physicists turned to string theory as a unified theory of particle physics and quantum gravity. Unlike supergravity theory, string theory was able to accommodate the chirality of the standard model, and it provided a theory of gravity consistent with quantum effects.[38] Another feature of string theory that many physicists were drawn to in the 1980s and 1990s was its high degree of uniqueness. In ordinary particle theories, one can consider any collection of elementary particles whose classical behavior is described by an arbitrary Lagrangian. In string theory, the possibilities are much more constrained: by the 1990s, physicists had argued that there were only five consistent supersymmetric versions of the theory.[38]

Although there were only a handful of consistent superstring theories, it remained a mystery why there was not just one consistent formulation.[38] However, as physicists began to examine string theory more closely, they realized that these theories are related in intricate and nontrivial ways. They found that a system of strongly interacting strings can, in some cases, be viewed as a system of weakly interacting strings. This phenomenon is known as S-duality. It was studied by Ashoke Sen in the context of heterotic strings in four dimensions[39][40] and by Chris Hull and Paul Townsend in the context of the type IIB theory.[41] Theorists also found that different string theories may be related by T-duality. This duality implies that strings propagating on completely different spacetime geometries may be physically equivalent.[42]

At around the same time, as many physicists were studying the properties of strings, a small group of physicists was examining the possible applications of higher dimensional objects. In 1987, Eric Bergshoeff, Ergin Sezgin, and Paul Townsend showed that eleven-dimensional supergravity includes two-dimensional branes.[43] Intuitively, these objects look like sheets or membranes propagating through the eleven-dimensional spacetime. Shortly after this discovery, Michael Duff, Paul Howe, Takeo Inami, and Kellogg Stelle considered a particular compactification of eleven-dimensional supergravity with one of the dimensions curled up into a circle.[44] In this setting, one can imagine the membrane wrapping around the circular dimension. If the radius of the circle is sufficiently small, then this membrane looks just like a string in ten-dimensional spacetime. In fact, Duff and his collaborators showed that this construction reproduces exactly the strings appearing in type IIA superstring theory.[45]

Speaking at a string theory conference in 1995, Edward Witten made the surprising suggestion that all five superstring theories were in fact just different limiting cases of a single theory in eleven spacetime dimensions. Witten's announcement drew together all of the previous results on S- and T-duality and the appearance of higher dimensional branes in string theory.[46] In the months following Witten's announcement, hundreds of new papers appeared on the Internet confirming different parts of his proposal.[47] Today this flurry of work is known as the second superstring revolution.[48]

Initially, some physicists suggested that the new theory was a fundamental theory of membranes, but Witten was skeptical of the role of membranes in the theory. In a paper from 1996, Hořava and Witten wrote "As it has been proposed that the eleven-dimensional theory is a supermembrane theory but there are some reasons to doubt that interpretation, we will non-committally call it the M-theory, leaving to the future the relation of M to membranes."[49] In the absence of an understanding of the true meaning and structure of M-theory, Witten has suggested that the *M* should stand for "magic", "mystery", or "membrane" according to taste, and the true meaning of the title should be decided when a more fundamental formulation of the theory is known.[50]

Matrix theory

Main article: Matrix theory (physics)

In mathematics, a matrix is a rectangular array of numbers or other data. In physics, a matrix model is a particular kind of physical theory whose mathematical formulation involves the notion of a matrix in an important way. A matrix model describes the behavior of a set of matrices within the framework of quantum mechanics.[51]

One important example of a matrix model is the BFSS matrix model proposed by Tom Banks, Willy Fischler, Stephen Shenker, and Leonard Susskind in 1997. This theory describes the behavior of a set of nine large matrices. In their original paper, these authors showed, among other things, that the low energy limit of this matrix model is described by eleven-dimensional supergravity. These calculations led them to propose that the BFSS matrix model is exactly equivalent to M-theory. The BFSS matrix model can therefore be used as a prototype for a correct formulation of M-theory and a tool for investigating the properties of M-theory in a relatively simple setting.[51]

The development of the matrix model formulation of M-theory has led physicists to consider various connections between string theory and a branch of mathematics called noncommutative geometry. This subject is a generalization of ordinary geometry in which mathematicians define new geometric notions using tools from noncommutative algebra.[52] In a paper from 1998, Alain Connes, Michael R. Douglas, and Albert Schwarz showed that some aspects of matrix models and M-theory are described by a noncommutative quantum field theory, a special kind of physical theory in which spacetime is described mathematically using noncommutative geometry.[53] This established a link between matrix models and M-theory on the one hand, and noncommutative geometry on the other hand. It quickly led to the discovery of other important links between noncommutative geometry and various physical theories.[154][155]

3.2.3 Black holes

In general relativity, a black hole is defined as a region of spacetime in which the gravitational field is so strong that no particle or radiation can escape. In the currently accepted models of stellar evolution, black holes are thought to arise when massive stars undergo gravitational collapse, and many galaxies are thought to contain supermassive black holes at their centers. Black holes are also important for theoretical reasons, as they present profound challenges for theorists attempting to understand the quantum aspects of gravity. String theory has proved to be an important tool for investigating the theoretical properties of black holes because it provides a framework in which theorists can study their thermodynamics.[156]

Bekenstein–Hawking formula

In the branch of physics called statistical mechanics, entropy is a measure of the randomness or disorder of a physical system. This concept was studied in the 1870s by the Austrian physicist Ludwig Boltzmann, who showed that the thermodynamic properties of a gas could be derived from the combined properties of its many constituent molecules. Boltzmann argued that by averaging the behaviors of all the different molecules in a gas, one can understand macroscopic properties such as volume, temperature, and pressure. In addition, this perspective led him to give a precise definition of entropy as the natural logarithm of the number of different states of the molecules (also called *microstates*) that give rise to the same macroscopic features.[157]

In the twentieth century, physicists began to apply the same concepts to black holes. In most systems such as gases, the entropy scales with the volume. In the 1970s, the physicist Jacob Bekenstein suggested that the entropy of a black hole is instead proportional to the *surface area* of its event horizon, the boundary beyond which matter and radiation is lost to its gravitational attraction.[158] When combined with ideas of the physicist Stephen Hawking,[159] Bekenstein's work yielded a precise formula for the entropy of a black hole. The formula expresses the entropy S as

$$S = \frac{c^3 k A}{4 \hbar G}$$

where c is the speed of light, k is Boltzmann's constant, $\hbar$ is the reduced Planck constant, G is Newton's constant, and A

is the surface area of the event horizon.[160]

Like any physical system, a black hole has an entropy defined in terms of the number of different microstates that lead to the same macroscopic features. The Bekenstein–Hawking entropy formula gives the expected value of the entropy of a black hole, but by the 1990s, physicists still lacked a derivation of this formula by counting microstates in a theory of quantum gravity. Finding such a derivation of this formula was considered an important test of the viability of any theory of quantum gravity such as string theory.[161]

Derivation within string theory

In a paper from 1996, Andrew Strominger and Cumrun Vafa showed how to derive the Beckenstein–Hawking formula for certain black holes in string theory.[162] Their calculation was based on the observation that D-branes—which look like fluctuating membranes when they are weakly interacting—become dense, massive objects with event horizons when the interactions are strong. In other words, a system of strongly interacting D-branes in string theory is indistinguishable from a black hole. Strominger and Vafa analyzed such D-brane systems and calculated the number of different ways of placing D-branes in spacetime so that their combined mass and charge is equal to a given mass and charge for the resulting black hole. Their calculation reproduced the Bekenstein–Hawking formula exactly, including the factor of 1/4.[163] Subsequent work by Strominger, Vafa, and others refined the original calculations and gave the precise values of the "quantum corrections" needed to describe very small black holes.[164][165]

The black holes that Strominger and Vafa considered in their original work were quite different from real astrophysical black holes. One difference was that Strominger and Vafa considered only extremal black holes in order to make the calculation tractable. These are defined as black holes with the lowest possible mass compatible with a given charge.[166] Strominger and Vafa also restricted attention to black holes in five-dimensional spacetime with unphysical supersymmetry.[167]

Although it was originally developed in this very particular and physically unrealistic context in string theory, the entropy calculation of Strominger and Vafa has led to a qualitative understanding of how black hole entropy can be accounted for in any theory of quantum gravity. Indeed, in 1998, Strominger argued that the original result could be generalized to an arbitrary consistent theory of quantum gravity without relying on strings or supersymmetry.[168] In collaboration with several other authors in 2010, he showed that some results on black hole entropy could be extended to non-extremal astrophysical black holes.[169][170]

3.2.4   AdS/CFT correspondence

Main article: AdS/CFT correspondence

One approach to formulating string theory and studying
its properties is provided by the anti-de Sitter/conformal
field theory (AdS/CFT) correspondence. This is a theoretical result which implies that string theory is in some
cases equivalent to a quantum field theory. In addition
to providing insights into the mathematical structure of
string theory, the AdS/CFT correspondence has shed light
on many aspects of quantum field theory in regimes where
traditional calculational techniques are ineffective.[6] The
AdS/CFT correspondence was first proposed by Juan Maldacena in late 1997.[71] Important aspects of the correspondence were elaborated in articles by Steven Gubser, Igor
Klebanov, and Alexander Markovich Polyakov,[72] and by
Edward Witten.[73] By 2010, Maldacena's article had over
7000 citations, becoming the most highly cited article in the
field of high energy physics.[lower-alpha 3]

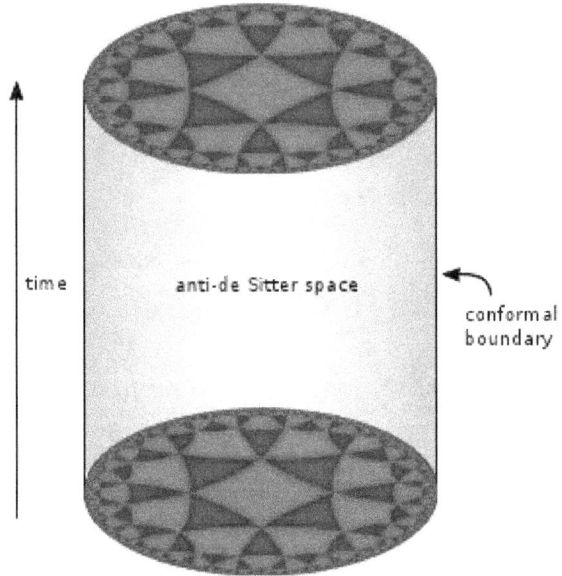

*Three-dimensional anti-de Sitter space is like a stack of hyperbolic
disks, each one representing the state of the universe at a given time.
The resulting spacetime looks like a solid cylinder.*

Overview of the correspondence

In the AdS/CFT correspondence, the geometry of spacetime is described in terms of a certain vacuum solution of
Einstein's equation called anti-de Sitter space.[74] In very
elementary terms, anti-de Sitter space is a mathematical
model of spacetime in which the notion of distance between points (the metric) is different from the notion of
distance in ordinary Euclidean geometry. It is closely related to hyperbolic space, which can be viewed as a disk as
illustrated on the left.[75] This image shows a tessellation of
a disk by triangles and squares. One can define the distance
between points of this disk in such a way that all the triangles and squares are the same size and the circular outer
boundary is infinitely far from any point in the interior.[76]

One can imagine a stack of hyperbolic disks where each
disk represents the state of the universe at a given time.
The resulting geometric object is three-dimensional anti-de
Sitter space.[75] It looks like a solid cylinder in which any
cross section is a copy of the hyperbolic disk. Time runs
along the vertical direction in this picture. The surface of
this cylinder plays an important role in the AdS/CFT correspondence. As with the hyperbolic plane, anti-de Sitter
space is curved in such a way that any point in the interior
is actually infinitely far from this boundary surface.[76]

This construction describes a hypothetical universe with
only two space dimensions and one time dimension, but it
can be generalized to any number of dimensions. Indeed,
hyperbolic space can have more than two dimensions and
one can "stack up" copies of hyperbolic space to get higher-
dimensional models of anti-de Sitter space.[75]

An important feature of anti-de Sitter space is its boundary
(which looks like a cylinder in the case of three-dimensional
anti-de Sitter space). One property of this boundary is that,
within a small region on the surface around any given point,
it looks just like Minkowski space, the model of spacetime
used in nongravitational physics.[77] One can therefore consider an auxiliary theory in which "spacetime" is given by
the boundary of anti-de Sitter space. This observation is the
starting point for AdS/CFT correspondence, which states
that the boundary of anti-de Sitter space can be regarded
as the "spacetime" for a quantum field theory. The claim is
that this quantum field theory is equivalent to a gravitational
theory, such as string theory, in the bulk anti-de Sitter space
in the sense that there is a "dictionary" for translating entities and calculations in one theory into their counterparts
in the other theory. For example, a single particle in the
gravitational theory might correspond to some collection of
particles in the boundary theory. In addition, the predictions in the two theories are quantitatively identical so that
if two particles have a 40 percent chance of colliding in the
gravitational theory, then the corresponding collections in
the boundary theory would also have a 40 percent chance
of colliding.[78]

Applications to quantum gravity

The discovery of the AdS/CFT correspondence was a major advance in physicists' understanding of string theory and
quantum gravity. One reason for this is that the correspon-

dence provides a formulation of string theory in terms of quantum field theory, which is well understood by comparison. Another reason is that it provides a general framework in which physicists can study and attempt to resolve the paradoxes of black holes.[56]

In 1975, Stephen Hawking published a calculation which suggested that black holes are not completely black but emit a dim radiation due to quantum effects near the event horizon.[59] At first, Hawking's result posed a problem for theorists because it suggested that black holes destroy information. More precisely, Hawking's calculation seemed to conflict with one of the basic postulates of quantum mechanics, which states that physical systems evolve in time according to the Schrödinger equation. This property is usually referred to as unitarity of time evolution. The apparent contradiction between Hawking's calculation and the unitarity postulate of quantum mechanics came to be known as the black hole information paradox.[79]

The AdS/CFT correspondence resolves the black hole information paradox, at least to some extent, because it shows how a black hole can evolve in a manner consistent with quantum mechanics in some contexts. Indeed, one can consider black holes in the context of the AdS/CFT correspondence, and any such black hole corresponds to a configuration of particles on the boundary of anti-de Sitter space.[80] These particles obey the usual rules of quantum mechanics and in particular evolve in a unitary fashion, so the black hole must also evolve in a unitary fashion, respecting the principles of quantum mechanics.[81] In 2005, Hawking announced that the paradox had been settled in favor of information conservation by the AdS/CFT correspondence, and he suggested a concrete mechanism by which black holes might preserve information.[82]

Applications to quantum field theory

Main articles: AdS/QCD correspondence and AdS/CMT correspondence

In addition to its applications to theoretical problems in quantum gravity, the AdS/CFT correspondence has been applied to a variety of problems in quantum field theory. One physical system that has been studied using the AdS/CFT correspondence is the quark–gluon plasma, an exotic state of matter produced in particle accelerators. This state of matter arises for brief instants when heavy ions such as gold or lead nuclei are collided at high energies. Such collisions cause the quarks that make up atomic nuclei to deconfine at temperatures of approximately two trillion kelvins, conditions similar to those present at around 10^{-11} seconds after the Big Bang.[83]

The physics of the quark–gluon plasma is governed by a theory called quantum chromodynamics, but this the-

A magnet levitating above a high-temperature superconductor. Today some physicists are working to understand high-temperature superconductivity using the AdS/CFT correspondence.[7]

ory is mathematically intractable in problems involving the quark–gluon plasma.[lower-alpha 4] In an article appearing in 2005, Đàm Thanh Sơn and his collaborators showed that the AdS/CFT correspondence could be used to understand some aspects of the quark–gluon plasma by describing it in the language of string theory.[84] By applying the AdS/CFT correspondence, Sơn and his collaborators were able to describe the quark gluon plasma in terms of black holes in five-dimensional spacetime. The calculation showed that the ratio of two quantities associated with the quark–gluon plasma, the shear viscosity and volume density of entropy, should be approximately equal to a certain universal constant. In 2008, the predicted value of this ratio for the quark–gluon plasma was confirmed at the Relativistic Heavy Ion Collider at Brookhaven National Laboratory.[85][86]

The AdS/CFT correspondence has also been used to study aspects of condensed matter physics. Over the decades, experimental condensed matter physicists have discovered a number of exotic states of matter, including superconductors and superfluids. These states are described using the formalism of quantum field theory, but some phenomena are difficult to explain using standard field theoretic techniques. Some condensed matter theorists including Subir Sachdev hope that the AdS/CFT correspondence will make it possible to describe these systems in the language of string theory and learn more about their behavior.[85]

So far some success has been achieved in using string theory methods to describe the transition of a superfluid to an insulator. A superfluid is a system of electrically neutral atoms that flows without any friction. Such systems are often produced in the laboratory using liquid helium, but recently experimentalists have developed new ways of producing artificial superfluids by pouring trillions of cold atoms into a lattice of criss-crossing lasers. These atoms initially behave as a superfluid, but as experimentalists increase the intensity of the lasers, they become less mobile and then suddenly transition to an insulating state. During the transition, the atoms behave in an unusual way. For example, the atoms slow to a halt at a rate that depends on the temperature and on Planck's constant, the fundamental parameter of quantum mechanics, which does not enter into the description of the other phases. This behavior has recently been understood by considering a dual description where properties of the fluid are described in terms of a higher dimensional black hole.[87]

3.2.5 Phenomenology

Main article: String phenomenology

In addition to being an idea of considerable theoretical interest, string theory provides a framework for constructing models of real world physics that combine general relativity and particle physics. Phenomenology is the branch of theoretical physics in which physicists construct realistic models of nature from more abstract theoretical ideas. String phenomenology is the part of string theory that attempts to construct realistic or semi-realistic models based on string theory.

Partly because of theoretical and mathematical difficulties and partly because of the extremely high energies needed to test these theories experimentally, there is so far no experimental evidence that would unambiguously point to any of these models being a correct fundamental description of nature. This has led some in the community to criticize these approaches to unification and question the value of continued research on these problems.[12]

Particle physics

The currently accepted theory describing elementary particles and their interactions is known as the standard model of particle physics. This theory provides a unified description of three of the fundamental forces of nature: electromagnetism and the strong and weak nuclear forces. Despite its remarkable success in explaining a wide range of physical phenomena, the standard model cannot be a complete description of reality. This is because the standard model fails to incorporate the force of gravity and because of problems such as the hierarchy problem and the inability to explain the structure of fermion masses or dark matter.

String theory has been used to construct a variety of models of particle physics going beyond the standard model. Typically, such models are based on the idea of compactification. Starting with the ten- or eleven-dimensional space-

time of string or M-theory, physicists postulate a shape for the extra dimensions. By choosing this shape appropriately, they can construct models roughly similar to the standard model of particle physics, together with additional undiscovered particles.[88] One popular way of deriving realistic physics from string theory is to start with the heterotic theory in ten dimensions and assume that the six extra dimensions of spacetime are shaped like a six-dimensional Calabi–Yau manifold. Such compactifications offer many ways of extracting realistic physics from string theory. Other similar methods can be used to construct realistic or semi-realistic models of our four-dimensional world based on M-theory.[89]

Cosmology

Main article: String cosmology

The Big Bang theory is the prevailing cosmological model

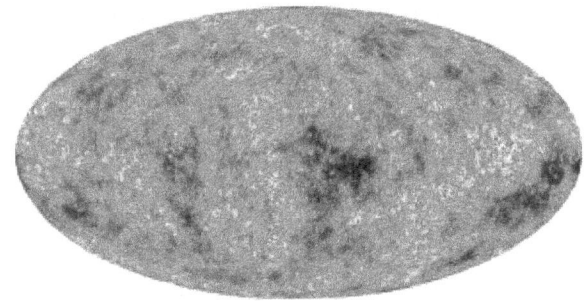

A map of the cosmic microwave background produced by the Wilkinson Microwave Anisotropy Probe

for the universe from the earliest known periods through its subsequent large-scale evolution. Despite its success in explaining many observed features of the universe including galactic redshifts, the relative abundance of light elements such as hydrogen and helium, and the existence of a cosmic microwave background, there are several questions that remain unanswered. For example, the standard Big Bang model does not explain why the universe appears to be same in all directions, why it appears flat on very large distance scales, or why certain hypothesized particles such as magnetic monopoles are not observed in experiments.[90]

Currently, the leading candidate for a theory going beyond the Big Bang is the theory of cosmic inflation. Developed by Alan Guth and others in the 1980s, inflation postulates a period of extremely rapid accelerated expansion of the universe prior to the expansion described by the standard Big Bang theory. The theory of cosmic inflation preserves the successes of the Big Bang while providing a natural explanation for some of the mysterious features of the universe.[91] The theory has also received striking support from observations of the cosmic microwave background, the radiation that has filled the sky since around 380,000 years after the Big Bang.[92]

In the theory of inflation, the rapid initial expansion of the universe is caused by a hypothetical particle called the inflaton. The exact properties of this particle are not fixed by the theory but should ultimately be derived from a more fundamental theory such as string theory.[93] Indeed, there have been a number of attempts to identify an inflaton within the spectrum of particles described by string theory, and to study inflation using string theory. While these approaches might eventually find support in observational data such as measurements of the cosmic microwave background, the application of string theory to cosmology is still in its early stages.[94]

3.2.6 Connections to mathematics

In addition to influencing research in theoretical physics, string theory has stimulated a number of major developments in pure mathematics. Like many developing ideas in theoretical physics, string theory does not at present have a mathematically rigorous formulation in which all of its concepts can be defined precisely. As a result, physicists who study string theory are often guided by physical intuition to conjecture relationships between the seemingly different mathematical structures that are used to formalize different parts of the theory. These conjectures are later proved by mathematicians, and in this way, string theory serves as a source of new ideas in pure mathematics.[95]

Mirror symmetry

Main article: Mirror symmetry (string theory)

After Calabi–Yau manifolds had entered physics as a way to compactify extra dimensions in string theory, many physicists began studying these manifolds. In the late 1980s, several physicists noticed that given such a compactification of string theory, it is not possible to reconstruct uniquely a corresponding Calabi–Yau manifold.[96] Instead, two different versions of string theory, type IIA and type IIB, can be compactified on completely different Calabi–Yau manifolds giving rise to the same physics. In this situation, the manifolds are called mirror manifolds, and the relationship between the two physical theories is called mirror symmetry.[97]

Regardless of whether Calabi–Yau compactifications of string theory provide a correct description of nature, the existence of the mirror duality between different string theories has significant mathematical consequences. The Calabi–Yau manifolds used in string theory are of interest in pure mathematics, and mirror symmetry allows mathematicians to solve problems in enumerative geometry, a

The Clebsch cubic is an example of a kind of geometric object called an algebraic variety. A classical result of enumerative geometry states that there are exactly 27 straight lines that lie entirely on this surface.

branch of mathematics concerned with counting the numbers of solutions to geometric questions.[31][98]

Enumerative geometry studies a class of geometric objects called algebraic varieties which are defined by the vanishing of polynomials. For example, the Clebsch cubic illustrated on the right is an algebraic variety defined using a certain polynomial of degree three in four variables. A celebrated result of nineteenth-century mathematicians Arthur Cayley and George Salmon states that there are exactly 27 straight lines that lie entirely on such a surface.[99]

Generalizing this problem, one can ask how many lines can be drawn on a quintic Calabi–Yau manifold, such as the one illustrated above, which is defined by a polynomial of degree five. This problem was solved by the nineteenth-century German mathematician Hermann Schubert, who found that there are exactly 2,875 such lines. In 1986, geometer Sheldon Katz proved that the number of curves, such as circles, that are defined by polynomials of degree two and lie entirely in the quintic is 609,250.[100]

By the year 1991, most of the classical problems of enumerative geometry had been solved and interest in enumerative geometry had begun to diminish.[101] The field was reinvigorated in May 1991 when physicists Philip Candelas, Xenia de la Ossa, Paul Green, and Linda Parks showed that mirror symmetry could be used to translate difficult mathematical questions about one Calabi–Yau manifold into easier questions about its mirror.[102] In particular, they used mirror symmetry to show that a six-dimensional Calabi–Yau manifold can contain exactly 317,206,375 curves of degree three.[101] In addition to counting degree-three curves, Candelas and his collaborators obtained a number of more general results for counting rational curves which went far beyond the results obtained by mathematicians.[103]

Originally, these results of Candelas were justified on physical grounds. However, mathematicians generally prefer rigorous proofs that do not require an appeal to physical intuition. Inspired by physicists' work on mirror symmetry, mathematicians have therefore constructed their own arguments proving the enumerative predictions of mirror symmetry.[lower-alpha 5] Today mirror symmetry is an active area of research in mathematics, and mathematicians are working to develop a more complete mathematical understanding of mirror symmetry based on physicists' intuition.[104] Major approaches to mirror symmetry include the homological mirror symmetry program of Maxim Kontsevich[32] and the SYZ conjecture of Andrew Strominger, Shing-Tung Yau, and Eric Zaslow.[105]

Monstrous moonshine

Main article: Monstrous moonshine
Group theory is the branch of mathematics that studies the

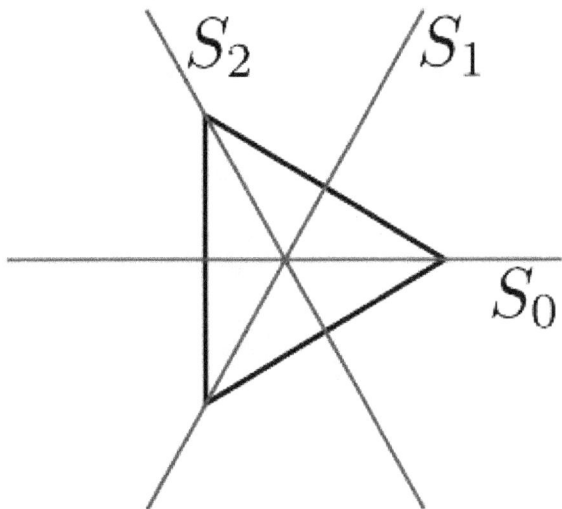

An equilateral triangle can be rotated through 120°, 240°, or 360°, or reflected in any of the three lines pictured without changing its shape.

concept of symmetry. For example, one can consider a geometric shape such as an equilateral triangle. There are various operations that one can perform on this triangle without changing its shape. One can rotate it through 120°, 240°, or 360°, or one can reflect in any of the lines labeled S_0, S_1, or S_2 in the picture. Each of these operations is called a *symmetry*, and the collection of these symmetries satisfies certain technical properties making it into what mathematicians call a group. In this particular example, the group is

known as the dihedral group of order 6 because it has six elements. A general group may describe finitely many or infinitely many symmetries; if there are only finitely many symmetries, it is called a finite group.[106]

Mathematicians often strive for a classification (or list) of all mathematical objects of a given type. It is generally believed that finite groups are too diverse to admit a useful classification. A more modest but still challenging problem is to classify all finite *simple* groups. These are finite groups which may be used as building blocks for constructing arbitrary finite groups in the same way that prime numbers can be used to construct arbitrary whole numbers by taking products.[lower-alpha 6] One of the major achievements of contemporary group theory is the classification of finite simple groups, a mathematical theorem which provides a list of all possible finite simple groups.[107]

This classification theorem identifies several infinite families of groups as well as 26 additional groups which do not fit into any family. The latter groups are called the "sporadic" groups, and each one owes its existence to a remarkable combination of circumstances. The largest sporadic group, the so-called monster group, has over 10^{53} elements, more than a thousand times the number of atoms in the Earth.[108]

A graph of the j-function in the complex plane

A seemingly unrelated construction is the *j*-function of number theory. This object belongs to a special class of functions called modular functions, whose graphs form a certain kind of repeating pattern.[109] Although this function appears in a branch of mathematics which seems very different from the theory of finite groups, the two subjects turn out to be intimately related. In the late 1970s, mathematicians John McKay and John Thompson noticed that certain numbers arising in the analysis of the monster group (namely, the dimensions of its irreducible representations) are related to numbers that appear in a formula for the *j*-function (namely, the coefficients of its Fourier series).[110] This relationship was further developed by John Horton

Conway and Simon Norton[111] who called it monstrous moonshine because it seemed so far fetched.[112]

In 1992, Richard Borcherds constructed a bridge between the theory of modular functions and finite groups and, in the process, explained the observations of McKay and Thompson.[113][114] Borcherds' work used ideas from string theory in an essential way, extending earlier results of Igor Frenkel, James Lepowsky, and Arne Meurman, who had realized the monster group as the symmetries of a particular version of string theory.[115] In 1998, Borcherds was awarded the Fields medal for his work.[116]

Since the 1990s, the connection between string theory and moonshine has led to further results in mathematics and physics.[108] In 2010, physicists Tohru Eguchi, Hirosi Ooguri, and Yuji Tachikawa discovered connections between a different sporadic group, the Mathieu group M_{24}, and a certain version of string theory.[117] Miranda Cheng, John Duncan, and Jeffrey A. Harvey proposed a generalization of this moonshine phenomenon called umbral moonshine,[118] and their conjecture was proved mathematically by Duncan, Michael Griffin, and Ken Ono.[119] Witten has also speculated that the version of string theory appearing in monstrous moonshine might be related to a certain simplified model of gravity in three spacetime dimensions.[120]

3.2.7 History

Main article: History of string theory

Early results

Some of the structures reintroduced by string theory arose for the first time much earlier as part of the program of classical unification started by Albert Einstein. The first person to add a fifth dimension to a theory of gravity was Gunnar Nordström in 1914, who noted that gravity in five dimensions describes both gravity and electromagnetism in four. Nordström attempted to unify electromagnetism with his theory of gravitation, which was however superseded by Einstein's general relativity in 1919. Thereafter, German mathematician Theodor Kaluza combined the fifth dimension with general relativity, and only Kaluza is usually credited with the idea. In 1926, the Swedish physicist Oskar Klein gave a physical interpretation of the unobservable extra dimension—it is wrapped into a small circle. Einstein introduced a non-symmetric metric tensor, while much later Brans and Dicke added a scalar component to gravity. These ideas would be revived within string theory, where they are demanded by consistency conditions.

String theory was originally developed during the late 1960s

Leonard Susskind

and early 1970s as a never completely successful theory of hadrons, the subatomic particles like the proton and neutron that feel the strong interaction. In the 1960s, Geoffrey Chew and Steven Frautschi discovered that the mesons make families called Regge trajectories with masses related to spins in a way that was later understood by Yoichiro Nambu, Holger Bech Nielsen and Leonard Susskind to be the relationship expected from rotating strings. Chew advocated making a theory for the interactions of these trajectories that did not presume that they were composed of any fundamental particles, but would construct their interactions from self-consistency conditions on the S-matrix. The S-matrix approach was started by Werner Heisenberg in the 1940s as a way of constructing a theory that did not rely on the local notions of space and time, which Heisenberg believed break down at the nuclear scale. While the scale was off by many orders of magnitude, the approach he advocated was ideally suited for a theory of quantum gravity.

Working with experimental data, R. Dolen, D. Horn and C. Schmid developed some sum rules for hadron exchange. When a particle and antiparticle scatter, virtual particles can be exchanged in two qualitatively different ways. In the s-channel, the two particles annihilate to make temporary intermediate states that fall apart into the final state particles. In the t-channel, the particles exchange intermediate states by emission and absorption. In field theory, the two contributions add together, one giving a continuous background contribution, the other giving peaks at certain energies. In the data, it was clear that the peaks were stealing from the background—the authors interpreted this as saying that the t-channel contribution was dual to the s-channel one, meaning both described the whole amplitude and included the other.

Gabriele Veneziano

The result was widely advertised by Murray Gell-Mann, leading Gabriele Veneziano to construct a scattering amplitude that had the property of Dolen-Horn-Schmid duality, later renamed world-sheet duality. The amplitude needed poles where the particles appear, on straight line trajectories, and there is a special mathematical function whose poles are evenly spaced on half the real line— the Gamma function— which was widely used in Regge theory. By manipulating combinations of Gamma functions, Veneziano was able to find a consistent scattering amplitude with poles on straight lines, with mostly positive residues, which obeyed duality and had the appropriate Regge scaling at high energy. The amplitude could fit near-beam scattering data as well as other Regge type fits, and had a suggestive integral representation that could be used for generalization.

Over the next years, hundreds of physicists worked to complete the bootstrap program for this model, with many surprises. Veneziano himself discovered that for the scattering

amplitude to describe the scattering of a particle that appears in the theory, an obvious self-consistency condition, the lightest particle must be a tachyon. Miguel Virasoro and Joel Shapiro found a different amplitude now understood to be that of closed strings, while Ziro Koba and Holger Nielsen generalized Veneziano's integral representation to multiparticle scattering. Veneziano and Sergio Fubini introduced an operator formalism for computing the scattering amplitudes that was a forerunner of world-sheet conformal theory, while Virasoro understood how to remove the poles with wrong-sign residues using a constraint on the states. Claud Lovelace calculated a loop amplitude, and noted that there is an inconsistency unless the dimension of the theory is 26. Charles Thorn, Peter Goddard and Richard Brower went on to prove that there are no wrong-sign propagating states in dimensions less than or equal to 26.

In 1969, Yoichiro Nambu, Holger Bech Nielsen, and Leonard Susskind recognized that the theory could be given a description in space and time in terms of strings. The scattering amplitudes were derived systematically from the action principle by Peter Goddard, Jeffrey Goldstone, Claudio Rebbi, and Charles Thorn, giving a space-time picture to the vertex operators introduced by Veneziano and Fubini and a geometrical interpretation to the Virasoro conditions.

In 1970, Pierre Ramond added fermions to the model, which led him to formulate a two-dimensional supersymmetry to cancel the wrong-sign states. John Schwarz and André Neveu added another sector to the fermi theory a short time later. In the fermion theories, the critical dimension was 10. Stanley Mandelstam formulated a world sheet conformal theory for both the bose and fermi case, giving a two-dimensional field theoretic path-integral to generate the operator formalism. Michio Kaku and Keiji Kikkawa gave a different formulation of the bosonic string, as a string field theory, with infinitely many particle types and with fields taking values not on points, but on loops and curves.

In 1974, Tamiaki Yoneya discovered that all the known string theories included a massless spin-two particle that obeyed the correct Ward identities to be a graviton. John Schwarz and Joel Scherk came to the same conclusion and made the bold leap to suggest that string theory was a theory of gravity, not a theory of hadrons. They reintroduced Kaluza–Klein theory as a way of making sense of the extra dimensions. At the same time, quantum chromodynamics was recognized as the correct theory of hadrons, shifting the attention of physicists and apparently leaving the bootstrap program in the dustbin of history.

String theory eventually made it out of the dustbin, but for the following decade all work on the theory was completely ignored. Still, the theory continued to develop at a steady pace thanks to the work of a handful of devotees. Ferdinando Gliozzi, Joel Scherk, and David Olive realized in 1976 that the original Ramond and Neveu Schwarz-strings were separately inconsistent and needed to be combined. The resulting theory did not have a tachyon, and was proven to have space-time supersymmetry by John Schwarz and Michael Green in 1981. The same year, Alexander Polyakov gave the theory a modern path integral formulation, and went on to develop conformal field theory extensively. In 1979, Daniel Friedan showed that the equations of motions of string theory, which are generalizations of the Einstein equations of General Relativity, emerge from the Renormalization group equations for the two-dimensional field theory. Schwarz and Green discovered T-duality, and constructed two superstring theories—IIA and IIB related by T-duality, and type I theories with open strings. The consistency conditions had been so strong, that the entire theory was nearly uniquely determined, with only a few discrete choices.

First superstring revolution

Edward Witten

In the early 1980s, Edward Witten discovered that most theories of quantum gravity could not accommodate chiral fermions like the neutrino. This led him, in collaboration with Luis Álvarez-Gaumé to study violations of the conservation laws in gravity theories with anomalies, concluding that type I string theories were inconsistent. Green and Schwarz discovered a contribution to the anomaly that Witten and Alvarez-Gaumé had missed, which restricted the gauge group of the type I string theory to be SO(32). In coming to understand this calculation, Edward Witten became convinced that string theory was truly a consistent theory of gravity, and he became a high-profile advocate. Fol-

lowing Witten's lead, between 1984 and 1986, hundreds of physicists started to work in this field, and this is sometimes called the first superstring revolution.

During this period, David Gross, Jeffrey Harvey, Emil Martinec, and Ryan Rohm discovered heterotic strings. The gauge group of these closed strings was two copies of E8, and either copy could easily and naturally include the standard model. Philip Candelas, Gary Horowitz, Andrew Strominger and Edward Witten found that the Calabi–Yau manifolds are the compactifications that preserve a realistic amount of supersymmetry, while Lance Dixon and others worked out the physical properties of orbifolds, distinctive geometrical singularities allowed in string theory. Cumrun Vafa generalized T-duality from circles to arbitrary manifolds, creating the mathematical field of mirror symmetry. Daniel Friedan, Emil Martinec and Stephen Shenker further developed the covariant quantization of the superstring using conformal field theory techniques. David Gross and Vipul Periwal discovered that string perturbation theory was divergent. Stephen Shenker showed it diverged much faster than in field theory suggesting that new non-perturbative objects were missing.

Joseph Polchinski

In the 1990s, Joseph Polchinski discovered that the theory requires higher-dimensional objects, called D-branes and identified these with the black-hole solutions of supergravity. These were understood to be the new objects suggested by the perturbative divergences, and they opened up a new field with rich mathematical structure. It quickly became clear that D-branes and other p-branes, not just strings, formed the matter content of the string theories, and the physical interpretation of the strings and branes was revealed—they are a type of black hole. Leonard Susskind had incorporated the holographic principle of Gerardus 't Hooft into string theory, identifying the long highly excited string states with ordinary thermal black hole states. As suggested by 't Hooft, the fluctuations of the black hole horizon, the world-sheet or world-volume theory, describes not only the degrees of freedom of the black hole, but all nearby objects too.

Second superstring revolution

In 1995, at the annual conference of string theorists at the University of Southern California (USC), Edward Witten gave a speech on string theory that in essence united the five string theories that existed at the time, and giving birth to a new 11-dimensional theory called M-theory. M-theory was also foreshadowed in the work of Paul Townsend at approximately the same time. The flurry of activity that began at this time is sometimes called the second superstring revolution.[34]

Juan Maldacena

During this period, Tom Banks, Willy Fischler, Stephen Shenker and Leonard Susskind formulated matrix theory,

a full holographic description of M-theory using IIA D0 branes.[51] This was the first definition of string theory that was fully non-perturbative and a concrete mathematical realization of the holographic principle. It is an example of a gauge-gravity duality and is now understood to be a special case of the AdS/CFT correspondence. Andrew Strominger and Cumrun Vafa calculated the entropy of certain configurations of D-branes and found agreement with the semi-classical answer for extreme charged black holes.[62] Petr Hořava and Witten found the eleven-dimensional formulation of the heterotic string theories, showing that orbifolds solve the chirality problem. Witten noted that the effective description of the physics of D-branes at low energies is by a supersymmetric gauge theory, and found geometrical interpretations of mathematical structures in gauge theory that he and Nathan Seiberg had earlier discovered in terms of the location of the branes.

In 1997, Juan Maldacena noted that the low energy excitations of a theory near a black hole consist of objects close to the horizon, which for extreme charged black holes looks like an anti-de Sitter space.[71] He noted that in this limit the gauge theory describes the string excitations near the branes. So he hypothesized that string theory on a near-horizon extreme-charged black-hole geometry, an anti-deSitter space times a sphere with flux, is equally well described by the low-energy limiting gauge theory, the N = 4 supersymmetric Yang–Mills theory. This hypothesis, which is called the AdS/CFT correspondence, was further developed by Steven Gubser, Igor Klebanov and Alexander Polyakov,[72] and by Edward Witten,[73] and it is now well-accepted. It is a concrete realization of the holographic principle, which has far-reaching implications for black holes, locality and information in physics, as well as the nature of the gravitational interaction.[56] Through this relationship, string theory has been shown to be related to gauge theories like quantum chromodynamics and this has led to more quantitative understanding of the behavior of hadrons, bringing string theory back to its roots.[84]

3.2.8   Criticism

Number of solutions

Main article: String theory landscape

To construct models of particle physics based on string theory, physicists typically begin by specifying a shape for the extra dimensions of spacetime. Each of these different shapes corresponds to a different possible universe, or "vacuum state", with a different collection of particles and forces. String theory as it is currently understood has an enormous number of vacuum states, typically estimated to be around 10^{500}, and these might be sufficiently diverse to accommodate almost any phenomena that might be observed at low energies.[121]

Many critics of string theory have expressed concerns about the large number of possible universes described by string theory. In his book *Not Even Wrong*, Peter Woit, a lecturer in the mathematics department at Columbia University, has argued that the large number of different physical scenarios renders string theory vacuous as a framework for constructing models of particle physics. According to Woit,

> The possible existence of, say, 10^{500} consistent different vacuum states for superstring theory probably destroys the hope of using the theory to predict anything. If one picks among this large set just those states whose properties agree with present experimental observations, it is likely there still will be such a large number of these that one can get just about whatever value one wants for the results of any new observation.[122]

Some physicists believe this large number of solutions is actually a virtue because it may allow a natural anthropic explanation of the observed values of physical constants, in particular the small value of the cosmological constant.[122] The anthropic principle is the idea that some of the numbers appearing in the laws of physics are not fixed by any fundamental principle but must be compatible with the evolution of intelligent life. In 1987, Steven Weinberg published an article in which he argued that the cosmological constant could not have been too large, or else galaxies and intelligent life would not have been able to develop.[123] Weinberg suggested that there might be a huge number of possible consistent universes, each with a different value of the cosmological constant, and observations indicate a small value of the cosmological constant only because humans happen to live in a universe that has allowed intelligent life, and hence observers, to exist.[124]

String theorist Leonard Susskind has argued that string theory provides a natural anthropic explanation of the small value of the cosmological constant.[125] According to Susskind, the different vacuum states of string theory might be realized as different universes within a larger multiverse. The fact that the observed universe has a small cosmological constant is just a tautological consequence of the fact that a small value is required for life to exist.[126] Many prominent theorists and critics have disagreed with Susskind's conclusions.[127] According to Woit, "in this case [anthropic reasoning] is nothing more than an excuse for failure. Speculative scientific ideas fail not just when they make incorrect predictions, but also when they turn out to be vacuous and incapable of predicting anything."[128]

Background independence

Main article: Background independence

One of the fundamental properties of Einstein's general theory of relativity is that it is background independent, meaning that the formulation of the theory does not in any way privilege a particular spacetime geometry.[129]

One of the main criticisms of string theory from early on is that it is not manifestly background independent. In string theory, one must typically specify a fixed reference geometry for spacetime, and all other possible geometries are described as perturbations of this fixed one. In his book *The Trouble With Physics*, physicist Lee Smolin of the Perimeter Institute for Theoretical Physics claims that this is the principal weakness of string theory as a theory of quantum gravity, saying that string theory has failed to incorporate this important insight from general relativity.[130]

Others have disagreed with Smolin's characterization of string theory. In a review of Smolin's book, string theorist Joseph Polchinski writes

> [Smolin] is mistaking an aspect of the mathematical language being used for one of the physics being described. New physical theories are often discovered using a mathematical language that is not the most suitable for them... In string theory it has always been clear that the physics is background-independent even if the language being used is not, and the search for more suitable language continues. Indeed, as Smolin belatedly notes, [AdS/CFT] provides a solution to this problem, one that is unexpected and powerful.[131]

Polchinski notes that an important open problem in quantum gravity is to develop holographic descriptions of gravity which do not require the gravitational field to be asymptotically anti-de Sitter.[131]

Smolin responded that the claims about background-independence, which Polchinski presents as "clear", are in fact only an unproven hope for future results, and Smolin is skeptical about them being true at all because of fundamental reasons: "If the strong form of the AdS/CFT conjecture is shown to be correct, then a very weak, and limited form of background will have been achieved. But ... this is still a big if". Smolin points out that current results about the [AdS/CFT] conjecture rely on global super-symmetry as perturbative physics, "but the whole point of general relativity and quantum gravity is that the generic solutions are governed by no global symmetries because the geometry of spacetime is completely dynamical", which "makes it very non-trivial to show the strong form of the [AdS/CFT] conjecture, because it must extend to solutions of supergravity arbitrarily far from those with global symmetries in the bulk".[132] Smolin summarizes:

> It would be more accurate to say, "Some string theorists believe that the formulations of perturbative string theories and dualities between them that they study concretely are approximations to a deeper, background independent formulation. This missing background independent formulation is not just a different language for the theory, it is hoped to be the statement of the principles and laws that define the theory, from which everything studied so far would be derived as an approximation."[132]

Sociological issues

Since the superstring revolutions of the 1980s and 1990s, string theory has become the dominant paradigm of high energy theoretical physics.[133] Some string theorists have expressed the view that there does not exist an equally successful alternative theory addressing the deep questions of fundamental physics. In an interview from 1987, Nobel laureate David Gross made the following controversial comments about the reasons for the popularity of string theory:

> The most important [reason] is that there are no other good ideas around. That's what gets most people into it. When people started to get interested in string theory they didn't know anything about it. In fact, the first reaction of most people is that the theory is extremely ugly and unpleasant, at least that was the case a few years ago when the understanding of string theory was much less developed. It was difficult for people to learn about it and to be turned on. So I think the real reason why people have got attracted by it is because there is no other game in town. All other approaches of constructing grand unified theories, which were more conservative to begin with, and only gradually became more and more radical, have failed, and this game hasn't failed yet.[134]

Several other high profile theorists and commentators have expressed similar views, suggesting that there are no viable alternatives to string theory.[135]

Many critics of string theory have commented on this state of affairs. In his book criticizing string theory, Peter Woit views the status of string theory research as unhealthy and

detrimental to the future of fundamental physics. He argues that the extreme popularity of string theory among theoretical physicists is partly a consequence of the financial structure of academia and the fierce competition for scarce resources.[136] In his book *The Road to Reality*, mathematical physicist Roger Penrose expresses similar views, stating "The often frantic competitiveness that this ease of communication engenders leads to 'bandwagon' effects, where researchers fear to be left behind if they do not join in."[137] Penrose also claims that the technical difficulty of modern physics forces young scientists to rely on the preferences of established researchers, rather than forging new paths of their own.[138] Lee Smolin expresses a slightly different position in his critique, claiming that string theory grew out of a tradition of particle physics which discourages speculation about the foundations of physics, while his preferred approach, loop quantum gravity, encourages more radical thinking. According to Smolin,

> String theory is a powerful, well-motivated idea and deserves much of the work that has been devoted to it. If it has so far failed, the principal reason is that its intrinsic flaws are closely tied to its strengths—and, of course, the story is unfinished, since string theory may well turn out to be part of the truth. The real question is not why we have expended so much energy on string theory but why we haven't expended nearly enough on alternative approaches.[139]

Smolin goes on to offer a number of prescriptions for how scientists might encourage a greater diversity of approaches to quantum gravity research.[140]

3.2.9   References

Notes

[1] For example, physicists are still working to understand the phenomenon of quark confinement, the paradoxes of black holes, and the origin of dark energy.

[2] For example, in the context of the AdS/CFT correspondence, theorists often formulate and study theories of gravity in unphysical numbers of spacetime dimensions.

[3] "Top Cited Articles during 2010 in hep-th". Retrieved 25 July 2013.

[4] More precisely, one cannot apply the methods of perturbative quantum field theory.

[5] Two independent mathematical proofs of mirror symmetry were given by Givental 1996, 1998 and Lian, Liu, Yau 1997, 1999, 2000.

[6] More precisely, a nontrivial group is called *simple* if its only normal subgroups are the trivial group and the group itself. The Jordan–Hölder theorem exhibits finite simple groups as the building blocks for all finite groups.

Citations

[1] Becker, Becker, and Schwarz 2007, p. 1

[2] Zwiebach 2009, p. 6

[3] Becker, Becker, and Schwarz 2007, pp. 2–3

[4] Becker, Becker, and Schwarz 2007, pp. 9–12

[5] Becker, Becker, and Schwarz 2007, pp. 14–15

[6] Klebanov and Maldacena 2009

[7] Merali 2011

[8] Sachdev 2013

[9] Becker, Becker, and Schwarz 2007, pp. 3, 15–16

[10] Becker, Becker, and Schwarz 2007, p. 8

[11] Becker, Becker, and Schwarz 13–14

[12] Woit 2006

[13] Zee 2010

[14] Becker, Becker, and Schwarz 2007, p. 2

[15] Becker, Becker, and Schwarz 2007, p. 6

[16] Zwiebach 2009, p. 12

[17] Becker, Becker, and Schwarz 2007, p. 4

[18] Zwiebach 2009, p. 324

[19] Wald 1984, p. 4

[20] Zee 2010, Parts V and VI

[21] Zwiebach 2009, p. 9

[22] Zwiebach 2009, p. 8

[23] Yau and Nadis 2010, Ch. 6

[24] Greene 2000, p. 186

[25] Yau and Nadis 2010, p. ix

[26] Randall and Sundrum 1999

[27] Becker, Becker, and Schwarz 2007

[28] Zwiebach 2009, p. 376

[29] Moore 2005, p. 214

[30] Moore 2005, p. 215

[31] Aspinwall et al. 2009

[32] Kontsevich 1995

[33] Kapustin and Witten 2007

[34] Duff 1998

[35] Duff 1998, p. 64

[36] Nahm 1978

[37] Cremmer, Julia, and Scherk 1978

[38] Duff 1998, p. 65

[39] Sen 1994a

[40] Sen 1994b

[41] Hull and Townsend 1995

[42] Duff 1998, p. 67

[43] Bergshoeff, Sezgin, and Townsend 1987

[44] Duff et al. 1987

[45] Duff 1998, p. 66

[46] Witten 1995

[47] Duff 1998, pp. 67–68

[48] Becker, Becker, and Schwarz 2007, p. 296

[49] Hořava and Witten 1996

[50] Duff 1996, sec. 1

[51] Banks et al. 1997

[52] Connes 1994

[53] Connes, Douglas, and Schwarz 1998

[54] Nekrasov and Schwarz 1998

[55] Seiberg and Witten 1999

[56] de Haro et al. 2013, p. 2

[57] Yau and Nadis 2010, p. 187–188

[58] Bekenstein 1973

[59] Hawking 1975

[60] Wald 1984, p. 417

[61] Yau and Nadis 2010, p. 189

[62] Strominger and Vafa 1996

[63] Yau and Nadis 2010, pp. 190–192

[64] Maldacena, Strominger, and Witten 1997

[65] Ooguri, Strominger, and Vafa 2004

[66] Yau and Nadis 2010, pp. 192–193

[67] Yau and Nadis 2010, pp. 194–195

[68] Strominger 1998

[69] Guica et al. 2009

[70] Castro, Maloney, and Strominger 2010

[71] Maldacena 1998

[72] Gubser, Klebanov, and Polyakov 1998

[73] Witten 1998

[74] Klebanov and Maldacena 2009, p. 28

[75] Maldacena 2005, p. 60

[76] Maldacena 2005, p. 61

[77] Zwiebach 2009, p. 552

[78] Maldacena 2005, pp. 61–62

[79] Susskind 2008

[80] Zwiebach 2009, p. 554

[81] Maldacena 2005, p. 63

[82] Hawking 2005

[83] Zwiebach 2009, p. 559

[84] Kovtun, Son, and Starinets 2001

[85] Merali 2011, p. 303

[86] Luzum and Romatschke 2008

[87] Sachdev 2013, p. 51

[88] Candelas et al. 1985

[89] Yau and Nadis 2010, pp. 147–150

[90] Becker, Becker, and Schwarz 2007, pp. 530–531

[91] Becker, Becker, and Schwarz 2007, p. 531

[92] Becker, Becker, and Schwarz 2007, p. 538

[93] Becker, Becker, and Schwarz 2007, p. 533

[94] Becker, Becker, and Schwarz 2007, pp. 539–543

[95] Deligne et al. 1999, p. 1

[96] Hori et al. 2003, p. xvii

[97] Aspinwall et al. 2009, p. 13

[98] Hori et al. 2003

[99] Yau and Nadis 2010, p. 167

[100] Yau and Nadis 2010, p. 166

[101] Yau and Nadis 2010, p. 169

[102] Candelas et al. 1991

[103] Yau and Nadis 2010, p. 171

[104] Hori et al. 2003, p. xix

[105] Strominger, Yau, and Zaslow 1996

[106] Dummit and Foote 2004

[107] Dummit and Foote 2004, pp. 102–103

[108] Klarreich 2015

[109] Gannon 2006, p. 2

[110] Gannon 2006, p. 4

[111] Conway and Norton 1979

[112] Gannon 2006, p. 5

[113] Gannon 2006, p. 8

[114] Borcherds 1992

[115] Frenkel, Lepowsky, and Meurman 1988

[116] Gannon 2006, p. 11

[117] Eguchi, Ooguri, and Tachikawa 2010

[118] Cheng, Duncan, and Harvey 2013

[119] Duncan, Griffin, and Ono 2015

[120] Witten 2007

[121] Woit 2006, pp. 240–242

[122] Woit 2006, p. 242

[123] Weinberg 1987

[124] Woit 2006, p. 243

[125] Susskind 2005

[126] Woit 2006, pp. 242–243

[127] Woit 2006, p. 240

[128] Woit 2006, p. 249

[129] Smolin 2006, p. 81

[130] Smolin 2006, p. 184

[131] Polchinski 2007

[132] Lee Smolin, April 2007:"Archived copy". Archived from the original on November 5, 2015. Retrieved December 31, 2015. Response to review of The Trouble with Physics by Joe Polchinski

[133] Penrose 2004, p. 1017

[134] Woit 2006, pp. 224–225

[135] Woit 2006, Ch. 16

[136] Woit 2006, p. 239

[137] Penrose 2004, p. 1018

[138] Penrose 2004, pp. 1019–1020

[139] Smolin 2006, p. 349

[140] Smolin 2006, Ch. 20

Bibliography

- Aspinwall, Paul; Bridgeland, Tom; Craw, Alastair; Douglas, Michael; Gross, Mark; Kapustin, Anton; Moore, Gregory; Segal, Graeme; Szendröi, Balázs; Wilson, P.M.H., eds. (2009). *Dirichlet Branes and Mirror Symmetry*. American Mathematical Society. ISBN 978-0-8218-3848-8.

- Banks, Tom; Fischler, Willy; Schenker, Stephen; Susskind, Leonard (1997). "M theory as a matrix model: A conjecture". *Physical Review D* **55** (8): 5112–5128. arXiv:hep-th/9610043. Bibcode:1997PhRvD..55.5112B. doi:10.1103/physrevd.55.5112.

- Becker, Katrin; Becker, Melanie; Schwarz, John (2007). *String theory and M-theory: A modern introduction*. Cambridge University Press. ISBN 978-0-521-86069-7.

- Bekenstein, Jacob (1973). "Black holes and entropy". *Physical Review D* **7** (8): 2333–2346. Bibcode:1973PhRvD...7.2333B. doi:10.1103/PhysRevD.7.2333.

- Bergshoeff, Eric; Sezgin, Ergin; Townsend, Paul (1987). "Supermembranes and eleven-dimensional supergravity". *Physics Letters B* **189** (1): 75–78. Bibcode:1987PhLB..189...75B. doi:10.1016/0370-2693(87)91272-X.

- Borcherds, Richard (1992). "Monstrous moonshine and Lie superalgebras". *Inventiones Mathematicae* **109** (1): 405–444. Bibcode:1992InMat.109..405B. doi:10.1007/BF01232032.

- Candelas, Philip; de la Ossa, Xenia; Green, Paul; Parks, Linda (1991). "A pair of Calabi–Yau manifolds as an exactly soluble superconformal field theory". *Nuclear Physics B* **359** (1): 21–74. Bibcode:1991NuPhB.359...21C. doi:10.1016/0550-3213(91)90292-6.

- Candelas, Philip; Horowitz, Gary; Strominger, Andrew; Witten, Edward (1985). "Vacuum configurations for superstrings". *Nuclear Physics B* **258**: 46–74. Bibcode:1985NuPhB.258...46C. doi:10.1016/0550-3213(85)90602-9.

- Castro, Alejandra; Maloney, Alexander; Strominger, Andrew (2010). "Hidden conformal symmetry of the Kerr black hole". *Physical Review D* **82** (2). arXiv:1004.0996. Bibcode:2010PhRvD..82b4008C. doi:10.1103/PhysRevD.82.024008.

- Cheng, Miranda; Duncan, John; Harvey, Jeffrey (2013). "Umbral Moonshine". arXiv:1204.2779.

- Connes, Alain (1994). *Noncommutative Geometry*. Academic Press. ISBN 978-0-12-185860-5.

- Connes, Alain; Douglas, Michael; Schwarz, Albert (1998). "Noncommutative geometry and matrix theory". *Journal of High Energy Physics*. 19981 (2): 003. arXiv:hep-th/9711162. Bibcode:1998JHEP...02..003C. doi:10.1088/1126-6708/1998/02/003.

- Conway, John; Norton, Simon (1979). "Monstrous moonshine". *Bull. London Math. Soc.* **11** (3): 308–339. doi:10.1112/blms/11.3.308.

- Cremmer, Eugene; Julia, Bernard; Scherk, Joel (1978). "Supergravity theory in eleven dimensions". *Physics Letters B* **76** (4): 409–412. Bibcode:1978PhLB...76..409C. doi:10.1016/0370-2693(78)90894-8.

- de Haro, Sebastian; Dieks, Dennis; 't Hooft, Gerard; Verlinde, Erik (2013). "Forty Years of String Theory Reflecting on the Foundations". *Foundations of Physics* **43** (1): 1–7. Bibcode:2013FoPh...43....1D. doi:10.1007/s10701-012-9691-3.

- Deligne, Pierre; Etingof, Pavel; Freed, Daniel; Jeffery, Lisa; Kazhdan, David; Morgan, John; Morrison, David; Witten, Edward, eds. (1999). *Quantum Fields and Strings: A Course for Mathematicians* **1**. American Mathematical Society. ISBN 978-0821820124.

- Duff, Michael (1996). "M-theory (the theory formerly known as strings)". *International Journal of Modern Physics A* **11** (32): 6523–41. arXiv:hep-th/9608117. Bibcode:1996IJMPA..11.5623D. doi:10.1142/S0217751X96002583.

- Duff, Michael (1998). "The theory formerly known as strings". *Scientific American* **278** (2): 64–9. doi:10.1038/scientificamerican0298-64.

- Duff, Michael; Howe, Paul; Inami, Takeo; Stelle, Kellogg (1987). "Superstrings in $D=10$ from supermembranes in $D=11$". *Nuclear Physics B* **191** (1): 70–74. Bibcode:1987PhLB..191...70D. doi:10.1016/0370-2693(87)91323-2.

- Dummit, David; Foote, Richard (2004). *Abstract Algebra*. Wiley. ISBN 978-0-471-43334-7.

- Duncan, John; Griffin, Michael; Ono, Ken (2015). "Proof of the Umbral Moonshine Conjecture". arXiv:1503.01472.

- Eguchi, Tohru; Ooguri, Hirosi; Tachikawa, Yuji (2011). "Notes on the K3 surface and the Mathieu group M_{24}". *Experimental Mathematics* **20** (1): 91–96. doi:10.1080/10586458.2011.544585.

- Frenkel, Igor; Lepowsky, James; Meurman, Arne (1988). *Vertex Operator Algebras and the Monster*. Pure and Applied Mathematics **134**. Academic Press. ISBN 0-12-267065-5.

- Gannon, Terry. *Moonshine Beyond the Monster: The Bridge Connecting Algebra, Modular Forms, and Physics*. Cambridge University Press.

- Givental, Alexander (1996). "Equivariant Gromov-Witten invariants". *International Mathematics Research Notices* **1996** (13): 613–663. doi:10.1155/S1073792896000414.

- Givental, Alexander (1998). "A mirror theorem for toric complete intersections". *Topological field theory, primitive forms and related topics*: 141–175. doi:10.1007/978-1-4612-0705-4_5. ISBN 978-1-4612-6874-1.

- Gubser, Steven; Klebanov, Igor; Polyakov, Alexander (1998). "Gauge theory correlators from non-critical string theory". *Physics Letters B* **428**: 105–114. arXiv:hep-th/9802109. Bibcode:1998PhLB..428..105G. doi:10.1016/S0370-2693(98)00377-3.

- Guica, Monica; Hartman, Thomas; Song, Wei; Strominger, Andrew (2009). "The Kerr/CFT Correspondence". *Physical Review D* **80** (12). arXiv:0809.4266. Bibcode:2009PhRvD..80l4008G. doi:10.1103/PhysRevD.80.124008.

- Hawking, Stephen (1975). "Particle creation by black holes". *Communications in Mathematical Physics* **43** (3): 199–220. Bibcode:1975CMaPh..43..199H. doi:10.1007/BF02345020.

- Hawking, Stephen (2005). "Information loss in black holes". *Physical Review D* **72** (8). arXiv:hep-th/0507171. Bibcode:2005PhRvD..72h4013H. doi:10.1103/PhysRevD.72.084013.

- Hořava, Petr; Witten, Edward (1996). "Heterotic and Type I string dynamics from eleven dimensions". *Nuclear Physics B* **460** (3): 506–524. arXiv:hep-th/9510209. Bibcode:1996NuPhB.460..506H. doi:10.1016/0550-3213(95)00621-4.

- Hori, Kentaro; Katz, Sheldon; Klemm, Albrecht; Pandharipande, Rahul; Thomas, Richard; Vafa, Cumrun; Vakil, Ravi; Zaslow, Eric, eds. (2003). *Mirror Symmetry* (PDF). American Mathematical Society. ISBN 0-8218-2955-6.

- Hull, Chris; Townsend, Paul (1995). "Unity of superstring dualities". *Nuclear Physics B* **4381** (1): 109–137. arXiv:hep-th/9410167. Bibcode:1995NuPhB.438..109H. doi:10.1016/0550-3213(94)00559-W.

- Kapustin, Anton; Witten, Edward (2007). "Electric-magnetic duality and the geometric Langlands program". *Communications in Number Theory and Physics* **1** (1): 1–236. arXiv:hep-th/0604151. Bibcode:2007CNTP....1....1K. doi:10.4310/cntp.2007.v1.n1.a1.

- Klarreich, Erica. "Mathematicians chase moonshine's shadow". *Quanta Magazine*. Retrieved March 2015.

- Klebanov, Igor; Maldacena, Juan (2009). "Solving Quantum Field Theories via Curved Spacetimes" (PDF). *Physics Today* **62**: 28–33. Bibcode:2009PhT....62a..28K. doi:10.1063/1.3074260. Archived from the original (PDF) on July 2, 2013. Retrieved May 2013.

- Kontsevich, Maxim (1995). "Homological algebra of mirror symmetry". *Proceedings of the International Congress of Mathematicians*: 120–139. arXiv:alg-geom/9411018. Bibcode:1994alg.geom.11018K.

- Kovtun, P. K.; Son, Dam T.; Starinets, A. O. (2001). "Viscosity in strongly interacting quantum field theories from black hole physics". *Physical Review Letters* **94** (11): 111601. arXiv:hep-th/0405231. Bibcode:2005PhRvL..94k1601K. doi:10.1103/PhysRevLett.94.111601. PMID 15903845.

- Lian, Bong; Liu, Kefeng; Yau, Shing-Tung (1997). "Mirror principle, I". *Asian Journal of Mathematics* **1**: 729–763. arXiv:alg-geom/9712011. Bibcode:1997alg.geom.12011L.

- Lian, Bong; Liu, Kefeng; Yau, Shing-Tung (1999a). "Mirror principle, II". *Asian Journal of Mathematics* **3**: 109–146. arXiv:math/9905006. Bibcode:1999math......5006L.

- Lian, Bong; Liu, Kefeng; Yau, Shing-Tung (1999b). "Mirror principle, III". *Asian Journal of Mathematics* **3**: 771–800. arXiv:math/9912038. Bibcode:1999math.....12038L.

- Lian, Bong; Liu, Kefeng; Yau, Shing-Tung (2000). "Mirror principle, IV". *Surveys in Differential Geometry* **7**: 475–496. arXiv:math/0007104. Bibcode:2000math......7104L. doi:10.4310/sdg.2002.v7.n1.a15.

- Luzum, Matthew; Romatschke, Paul (2008). "Conformal relativistic viscous hydrodynamics: Applications to RHIC results at √sNN=200 GeV". *Physical Review C* **78** (3). arXiv:0804.4015. doi:10.1103/PhysRevC.78.034915.

- Maldacena, Juan (1998). "The Large *N* limit of superconformal field theories and supergravity". *Advances in Theoretical and Mathematical Physics* **2**: 231–252. arXiv:hep-th/9711200. Bibcode:1998AdTMP...2..231M. doi:10.1063/1.59653.

- Maldacena, Juan (2005). "The Illusion of Gravity" (PDF). *Scientific American* **293** (5): 56–63. Bibcode:2005SciAm.293e..56M. doi:10.1038/scientificamerican1105-56. PMID 16318027. Archived from the original (PDF) on November 1, 2014. Retrieved July 2013.

- Maldacena, Juan; Strominger, Andrew; Witten, Edward (1997). "Black hole entropy in M-theory". *Journal of High Energy Physics* **1997** (12). arXiv:hep-th/9711053. Bibcode:1997JHEP...12..002M. doi:10.1088/1126-6708/1997/12/002.

- Merali, Zeeya (2011). "Collaborative physics: string theory finds a bench mate". *Nature* **478** (7369): 302–304. Bibcode:2011Natur.478..302M. doi:10.1038/478302a. PMID 22012369.

- Moore, Gregory (2005). "What is ... a Brane?" (PDF). *Notices of the AMS* **52**: 214. Retrieved June 2013.

- Nahm, Walter (1978). "Supersymmetries and their representations". *Nuclear Physics B* **135** (1): 149–166. Bibcode:1978NuPhB.135..149N. doi:10.1016/0550-3213(78)90218-3.

- Nekrasov, Nikita; Schwarz, Albert (1998). "Instantons on noncommutative $\mathbf{R}^4$ and (2,0) superconformal six dimensional theory". *Communications in Mathematical Physics* **198** (3): 689–703. arXiv:hep-th/9802068. Bibcode:1998CMaPh.198..689N. doi:10.1007/s002200050490.

- Ooguri, Hirosi; Strominger, Andrew; Vafa, Cumrun (2004). "Black hole attractors and the topological string". *Physical Review D* **70** (10). arXiv:hep-th/0405146. Bibcode:2004PhRvD..70j6007O. doi:10.1103/physrevd.70.106007.

- Polchinski, Joseph (2007). "All Strung Out?". *American Scientist*. Retrieved April 2015.

- Penrose, Roger (2005). *The Road to Reality: A Complete Guide to the Laws of the Universe*. Knopf. ISBN 0-679-45443-8.

- Randall, Lisa; Sundrum, Raman (1999). "An alternative to compactification". *Physical Review Letters* **83** (23): 4690–4693. arXiv:hep-th/9906064. Bibcode:1999PhRvL..83.4690R. doi:10.1103/PhysRevLett.83.4690.

- Sachdev, Subir (2013). "Strange and stringy". *Scientific American* **308** (44): 44–51. Bibcode:2012SciAm.308a..44S. doi:10.1038/scientificamerican0113-44.

- Seiberg, Nathan; Witten, Edward (1999). "String Theory and Noncommutative Geometry". *Journal of High Energy Physics* **1999** (9): 032. arXiv:hep-th/9908142. Bibcode:1999JHEP...09..032S. doi:10.1088/1126-6708/1999/09/032.

- Sen, Ashoke (1994a). "Strong-weak coupling duality in four-dimensional string theory". *International Journal of Modern Physics A* **9** (21): 3707–3750. arXiv:hep-th/9402002. Bibcode:1994IJMPA...9.3707S. doi:10.1142/S0217751X94001497.

- Sen, Ashoke (1994b). "Dyon-monopole bound states, self-dual harmonic forms on the multi-monopole moduli space, and $SL(2,\mathbf{Z})$ invariance in string theory". *Physics Letters B* **329** (2): 217–221. arXiv:hep-th/9402032. Bibcode:1994PhLB..329..217S. doi:10.1016/0370-2693(94)90763-3.

- Smolin, Lee (2006). *The Trouble with Physics: The Rise of String Theory, the Fall of a Science, and What Comes Next*. New York: Houghton Mifflin Co. ISBN 0-618-55105-0.

- Strominger, Andrew (1998). "Black hole entropy from near-horizon microstates". *Journal of High Energy Physics* **1998** (2): 009. arXiv:hep-th/9712251. Bibcode:1998JHEP...02..009S. doi:10.1088/1126-6708/1998/02/009.

- Strominger, Andrew; Vafa, Cumrun (1996). "Microscopic origin of the Bekenstein–Hawking entropy". *Physics Letters B* **379** (1): 99–104. arXiv:hep-th/9601029. Bibcode:1996PhLB..379...99S. doi:10.1016/0370-2693(96)00345-0.

- Strominger, Andrew; Yau, Shing-Tung; Zaslow, Eric (1996). "Mirror symmetry is T-duality". *Nuclear Physics B* **479** (1): 243–259. arXiv:hep-th/9606040. Bibcode:1996NuPhB.479..243S. doi:10.1016/0550-3213(96)00434-8.

- Susskind, Leonard (2005). *The Cosmic Landscape: String Theory and the Illusion of Intelligent Design*. Back Bay Books. ISBN 978-0316013338.

- Susskind, Leonard (2008). *The Black Hole War: My Battle with Stephen Hawking to Make the World Safe for Quantum Mechanics*. Little, Brown and Company. ISBN 978-0-316-01641-4.

- Wald, Robert (1984). *General Relativity*. University of Chicago Press. ISBN 978-0-226-87033-5.

- Weinberg, Steven (1987). *Anthropic bound on the cosmological constant* **59**. Physical Review Letters. p. 2607.

- Witten, Edward (1995). "String theory dynamics in various dimensions". *Nuclear Physics B* **443** (1): 85–126. arXiv:hep-th/9503124. Bibcode:1995NuPhB.443...85W. doi:10.1016/0550-3213(95)00158-O.

- Witten, Edward (1998). "Anti-de Sitter space and holography". *Advances in Theoretical and Mathematical Physics* **2**: 253–291. arXiv:hep-th/9802150. Bibcode:1998AdTMP...2..253W.

- Witten, Edward (2007). "Three-dimensional gravity revisited". arXiv:0706.3359 [hep-th].

- Woit, Peter (2006). *Not Even Wrong: The Failure of String Theory and the Search for Unity in Physical Law*. Basic Books. p. 105. ISBN 0-465-09275-6.

- Yau, Shing-Tung; Nadis, Steve (2010). *The Shape of Inner Space: String Theory and the Geometry of the Universe's Hidden Dimensions*. Basic Books. ISBN 978-0-465-02023-2.

- Zee, Anthony (2010). *Quantum Field Theory in a Nutshell* (2nd ed.). Princeton University Press. ISBN 978-0-691-14034-6.

- Zwiebach, Barton (2009). *A First Course in String Theory*. Cambridge University Press. ISBN 978-0-521-88032-9.

3.2.10   Further reading

Popularizations

General

- Greene, Brian (2003). *The Elegant Universe: Superstrings, Hidden Dimensions, and the Quest for the Ultimate Theory*. New York: W.W. Norton & Company. ISBN 0-393-05858-1.

- Greene, Brian (2004). *The Fabric of the Cosmos: Space, Time, and the Texture of Reality*. New York: Alfred A. Knopf. ISBN 0-375-41288-3.

Critical

- Penrose, Roger (2005). *The Road to Reality: A Complete Guide to the Laws of the Universe*. Knopf. ISBN 0-679-45443-8.

- Smolin, Lee (2006). *The Trouble with Physics: The Rise of String Theory, the Fall of a Science, and What Comes Next*. New York: Houghton Mifflin Co. ISBN 0-618-55105-0.

- Woit, Peter (2006). *Not Even Wrong: The Failure of String Theory And the Search for Unity in Physical Law*. London: Jonathan Cape &: New York: Basic Books. ISBN 978-0-465-09275-8.

Textbooks

For physicists

- Becker, Katrin; Becker, Melanie; Schwarz, John (2007). *String Theory and M-theory: A Modern Introduction*. Cambridge University Press. ISBN 978-0-521-86069-7.

- Green, Michael; Schwarz, John; Witten, Edward (2012). *Superstring theory. Vol. 1: Introduction*. Cambridge University Press. ISBN 978-1107029118.

- Green, Michael; Schwarz, John; Witten, Edward (2012). *Superstring theory. Vol. 2: Loop amplitudes, anomalies and phenomenology*. Cambridge University Press. ISBN 978-1107029132.

- Polchinski, Joseph (1998). *String Theory Vol. 1: An Introduction to the Bosonic String*. Cambridge University Press. ISBN 0-521-63303-6.

- Polchinski, Joseph (1998). *String Theory Vol. 2: Superstring Theory and Beyond*. Cambridge University Press. ISBN 0-521-63304-4.

- Zwiebach, Barton (2009). *A First Course in String Theory*. Cambridge University Press. ISBN 978-0-521-88032-9.

For mathematicians

- Deligne, Pierre; Etingof, Pavel; Freed, Daniel; Jeffery, Lisa; Kazhdan, David; Morgan, John; Morrison, David; Witten, Edward, eds. (1999). *Quantum Fields and Strings: A Course for Mathematicians, Vol. 2*. American Mathematical Society. ISBN 978-0821819883.

3.2.11 External links

- *The Elegant Universe*—A three-hour miniseries with Brian Greene by *NOVA* (original PBS Broadcast Dates: October 28, 8–10 p.m. and November 4, 8–9 p.m., 2003). Various images, texts, videos and animations explaining string theory.

- Not Even Wrong—A blog critical of string theory

- The Official String Theory Web Site

- Why String Theory—An introduction to string theory.

3.3 Superstring theory

"Superstring" redirects here. For the converse relation of "substring", see Superstring (formal languages). For the bundle of firecrackers, see Superstring (fireworks).

Superstring theory is an attempt to explain all of the particles and fundamental forces of nature in one theory by modelling them as vibrations of tiny supersymmetric strings.

'Superstring theory' is a shorthand for **supersymmetric string theory** because unlike bosonic string theory, it is the version of string theory that incorporates fermions and supersymmetry.

Since the second superstring revolution, the five superstring theories are regarded as different limits of a single theory tentatively called M-theory, or simply string theory.

3.3.1 Background

The deepest problem in theoretical physics is harmonizing the theory of general relativity, which describes gravitation and applies to large-scale structures (stars, galaxies, super clusters), with quantum mechanics, which describes the other three fundamental forces acting on the atomic scale.

The development of a quantum field theory of a force invariably results in infinite possibilities. Physicists have developed mathematical techniques (renormalization) to eliminate these infinities that work for three of the four fundamental forces—electromagnetic, strong nuclear and weak nuclear forces—but not for gravity. The development of a quantum theory of gravity must therefore come about by different means than those used for the other forces.[1]

According to the theory, the fundamental constituents of reality are strings of the Planck length (about 10^{-33} cm) that vibrate at resonant frequencies. Every string, in theory, has a unique resonance, or harmonic. Different harmonics determine different fundamental particles. The tension in a string is on the order of the Planck force (10^{44} newtons). The graviton (the proposed messenger particle of the gravitational force), for example, is predicted by the theory to be a string with wave amplitude zero.

3.3.2 History

Main article History of string theory

Since its beginnings in late sixties, the theory was developed through several decades of intense research and combined effort of numerous scientists. It has developed into a broad and varied subject with connections to quantum gravity, particle and condensed matter physics, cosmology, and pure mathematics.

3.3.3 Lack of experimental evidence

Superstring theory is based on supersymmetry. No supersymmetric particles have been discovered and recent research at LHC and Tevatron has excluded some of the ranges.[2][3][4][5] For instance, the mass constraint of the Minimal Supersymmetric Standard Model squarks has been up to 1.1 TeV, and gluinos up to 500 GeV.[6] No report on suggesting large extra dimensions has been delivered from LHC. There have been no principles so far to limit the number of vacua in the concept of a landscape of vacua.[7]

Some particle physicists became disappointed[8] by the lack of experimental verification of supersymmetry, and some have already discarded it; Jon Butterworth at the University College London said that we had no sign of supersymmetry, even in higher energy region, excluding the superpartners of the top quark up to a few TeV. Ben Allanach at the University of Cambridge states that if we do not discover any new particles in the next trial at the LHC, then we can say it is unlikely to discover supersymmetry at CERN in the foreseeable future.[8]

3.3.4 Extra dimensions

See also: Why does consistency require 10 dimensions?

Our physical space is observed to have three large spatial dimensions and, along with time, is a boundless four-dimensional continuum known as spacetime. However, nothing prevents a theory from including more than 4 dimensions. In the case of string theory, consistency requires spacetime to have 10 (3+1+6) dimensions. The fact that we see only 3 dimensions of space can be explained by one of two mechanisms: either the extra dimensions are compactified on a very small scale, or else our world may live on a 3-dimensional submanifold corresponding to a brane, on which all known particles besides gravity would be restricted.

If the extra dimensions are compactified, then the extra six dimensions must be in the form of a Calabi–Yau manifold. Within the more complete framework of M-theory, they would have to take form of a G2 manifold. Calabi-Yaus are interesting mathematical spaces in their own right. A particular exact symmetry of string/M-theory called T-duality (which exchanges momentum modes for winding number and sends compact dimensions of radius R to radius 1/R),[9] has led to the discovery of equivalences between different Calabi-Yaus called Mirror Symmetry.

Superstring theory is not the first theory to propose extra spatial dimensions. It can be seen as building upon the Kaluza–Klein theory, which proposed a 4+1-dimensional theory of gravity. When compactified on a circle, the gravity in the extra dimension precisely describes electromagnetism from the perspective of the 3 remaining large space dimensions. Thus the original Kaluza–Klein theory is a prototype for the unification of gauge and gravity interactions, at least at the classical level, however it is known to be insufficient to describe nature for a variety of

reasons (missing weak and strong forces, lack of parity violation, etc.) A more complex compact geometry is needed to reproduce the known gauge forces. Also, to obtain a consistent, fundamental, quantum theory requires the upgrade to string theory—not just the extra dimensions.

3.3.5 Number of superstring theories

Theoretical physicists were troubled by the existence of five separate string theories. A possible solution for this dilemma was suggested at the beginning of what is called the second superstring revolution in the 1990s, which suggests that the five string theories might be different limits of a single underlying theory, called M-theory. This remains a conjecture.[10]

The five consistent superstring theories are:

- The type I string has one supersymmetry in the ten-dimensional sense (16 supercharges). This theory is special in the sense that it is based on unoriented open and closed strings, while the rest are based on oriented closed strings.

- The type II string theories have two supersymmetries in the ten-dimensional sense (32 supercharges). There are actually two kinds of type II strings called type IIA and type IIB. They differ mainly in the fact that the IIA theory is non-chiral (parity conserving) while the IIB theory is chiral (parity violating).

- The heterotic string theories are based on a peculiar hybrid of a type I superstring and a bosonic string. There are two kinds of heterotic strings differing in their ten-dimensional gauge groups: the heterotic $E_8 \times E_8$ string and the heterotic SO(32) string. (The name heterotic SO(32) is slightly inaccurate since among the SO(32) Lie groups, string theory singles out a quotient $Spin(32)/Z_2$ that is not equivalent to SO(32).)

Chiral gauge theories can be inconsistent due to anomalies. This happens when certain one-loop Feynman diagrams cause a quantum mechanical breakdown of the gauge symmetry. The anomalies were canceled out via the Green–Schwarz mechanism.

Even though there are only five superstring theories, making detailed predictions for real experiments requires information about exactly what physical configuration the theory is in. This considerably complicates efforts to test string theory because there is an astronomically high number – 10^{500} or more – of configurations that meet some of the basic requirements to be consistent with our world. Along with the extreme remoteness of the Planck scale, this is the other major reason it is hard to test superstring theory.

Another approach to the number of superstring theories refers to the mathematical structure called composition algebra. In the findings of abstract algebra there are just seven composition algebras over the field of real numbers. In 1990 physicists R. Foot and G.C. Joshi in Australia stated that "the seven classical superstring theories are in one-to-one correspondence to the seven composition algebras."[11]

3.3.6 Integrating general relativity and quantum mechanics

General relativity typically deals with situations involving large mass objects in fairly large regions of spacetime whereas quantum mechanics is generally reserved for scenarios at the atomic scale (small spacetime regions). The two are very rarely used together, and the most common case that combines them is in the study of black holes. Having *peak density*, or the maximum amount of matter possible in a space, and very small area, the two must be used in synchrony to predict conditions in such places. Yet, when used together, the equations fall apart, spitting out impossible answers, such as imaginary distances and less than one dimension.

The major problem with their congruence is that, at Planck scale (a fundamental small unit of length) lengths, general relativity predicts a smooth, flowing surface, while quantum mechanics predicts a random, warped surface, neither of which are anywhere near compatible. Superstring theory resolves this issue, replacing the classical idea of point particles with strings. These strings have an average diameter of the Planck length, with extremely small variances, which completely ignores the quantum mechanical predictions of Planck-scale length dimensional warping. Also, these surfaces can be mapped as branes. These branes can be viewed as objects with a morphism between them. In this case, the morphism will be state of a string that stretches between brane A and brane B.

Singularities are avoided because the observed consequences of "Big Crunches" never reach zero size. In fact, should the universe begin a "big crunch" sort of process, string theory dictates that the universe could never be smaller than the size of one string, at which point it would actually begin expanding.

3.3.7 Mathematics

D-branes

D-branes are membrane-like objects in 10D string theory. They can be thought of as occurring as a result of a Kaluza–Klein compactification of 11D M-theory that contains membranes. Because compactification of a geometric theory produces extra vector fields the D-branes can be included in the action by adding an extra U(1) vector field to the string action.

$$\partial_z \to \partial_z + iA_z(z, \bar{z})$$

In **type I** open string theory, the ends of open strings are always attached to D-brane surfaces. A string theory with more gauge fields such as SU(2) gauge fields would then correspond to the compactification of some higher-dimensional theory above 11 dimensions, which is not thought to be possible to date. Furthermore, the tachyons attached to the D-branes, show, the instability of those d-branes with respect to the annihilation. We will consider that tachyon total energy is (or reflects) the total energy of the D-branes.

Why five superstring theories?

For a 10 dimensional supersymmetric theory we are allowed a 32-component Majorana spinor. This can be decomposed into a pair of 16-component Majorana-Weyl (chiral) spinors. There are then various ways to construct an invariant depending on whether these two spinors have the same or opposite chiralities:

The heterotic superstrings come in two types SO(32) and $E_8 \times E_8$ as indicated above and the type I superstrings include open strings.

3.3.8 Beyond superstring theory

It is conceivable that the five superstring theories are approximated to a theory in higher dimensions possibly involving membranes. Because the action for this involves quartic terms and higher so is not Gaussian, the functional integrals are very difficult to solve and so this has confounded the top theoretical physicists. Edward Witten has popularised the concept of a theory in 11 dimensions M-theory involving membranes interpolating from the known symmetries of superstring theory. It may turn out that there exist membrane models or other non-membrane models in higher dimensions—which may become acceptable when we find new unknown symmetries of nature, such as non-commutative geometry. It is thought, however, that 16 is probably the maximum since O(16) is a maximal subgroup of E8 the largest exceptional lie group and also is more than large enough to contain the Standard Model. Quartic integrals of the non-functional kind are easier to solve so there is hope for the future. This is the series solution, which is always convergent when a is non-zero and negative:

$$\int_{-\infty}^{\infty} \exp(ax^4 + bx^3 + cx^2 + dx + f)\,dx = e^f$$
$$\times \sum_{n,m,p=0}^{\infty} \frac{b^{4n}}{(4n)!} \frac{c^{2m}}{(2m)!} \frac{d^{4p}}{(4p)!} \frac{\Gamma(3n + m + p + \frac{1}{4})}{a^{3n+m+p+\frac{1}{4}}}$$

In the case of membranes the series would correspond to sums of various membrane interactions that are not seen in string theory.

Compactification

Investigating theories of higher dimensions often involves looking at the 10 dimensional superstring theory and interpreting some of the more obscure results in terms of compactified dimensions. For example, D-branes are seen as compactified membranes from 11D M-theory. Theories of higher dimensions such as 12D F-theory and beyond produce other effects, such as gauge terms higher than $U(1)$. The components of the extra vector fields (A) in the D-brane actions can be thought of as extra coordinates (X) in disguise. However, the *known* symmetries including supersymmetry currently restrict the spinors to 32-components—which limits the number of dimensions to 11 (or 12 if you include two time dimensions.) Some commentators (e.g., John Baez et al.) have speculated that the exceptional lie groups E_6, E_7 and E_8 having maximum orthogonal subgroups O(10), O(12) and O(16) may be related to theories in 10, 12 and 16 dimensions; 10 dimensions corresponding to string theory and the 12 and 16 dimensional theories being yet undiscovered but would be theories based on 3-branes and 7-branes respectively. However this is a minority view within the string community. Since E_7 is in some sense F_4 quaternified and E_8 is F_4 octonified, then the 12 and 16 dimensional theories, if they did exist, may involve the noncommutative geometry based on the quaternions and octonions respectively. From the above discussion, it can be seen that physicists have many ideas for extending superstring theory beyond the current 10 dimensional theory, but so far none have been successful.

Kac–Moody algebras

Since strings can have an infinite number of modes, the symmetry used to describe string theory is based on infinite dimensional Lie algebras. Some Kac–Moody algebras that have been considered as symmetries for M-theory have been E_{10} and E_{11} and their supersymmetric extensions.

3.3.9 See also

- AdS/CFT correspondence
- dS/CFT correspondence
- Grand unification theory
- Large Hadron Collider
- List of string theory topics
- Quantum gravity
- String field theory

3.3.10 Notes

[1] Polchinski, Joseph. *String Theory: Volume I*. Cambridge University Press, p. 4.

[2] Woit, Peter (February 22, 2011). "Implications of Initial LHC Searches for Supersymmetry".

[3] Cassel, S.; Ghilencea, D. M.; Kraml, S.; Lessa, A.; Ross, G. G. (2011). "Fine-tuning implications for complementary dark matter and LHC SUSY searches". *Journal of High Energy Physics* **2011** (5): 120. arXiv:1101.4664. Bibcode:2011JHEP...05..120C. doi:10.1007/JHEP05(2011)120.

[4] Falkowski, Adam (Jester) (February 16, 2011). "What LHC tells about SUSY". *resonaances.blogspot.com*. Archived from the original on March 22, 2014. Retrieved March 22, 2014.

[5] Tapper, Alex (24 March 2010). "Early SUSY searches at the LHC" (PDF). Imperial College London.

[6] CMS Collaboration (2011). "Search for Supersymmetry at the LHC in Events with Jets and Missing Transverse Energy". *Physical Review Letters* **107** (22): 221804. arXiv:1109.2352. Bibcode:2011PhRvL.107v1804C. doi:10.1103/PhysRevLett.107.221804. PMID 22182023.

[7] Shifman, M. (2012). "Frontiers Beyond the Standard Model: Reflections and Impressionistic Portrait of the Conference". *Modern Physics Letters A* **27** (40): 1230043. Bibcode:2012MPLA...2730043S. doi:10.1142/S0217732312300431.

[8] Jha, Alok (August 6, 2013). "One year on from the Higgs boson find, has physics hit the buffers?". *The Guardian*. photograph: Harold Cunningham/Getty Images (London: GMG). ISSN 0261-3077. OCLC 60623878. Archived from the original on March 22, 2014. Retrieved March 22, 2014.

[9] Polchinski, Joseph. *String Theory: Volume I*. Cambridge University Press, p. 247.

[10] Polchinski, Joseph. *String Theory: Volume II*. Cambridge University Press, p. 198.

[11] Foot, R.; Joshi, G. C. (1990). "Nonstandard signature of spacetime, superstrings, and the split composition algebras". *Letters in Mathematical Physics* **19**: 65–71. Bibcode:1990LMaPh..19...65F. doi:10.1007/BF00402262.

3.3.11 References

- Kaku, Michio (1999). *Introduction to Superstring and M-Theory* (2nd ed.). New York, USA: Springer-Verlag.

- Shen, Sinyan (1982). *Introduction to Superfluidity* (2nd ed.). Beijing, China: Science Press.

- Greene, Brian (2000). *The Elegant Universe: Superstrings, Hidden Dimensions, and the Quest for the Ultimate Theory*. Random House Inc.

3.3.12 External links

- Wellcome Collection video on superstring theory

- The Official Superstring theory website: http://superstringtheory.com/index.html

3.4 Supergravity

In theoretical physics, **supergravity** (**supergravity theory**; **SUGRA** for short) is a field theory that combines the principles of supersymmetry and general relativity. Together, these imply that, in supergravity, the supersymmetry is a local symmetry (in contrast to non-gravitational supersymmetric theories, such as the Minimal Supersymmetric Standard Model). Since the generators of supersymmetry (SUSY) are convoluted with the Poincaré group to form a super-Poincaré algebra, it can be seen that supergravity follows naturally from supersymmetry.[1] All traditional literature on supergravity is generally written in terms of Cartan connections.[2]

3.4.1 Gravitons

Like any field theory of gravity, a supergravity theory contains a spin-2 field whose quantum is the graviton. Supersymmetry requires the graviton field to have a superpartner. This field has spin 3/2 and its quantum is the gravitino. The number of gravitino fields is equal to the number of supersymmetries.

3.4.2 History

Gauge supersymmetry

The first theory[3] of local supersymmetry was proposed in 1975 by Dick Arnowitt and Pran Nath and was called **gauge supersymmetry**.

SUGRA

SUGRA, or supergravity, was discovered in 1976 by Dan Freedman, Sergio Ferrara and Peter van Nieuwenhuizen,[4] but was quickly generalized to many different theories in various numbers of dimensions and additional (N) supersymmetry charges. Supergravity theories with N>1 are usually referred to as extended supergravity (SUEGRA). Some supergravity theories were shown to be equivalent to certain higher-dimensional supergravity theories via dimensional reduction (e.g. $N = 1$ **11-dimensional** supergravity is dimensionally reduced on S^7 to $N = 8$, $d = 4$ SUGRA). The resulting theories were sometimes referred to as Kaluza–Klein theories as Kaluza and Klein constructed in 1919 a 5-dimensional gravitational theory, that when dimensionally reduced on circle, its 4-dimensional non-massive modes describe electromagnetism coupled to gravity.

mSUGRA

mSUGRA means minimal SUper GRAvity. The construction of a realistic model of particle interactions within the $N = 1$ supergravity framework where supersymmetry (SUSY) is broken by a super Higgs mechanism was carried out by Ali Chamseddine, Richard Arnowitt and Pran Nath in 1982. In these classes of models collectively now known as minimal supergravity Grand Unification Theories (mSUGRA GUT), gravity mediates the breaking of SUSY through the existence of a hidden sector. mSUGRA naturally generates the Soft SUSY breaking terms which are a consequence of the Super Higgs effect. Radiative breaking of electroweak symmetry through Renormalization Group Equations (RGEs) follows as an immediate consequence. mSUGRA is one of the most widely investigated models of particle physics due to its predictive power—requiring only four input parameters and a sign to determine the low energy phenomenology from the scale of Grand Unification.

See also: Gravity-Mediated Supersymmetry Breaking in the MSSM

11d: the maximal SUGRA

One of these supergravities, the 11-dimensional theory, generated considerable excitement as the first potential candidate for the theory of everything. This excitement was built on four pillars, two of which have now been largely discredited:

- Werner Nahm showed[5] that 11 dimensions was the largest number of dimensions consistent with a single graviton, and that a theory with more dimensions would also have particles with spins greater than 2. These problems are avoided in 12 dimensions if two of these dimensions are timelike, as has been often emphasized by Itzhak Bars.

- In 1981, Ed Witten showed[6] that 11 was the smallest number of dimensions that was big enough to contain the gauge groups of the Standard Model, namely SU(3) for the strong interactions and SU(2) times U(1) for the electroweak interactions. Today many techniques exist to embed the standard model gauge group in supergravity in any number of dimensions. For example, in the mid and late 1980s, the obligatory gauge symmetry in type I and heterotic string theories was often used. In type II string theory they could also be obtained by compactifying on certain Calabi–Yau manifolds. Today one may also use D-branes to engineer gauge symmetries.

- In 1978, Eugène Cremmer, Bernard Julia and Joël Scherk (CJS) found[7] the classical action for an 11-dimensional supergravity theory. This remains today the only known classical 11-dimensional theory with local supersymmetry and no fields of spin higher than two. Other 11-dimensional theories are known that are quantum-mechanically inequivalent to the CJS theory, but classically equivalent (that is, they reduce to the CJS theory when one imposes the classical equations of motion). For example, in the mid 1980s Bernard de Wit and Hermann Nicolai found an alternate theory in D=11 Supergravity with Local SU(8) Invariance. This theory, while not manifestly Lorentz-invariant, is in many ways superior to the CJS theory in that, for example, it dimensionally-reduces to the 4-dimensional theory without recourse to the classical equations of motion.

- In 1980, Peter Freund and M. A. Rubin showed that compactification from 11 dimensions preserving all the SUSY generators could occur in two ways, leaving only 4 or 7 macroscopic dimensions (the other 7 or 4 being compact).[8] Unfortunately, the noncompact dimensions have to form an anti-de Sitter space. Today it is understood that there are many possible compactifications, but that the Freund-Rubin compactifications are invariant under all of the supersymmetry transformations that preserve the action.

Thus, the first two results appeared to establish 11 dimensions uniquely, the third result appeared to specify the theory, and the last result explained why the observed universe appears to be four-dimensional.

Many of the details of the theory were fleshed out by Peter van Nieuwenhuizen, Sergio Ferrara and Daniel Z. Freedman.

The end of the SUGRA era

The initial excitement over 11-dimensional supergravity soon waned, as various failings were discovered, and attempts to repair the model failed as well. Problems included:

- The compact manifolds which were known at the time and which contained the standard model were not compatible with supersymmetry, and could not hold quarks or leptons. One suggestion was to replace the compact dimensions with the 7-sphere, with the symmetry group SO(8), or the squashed 7-sphere, with symmetry group SO(5) times SU(2).

- Until recently, the physical neutrinos seen in experiments were believed to be massless, and appeared to be left-handed, a phenomenon referred to as the chirality of the Standard Model. It was very difficult to construct a chiral fermion from a compactification — the compactified manifold needed to have singularities, but physics near singularities did not begin to be understood until the advent of orbifold conformal field theories in the late 1980s.

- Supergravity models generically result in an unrealistically large cosmological constant in four dimensions, and that constant is difficult to remove, and so require fine-tuning. This is still a problem today.

- Quantization of the theory led to quantum field theory gauge anomalies rendering the theory inconsistent. In the intervening years physicists have learned how to cancel these anomalies.

Some of these difficulties could be avoided by moving to a 10-dimensional theory involving superstrings. However, by moving to 10 dimensions one loses the sense of uniqueness of the 11-dimensional theory.

The core breakthrough for the 10-dimensional theory, known as the first superstring revolution, was a demonstration by Michael B. Green, John H. Schwarz and David Gross that there are only three supergravity models in 10 dimensions which have gauge symmetries and in which all of the gauge and gravitational anomalies cancel. These were theories built on the groups $SO(32)$ and $E_8 \times E_8$, the direct product of two copies of E_8. Today we know that, using D-branes for example, gauge symmetries can be introduced in other 10-dimensional theories as well.[9]

The second superstring revolution

Initial excitement about the 10-dimensional theories, and the string theories that provide their quantum completion, died by the end of the 1980s. There were too many Calabi–Yaus to compactify on, many more than Yau had estimated, as he admitted in December 2005 at the 23rd International Solvay Conference in Physics. None quite gave the standard model, but it seemed as though one could get close with enough effort in many distinct ways. Plus no one understood the theory beyond the regime of applicability of string perturbation theory.

There was a comparatively quiet period at the beginning of the 1990s; however, several important tools were developed. For example, it became apparent that the various superstring theories were related by "string dualities", some of which relate weak string-coupling (i.e. perturbative) physics in one model with strong string-coupling (i.e. non-perturbative) in another.

Then it all changed, in what is known as the second superstring revolution. Joseph Polchinski realized that obscure string theory objects, called D-branes, which he had discovered six years earlier, are stringy versions of the p-branes that were known in supergravity theories. The treatment of these p-branes was not restricted by string perturbation theory; in fact, thanks to supersymmetry, p-branes in supergravity were understood well beyond the limits in which string theory was understood.

Armed with this new nonperturbative tool, Edward Witten and many others were able to show that all of the perturbative string theories were descriptions of different states in a single theory which Witten named M-theory. Furthermore, he argued that M-theory's long wavelength limit (i.e. when the quantum wavelength associated to objects in the theory are much larger than the size of the 11th dimension) should be described by the 11-dimensional supergravity that had fallen out of favor with the first superstring revolution 10 years earlier, accompanied by the 2- and 5-branes.

Historically, then, supergravity has come "full circle". It is a commonly used framework in understanding features of string theories, M-theory and their compactifications to lower spacetime dimensions.

3.4.3 Relation to superstrings

Particular 10-dimensional supergravity theories are considered "low energy limits" of the 10-dimensional superstring theories; more precisely, these arise as the massless, tree-level approximation of string theories. True effective field theories of string theories, rather than truncations, are rarely available. Due to string dualities, the conjectured 11-dimensional M-theory is required to have 11-dimensional supergravity as a "low energy limit". However, this doesn't necessarily mean that string theory/M-theory is the only possible UV completion of supergravity; supergravity research is useful independent of those relations.

3.4.4   4D $N = 1$ SUGRA

Before we move on to SUGRA proper, let's recapitulate some important details about general relativity. We have a 4D differentiable manifold M with a Spin(3,1) principal bundle over it. This principal bundle represents the local Lorentz symmetry. In addition, we have a vector bundle T over the manifold with the fiber having four real dimensions and transforming as a vector under Spin(3,1). We have an invertible linear map from the tangent bundle TM to T. This map is the vierbein. The local Lorentz symmetry has a gauge connection associated with it, the spin connection.

The following discussion will be in superspace notation, as opposed to the component notation, which isn't manifestly covariant under SUSY. There are actually *many* different versions of SUGRA out there which are inequivalent in the sense that their actions and constraints upon the torsion tensor are different, but ultimately equivalent in that we can always perform a field redefinition of the supervierbeins and spin connection to get from one version to another.

In 4D N=1 SUGRA, we have a 4|4 real differentiable supermanifold M, i.e. we have 4 real bosonic dimensions and 4 real fermionic dimensions. As in the nonsupersymmetric case, we have a Spin(3,1) principal bundle over M. We have an $\mathbf{R}^{4|4}$ vector bundle T over M. The fiber of T transforms under the local Lorentz group as follows: the four real bosonic dimensions transform as a vector and the four real fermionic dimensions transform as a Majorana spinor. This Majorana spinor can be reexpressed as a complex left-handed Weyl spinor and its complex conjugate right-handed Weyl spinor (they're not independent of each other). We also have a spin connection as before.

We will use the following conventions: the spatial (both bosonic and fermionic) indices will be indicated by M, N,

... . The bosonic spatial indices will be indicated by μ, ν, ... the left-handed Weyl spatial indices by α, β...., and the right-handed Weyl spatial indices by $\dot{\alpha}$, $\dot{\beta}$, ... The indices for the fiber of T will follow a similar notation, except that they will be hatted like this: $\hat{M}, \hat{\alpha}$. See van der Waerden notation for more details. $M = (\mu, \alpha, \dot{\alpha})$. The supervierbein is denoted by e_N^M , and the spin connection by ω_{MNP} . The *inverse* supervierbein is denoted by E_M^N .

The supervierbein and spin connection are real in the sense that they satisfy the reality conditions

$$e_N^M(x, \overline{\theta}, \theta)^* = e_{N^*}^{M^*}(x, \theta, \overline{\theta}) \text{ where } \mu^* = \mu , \alpha^* = \dot{\alpha} , \text{ and } \dot{\alpha}^* = \alpha \text{ and } \omega(x, \overline{\theta}, \theta)^* = \omega(x, \theta, \overline{\theta}) .$$

The covariant derivative is defined as

$$D_M f = E_M^N \left(\partial_N f + \omega_N [f] \right)$$

The covariant exterior derivative as defined over supermanifolds needs to be super graded. This means that every time we interchange two fermionic indices, we pick up a $+1$ sign factor, instead of -1.

The presence or absence of R symmetries is optional, but if R-symmetry exists, the integrand over the full superspace has to have an R-charge of 0 and the integrand over chiral superspace has to have an R-charge of 2.

A chiral superfield X is a superfield which satisfies $\overline{D}_{\dot{\alpha}} X = 0$. In order for this constraint to be consistent, we require the integrability conditions that $\left\{ \overline{D}_{\dot{\alpha}}, \overline{D}_{\dot{\beta}} \right\} = c_{\dot{\alpha}\dot{\beta}}^{\dot{\gamma}} \overline{D}_{\dot{\gamma}}$ for some coefficients c.

Unlike nonSUSY GR, the torsion has to be nonzero, at least with respect to the fermionic directions. Already, even in flat superspace, $D_\alpha e_{\dot{\alpha}} + \overline{D}_{\dot{\alpha}} e_\alpha \neq 0$. In one version of SUGRA (but certainly not the only one), we have the following constraints upon the torsion tensor:

$$T_{\underline{\alpha}\beta}^{\hat{\gamma}} = 0$$

$$T_{\dot{\alpha}\beta}^{\mu} = 0$$

$$T_{\dot{\alpha}\beta}^{\mu} = 0$$

$$T_{\dot{\alpha}\beta}^{\mu} = 2i\sigma_{\dot{\alpha}\beta}^{\mu}$$

$$T_{\mu\dot{\alpha}}^{\nu} = 0$$

$$T_{\mu\nu}^{\hat{\rho}} = 0$$

Here, $\underline{\alpha}$ is a shorthand notation to mean the index runs over either the left or right Weyl spinors.

The superdeterminant of the supervierbein, $|e|$, gives us the volume factor for M. Equivalently, we have the volume 4|4-superform $e^{\hat{\mu}=0} \wedge \cdots \wedge e^{\hat{\mu}=3} \wedge e^{\hat{\alpha}=1} \wedge e^{\hat{\alpha}=2} \wedge e^{\dot{\hat{\alpha}}=1} \wedge e^{\dot{\hat{\alpha}}=2}$

If we complexify the superdiffeomorphisms, there is a gauge where $E_{\dot{\alpha}}^\mu = 0$, $E_{\dot{\alpha}}^\beta = 0$ and $E_{\dot{\alpha}}^{\dot{\beta}} = \delta_{\dot{\alpha}}^{\dot{\beta}}$. The resulting chiral superspace has the coordinates x and Θ.

R is a scalar valued chiral superfield derivable from the supervielbeins and spin connection. If f is any superfield, $\left(D^2 - 8R \right) f$ is always a chiral superfield.

The action for a SUGRA theory with chiral superfields X, is given by

$$S = \int d^4x d^2\Theta 2\mathcal{E} \left[\frac{3}{8} \left(\overline{D}^2 - 8R \right) e^{-K(X,\overline{X})/3} + W(X) \right] + c.c.$$

where K is the Kähler potential and W is the superpotential, and $\mathcal{E}$ is the chiral volume factor.

Unlike the case for flat superspace, adding a constant to either the Kähler or superpotential is now physical. A constant shift to the Kähler potential changes the effective Planck constant, while a constant shift to the superpotential changes the effective cosmological constant. As the effective Planck constant now depends upon the value of the chiral superfield X, we need to rescale the supervierbeins (a field redefinition) to get a constant Planck constant. This is called the **Einstein frame**.

3.4.5 N = 8 supergravity in 4 dimensions

N=8 Supergravity is the most symmetric quantum field theory which involves gravity and a finite number of fields. It can be found from a dimensional reduction of 11D supergravity by making the size of 7 of the dimensions go to zero. It has 8 supersymmetries which is the most any gravitational theory can have since there are 8 half-steps between spin 2 and spin -2. (A graviton has the highest spin in this theory which is a spin 2 particle). More supersymmetries would mean the particles would have superpartners with spins higher than 2. The only theories with spins higher than 2 which are consistent involve an infinite number of particles (such as String Theory and Higher-Spin Theories). Stephen Hawking in his *A Brief History of Time* speculated that this theory could be the Theory of Everything. However, in later years this was abandoned in favour of String Theory. There has been renewed interest in the 21st century with the possibility that this theory may be finite.

3.4.6 Higher-dimensional SUGRA

Main article: Higher-dimensional supergravity

Higher-dimensional SUGRA is the higher-dimensional, supersymmetric generalization of general relativity. Supergravity can be formulated in any number of dimensions up to eleven. Higher-dimensional SUGRA focuses upon supergravity in greater than four dimensions.

The number of supercharges in a spinor depends on the dimension and the signature of spacetime. The supercharges occur in spinors. Thus the limit on the number of supercharges cannot be satisfied in a spacetime of arbitrary dimension. Some theoretical examples in which this is satisfied are:

- 12-dimensional two-time theory

- 11-dimensional maximal SUGRA

- 10-dimensional SUGRA theories

  - Type IIA SUGRA: N = (1, 1)

  - IIA SUGRA from 11d SUGRA

  - Type IIB SUGRA: N = (2, 0)

  - Type I gauged SUGRA: N = (1, 0)

- 9d SUGRA theories

  - Maximal 9d SUGRA from 10d

  - T-duality

  - N = 1 Gauged SUGRA

The supergravity theories that have attracted the most interest contain no spins higher than two. This means, in particular, that they do not contain any fields that transform as symmetric tensors of rank higher than two under Lorentz transformations. The consistency of interacting higher spin field theories is, however, presently a field of very active interest.

3.4.7 See also

3.4.8 Notes

[1] P. van Nieuwenhuizen, Phys. Rep. 68, 189 (1981)

[2] "supergravity in nLab". *ncatlab.org*. Retrieved 2015-10-05.

[3] P. Nath and R. Arnowitt. "Generalized Super-Gauge Symmetry as a New Framework for Unified Gauge Theories", *Physics Letters B* **56** (1975) 177

[4] D.Z. Freedman, P. van Nieuwenhuizen and S. Ferrara, "Progress Toward A Theory Of Supergravity", *Physical Review* **D13** (1976) pp 3214–3218.

[5] Werner Nahm, "Supersymmetries and their representations". *Nuclear Physics B* **135** no 1 (1978) pp 149-166, doi:10.1016/0550-3213(78)90218-3

[6] Ed Witten, "Search for a realistic Kaluza-Klein theory". *Nuclear Physics B* **186** no 3 (1981) pp 412-428, doi:10.1016/0550-3213(81)90021-3

[7] E. Cremmer, B. Julia and J. Scherk, "Supergravity theory in eleven dimensions", *Physics Letters* **B76** (1978) pp 409-412.

[8] Peter G.O. Freund; Mark A. Rubin (1980). "Dynamics of dimensional reduction". *Physics Letters B* **97** (2): 233–235. Bibcode:1980PhLB...97..233F. doi:10.1016/0370-2693(80)90590-0.

[9] Blumenhagen, R.; Cvetic, M.; Langacker, P.; Shiu, G. (2005). "Toward Realistic Intersecting D-Brane Models". arXiv:hep-th/0502005 [hep-th].

3.4.9 References

Historical

- P. Nath and R. Arnowitt, "Generalized Super-Gauge Symmetry as a New Framework for Unified Gauge Theories", *Physics Letters B '56* (1975) 177.

- D.Z. Freedman, P. van Nieuwenhuizen and S. Ferrara, "Progress Toward A Theory Of Supergravity", *Physical Review* **D13** (1976) pp 3214–3218.

- E. Cremmer, B. Julia and J. Scherk, "Supergravity theory in eleven dimensions", *Physics Letters* **B76** (1978) pp 409–412. scanned version

- P. Freund and M. Rubin, "Dynamics of dimensional reduction", *Physics Letters* **B97** (1980) pp 233–235.

- Ali H. Chamseddine, R. Arnowitt, Pran Nath, "Locally Supersymmetric Grand Unification", " Phys. Rev.Lett.49:970,1982"

- Michael B. Green, John H. Schwarz, "Anomaly Cancellation in Supersymmetric D=10 Gauge Theory and Superstring Theory". *Physics Letters* **B149** (1984) pp117–122.

General

- Bernard de Wit(2002) Supergravity

- A Supersymmetry Primer (1998); updated in (2006).

- Adel Bilal, Introduction to supersymmetry (2001) ArXiv hep-th/0101055, (*a comprehensive introduction to supersymmetry*).

- Friedemann Brandt, Lectures on supergravity (2002) ArXiv hep-th/0204035, (*an introduction to 4-dimensional N = 1 supergravity*).

- Wess, Julius; Bagger, Jonathan (1992). *Supersymmetry and Supergravity.* Princeton University Press. p. 260. ISBN 0-691-02530-4.

3.5   M-theory

For a more accessible and less technical introduction to this topic, see Introduction to M-theory.

M-theory is a theory in physics that unifies all consistent versions of superstring theory. The existence of such a theory was first conjectured by Edward Witten at a string theory conference at the University of Southern California in the spring of 1995. Witten's announcement initiated a flurry of research activity known as the second superstring revolution.

Prior to Witten's announcement, string theorists had identified five versions of superstring theory. Although these theories appeared at first to be very different, work by several physicists showed that the theories were related in intricate and nontrivial ways. In particular, physicists found that apparently distinct theories could be unified by mathematical transformations called S-duality and T-duality. Witten's conjecture was based in part on the existence of these dualities and in part on the relationship of the string theories to a field theory called eleven-dimensional supergravity.

Although a complete formulation of M-theory is not known, the theory should describe two- and five-dimensional objects called branes and should be approximated by eleven-dimensional supergravity at low energies. Modern attempts to formulate M-theory are typically based on matrix theory or the AdS/CFT correspondence.

According to Witten, M should stand for "magic", "mystery", or "membrane" according to taste, and the true meaning of the title should be decided when a more fundamental formulation of the theory is known.[1]

Investigations of the mathematical structure of M-theory have spawned important theoretical results in physics and mathematics. More speculatively, M-theory may provide a framework for developing a unified theory of all of the fundamental forces of nature. Attempts to connect M-theory to experiment typically focus on compactifying its extra dimensions to construct candidate models of our four-dimensional world, although so far none have been verified to give rise to physics as observed at, for instance, the Large Hadron Collider.

3.5.1   Background

Quantum gravity and strings

Main articles: Quantum gravity and String theory

One of the deepest problems in modern physics is the

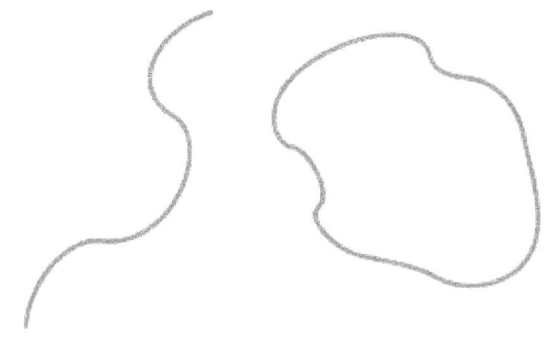

The fundamental objects of string theory are open and closed strings.

problem of quantum gravity. The current understanding of gravity is based on Albert Einstein's general theory of relativity, which is formulated within the framework of classical physics. However, nongravitational forces are described within the framework of quantum mechanics, a radically different formalism for describing physical phenomena based on probability.[lower-alpha 1] A quantum theory of gravity is needed in order to reconcile general relativity with the principles of quantum mechanics,[lower-alpha 2] but difficulties arise when one attempts to apply the usual prescriptions of quantum theory to the force of gravity.[lower-alpha 3]

String theory is a theoretical framework that attempts to reconcile gravity and quantum mechanics. In string theory, the point-like particles of particle physics are replaced by one-dimensional objects called strings.

String theory describes how strings propagate through space and interact with each other. In a given version of string theory, there is only one kind of string, which may look like a small loop or segment of ordinary string, and it can vibrate in different ways. On distance scales larger than the string scale, a string will look just like an ordinary particle, with its mass, charge, and other properties determined by the vibrational state of the string. In this way, all of the different elementary particles may be viewed as vibrating strings. One of the vibrational states of a string gives rise to the graviton, a quantum mechanical particle that carries gravitational force.[lower-alpha 4]

There are several versions of string theory: type I, type IIA, type IIB, and two flavors of heterotic string theory ($SO(32)$ and $E_8 \times E_8$). The different theories allow different types of strings, and the particles that arise at low energies exhibit different symmetries. For example, the type I theory includes both open strings (which are segments with endpoints) and closed strings (which form closed loops), while types IIA and IIB include only closed strings.[2] Each of these five string theories arises as a special limiting case of M-theory. This theory, like its string theory predecessors, is an example of a quantum theory of gravity. It describes a force just like the familiar gravitational force subject to the rules of quantum mechanics.[3]

Number of dimensions

Main article: Compactification (physics)
In everyday life, there are three familiar dimensions of

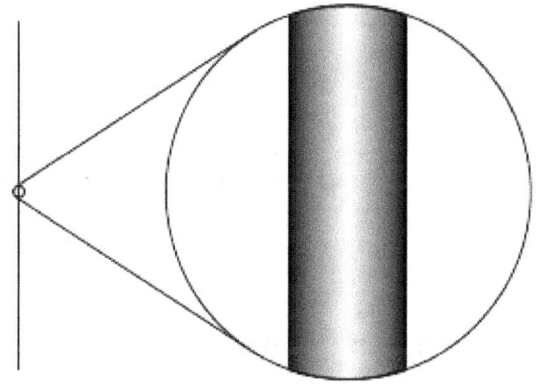

An example of compactification: At large distances, a two dimensional surface with one circular dimension looks one-dimensional.

space: height, width and depth. Einstein's general theory of relativity treats time as a dimension on par with the three spatial dimensions; in general relativity, space and time are not modeled as separate entities but are instead unified to a four-dimensional spacetime. In this framework, the phenomenon of gravity is viewed as a consequence of the geometry of spacetime.[4]

In spite of the fact that the universe is well described by four-dimensional spacetime, there are several reasons why physicists consider theories in other dimensions. In some cases, by modeling spacetime in a different number of dimensions, a theory becomes more mathematically tractable, and one can perform calculations and gain general insights more easily.[lower-alpha 5] There are also situations where theories in two or three spacetime dimensions are useful for describing phenomena in condensed matter physics.[5] Finally, there exist scenarios in which there could actually be more

than four dimensions of spacetime which have nonetheless managed to escape detection.[6]

One notable feature of string theory and M-theory is that these theories require extra dimensions of spacetime for their mathematical consistency. In string theory, spacetime is ten-dimensional, while in M-theory it is eleven-dimensional. In order to describe real physical phenomena using these theories, one must therefore imagine scenarios in which these extra dimensions would not be observed in experiments.[7]

Compactification is one way of modifying the number of dimensions in a physical theory.[lower-alpha 6] In compactification, some of the extra dimensions are assumed to "close up" on themselves to form circles.[8] In the limit where these curled up dimensions become very small, one obtains a theory in which spacetime has effectively a lower number of dimensions. A standard analogy for this is to consider a multidimensional object such as a garden hose. If the hose is viewed from a sufficient distance, it appears to have only one dimension, its length. However, as one approaches the hose, one discovers that it contains a second dimension, its circumference. Thus, an ant crawling on the surface of the hose would move in two dimensions.[lower-alpha 7]

Dualities

Main articles: S-duality and T-duality
Theories that arise as different limits of M-theory turn out to be related in highly nontrivial ways. One of the relationships that can exist between these different physical theories is called S-duality. This is a relationship which says that a collection of strongly interacting particles in one theory can, in some cases, be viewed as a collection of weakly interacting particles in a completely different theory. Roughly speaking, a collection of particles is said to be strongly interacting if they combine and decay often and weakly interacting if they do so infrequently. Type I string theory turns out to be equivalent by S-duality to the $SO(32)$ heterotic string theory. Similarly, type IIB string theory is related to itself in a nontrivial way by S-duality.[10]

Another relationship between different string theories is T-duality. Here one considers strings propagating around a circular extra dimension. T-duality states that a string propagating around a circle of radius R is equivalent to a string propagating around a circle of radius $1/R$ in the sense that all observable quantities in one description are identified with quantities in the dual description. For example, a string has momentum as it propagates around a circle, and it can also wind around the circle one or more times. The number of times the string winds around a circle is called the winding number. If a string has momentum p and winding number n in one description, it will have momentum n and winding

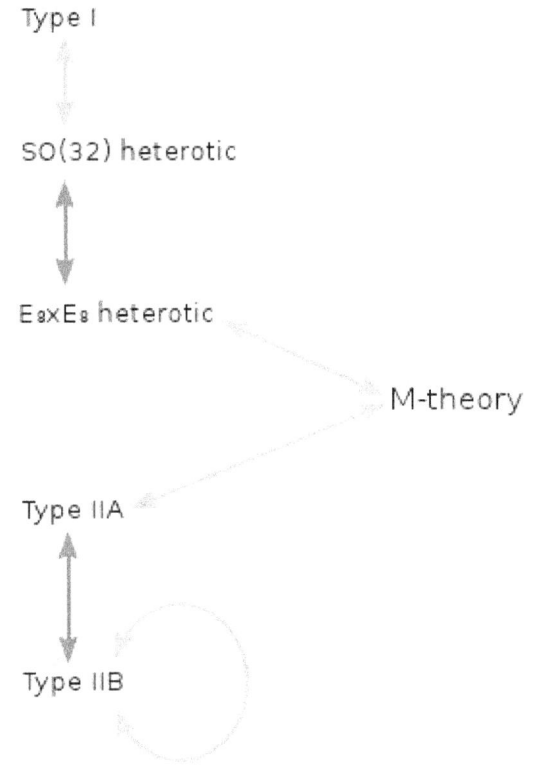

Type I

SO(32) heterotic

E₈×E₈ heterotic

M-theory

Type IIA

Type IIB

A diagram of string theory dualities. Yellow arrows indicate S-duality. Blue arrows indicate T-duality. These dualities may be combined to obtain equivalences of any of the five theories with M-theory.[9]

number p in the dual description. For example, type IIA string theory is equivalent to type IIB string theory via T-duality, and the two versions of heterotic string theory are also related by T-duality.[10]

In general, the term *duality* refers to a situation where two seemingly different physical systems turn out to be equivalent in a nontrivial way. If two theories are related by a duality, it means that one theory can be transformed in some way so that it ends up looking just like the other theory. The two theories are then said to be *dual* to one another under the transformation. Put differently, the two theories are mathematically different descriptions of the same phenomena.[11]

Supersymmetry

Main article: Supersymmetry

Another important theoretical idea that plays a role in M-theory is supersymmetry. This is a mathematical relation that exists in certain physical theories between a class of particles called bosons and a class of particles called fermions. Roughly speaking, fermions are the constituents of matter, while bosons mediate interactions between particles. In theories with supersymmetry, each boson has a counterpart which is a fermion, and vice versa. When supersymmetry is imposed as a local symmetry, one automatically obtains a quantum mechanical theory that includes gravity. Such a theory is called a supergravity theory.[12]

A theory of strings that incorporates the idea of supersymmetry is called a superstring theory. There are several different versions of superstring theory which are all subsumed within the M-theory framework. At low energies, the superstring theories are approximated by supergravity in ten spacetime dimensions. Similarly, M-theory is approximated at low energies by supergravity in eleven dimensions.[3]

Branes

Main article: Brane

In string theory and related theories such as supergravity theories, a brane is a physical object that generalizes the notion of a point particle to higher dimensions. For example, a point particle can be viewed as a brane of dimension zero, while a string can be viewed as a brane of dimension one. It is also possible to consider higher-dimensional branes. In dimension p, these are called p-branes. Branes are dynamical objects which can propagate through spacetime according to the rules of quantum mechanics. They can have mass and other attributes such as charge. A p-brane sweeps out a $(p+1)$-dimensional volume in spacetime called its *worldvolume*. Physicists often study fields analogous to the electromagnetic field which live on the worldvolume of a brane. The word brane comes from the word "membrane" which refers to a two-dimensional brane.[13]

In string theory, the fundamental objects that give rise to elementary particles are the one-dimensional strings. Although the physical phenomena described by M-theory are still poorly understood, physicists know that the theory describes two- and five-dimensional branes. Much of the current research in M-theory attempts to better understand the properties of these branes.[lower-alpha 8]

3.5.2 History and development

Kaluza–Klein theory

Main article: Kaluza–Klein theory

In the early 20th century, physicists and mathematicians

including Albert Einstein and Hermann Minkowski pioneered the use of four-dimensional geometry for describing the physical world.[14] These efforts culminated in the formulation of Einstein's general theory of relativity, which relates gravity to the geometry of four-dimensional spacetime.[15]

The success of general relativity led to efforts to apply higher dimensional geometry to explain other forces. In 1919, work by Theodor Kaluza showed that by passing to five-dimensional spacetime, one can unify gravity and electromagnetism into a single force.[15] This idea was improved by physicist Oskar Klein, who suggested that the additional dimension proposed by Kaluza could take the form of a circle with radius around 10^{-30} cm.[16]

The Kaluza–Klein theory and subsequent attempts by Einstein to develop unified field theory were never completely successful. In part this was because Kaluza–Klein theory predicted a particle that has never been shown to exist, and in part because it was unable to correctly predict the ratio of an electron's mass to its charge. In addition, these theories were being developed just as other physicists were beginning to discover quantum mechanics, which would ultimately prove successful in describing known forces such as electromagnetism, as well as new nuclear forces that were being discovered throughout the middle part of the century. Thus it would take almost fifty years for the idea of new dimensions to be taken seriously again.[17]

In the 1980s, Edward Witten contributed to the understanding of supergravity theories. In 1995, he introduced M-theory, sparking the second superstring revolution.

Early work on supergravity

Main article: Supergravity

New concepts and mathematical tools provided fresh insights into general relativity, giving rise to a period in the 1960s and 70s now known as the golden age of general relativity.[18] In the mid-1970s, physicists began studying higher-dimensional theories combining general relativity with supersymmetry, the so-called supergravity theories.[19]

General relativity does not place any limits on the possible dimensions of spacetime. Although the theory is typically formulated in four dimensions, one can write down the same equations for the gravitational field in any number of dimensions. Supergravity is more restrictive because it places an upper limit on the number of dimensions.[12] In 1978, work by Werner Nahm showed that the maximum spacetime dimension in which one can formulate a consistent supersymmetric theory is eleven.[20] In the same year, Eugene Cremmer, Bernard Julia, and Joel Scherk of the École Normale Supérieure showed that supergravity not only permits up to eleven dimensions but is in fact most elegant in this maximal number of dimensions.[21][22]

Initially, many physicists hoped that by compactifying eleven-dimensional supergravity, it might be possible to construct realistic models of our four-dimensional world. The hope was that such models would provide a unified description of the four fundamental forces of nature: electromagnetism, the strong and weak nuclear forces, and gravity. Interest in eleven-dimensional supergravity soon waned as various flaws in this scheme were discovered. One of the problems was that the laws of physics appear to distinguish between clockwise and counterclockwise, a phenomenon known as chirality. Edward Witten and others observed this chirality property cannot be readily derived by compactifying from eleven dimensions.[22]

In the first superstring revolution in 1984, many physicists turned to string theory as a unified theory of particle physics and quantum gravity. Unlike supergravity theory, string theory was able to accommodate the chirality of the standard model, and it provided a theory of gravity consistent with quantum effects.[22] Another feature of string theory that many physicists were drawn to in the 1980s and 1990s was its high degree of uniqueness. In ordinary particle theories, one can consider any collection of elementary particles whose classical behavior is described by an arbitrary Lagrangian. In string theory, the possibilities are much more constrained: by the 1990s, physicists had argued that there were only five consistent supersymmetric versions of the theory.[22]

Relationships between string theories

Although there were only a handful of consistent superstring theories, it remained a mystery why there was not just one consistent formulation.[22] However, as physicists began to examine string theory more closely, they realized that these theories are related in intricate and nontrivial ways.[23]

In the late 1970s, Claus Montonen and David Olive had conjectured a special property of certain physical theories.[24] A sharpened version of their conjecture concerns a theory called $N=4$ supersymmetric Yang–Mills theory, which describes particles similar to the quarks and gluons that make up atomic nuclei. The strength with which the particles of this theory interact is measured by a number called the coupling constant. The result of Montonen and Olive, now known as Montonen–Olive duality, states that $N=4$ supersymmetric Yang–Mills theory with coupling constant g is equivalent to the same theory with coupling constant $1/g$. In other words, a system of strongly interacting particles (large coupling constant) has an equivalent description as a system of weakly interacting particles (small coupling constant) and vice versa.[25]

In the 1990s, several theorists generalized Montonen–Olive duality to the S-duality relationship, which connects different string theories. Ashoke Sen studied S-duality in the context of heterotic strings in four dimensions.[26][27] Chris Hull and Paul Townsend showed that type IIB string theory with a large coupling constant is equivalent via S-duality to the same theory with small coupling constant.[28] Theorists also found that different string theories may be related by T-duality. This duality implies that strings propagating on completely different spacetime geometries may be physically equivalent.[29]

Membranes and fivebranes

String theory extends ordinary particle physics by promoting zero-dimensional point particles to one-dimensional objects called strings. In the late 1980s, it was natural for theorists to attempt to formulate other extensions in which particles are replaced by two-dimensional supermembranes or by higher-dimensional objects called branes. Such objects had been considered as early as 1962 by Paul Dirac,[30] and they were reconsidered by a small but enthusiastic group of physicists in the 1980s.[22]

Supersymmetry severely restricts the possible number of dimensions of a brane. In 1987, Eric Bergshoeff, Ergin Sezgin, and Paul Townsend showed that eleven-dimensional supergravity includes two-dimensional branes.[31] Intuitively, these objects look like sheets or membranes propagating through the eleven-dimensional spacetime. Shortly after this discovery, Michael Duff, Paul Howe, Takeo Inami, and Kellogg Stelle considered a particular compactification of eleven-dimensional supergravity with one of the dimensions curled up into a circle.[32] In this setting, one can imagine the membrane wrapping around the circular dimension. If the radius of the circle is sufficiently small, then this membrane looks just like a string in ten-dimensional spacetime. In fact, Duff and his collaborators showed that this construction reproduces exactly the strings appearing in type IIA superstring theory.[25]

In 1990, Andrew Strominger published a similar result which suggested that strongly interacting strings in ten dimensions might have an equivalent description in terms of weakly interacting five-dimensional branes.[33] Initially, physicists were unable to prove this relationship for two important reasons. On the one hand, the Montonen–Olive duality was still unproven, and so Strominger's conjecture was even more tenuous. On the other hand, there were many technical issues related to the quantum properties of five-dimensional branes.[34] The first of these problems was solved in 1993 when Ashoke Sen established that certain physical theories require the existence of objects with both electric and magnetic charge which were predicted by the work of Montonen and Olive.[35]

In spite of this progress, the relationship between strings and five-dimensional branes remained conjectural because theorists were unable to quantize the branes. Starting in 1991, a team of researchers including Michael Duff, Ramzi Khuri, Jianxin Lu, and Ruben Minasian considered a special compactification of string theory in which four of the ten dimensions curl up. If one considers a five-dimensional brane wrapped around these extra dimensions, then the brane looks just like a one-dimensional string. In this way, the conjectured relationship between strings and branes was reduced to a relationship between strings and strings, and the latter could be tested using already established theoretical techniques.[29]

Second superstring revolution

Main article: Second superstring revolution

Speaking at the string theory conference at the University of Southern California in 1995, Edward Witten of the Institute for Advanced Study made the surprising suggestion that all five superstring theories were in fact just different limiting cases of a single theory in eleven spacetime dimensions. Witten's announcement drew together all of the previous results on S- and T-duality and the appearance of two- and five-dimensional branes in string theory.[36] In the months following Witten's announcement, hundreds of new papers appeared on the Internet confirming that the new theory involved membranes in an important way.[37] Today this flurry

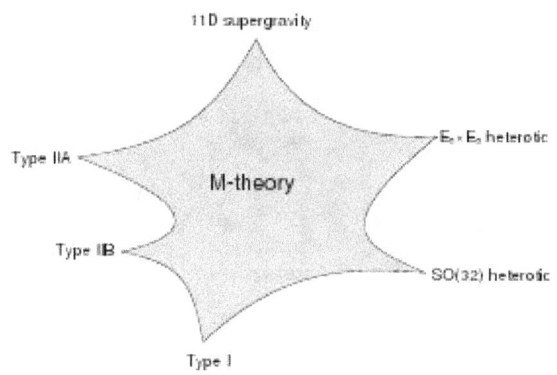

A schematic illustration of the relationship between M-theory, the five superstring theories, and eleven-dimensional supergravity. The shaded region represents a family of different physical scenarios that are possible in M-theory. In certain limiting cases corresponding to the cusps, it is natural to describe the physics using one of the six theories labeled there.

of work is known as the second superstring revolution.[38]

One of the important developments following Witten's announcement was Witten's work in 1996 with string theorist Petr Hořava.[39][40] Witten and Hořava studied M-theory on a special spacetime geometry with two ten-dimensional boundary components. Their work shed light on the mathematical structure of M-theory and suggested possible ways of connecting M-theory to real world physics.[41]

Origin of the term

Initially, some physicists suggested that the new theory was a fundamental theory of membranes, but Witten was skeptical of the role of membranes in the theory. In a paper from 1996, Hořava and Witten wrote

> As it has been proposed that the eleven-dimensional theory is a supermembrane theory but there are some reasons to doubt that interpretation, we will non-committally call it the M-theory, leaving to the future the relation of M to membranes.[39]

In the absence of an understanding of the true meaning and structure of M-theory, Witten has suggested that the *M* should stand for "magic", "mystery", or "membrane" according to taste, and the true meaning of the title should be decided when a more fundamental formulation of the theory is known.[1]

3.5.3 Matrix theory

BFSS matrix model

Main article: Matrix theory (physics)

In mathematics, a matrix is a rectangular array of numbers or other data. In physics, a matrix model is a particular kind of physical theory whose mathematical formulation involves the notion of a matrix in an important way. A matrix model describes the behavior of a set of matrices within the framework of quantum mechanics.[42][43]

One important example of a matrix model is the BFSS matrix model proposed by Tom Banks, Willy Fischler, Stephen Shenker, and Leonard Susskind in 1997. This theory describes the behavior of a set of nine large matrices. In their original paper, these authors showed, among other things, that the low energy limit of this matrix model is described by eleven-dimensional supergravity. These calculations led them to propose that the BFSS matrix model is exactly equivalent to M-theory. The BFSS matrix model can therefore be used as a prototype for a correct formulation of M-theory and a tool for investigating the properties of M-theory in a relatively simple setting.[42]

Noncommutative geometry

Main articles: Noncommutative geometry and Noncommutative quantum field theory

In geometry, it is often useful to introduce coordinates. For example, in order to study the geometry of the Euclidean plane, one defines the coordinates x and y as the distances between any point in the plane and a pair of axes. In ordinary geometry, the coordinates of a point are numbers, so they can be multiplied, and the product of two coordinates does not depend on the order of multiplication. That is, $xy = yx$. This property of multiplication is known as the commutative law, and this relationship between geometry and the commutative algebra of coordinates is the starting point for much of modern geometry.[44]

Noncommutative geometry is a branch of mathematics that attempts to generalize this situation. Rather than working with ordinary numbers, one considers some similar objects, such as matrices, whose multiplication does not satisfy the commutative law (that is, objects for which xy is not necessarily equal to yx). One imagines that these noncommuting objects are coordinates on some more general notion of "space" and proves theorems about these generalized spaces by exploiting the analogy with ordinary geometry.[45]

In a paper from 1998, Alain Connes, Michael R. Douglas, and Albert Schwarz showed that some aspects of matrix models and M-theory are described by a noncommutative quantum field theory, a special kind of physical theory in which the coordinates on spacetime do not satisfy the commutativity property.[43] This established a link between matrix models and M-theory on the one hand, and noncommutative geometry on the other hand. It quickly led to the discovery of other important links between noncommutative geometry and various physical theories.[46][47]

3.5.4 AdS/CFT correspondence

Overview

Main article: AdS/CFT correspondence

The application of quantum mechanics to physical objects such as the electromagnetic field, which are extended in space and time, is known as quantum field theory.[lower-alpha 9] In particle physics, quantum field theories form the basis for our understanding of elementary particles, which are modeled as excitations in the fundamental fields. Quantum field theories are also used throughout condensed matter physics to model particle-like objects called quasiparticles.[lower-alpha 10]

One approach to formulating M-theory and studying its properties is provided by the anti-de Sitter/conformal field theory (AdS/CFT) correspondence. Proposed by Juan Maldacena in late 1997, the AdS/CFT correspondence is a theoretical result which implies that M-theory is in some cases equivalent to a quantum field theory.[48] In addition to providing insights into the mathematical structure of string and M-theory, the AdS/CFT correspondence has shed light on many aspects of quantum field theory in regimes where traditional calculational techniques are ineffective.[49]

In the AdS/CFT correspondence, the geometry of spacetime is described in terms of a certain vacuum solution of Einstein's equation called anti-de Sitter space.[50] In very elementary terms, anti-de Sitter space is a mathematical model of spacetime in which the notion of distance between points (the metric) is different from the notion of distance in ordinary Euclidean geometry. It is closely related to hyperbolic space, which can be viewed as a disk as illustrated on the left.[51] This image shows a tessellation of a disk by triangles and squares. One can define the distance between points of this disk in such a way that all the triangles and squares are the same size and the circular outer boundary is infinitely far from any point in the interior.[52]

Now imagine a stack of hyperbolic disks where each disk represents the state of the universe at a given time. The resulting geometric object is three-dimensional anti-de Sitter space.[51] It looks like a solid cylinder in which any cross section is a copy of the hyperbolic disk. Time runs along the vertical direction in this picture. The surface of this

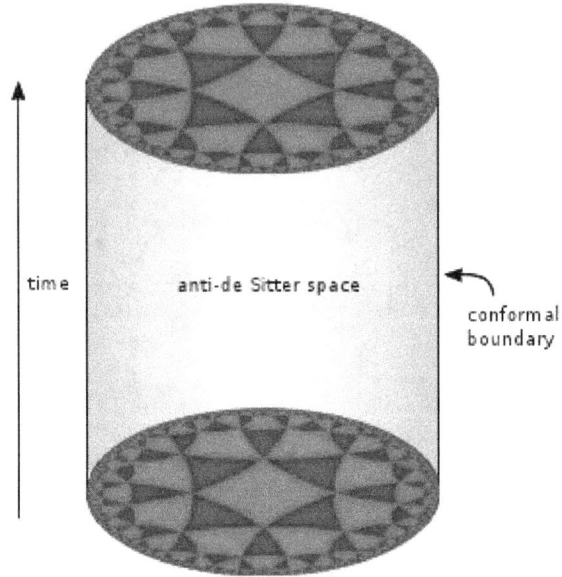

Three-dimensional anti-de Sitter space is like a stack of hyperbolic disks, each one representing the state of the universe at a given time. One can study theories of quantum gravity such as M-theory in the resulting spacetime.

cylinder plays an important role in the AdS/CFT correspondence. As with the hyperbolic plane, anti-de Sitter space is curved in such a way that any point in the interior is actually infinitely far from this boundary surface.[52]

This construction describes a hypothetical universe with only two space dimensions and one time dimension, but it can be generalized to any number of dimensions. Indeed, hyperbolic space can have more than two dimensions and one can "stack up" copies of hyperbolic space to get higher-dimensional models of anti-de Sitter space.[51]

An important feature of anti-de Sitter space is its boundary (which looks like a cylinder in the case of three-dimensional anti-de Sitter space). One property of this boundary is that, within a small region on the surface around any given point, it looks just like Minkowski space, the model of spacetime used in nongravitational physics.[53] One can therefore consider an auxiliary theory in which "spacetime" is given by the boundary of anti-de Sitter space. This observation is the starting point for AdS/CFT correspondence, which states that the boundary of anti-de Sitter space can be regarded as the "spacetime" for a quantum field theory. The claim is that this quantum field theory is equivalent to the gravitational theory on the bulk anti-de Sitter space in the sense that there is a "dictionary" for translating entities and calculations in one theory into their counterparts in the other theory. For example, a single particle in the gravitational theory might correspond to some collection of particles in the boundary theory. In addition, the predictions in the two theories are quantitatively identical so that if two particles have a 40 percent chance of colliding in the gravitational theory, then the corresponding collections in the boundary theory would also have a 40 percent chance of colliding.[54]

6D (2,0) superconformal field theory

Main article: 6D (2,0) superconformal field theory
 One particular realization of the AdS/CFT correspondence

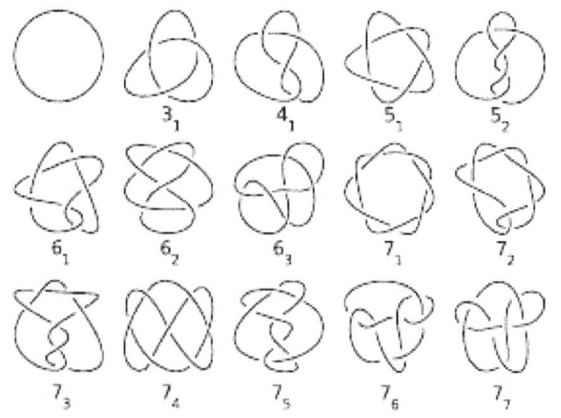

The six-dimensional (2,0)-theory has been used to understand results from the mathematical theory of knots.

states that M-theory on the product space $AdS_7 \times S^4$ is equivalent to the so-called (2,0)-theory on the six-dimensional boundary.[48] Here "(2,0)" refers to the particular type of supersymmetry that appears in the theory. In this example, the spacetime of the gravitational theory is effectively seven-dimensional (hence the notation AdS_7), and there are four additional "compact" dimensions (encoded by the S^4 factor). In the real world, spacetime is four-dimensional, at least macroscopically, so this version of the correspondence does not provide a realistic model of gravity. Likewise, the dual theory is not a viable model of any real-world system since it describes a world with six spacetime dimensions.[lower-alpha 11]

Nevertheless, the (2,0)-theory has proven to be important for studying the general properties of quantum field theories. Indeed, this theory subsumes many mathematically interesting effective quantum field theories and points to new dualities relating these theories. For example, Luis Alday, Davide Gaiotto, and Yuji Tachikawa showed that by compactifying this theory on a surface, one obtains a four-dimensional quantum field theory, and there is a duality known as the AGT correspondence which relates the physics of this theory to certain physical concepts associated with the surface itself.[55] More recently, theorists have extended these ideas to study the theories obtained by compactifying down to three dimensions.[56]

In addition to its applications in quantum field theory, the (2,0)-theory has spawned important results in pure mathematics. For example, the existence of the (2,0)-theory was used by Witten to give a "physical" explanation for a conjectural relationship in mathematics called the geometric Langlands correspondence.[57] In subsequent work, Witten showed that the (2,0)-theory could be used to understand a concept in mathematics called Khovanov homology.[58] Developed by Mikhail Khovanov around 2000, Khovanov homology provides a tool in knot theory, the branch of mathematics that studies and classifies the different shapes of knots.[59] Another application of the (2,0)-theory in mathematics is the work of Davide Gaiotto, Greg Moore, and Andrew Neitzke, which used physical ideas to derive new results in hyperkähler geometry.[60]

ABJM superconformal field theory

Main article: ABJM superconformal field theory

Another realization of the AdS/CFT correspondence states that M-theory on $AdS_4 \times S^7$ is equivalent to a quantum field theory called the ABJM theory in three dimensions. In this version of the correspondence, seven of the dimensions of M-theory are curled up, leaving four non-compact dimensions. Since the spacetime of our universe is four-dimensional, this version of the correspondence provides a somewhat more realistic description of gravity.[61]

The ABJM theory appearing in this version of the correspondence is also interesting for a variety of reasons. Introduced by Aharony, Bergman, Jafferis, and Maldacena, it is closely related to another quantum field theory called Chern–Simons theory. The latter theory was popularized by Witten in the late 1980s because of its applications to knot theory.[62] In addition, the ABJM theory serves as a semi-realistic simplified model for solving problems that arise in condensed matter physics.[61]

3.5.5 Phenomenology

Overview

Main article: String phenomenology
 In addition to being an idea of considerable theoretical interest, M-theory provides a framework for constructing models of real world physics that combine general relativity with the standard model of particle physics. Phenomenology is the branch of theoretical physics in which physicists construct realistic models of nature from more abstract theoretical ideas. String phenomenology is the part of string theory that attempts to construct realistic models of particle physics based on string and M-theory.[63]

A cross section of a Calabi–Yau manifold

Typically, such models are based on the idea of compactification.[lower-alpha 12] Starting with the ten- or eleven-dimensional spacetime of string or M-theory, physicists postulate a shape for the extra dimensions. By choosing this shape appropriately, they can construct models roughly similar to the standard model of particle physics, together with additional undiscovered particles,[64] usually supersymmetric partners to analogues of known particles. One popular way of deriving realistic physics from string theory is to start with the heterotic theory in ten dimensions and assume that the six extra dimensions of spacetime are shaped like a six-dimensional Calabi–Yau manifold. This is a special kind of geometric object named after mathematicians Eugenio Calabi and Shing-Tung Yau.[65] Calabi–Yau manifolds offer many ways of extracting realistic physics from string theory. Other similar methods can be used to construct models with physics resembling to some extent that of our four-dimensional world based on M-theory.[66]

Partly because of theoretical and mathematical difficulties and partly because of the extremely high energies (beyond what is technologically possible for the foreseeable future) needed to test these theories experimentally, there is so far no experimental evidence that would unambiguously point to any of these models being a correct fundamental description of nature. This has led some in the community to criticize these approaches to unification and question the value of continued research on these problems.[67]

Compactification on G_2 manifolds

In one approach to M-theory phenomenology, theorists assume that the seven extra dimensions of M-theory are shaped like a G_2 manifold. This is a special kind of seven-dimensional shape constructed by mathematician Dominic Joyce of the University of Oxford.[68] These G_2 manifolds are still poorly understood mathematically, and this fact has made it difficult for physicists to fully develop this approach to phenomenology.[69]

For example, physicists and mathematicians often assume that space has a mathematical property called smoothness, but this property cannot be assumed in the case of a G_2 manifold if one wishes to recover the physics of our four-dimensional world. Another problem is that G_2 manifolds are not complex manifolds, so theorists are unable to use tools from the branch of mathematics known as complex analysis. Finally, there are many open questions about the existence, uniqueness, and other mathematical properties of G_2 manifolds, and mathematicians lack a systematic way of searching for these manifolds.[69]

Heterotic M-theory

Because of the difficulties with G_2 manifolds, most attempts to construct realistic theories of physics based on M-theory have taken a more indirect approach to compactifying eleven-dimensional spacetime. One approach, pioneered by Witten, Hořava, Burt Ovrut, and others, is known as heterotic M-theory. In this approach, one imagines that one of the eleven dimensions of M-theory is shaped like a circle. If this circle is very small, then the spacetime becomes effectively ten-dimensional. One then assumes that six of the ten dimensions form a Calabi–Yau manifold. If this Calabi–Yau manifold is also taken to be small, one is left with a theory in four-dimensions.[69]

Heterotic M-theory has been used to construct models of brane cosmology in which the observable universe is thought to exist on a brane in a higher dimensional ambient space. It has also spawned alternative theories of the early universe that do not rely on the theory of cosmic inflation.[69]

3.5.6   References

Notes

[1] For a standard introduction to quantum mechanics, see Griffiths 2004.

[2] The necessity of a quantum mechanical description of gravity follows from the fact that one cannot consistently couple a classical system to a quantum one. See Wald 1984, p. 382.

[3] From a technical point of view, the problem is that the theory one gets in this way is not renormalizable and therefore cannot be used to make meaningful physical predictions. See Zee 2010, p. 72 for a discussion of this issue.

[4] For an accessible introduction to string theory, see Greene 2000.

[5] For example, in the context of the AdS/CFT correspondence, theorists often formulate and study theories of gravity in unphysical numbers of spacetime dimensions.

[6] Dimensional reduction is another way of modifying the number of dimensions.

[7] This analogy is used for example in Greene 2000, p. 186.

[8] For example, see the subsections on the 6D (2,0) superconformal field theory and ABJM superconformal field theory.

[9] A standard text is Peskin and Schroeder 1995.

[10] For an introduction to the applications of quantum field theory to condensed matter physics, see Zee 2010.

[11] For a review of the (2,0)-theory, see Moore 2012.

[12] Brane world scenarios provide an alternative way of recovering real world physics from string theory. See Randall and Sundrum 1999.

Citations

[1] Duff 1996, sec. 1

[2] Zwiebach 2009, p. 324

[3] Becker, Becker, and Schwarz 2007, p. 12

[4] Wald 1984, p. 4

[5] Zee 2010, Parts V and VI

[6] Zwiebach 2009, p. 9

[7] Zwiebach 2009, p. 8

[8] Yau and Nadis 2010, Ch. 6

[9] Becker, Becker, and Schwarz 2007, pp. 339–347

[10] Becker, Becker, and Schwarz 2007

[11] Zwiebach 2009, p. 376

[12] Duff 1998, p. 64

[13] Moore 2005

[14] Yau and Nadis 2010, p. 9

[15] Yau and Nadis 2010, p. 10

[16] Yau and Nadis 2010, p. 12

[17] Yau and Nadis 2010, p. 13

[18] Wald 1984, p. 3

[19] van Nieuwenhuizen 1981

[20] Nahm 1978

[21] Cremmer, Julia, and Scherk 1978

[22] Duff 1998, p. 65

[23] Duff 1998

[24] Montonen and Olive 1977

[25] Duff 1998, p. 66

[26] Sen 1994a

[27] Sen 1994b

[28] Hull and Townsend 1995

[29] Duff 1998, p. 67

[30] Dirac 1962

[31] Bergshoeff, Sezgin, and Townsend 1987

[32] Duff et al. 1987

[33] Strominger 1990

[34] Duff 1998, pp 66–67

[35] Sen 1993

[36] Witten 1995

[37] Duff 1998, pp. 67–68

[38] Becker, Becker, and Schwarz 2007, p. 296

[39] Hořava and Witten 1996a

[40] Hořava and Witten 1996b

[41] Duff 1998, p. 68

[42] Banks et al. 1997

[43] Connes, Douglas, and Schwarz 1998

[44] Connes 1994, p. 1

[45] Connes 1994

[46] Nekrasov and Schwarz 1998

[47] Seiberg and Witten 1999

[48] Maldacena 1998

[49] Klebanov and Maldacena 2009

[50] Klebanov and Maldacena 2009, p. 28

[51] Maldacena 2005, p. 60

[52] Maldacena 2005, p. 61

[53] Zwiebach 2009, p. 552

[54] Maldacena 2005, pp. 61–62

[55] Alday, Gaiotto, and Tachikawa 2010

[56] Dimotte, Gaiotto, and Gukov 2010

[57] Witten 2009

[58] Witten 2012

[59] Khovanov 2000

[60] Gaiotto, Moore, and Neitzke 2013

[61] Aharony et al. 2008

[62] Witten 1989

[63] Dine 2000

[64] Candelas et al. 1985

[65] Yau and Nadis 2010, p. ix

[66] Yau and Nadis 2010, pp. 147–150

[67] Woit 2006

[68] Yau and Nadis 2010, p. 149

[69] Yau and Nadis 2010, p. 150

Bibliography

- Aharony, Ofer; Bergman, Oren; Jafferis, Daniel Louis; Maldacena, Juan (2008). "$N=6$ superconformal Chern-Simons-matter theories, M2-branes and their gravity duals". *Journal of High Energy Physics* **2008** (10): 091. arXiv:0806.1218. Bibcode:2008JHEP...10..091A. doi:10.1088/1126-6708/2008/10/091.

- Alday, Luis; Gaiotto, Davide; Tachikawa, Yuji (2010). "Liouville correlation functions from four-dimensional gauge theories". *Letters in Mathematical Physics* **91** (2): 167–197. arXiv:0906.3219. Bibcode:2010LMaPh..91..167A. doi:10.1007/s11005-010-0369-5.

- Banks, Tom; Fischler, Willy; Schenker, Stephen; Susskind, Leonard (1997). "M theory as a matrix model: A conjecture". *Physical Review D* **55** (8): 5112. arXiv:hep-th/9610043. Bibcode:1997PhRvD..55.5112B. doi:10.1103/physrevd.55.5112.

- Becker, Katrin; Becker, Melanie; Schwarz, John (2007). *String theory and M-theory: A modern introduction*. Cambridge University Press. ISBN 978-0-521-86069-7.

- Bergshoeff, Eric; Sezgin, Ergin; Townsend, Paul (1987). "Supermembranes and eleven-dimensional supergravity". *Physics Letters B* **189** (1): 75–78. Bibcode:1987PhLB..189...75B. doi:10.1016/0370-2693(87)91272-X.

- Candelas, Philip; Horowitz, Gary; Strominger, Andrew; Witten, Edward (1985). "Vacuum configurations for superstrings". *Nuclear Physics B* **258**: 46–74. Bibcode:1985NuPhB.258...46C. doi:10.1016/0550-3213(85)90602-9.

- Connes, Alain (1994). *Noncommutative Geometry*. Academic Press. ISBN 978-0-12-185860-5.

- Connes, Alain; Douglas, Michael; Schwarz, Albert (1998). "Noncommutative geometry and matrix theory". *Journal of High Energy Physics*. 19981 (2): 003. arXiv:hep-th/9711162. Bibcode:1998JHEP...02..003C. doi:10.1088/1126-6708/1998/02/003.

- Cremmer, Eugene; Julia, Bernard; Scherk, Joel (1978). "Supergravity theory in eleven dimensions". *Physics Letters B* **76** (4): 409–412. Bibcode:1978PhLB...76..409C. doi:10.1016/0370-2693(78)90894-8.

- Dimofte, Tudor; Gaiotto, Davide; Gukov, Sergei (2010). "Gauge theories labelled by three-manifolds". *Communications in Mathematical Physics* **325** (2): 367–419. Bibcode:2014CMaPh.325..367D. doi:10.1007/s00220-013-1863-2.

- Dine, Michael (2000). "TASI Lectures on M Theory Phenomenology". arXiv:hep-th/0003175.

- Dirac, Paul (1962). "An extensible model of the electron". *Proceedings of the Royal Society of London. A. Mathematical and Physical Sciences* **268** (1332): 57–67. Bibcode:1962RSPSA.268...57D. doi:10.1098/rspa.1962.0124.

- Duff, Michael (1996). "M-theory (the theory formerly known as strings)". *International Journal of Modern Physics A* **11** (32): 6523–41. arXiv:hep-th/9608117. Bibcode:1996IJMPA..11.5623D. doi:10.1142/S0217751X96002583.

- Duff, Michael (1998). "The theory formerly known as strings". *Scientific American* **278** (2): 64–9. doi:10.1038/scientificamerican0298-64.

- Duff, Michael; Howe, Paul; Inami, Takeo; Stelle, Kellogg (1987). "Superstrings in $D=10$ from supermembranes in $D=11$". *Nuclear Physics B* **191** (1): 70–74. Bibcode:1987PhLB..191...70D. doi:10.1016/0370-2693(87)91323-2.

- Gaiotto, Davide; Moore, Gregory; Neitzke, Andrew (2013). "Wall-crossing, Hitchin systems, and the WKB approximation". *Advances in Mathematics* **2341**: 239–403. arXiv:0907.3987. doi:10.1016/j.aim.2012.09.027.

- Greene, Brian (2000). *The Elegant Universe: Superstrings, Hidden Dimensions, and the Quest for the Ultimate Theory*. Random House. ISBN 978-0-9650888-0-0.

- Griffiths, David (2004). *Introduction to Quantum Mechanics*. Pearson Prentice Hall. ISBN 978-0-13-111892-8.

- Hořava, Petr; Witten, Edward (1996a). "Heterotic and Type I string dynamics from eleven dimensions". *Nuclear Physics B* **460** (3): 506–524. arXiv:hep-th/9510209. Bibcode:1996NuPhB.460..506H. doi:10.1016/0550-3213(95)00621-4.

- Hořava, Petr; Witten, Edward (1996b). "Eleven dimensional supergravity on a manifold with boundary". *Nuclear Physics B* **475** (1): 94–114. arXiv:hep-th/9603142. Bibcode:1996NuPhB.475...94H. doi:10.1016/0550-3213(96)00308-2.

- Hull, Chris; Townsend, Paul (1995). "Unity of superstring dualities". *Nuclear Physics B* **4381** (1): 109–137. arXiv:hep-th/9410167. Bibcode:1995NuPhB.438..109H. doi:10.1016/0550-3213(94)00559-W.

- Khovanov, Mikhail (2000). "A categorification of the Jones polynomial". *Duke Mathematical Journal* **1011** (3): 359–426. doi:10.1215/S0012-7094-00-10131-7.

- Klebanov, Igor; Maldacena, Juan (2009). "Solving Quantum Field Theories via Curved Spacetimes" (PDF). *Physics Today* **62**: 28. Bibcode:2009PhT....62a..28K. doi:10.1063/1.3074260. Retrieved May 2013.

- Maldacena, Juan (1998). "The Large N limit of superconformal field theories and supergravity". *Advances in Theoretical and Mathematical Physics* **2**: 231–252. arXiv:hep-th/9711200. Bibcode:1998AdTMP...2..231M. doi:10.1063/1.59653.

- Maldacena, Juan (2005). "The Illusion of Gravity" (PDF). *Scientific American* **293** (5): 56–63. Bibcode:2005SciAm.293e..56M. doi:10.1038/scientificamerican1105-56. PMID 16318027. Retrieved July 2013.

- Montonen, Claus; Olive, David (1977). "Magnetic monopoles as gauge particles?". *Physics Letters B* **72** (1): 117–120. Bibcode:1977PhLB..72..117M. doi:10.1016/0370-2693(77)90076-4.

- Moore, Gregory (2005). "What is ... a Brane?" (PDF). *Notices of the AMS* **52**: 214. Retrieved June 2013.

- Moore, Gregory (2012). "Lecture Notes for Felix Klein Lectures" (PDF). Retrieved 14 August 2013.

- Nahm, Walter (1978). "Supersymmetries and their representations". *Nuclear Physics B* **135** (1): 149–166. Bibcode:1978NuPhB.135..149N. doi:10.1016/0550-3213(78)90218-3.

- Nekrasov, Nikita; Schwarz, Albert (1998). "Instantons on noncommutative $\mathbf{R}^4$ and (2,0) superconformal six dimensional theory". *Communications in Mathematical Physics* **198** (3): 689–703. arXiv:hep-th/9802068. Bibcode:1998CMaPh.198..689N. doi:10.1007/s002200050490.

- Peskin, Michael; Schroeder, Daniel (1995). *An Introduction to Quantum Field Theory*. Westview Press. ISBN 978-0-201-50397-5.

- Randall, Lisa; Sundrum, Raman (1999). "An alternative to compactification". *Physical Review Letters* **83** (23): 4690. arXiv:hep-th/9906064. Bibcode:1999PhRvL..83.4690R. doi:10.1103/PhysRevLett.83.4690.

- Seiberg, Nathan; Witten, Edward (1999). "String Theory and Noncommutative Geometry". *Journal of High Energy Physics* **1999** (9): 032. arXiv:hep-th/9908142. Bibcode:1999JHEP...09..032S. doi:10.1088/1126-6708/1999/09/032.

- Sen, Ashoke (1993). "Electric-magnetic duality in string theory". *Nuclear Physics B* **404** (1): 109–126. arXiv:hep-th/9207053. Bibcode:1993NuPhB.404..109S. doi:10.1016/0550-3213(93)90475-5.

- Sen, Ashoke (1994a). "Strong-weak coupling duality in four-dimensional string theory". *International Journal of Modern Physics A* **9** (21): 3707–3750. arXiv:hep-th/9402002. Bibcode:1994IJMPA...9.3707S. doi:10.1142/S0217751X94001497.

- Sen, Ashoke (1994b). "Dyon-monopole bound states, self-dual harmonic forms on the multi-monopole moduli space, and $SL(2, \mathbf{Z})$ invariance in string theory". *Physics Letters B* **329** (2): 217–221. arXiv:hep-th/9402032. Bibcode:1994PhLB..329..217S. doi:10.1016/0370-2693(94)90763-3.

- Strominger, Andrew (1990). "Heterotic solitons". *Nuclear Physics B* **343** (1): 167–184. Bibcode:1990NuPhB.343..167S. doi:10.1016/0550-3213(90)90599-9.

- van Nieuwenhuizen, Peter (1981). "Supergravity". *Physics Reports* **68** (4): 189–398. Bibcode:1981PhR....68..189V. doi:10.1016/0370-1573(81)90157-5.

- Wald, Robert (1984). *General Relativity*. University of Chicago Press. ISBN 978-0-226-87033-5.

- Witten, Edward (1989). "Quantum Field Theory and the Jones Polynomial". *Communications in Mathematical Physics* **121** (3): 351–399. Bibcode:1989CMaPh.121..351W. doi:10.1007/BF01217730. MR 0990772.

- Witten, Edward (1995). "String theory dynamics in various dimensions". *Nuclear Physics B* **443** (1): 85–126. arXiv:hep-th/9503124. Bibcode:1995NuPhB.443...85W. doi:10.1016/0550-3213(95)00158-O.

- Witten, Edward (2009). "Geometric Langlands from six dimensions". arXiv:0905.2720 [hep-th].

- Witten, Edward (2012). "Fivebranes and knots". *Quantum Topology* **3** (1): 1–137. doi:10.4171/QT/26.

- Woit, Peter (2006). *Not Even Wrong: The Failure of String Theory and the Search for Unity in Physical Law*. Basic Books. p. 105. ISBN 0-465-09275-6.

- Yau, Shing-Tung; Nadis, Steve (2010). *The Shape of Inner Space: String Theory and the Geometry of the Universe's Hidden Dimensions*. Basic Books. ISBN 978-0-465-02023-2.

- Zee, Anthony (2010). *Quantum Field Theory in a Nutshell* (2nd ed.). Princeton University Press. ISBN 978-0-691-14034-6.

- Zwiebach, Barton (2009). *A First Course in String Theory*. Cambridge University Press. ISBN 978-0-521-88032-9.

3.5.7 External links

- The Elegant Universe—A three-hour miniseries with Brian Greene on the series *Nova* (original PBS broadcast dates: October 28, 8–10 p.m. and November 4, 8–9 p.m., 2003). Various images, texts, videos and animations explaining string theory and M-theory.

- Superstringtheory.com—The "Official String Theory Web Site", created by Patricia Schwarz. References on string theory and M-theory for the layperson and expert.

- Not Even Wrong—Peter Woit's blog on physics in general, and string theory in particular.

3.6 Minimal Supersymmetric Standard Model

The **Minimal Supersymmetric Standard Model** (**MSSM**) is an extension to the Standard Model that realizes supersymmetry. MSSM is the minimal supersymmetrical model as it considers only "the [minimum] number of new particle states and new interactions consistent with phenomenology".[1] Supersymmetry pairs bosons with fermions; therefore every Standard Model particle has a partner that has yet to be discovered. If the superparticles are found, it may be analogous to discovering dark matter [2] and depending on the details of what might be found, it could provide evidence for grand unification and might even, in principle, provide hints as to whether string theory describes nature. The failure to find evidence for supersymmetry using the Large Hadron Collider since 2010 has led to suggestions that the theory should be abandoned.[3]

3.6.1 Background

The MSSM was originally proposed in 1981 to stabilize the weak scale, solving the hierarchy problem.[4] The Higgs boson mass of the Standard Model is unstable to quantum corrections and the theory predicts that weak scale should be much weaker than what is observed to be. In the MSSM, the Higgs boson has a fermionic superpartner, the Higgsino, that has the same mass as it would if supersymmetry were an exact symmetry. Because fermion masses are radiatively stable, the Higgs mass inherits this stability. However, in MSSM there is a need for more than one Higgs field, as described below.

The only unambiguous way to claim discovery of supersymmetry is to produce superparticles in the laboratory. Because superparticles are expected to be 100 to 1000 times

3.6.2 Theoretical motivations

There are three principal motivations for the MSSM over other theoretical extensions of the Standard Model, namely:

- Naturalness

- Gauge coupling unification

- Dark Matter

These motivations come out without much effort and they are the primary reasons why the MSSM is the leading candidate for a new theory to be discovered at collider experiments such as the Tevatron or the LHC.

Naturalness

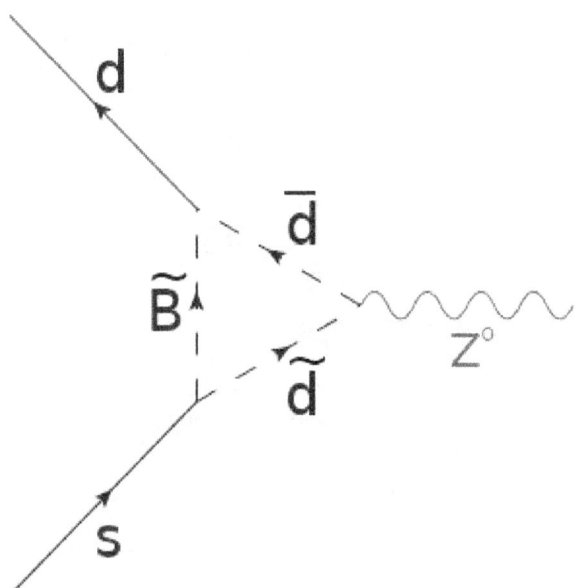

An example of a flavor changing neutral current process in MSSM. A strange quark emits a bino, turning into a sdown-type quark, which then emits a Z boson and reabsorbs the bino, turning into a down quark. If the MSSM squark masses are flavor violating, such a process can occur.

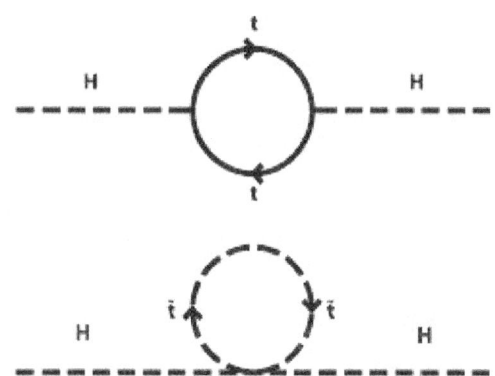

Cancellation of the Higgs boson quadratic mass renormalization between fermionic top quark loop and scalar top squark Feynman diagrams in a supersymmetric extension of the Standard Model

heavier than the proton, it requires a huge amount of energy to make these particles that can only be achieved at particle accelerators. The Tevatron was actively looking for evidence of the production of supersymmetric particles before it was shut down on 30 September 2011. Most physicists believe that supersymmetry must be discovered at the LHC if it is responsible for stabilizing the weak scale. There are five classes of particle that superpartners of the Standard Model fall into: squarks, gluinos, charginos, neutralinos, and sleptons. These superparticles have their interactions and subsequent decays described by the MSSM and each has characteristic signatures.

The MSSM imposes R-parity to explain the stability of the proton. It adds supersymmetry breaking by introducing explicit soft supersymmetry breaking operators into the Lagrangian that is communicated to it by some unknown (and unspecified) dynamics. This means that there are 120 new parameters in the MSSM. Most of these parameters lead to unnacceptable phenomenology such as large flavor changing neutral currents or large electric dipole moments for the neutron and electron. To avoid these problems, the MSSM takes all of the soft supersymmetry breaking to be diagonal in flavor space and for all of the new CP violating phases to vanish.

The original motivation for proposing the MSSM was to stabilize the Higgs mass to radiative corrections that are quadratically divergent in the Standard Model (hierarchy problem). In supersymmetric models, scalars are related to fermions and have the same mass. Since fermion masses are logarithmically divergent, scalar masses inherit the same radiative stability. The Higgs vacuum expectation value is related to the negative scalar mass in the Lagrangian. In order for the radiative corrections to the Higgs mass to not be dramatically larger than the actual value, the mass of the superpartners of the Standard Model should not be significantly heavier than the Higgs VEV—roughly 100 GeV. In 2012, the Higgs particle was discovered at the LHC, and its mass was found to be 125-126 GeV.

Gauge-coupling unification

If the superpartners of the Standard Model are near the TeV scale, then measured gauge couplings of the three gauge groups unify at high energies.[5] [6] [7] The beta-functions for the MSSM gauge couplings are given by

where α_1^{-1} is measured in SU(5) normalization—a factor of $\frac{3}{5}$ different than the Standard Model's normalization and predicted by Georgi–Glashow SU(5) .

The condition for gauge coupling unification at one loop is whether the following expression is satisfied $\frac{\alpha_3^{-1} - \alpha_2^{-1}}{\alpha_2^{-1} - \alpha_1^{-1}} = \frac{b_{0,3} - b_{0,2}}{b_{0,2} - b_{0,1}}$.

Remarkably, this is precisely satisfied to experimental errors in the values of $\alpha^{-1}(M_{Z^\circ})$. There are two loop corrections and both TeV-scale and GUT-scale threshold corrections that alter this condition on gauge coupling unification, and the results of more extensive calculations reveal that gauge coupling unification occurs to an accuracy of 1%, though this is about 3 standard deviations from the theoretical expectations.

This prediction is generally considered as indirect evidence for both the MSSM and SUSY GUTs.[8] It should be noted that gauge coupling unification does not necessarily imply grand unification and there exist other mechanisms to reproduce gauge coupling unification. However, if superpartners are found in the near future, the apparent success of gauge coupling unification would suggest that a supersymmetric grand unified theory is a promising candidate for high scale physics.

Dark matter

If R-parity is preserved, then the lightest superparticle (LSP) of the MSSM is stable and is a Weakly interacting massive particle (WIMP) — i.e. it does not have electromagnetic or strong interactions. This makes the LSP a good dark matter candidate and falls into the category of cold dark matter (CDM) particle.

3.6.3   Predictions of the MSSM regarding hadron colliders

The Tevatron and LHC have active experimental programs searching for supersymmetric particles. Since both of these machines are hadron colliders — proton antiproton for the Tevatron and proton proton for the LHC — they search best for strongly interacting particles. Therefore, most experimental signature involve production of squarks or gluinos. Since the MSSM has R-parity, the lightest supersymmetric particle is stable and after the squarks and gluinos decay

each decay chain will contain one LSP that will leave the detector unseen. This leads to the generic prediction that the MSSM will produce a 'missing energy' signal from these particles leaving the detector.

Neutralinos

There are four neutralinos that are fermions and are electrically neutral, the lightest of which is typically stable. They are typically labeled N0

1. N0
2. N0
3. N0

4 (although sometimes $\tilde{\chi}_1^0, \ldots, \tilde{\chi}_4^0$ is used instead). These four states are mixtures of the Bino and the neutral Wino (which are the neutral electroweak Gauginos), and the neutral Higgsinos. As the neutralinos are Majorana fermions, each of them is identical with its antiparticle. Because these particles only interact with the weak vector bosons, they are not directly produced at hadron colliders in copious numbers. They primarily appear as particles in cascade decays of heavier particles usually originating from colored supersymmetric particles such as squarks or gluinos.

In R-parity conserving models, the lightest neutralino is stable and all supersymmetric cascades decays end up decaying into this particle which leaves the detector unseen and its existence can only be inferred by looking for unbalanced momentum in a detector.

The heavier neutralinos typically decay through a Z0 to a lighter neutralino or through a W± to chargino. Thus a typical decay is

The mass splittings between the different Neutralinos will dictate which patterns of decays are allowed.

Charginos

There are two Charginos that are fermions and are electrically charged. They are typically labeled C $\tilde{\chi}$ ±
1 and C $\tilde{\chi}$ ±
2 (although sometimes $\tilde{\chi}_1^\pm$ and $\tilde{\chi}_2^\pm$ is used instead). The heavier chargino can decay through Z0 to the lighter chargino. Both can decay through a W± to neutralino.

Squarks

The squarks are the scalar superpartners of the quarks and there is one version for each Standard Model quark. Due to phenomenological constraints from flavor changing neutral currents, typically the lighter two generations of squarks have to be nearly the same in mass and therefore are not given distinct names. The superpartners of the top and bottom quark can be split from the lighter squarks and are called *stop* and *sbottom*.

On the other way, there may be a remarkable left-right mixing of the stops $\tilde{t}$ and of the sbottoms $\tilde{b}$ because of the high masses of the partner quarks top and bottom: [9]

- $\tilde{t}_1 = e^{+i\phi}\cos(\theta)\tilde{t}_L + \sin(\theta)\tilde{t}_R$

- $\tilde{t}_2 = e^{-i\phi}\cos(\theta)\tilde{t}_R - \sin(\theta)\tilde{t}_L$

Same holds for bottom $\tilde{b}$ with its own parameters ϕ and θ.

Squarks can be produced through strong interactions and therefore are easily produced at hadron colliders. They decay to quarks and neutralinos or charginos which further decay. In R-parity conserving scenarios, squarks are pair produced and therefore a typical signal is

- $\tilde{q}\tilde{q} \rightarrow q\tilde{N}_1^0 \bar{q}\tilde{N}_1^0 \rightarrow 2$ jets + missing energy

- $\tilde{q}\tilde{q} \rightarrow q\tilde{N}_2^0 \bar{q}\tilde{N}_1^0 \rightarrow q\tilde{N}_1^0 \ell\ell\bar{q}\tilde{N}_1^0 \rightarrow 2$ jets + 2 leptons + missing energy

Gluinos

Gluinos are Majorana fermionic partners of the gluon which means that they are their own antiparticles. They interact strongly and therefore can be produced significantly at the LHC. They can only decay to a quark and a squark and thus a typical gluino signal is

- $\tilde{g}\tilde{g} \rightarrow (q\tilde{\bar{q}})(\bar{q}\tilde{q}) \rightarrow (qq\tilde{N}_1^0)(\bar{q}\bar{q}\tilde{N}_1^0) \rightarrow 4$ jets + Missing energy

Because gluinos are Majorana, gluinos can decay to either a quark+anti-squark or an anti-quark+squark with equal probability. Therefore, pairs of gluinos can decay to

- $\tilde{g}\tilde{g} \rightarrow (\bar{q}\tilde{q})(\bar{q}\tilde{q}) \rightarrow (q\bar{q}\tilde{C}_1^+)(q\bar{q}\tilde{C}_1^+) \rightarrow (q\bar{q}W^+)(q\bar{q}W^+) \rightarrow 4$ jets+ $\ell^+\ell^+$ + Missing energy

This is a distinctive signature because it has same-sign di-leptons and has very little background in the Standard Model.

Sleptons

Sleptons are the scalar partners of the leptons of the Standard Model. They are not strongly interacting and therefore are not produced very often at hadron colliders unless they are very light.

Because of the high mass of the tau lepton there will be left-right mixing of the stau similar to that of stop and sbottom (see above).

Sfermions will typically be found in decays of a charginos and neutralinos if they are light enough to be a decay product

- $\tilde{C}^+ \rightarrow \tilde{\ell}^+\nu$

- $\tilde{N}^0 \rightarrow \tilde{\ell}^+\ell^-$

3.6.4 MSSM fields

Fermions have bosonic superpartners (called sfermions), and bosons have fermionic superpartners (called bosinos). For most of the Standard Model particles, doubling is very straightforward. However, for the Higgs boson, it is more complicated.

A single Higgsino (the fermionic superpartner of the Higgs boson) would lead to a gauge anomaly and would cause the theory to be inconsistent. However, if two Higgsinos are added, there is no gauge anomaly. The simplest theory is one with two Higgsinos and therefore two scalar Higgs doublets. Another reason for having two scalar Higgs doublets rather than one is in order to have Yukawa couplings between the Higgs and both down-type quarks and up-type quarks; these are the terms responsible for the quarks' masses. In the Standard Model the down-type quarks couple to the Higgs field (which has $Y=-1/2$) and the up-type quarks to its complex conjugate (which has $Y=+1/2$). However, in a supersymmetric theory this is not allowed, so two types of Higgs fields are needed.

MSSM superfields

In supersymmetric theories, every field and its superpartner can be written together as a superfield. The superfield formulation of supersymmetry is very convenient to write down manifestly supersymmetric theories (i.e. one does not have to tediously check that the theory is supersymmetric term by term in the Lagrangian). The MSSM contains vector superfields associated with the Standard Model gauge groups which contain the vector bosons and associated gauginos. It also contains chiral superfields for the Standard Model fermions and Higgs bosons (and their respective superpartners).

MSSM Higgs Mass

The MSSM Higgs Mass is a prediction of the Minimal Supersymmetric Standard Model. The mass of the lightest

Higgs boson is set by the Higgs *quartic coupling*. Quartic couplings are not soft supersymmetry-breaking parameters since they lead to a quadratic divergence of the Higgs mass. Furthermore, there are no supersymmetric parameters to make the Higgs mass a free parameter in the MSSM (though not in non-minimal extensions). This means that Higgs mass is a prediction of the MSSM. The LEP II and the IV experiments placed a lower limit on the Higgs mass of 114.4 GeV. This lower limit is significantly above where the MSSM would typically predict it to be, and while it does not rule out the MSSM, the discovery of the Higgs with a mass of 125 GeV makes proponents of the MSSM nervous.[10][11]

Formulas The only susy-preserving operator that creates a quartic coupling for the Higgs in the MSSM arise for the D-terms of the SU(2) and U(1) gauge sector and the magnitude of the quartic coupling is set by the size of the gauge couplings.

This leads to the prediction that the Standard Model-like Higgs mass (the scalar that couples approximately to the vev) is limited to be less than the Z mass

$$m_{h^0}^2 \leq m_{Z^0}^2 \cos^2 2\beta \ .$$

Since supersymmetry is broken, there are radiative corrections to the quartic coupling that can increase the Higgs mass. These dominantly arise from the 'top sector'

$$m_{h^0}^2 \leq m_{Z^0}^2 \cos^2 2\beta + \frac{3}{\pi^2} \frac{m_t^4 \sin^4 \beta}{v^2} \log \frac{m_{\tilde{t}}}{m_t}$$

where m_t is the top mass and $m_{\tilde{t}}$ is the mass of the top squark. This result can be interpreted as the RG running of the Higgs quartic coupling from the scale of supersymmetry to the top mass—however since the top squark mass should be relatively close to the top mass, this is usually a fairly modest contribution and increases the Higgs mass to roughly the LEP II bound of 114 GeV before the top squark becomes too heavy.

Finally there is a contribution from the top squark A-terms

$$\mathcal{L} = y_t \, m_{\tilde{t}} \, a \, h_u \tilde{q}_3 \tilde{u}_3^c$$

where a is a dimensionless number. This contributes an additional term to the Higgs mass at loop level, but is not logarithmically enhanced

$$m_{h^0}^2 \qquad \leq \qquad m_{Z^0}^2 \cos^2 2\beta \qquad +$$
$$\frac{3}{\pi^2} \frac{m_t^4 \sin^4 \beta}{v^2} \left(\log \frac{m_{\tilde{t}}}{m_t} + a^2(1 - a^2/12) \right)$$

by pushing $a \to \sqrt{6}$ (known as 'maximal mixing') it is possible to push the Higgs mass to 125 GeV without decoupling the top squark or adding new dynamics to the MSSM.

As the Higgs was found at around 125 GeV (along with no other superparticles) at the LHC, this strongly hints at new dynamics beyond the MSSM, such as the 'Next to Minimal Supersymmetric Standard Model' (NMSSM); and suggests some correlation to the little hierarchy problem.

3.6.5   The MSSM Lagrangian

The Lagrangian for the MSSM contains several pieces.

- The first is the Kähler potential for the matter and Higgs fields which produces the kinetic terms for the fields.

- The second piece is the gauge field superpotential that produces the kinetic terms for the gauge bosons and gauginos.

- The next term is the superpotential for the matter and Higgs fields. These produce the Yukawa couplings for the Standard Model fermions and also the mass term for the Higgsinos. After imposing R-parity, the renormalizable, gauge invariant operators in the superpotential are

$$W = \mu H_u H_d + y_u H_u Q U^c + y_d H_d Q D^c + y_l H_d L E^c$$

The constant term is unphysical in global supersymmetry (as opposed to supergravity).

Soft Susy breaking

Main article: Soft SUSY breaking

The last piece of the MSSM Lagrangian is the soft supersymmetry breaking Lagrangian. The vast majority of the parameters of the MSSM are in the susy breaking Lagrangian. The soft susy breaking are divided into roughly three pieces.

- The first are the gaugino masses

$$\mathcal{L} \supset m_{\frac{1}{2}} \bar{\lambda}\lambda + \text{h.c.}$$

Where $\bar{\lambda}$ are the gauginos and $m_{\frac{1}{2}}$ is different for the wino, bino and gluino.

- The next are the soft masses for the scalar fields

$$\mathcal{L} \supset m_0^2 \phi^\dagger \phi$$

where ϕ are any of the scalars in the MSSM and m_0 are 3×3 Hermitian matrices for the squarks and sleptons of a given set of gauge quantum numbers. The eigenvalues of these matrices are actually the masses squared, rather than the masses.

- There are the A and B terms which are given by

$$\mathcal{L} \supset B_\mu h_u h_d + A h_u \tilde{q} \tilde{u}^c + A h_d \tilde{q} \tilde{d}^c + A h_d \tilde{l} \tilde{e}^c + \text{h.c.}$$

The A terms are 3×3 complex matrices much as the scalar masses are.

- Although not often mentioned with regard to soft terms, to be consistent with observation, one must also include Gravitino and Goldstino soft masses given by

$$\mathcal{L} \supset m_{3/2} \Psi_\mu^\alpha (\sigma^{\mu\nu})_\alpha^\beta \Psi_\beta + m_{3/2} G^\alpha G_\alpha + \text{h.c.}$$

The reason these soft terms are not often mentioned are that they arise through local supersymmetry and not global supersymmetry, although they are required otherwise if the Goldstino were massless it would contradict observation. The Goldstino mode is eaten by the Gravitino to become massive, through a gauge shift, which also absorbs the would-be "mass" term of the Goldstino.

3.6.6 Problems with the MSSM

There are several problems with the MSSM — most of them falling into understanding the parameters.

- The mu problem: The Higgsino mass parameter μ appears as the following term in the superpotential: $\mu H_u H_d$. It should have the same order of magnitude as the electroweak scale, many orders of magnitude smaller than that of the Planck scale, which is the natural cutoff scale. The soft supersymmetry breaking terms should also be of the same order of magnitude as the electroweak scale. This brings about a problem of naturalness: why are these scales so much smaller than the cutoff scale yet happen to fall so close to each other?

- Flavor universality of soft masses and A-terms: since no flavor mixing additional to that predicted by the standard model has been discovered so far, the coefficients of the additional terms in the MSSM Lagrangian must be, at least approximately, flavor invariant (i.e. the same for all flavors).

- Smallness of CP violating phases: since no CP violation additional to that predicted by the standard model has been discovered so far, the additional terms in the MSSM Lagrangian must be, at least approximately, CP invariant, so that their CP violating phases are small.

3.6.7 Theories of supersymmetry breaking

A large amount of theoretical effort has been spent trying to understand the mechanism for soft supersymmetry breaking that produces the desired properties in the superpartner masses and interactions. The three most extensively studied mechanisms are:

Gravity-mediated supersymmetry breaking

Gravity-mediated supersymmetry breaking is a method of communicating supersymmetry breaking to the supersymmetric Standard Model through gravitational interactions. It was the first method proposed to communicate supersymmetry breaking. In gravity-mediated supersymmetry-breaking models, there is a part of the theory that only interacts with the MSSM through gravitational interaction. This hidden sector of the theory breaks supersymmetry. Through the supersymmetric version of the Higgs mechanism, the gravitino, the supersymmetric version of the graviton, acquires a mass. After the gravitino has a mass, gravitational radiative corrections to soft masses are incompletely cancelled beneath the gravitino's mass.

It is currently believed that it is not generic to have a sector completely decoupled from the MSSM and there should be higher dimension operators that couple different sectors together with the higher dimension operators suppressed by the Planck scale. These operators give as large of a contribution to the soft supersymmetry breaking masses as the gravitational loops; therefore, today people usually consider gravity mediation to be gravitational sized direct interactions between the hidden sector and the MSSM.

mSUGRA stands for minimal supergravity. The construction of a realistic model of interactions within $N = 1$ supergravity framework where supersymmetry breaking is communicated through the supergravity interactions was carried out by Ali Chamseddine, Richard Arnowitt, and Pran Nath in 1982.[12] mSUGRA is one of the most widely investigated models of particle physics due to its predictive power requiring only 4 input parameters and a sign, to determine the low energy phenomenology from the scale of Grand Unification. The most widely used set of parameters is:

Gravity-Mediated Supersymmetry Breaking was assumed to be flavor universal because of the universality of gravity; however, in 1986 Hall, Kostelecky, and Raby [13] showed that Planck-scale physics that are necessary to generate the Standard-Model Yukawa couplings spoil the universality of the supersymmetry breaking.

Gauge-mediated supersymmetry breaking (GMSB)

Gauge-mediated supersymmetry breaking is method of communicating supersymmetry breaking to the supersymmetric Standard Model through the Standard Model's gauge interactions. Typically a hidden sector breaks supersymmetry and communicates it to massive messenger fields that are charged under the Standard Model. These messenger fields induce a gaugino mass at one loop and then this is transmitted on to the scalar superpartners at two loops. Requiring stop squarks below 2 TeV, the maximum Higgs boson mass predicted is just 121.5GeV.[14] With the Higgs being discovered at 125GeV - this model requires stops above 2 TeV.

Anomaly-mediated supersymmetry breaking (AMSB)

Anomaly-mediated supersymmetry breaking is a special type of gravity mediated supersymmetry breaking that results in supersymmetry breaking being communicated to the supersymmetric Standard Model through the conformal anomaly.[15][16] Requiring stop squarks below 2 TeV, the maximum Higgs boson mass predicted is just 121.0GeV.[14] With the Higgs being discovered at 125GeV - this scenario requires stops heavier than 2 TeV.

3.6.8   Phenomenological MSSM (pMSSM)

The unconstrained MSSM has more than 100 parameters in addition to the Standard Model parameters. This makes any phenomenological analysis (e.g. finding regions in parameter space consistent with observed data) impractical. Under the following three assumptions:

- no new source of CP-violation

- no Flavour Changing Neutral Currents

- first and second generation universality

one can reduce the number of additional parameters to the following 19 quantities of the phenomenological MSSM (pMSSM):[17] The large parameter space of pMSSM makes searches in pMSSM extremely challenging and makes pMSSM difficult to exclude.

3.6.9   See also

- Desert (particle physics)

3.6.10   References

[1] Howard Baer: Xerxes Tata (2006). "8 - The Minimal Supersymmetric Standard Model". *Weak Scale Supersymmetry From Superfields to Scattering Events*. Cambridge: Cambridge University Press. p. 127. ISBN 9780511617270. It is minimal in the sense that it contains the smallest number of new particle states and new interactions consistent with phenomenology.

[2] Murayama, Hitoshi (2000). "Supersymmetry phenomenology". arXiv:hep-ph/0002232.

[3] Wolchover, Natalie (November 29, 2012). "Supersymmetry Fails Test, Forcing Physics to Seek New Ideas". *Scientific American*.

[4] S. Dimopoulos, H. Georgi; Georgi (1981). "Softly Broken Supersymmetry and SU(5)". *Nuclear Physics B* **193**: 150. Bibcode:1981NuPhB.193..150D. doi:10.1016/0550-3213(81)90522-8.

[5] S. Dimopoulos, S. Raby and F. Wilczek; Raby; Wilczek (1981). "Supersymmetry and the Scale of Unification". *Physical Review D* **24** (6): 1681–1683. Bibcode:1981PhRvD..24.1681D. doi:10.1103/PhysRevD.24.1681.

[6] L.E. Ibanez and G.G. Ross; Ross (1981). "Low-energy predictions in supersymmetric grand unified theories". *Physics Letters B* **105** (6): 439. Bibcode:1981PhLB..105..439I. doi:10.1016/0370-2693(81)91200-4.

[7] W.J. Marciano and G. Senjanovic; Senjanović (1982). "Predictions of supersymmetric grand unified theories". *Physical Review D* **25** (11): 3092. Bibcode:1982PhRvD..25.3092M. doi:10.1103/PhysRevD.25.3092.

[8] Gordon Kane. "The Dawn of Physics Beyond the Standard Model", *Scientific American*, June 2003, page 60 and *The frontiers of physics*, special edition, Vol 15, #3, page 8 "Indirect evidence for supersymmetry comes from the extrapolation of interactions to high energies."

[9] Bartl, A.; Hesselbach, S.; Hidaka, K.; Kernreiter, T.; Porod, W. (2003). "Impact of SUSY CP Phases on Stop and Sbottom Decays in the MSSM". arXiv:hep-ph/0306281 [hep-ph].

[10] Heinemeyer, S.; Stål, O.; Weiglein, G. (2012). "Interpreting the LHC Higgs search results in the MSSM". *Physics Letters B* **710**: 201. arXiv:1112.3026v3. Bibcode:2012PhLB..710..201H. doi:10.1016/j.physletb.2012.02.084.

[11] Carena, M.; Heinemeyer, S.; Wagner, C. E. M.; Weiglein, G. (2006). "MSSM Higgs boson searches at the evatron and the LHC: Impact of different benchmark scenarios" (PDF). *The European Physical Journal C* **45** (3): 797. arXiv:hep-ph/0511023. Bibcode:2006EPJC...45..797C. doi:10.1140/epjc/s2005-02470-y.

[12] A. Chamseddine. R. Arnowitt, P. Nath; Arnowitt; Nath (1982). "Locally Supersymmetric Grand Unification". *Physical Review Letters* **49** (14): 970–974. Bibcode:1982PhRvL..49..970C. doi:10.1103/PhysRevLett.49.970.

[13] L.J. Hall, V.A. Kostelecky, S. Raby; Kostelecky; Raby (1986). "New Flavor Violations in Supergravity Models". *Nuclear Physics B* **267** (2): 415. Bibcode:1986NuPhB.267..415H. doi:10.1016/0550-3213(86)90397-4.

[14] Arbey, A.; Battaglia, M.; Djouadi, A.; Mahmoudi. F.; Quevillon, J. (2011). "Implications of a 125 GeV Higgs for supersymmetric models". *Physics Letters B.* 3 **708** (2012): 162–169. arXiv:1112.3028. Bibcode:2012PhLB..708..162A. doi:10.1016/j.physletb.2012.01.053.

[15] L. Randall. R. Sundrum; Sundrum (1999). "Out of this world supersymmetry breaking". *Nuclear Physics B* **557**: 79–118. arXiv:hep-th/9810155. Bibcode:1999NuPhB.557...79R. doi:10.1016/S0550-3213(99)00359-4.

[16] G. Giudice, M. Luty, H. Murayama, R. Rattazzi; Rattazzi; Luty; Murayama (1998). "Gaugino mass without singlets". *Journal of High Energy Physics* **9812** (12): 027. arXiv:hep-ph/9810442. Bibcode:1998JHEP...12..027G. doi:10.1088/1126-6708/1998/12/027.

[17] Djouadi, A.; Rosier-Lees, S.; Bezouh, M.; Bizouard. M. A.; Boehm, C.; Borzumati, F.; Briot, C.; Carr, J.; Causse, M. B.; Charles, F.; Chereau, X.; Colas, P.; Duflot, L.; Dupperin, A.; Ealet, A.; El-Mamouni, H.; Ghodbane, N.; Gieres, F.; Gonzalez-Pineiro, B.; Gourmelen, S.; Grenier, G.; Gris, Ph.; Grivaz, J. -F.; Hebrard, C.; Ille, B.; Kneur, J. -L.; Kostantinidis, N.; Layssac, J.; Lebrun, P.; et al. (1999). "The Minimal Supersymmetric Standard Model: Group Summary Report". arXiv:hep-ph/9901246.

3.6.11   External links

- MSSM on arxiv.org

- Stephen P. Martin (1997). "A Supersymmetry Primer". arXiv:hep-ph/9709356.

- Particle Data Group review of MSSM and search for MSSM predicted particles

- Ian J R Aitchison (2005). "Supersymmetry and the MSSM: An Elementary Introduction". arXiv:hep-ph/0505105.

3.7   Next-to-Minimal Supersymmetric Standard Model

In particle physics. **NMSSM** is an acronym for **Next-to-Minimal Supersymmetric Standard Model**.[1][2][3][4][5] It is a supersymmetric extension to the Standard Model that adds an additional singlet chiral superfield to the MSSM and can be used to dynamically generate the μ term, solving the μ-problem. Articles about the NMSSM are available for review.[6][7]

The Minimal Supersymmetric Standard Model does not explain why the μ parameter in the superpotential term $\mu H_u H_d$ is at the electroweak scale. The idea behind the **Next to Minimal Supersymmetric Standard Model** is to promote the μ term to a gauge singlet, chiral superfield S. Note that the scalar superpartner of the singlino S is denoted by $\hat{S}$ and the spin-1/2 singlino superpartner by $\tilde{S}$ in the following. The superpotential for the NMSSM is given by

$$W_{\text{NMSSM}} = W_{\text{Yuk}} + \lambda S H_u H_d + \frac{\kappa}{3} S^3$$

where W_{Yuk} gives the Yukawa couplings for the Standard Model fermions. Since the superpotential has mass dimension three, the couplings λ and κ are dimensionless, hence the μ-problem of the MSSM is solved in the NMSSM – the superpotential of the NMSSM is scale invariant. The role of the λ term is to generate an effective μ term. This is done with the scalar component of the singlet $\hat{S}$ getting a vacuum-expectation value $\langle \hat{S} \rangle$, that is, we have

$$\mu_{\text{eff}} = \lambda \langle \hat{S} \rangle$$

Without the κ term the superpotential would have a U(1)' symmetry, so-called Peccei–Quinn symmetry; see Peccei–Quinn theory. This additional symmetry would alter the phenomenology completely. The role of the κ term is to break this U(1)' symmetry. The κ term is introduced trilinear such that κ is dimensionless. However there remains a discrete $\mathbb{Z}_3$ symmetry, which is moreover broken spontaneously.[8] In principle this leads to the domain wall problem. Introducing additional, but suppressed terms, the $\mathbb{Z}_3$ symmetry can be broken without changing phenomenology at the electroweak scale.[9] It is assumed that the domain wall problem is circumvented in this way without any modifications except far beyond the electroweak scale.

Also alternative models have been proposed which solve the μ-problem of the MSSM. One idea is to keep the κ term in the superpotential and take the U(1)' symmetry into account. Assuming this symmetry to be local an additional Z' gauge boson is predicted in this model, called UMSSM.

3.7.1 Phenomenology

Due to the additional singlet S the NMSSM alters in general the phenomenology of both the Higgs sector and the neutralino sector compared to the MSSM.

Higgs phenomenology

In the Standard Model we have one physical Higgs boson. In the MSSM we encounter five physical Higgs bosons. Due to the additional singlet $\hat{S}$ in the NMSSM we have two more Higgs bosons, that is, in total seven physical Higgs bosons. The Higgs sector is therefore much richer compared to the MSSM. In particular, the Higgs potential is in general no longer invariant under CP transformations; see CP violation. Typically, the Higgs bosons in the NMSSM are denoted in an order with increasing masses, that is, by $H_1, H_2, ..., H_7$ with H_1 the lightest Higgs boson. In the special case of a CP conserving Higgs potential we have three CP even Higgs bosons, H_1, H_2, H_3 , two CP odd ones, A_1, A_2 and a pair of charged Higgs bosons, H^+, H^- . In the MSSM, the lightest Higgs boson is always Standard Model-like, and therefore its production and decays are roughly known. In the NMSSM, the lightest Higgs can be very light (even of the order of 1 GeV) and may have escaped detection so far. In addition, in the CP-conserving case, the lightest CP-even Higgs boson turns out to have an enhanced lower bound compared to the MSSM. This is one of the reasons why the NMSSM deserves much attraction in recent years.

Neutralino phenomenology

The spin-1/2 singlino $\tilde{S}$ gives a fifth neutralino, compared to the four neutralinos of the MSSM. The singlino does not couple to gauge bosons, gauginos (the superpartners of the gauge bosons), leptons, sleptons (the superpartners of the leptons), quarks or squarks (the superpartners of the quarks). Suppose that a supersymmetric partner particle is produced at a collider, for instance at the LHC, the singlino is omitted in cascade decays and therefore escapes detection. However in case the singlino is the lightest supersymmetric particle (LSP) all supersymmetric partner particles eventually decay into the singlino. Due to R parity conservation this LSP is stable. In this way the singlino could be detected via missing transverse energy in the detector.

3.7.2 References

[1] Fayet, P. (1975). "Supergauge invariant extension of the Higgs mechanism and a model for the electron and its neutrino". *Nuclear Physics B* **90**: 104. Bibcode:1975NuPhB..90..104F.   doi:10.1016/0550-3213(75)90636-7.

[2] Dine, M.; Fischler, W.; Srednicki, M. (1981). "A simple solution to the strong CP problem with a harmless axion". *Physics Letters B* **104** (3): 199. Bibcode:1981PhLB..104..199D.   doi:10.1016/0370-2693(81)90590-6.

[3] Nilles, H. P.; Srednicki, M.; Wyler, D. (1983). "Weak interaction breakdown induced by supergravity". *Physics Letters B* **120** (4–6): 346. Bibcode:1983PhLB..120..346N. doi:10.1016/0370-2693(83)90460-4.

[4] Frere, J. M.; Jones, D. R. T.; Raby, S. (1983). "Fermion masses and induction of the weak scale by supergravity". *Nuclear Physics B* **222**: 11. Bibcode:1983NuPhB.222...11F.   doi:10.1016/0550-3213(83)90606-5.

[5] Derendinger, J. P.; Savoy, C. A. (1984). "Quantum effects and SU(2)×U(1) breaking in supergravity gauge theories". *Nuclear Physics B* **237** (2): 307. Bibcode:1984NuPhB.237..307D.   doi:10.1016/0550-3213(84)90162-7.

[6] Maniatis, M. (2010). "The Next-To-Minimal Supersymmetric Extension of the Standard Model Reviewed". *International Journal of Modern Physics A* **25** (18–19): 3505. arXiv:0906.0777. Bibcode:2010IJMPA..25.3505M. doi:10.1142/S0217751X10049827.

[7] Ellwanger, U.; Hugonie, C.; Teixeira, A. M. (2010). "The Next-to-Minimal Supersymmetric Standard Model". *Physics Reports* **496**: 1. arXiv:0910.1785.   Bibcode:2010PhR...496....1E. doi:10.1016/j.physrep.2010.07.001.

[8] Zeldovich, Ya. B.; Kobzarev, I. Y.; Okun, L. B. (1974). *Zhurnal Éksperimental'noĭ i Teoreticheskoĭ Fiziki* **67**: 3. Missing or empty |title= (help) Translated in *Soviet Physics JETP* **40**: 1. 1977. Bibcode:1975JETP...40....1Z. Missing or empty |title= (help)

[9] Panagiotakopoulos, P.; Tamvakis, K. (1999). "Stabilized NMSSM without domain walls". *Physics Letters B* **446** (3–4): 224. arXiv:hep-ph/9809475. Bibcode:1999PhLB..446..224P.   doi:10.1016/S0370-2693(98)01493-2.

3.8 Extra dimensions

In physics, **extra dimensions** are proposed additional space or time dimensions beyond the (3 + 1) typical of our observed space-time, such as the first attempts based on the Kaluza–Klein theory. Among theories proposing extra dimension are:[1]

1. Large extra dimension, mostly motivated by the ADD model, by Nima Arkani-Hamed, Savas Dimopoulos, and Gia Dvali in 1998, in an attempt to solve the hierarchy problem. This theory requires that the fields of the Standard Model are confined to a four-dimensional membrane, while gravity propagates in several additional spatial dimensions that are large compared to the Planck scale.[2]

2. Warped extra dimensions, such as those proposed by the Randall–Sundrum model (RS), based on warped geometry where our universe is a five-dimensional anti-de Sitter space and the elementary particles except for the graviton are localized on a $(3 + 1)$-dimensional brane or branes.[3]

3. Universal extra dimension, proposed and first studied in 2000, assume, at variance with the ADD and RS approaches, that all fields propagate universally in the extra dimensions.

4. Multiple time dimensions, i.e. the possibility that there might be more than one dimension of time, has occasionally been discussed in physics and philosophy, although those models have to deal with the problem of causality.

3.8.1 References

[1] Rizzo, Thomas G. (2004). "Pedagogical Introduction to Extra Dimensions". *SLAC Summer Institute*. Retrieved 2016.

[2] For a pedagogical introduction, see M. Shifman (2009). *Large Extra Dimensions: Becoming acquainted with an alternative paradigm*. Crossing the boundaries: Gauge dynamics at strong coupling. Singapore: World Scientific. arXiv:0907.3074.

[3] Randall, Lisa; Sundrum, Raman (1999). "Large Mass Hierarchy from a Small Extra Dimension". *Physical Review Letters* **83** (17): 3370–3373. arXiv:hep-ph/9905221. Bibcode:1999PhRvL..83.3370R. doi:10.1103/PhysRevLett.83.3370.

3.9 Technicolor (physics)

Technicolor theories are models of physics beyond the standard model that address electroweak gauge symmetry breaking, the mechanism through which W and Z bosons acquire masses. Early technicolor theories were modelled on quantum chromodynamics (QCD), the "color" theory of the strong nuclear force, which inspired their name.

Instead of introducing elementary Higgs bosons to explain observed phenomena, technicolor models hide electroweak symmetry and generate masses for the W and Z bosons through the dynamics of new gauge interactions. Although asymptotically free at very high energies, these interactions must become strong and confining (and hence unobservable) at lower energies that have been experimentally probed. This dynamical approach is natural and avoids issues of Quantum triviality and the hierarchy problem of the Standard Model.[1]

In order to produce quark and lepton masses, technicolor has to be "extended" by additional gauge interactions. Particularly when modelled on QCD, extended technicolor is challenged by experimental constraints on flavor-changing neutral current and precision electroweak measurements. It is not known what is the extended technicolor dynamics.

Much technicolor research focuses on exploring strongly interacting gauge theories other than QCD, in order to evade some of these challenges. A particularly active framework is "walking" technicolor, which exhibits nearly conformal behavior caused by an infrared fixed point with strength just above that necessary for spontaneous chiral symmetry breaking. Whether walking can occur and lead to agreement with precision electroweak measurements is being studied through non-perturbative lattice simulations.[2]

Experiments at the Large Hadron Collider are expected to discover the mechanism responsible for electroweak symmetry breaking, and will be critical for determining whether the technicolor framework provides the correct description of nature. In 2012 these experiments declared the discovery of a Higgs-like boson with mass approximately 125 GeV/c^2;[3][4][5] such a particle is not generically predicted by technicolor models, but can be accommodated by them.

3.9.1 Introduction

The mechanism for the breaking of electroweak gauge symmetry in the Standard Model of elementary particle interactions remains unknown. The breaking must be spontaneous, meaning that the underlying theory manifests the symmetry exactly (the gauge-boson fields are massless in the equations of motion), but the solutions (the ground state and the excited states) do not. In particular, the physical W and Z gauge bosons become massive. This phenomenon, in which the W and Z bosons also acquire an extra polarization state, is called the "Higgs mechanism". Despite the precise agreement of the electroweak theory with experiment at energies accessible so far, the necessary ingredients for the symmetry breaking remain hidden, yet to be revealed at higher energies.

The simplest mechanism of electroweak symmetry breaking introduces a single complex field and predicts the existence of the Higgs boson. Typically, the Higgs boson is "unnatural" in the sense that quantum mechanical fluctuations produce corrections to its mass that lift it to such high values that it cannot play the role for which it was introduced. Unless the Standard Model breaks down at energies less than a few TeV, the Higgs mass can be kept small only by a delicate fine-tuning of parameters.

Technicolor avoids this problem by hypothesizing a new gauge interaction coupled to new massless fermions. This interaction is asymptotically free at very high energies and becomes strong and confining as the energy decreases to the electroweak scale of 246 GeV. These strong forces spontaneously break the massless fermions' chiral symmetries, some of which are weakly gauged as part of the Standard Model. This is the dynamical version of the Higgs mechanism. The electroweak gauge symmetry is thus broken, producing masses for the W and Z bosons.

The new strong interaction leads to a host of new composite, short-lived particles at energies accessible at the Large Hadron Collider (LHC). This framework is natural because there are no elementary Higgs bosons and, hence, no fine-tuning of parameters. Quark and lepton masses also break the electroweak gauge symmetries, so they, too, must arise spontaneously. A mechanism for incorporating this feature is known as extended technicolor. Technicolor and extended technicolor face a number of phenomenological challenges, in particular issues of flavor-changing neutral currents, precision electroweak tests, and the top quark mass. Technicolor models also do not generically predict Higgs-like bosons as light as 125 GeV/c^2; such a particle was discovered by experiments at the Large Hadron Collider in 2012.[3][4][5] Some of these issues can be addressed with a class of theories known as walking technicolor.

3.9.2   Early technicolor

Technicolor is the name given to the theory of electroweak symmetry breaking by new strong gauge-interactions whose characteristic energy scale ΛTC is the weak scale itself, ΛTC $\cong$ FEW $\equiv$ 246 GeV. The guiding principle of technicolor is "naturalness": basic physical phenomena should not require fine-tuning of the parameters in the Lagrangian that describes them. What constitutes fine-tuning is to some extent a subjective matter, but a theory with elementary scalar particles typically is very finely tuned (unless it is supersymmetric). The quadratic divergence in the scalar's mass requires adjustments of a part in $\mathcal{O}\left(\frac{M_{bare}^2}{M_{physical}^2}\right)$, where M_{bare} is the cutoff of the theory, the energy scale at which the theory changes in some essential way. In the standard electroweak model with $M_{bare} \sim 10^{15}$ GeV (the grand-unification mass scale), and with the Higgs boson mass $M_{physical}$ = 100–500 GeV, the mass is tuned to at least a part in 10^{25}.

By contrast, a natural theory of electroweak symmetry breaking is an asymptotically free gauge theory with fermions as the only matter fields. The technicolor gauge group GTC is often assumed to be $SU(NTC)$. Based on analogy with quantum chromodynamics (QCD), it is assumed that there are one or more doublets of massless Dirac "technifermions" transforming vectorially under the same complex representation of GTC, T_iL,R = (U_i,D_i)L,R, i = 1,2, …, $N_f/2$. Thus, there is a chiral symmetry of these fermions, e.g., $SU(N_f)$L $\otimes$ $SU(N_f)$R, if they all transform according the same complex representation of GTC. Continuing the analogy with QCD, the running gauge coupling αTC(μ) triggers spontaneous chiral symmetry breaking, the technifermions acquire a dynamical mass, and a number of massless Goldstone bosons result. If the technifermions transform under $[SU(2) \otimes U(1)]$EW as left-handed doublets and right-handed singlets, three linear combinations of these Goldstone bosons couple to three of the electroweak gauge currents.

In 1973 Jackiw and Johnson[6] and Cornwall and Norton[7] studied the possibility that a (non-vectorial) gauge interaction of fermions can break itself; i.e., is strong enough to form a Goldstone boson coupled to the gauge current. Using Abelian gauge models, they showed that, *if* such a Goldstone boson is formed, it is "eaten" by the Higgs mechanism, becoming the longitudinal component of the now massive gauge boson. Technically, the polarization function $\Pi(p^2)$ appearing in the gauge boson propagator, $\Delta\mu\nu$ = $(p\mu\, p\nu/p^2$ - $g\mu\nu)/[p^2(1 - g^2\, \Pi(p^2))]$ develops a pole at p^2 = 0 with residue F^2, the square of the Goldstone boson's decay constant, and the gauge boson acquires mass $M \cong g\, F$. In 1973, Weinstein[8] showed that composite Goldstone bosons whose constituent fermions transform in the "standard" way under $SU(2) \otimes U(1)$ generate the weak boson masses

(1) $M_{W^\pm} = \frac{1}{2}gF_{EW}$ and $M_Z = \frac{1}{2}\sqrt{g^2 + g'^2}F_{EW} \equiv \frac{M_W}{\cos\theta_W}$.

This standard-model relation is achieved with elementary Higgs bosons in electroweak doublets; it is verified experimentally to better than 1%. Here, g and g' are $SU(2)$ and $U(1)$ gauge couplings and $\tan\theta$W = g'/g defines the weak mixing angle.

The important idea of a *new* strong gauge interaction of massless fermions at the electroweak scale FEW driving the spontaneous breakdown of its global chiral symmetry, of which an $SU(2) \otimes U(1)$ subgroup is weakly gauged, was first proposed in 1979 by S. Weinberg[9] and L. Susskind.[10] This "technicolor" mechanism is natural in that no fine-tuning of parameters is necessary.

3.9.3 Extended technicolor

Elementary Higgs bosons perform another important task. In the Standard Model, quarks and leptons are necessarily massless because they transform under $SU(2) \otimes U(1)$ as left-handed doublets and right-handed singlets. The Higgs doublet couples to these fermions. When it develops its vacuum expectation value, it transmits this electroweak breaking to the quarks and leptons, giving them their observed masses. (In general, electroweak-eigenstate fermions are not mass eigenstates, so this process also induces the mixing matrices observed in charged-current weak interactions.)

In technicolor, something else must generate the quark and lepton masses. The only natural possibility, one avoiding the introduction of elementary scalars, is to enlarge GTC to allow technifermions to couple to quarks and leptons. This coupling is induced by gauge bosons of the enlarged group. The picture, then, is that there is a large "extended technicolor" (ETC) gauge group $GETC \supset GTC$ in which technifermions, quarks, and leptons live in the same representations. At one or more high scales ΛETC, GETC is broken down to GTC, and quarks and leptons emerge as the TC-singlet fermions. When αTC(μ) becomes strong at scale ΛTC $\cong$ FEW, the fermionic condensate $\langle \bar{T}T \rangle_{TC} \cong 4\pi F_{EW}^3$ forms. (The condensate is the vacuum expectation value of the technifermion bilinear $\bar{T}T$. The estimate here is based on naive dimensional analysis of the quark condensate in QCD, expected to be correct as an order of magnitude.) Then, the transitions q_L(or ℓ_L) $\to T_L \to T_R \to q_R$ (or ℓ_R) can proceed through the technifermion's dynamical mass by the emission and reabsorption of ETC bosons whose masses METC $\cong g$ETC ΛETC are much greater than ΛTC. The quarks and leptons develop masses given approximately by

$$(2) \qquad m_{q,\ell}(M_{ETC}) \cong \frac{g_{ETC}^2 \langle \bar{T}T \rangle_{ETC}}{M_{ETC}^2} \cong \frac{4\pi F_{EW}^3}{\Lambda_{ETC}^2} .$$

Here, $\langle \bar{T}T \rangle_{ETC}$ is the technifermion condensate renormalized at the ETC boson mass scale,

$$(3) \qquad \langle \bar{T}T \rangle_{ETC} = \exp\left(\int_{\Lambda_{TC}}^{M_{ETC}} \frac{d\mu}{\mu} \gamma_m(\mu) \right) \langle \bar{T}T \rangle_{TC} ,$$

where $\gamma_m(\mu)$ is the anomalous dimension of the technifermion bilinear $\bar{T}T$ at the scale μ. The second estimate in Eq. (2) depends on the assumption that, as happens in QCD, αTC(μ) becomes weak not far above ΛTC, so that the anomalous dimension γ_m of $\bar{T}T$ is small there. Extended technicolor was introduced in 1979 by Dimopoulos and Susskind,[11] and by Eichten and Lane.[12] For a quark of mass $m_q \cong 1$ GeV, and with ΛTC $\cong 246$ GeV, one estimates ΛETC $\cong 15$ TeV. Therefore, assuming that $g_{ETC}^2 \gtrsim 1$, METC will be at least this large.

In addition to the ETC proposal for quark and lepton masses, Eichten and Lane observed that the size of the ETC representations required to generate all quark and lep-ton masses suggests that there will be more than one electroweak doublet of technifermions.[12] If so, there will be more (spontaneously broken) chiral symmetries and therefore more Goldstone bosons than are eaten by the Higgs mechanism. These must acquire mass by virtue of the fact that the extra chiral symmetries are also explicitly broken, by the standard-model interactions and the ETC interactions. These "pseudo-Goldstone bosons" are called technipions, πT. An application of Dashen's theorem[13] gives for the ETC contribution to their mass

$$(4) \qquad F_{EW}^2 \cdot M_{\pi T}^2 \cong \frac{g_{ETC}^2 \langle \bar{T}T\bar{T}T \rangle_{ETC}}{M_{ETC}^2} \cong \frac{16\pi^2 F_{EW}^6}{\Lambda_{ETC}^2} .$$

The second approximation in Eq. (4) assumes that $\langle \bar{T}T\bar{T}T \rangle_{ETC} \cong \langle \bar{T}T \rangle_{ETC}^2$. For FEW $\cong \Lambda$TC $\cong 246$ GeV and ΛETC $\cong 15$ TeV, this contribution to $M\pi$T is about 50 GeV. Since ETC interactions generate $m_{q,\ell}$ *and* the coupling of technipions to quark and lepton pairs, one expects the couplings to be Higgs-like; i.e., roughly proportional to the masses of the quarks and leptons. This means that technipions are expected to decay to the heaviest $\bar{q}q$ and $\bar{\ell}\ell$ pairs allowed.

Perhaps the most important restriction on the ETC framework for quark mass generation is that ETC interactions are likely to induce flavor-changing neutral current processes such as $\mu \to e\,\gamma$, $KL \to \mu\,e$, and $|\Delta S| = 2$ and $|\Delta B| = 2$ interactions that induce $K^0 \leftrightarrow \bar{K}^0$ and $B^0 \leftrightarrow \bar{B}^0$ mixing.[12] The reason is that the algebra of the ETC currents involved in $m_{q,\ell}$ generation imply $\bar{q}q'$ and $\bar{\ell}\ell'$ ETC currents which, when written in terms of fermion mass eigenstates, have no reason to conserve flavor. The strongest constraint comes from requiring that ETC interactions mediating K--$\bar{K}$ mixing contribute less than the Standard Model. This implies an effective ΛETC greater than 1000 TeV. The actual ΛETC may be reduced somewhat if CKM-like mixing angle factors are present. If these interactions are CP-violating, as they well may be, the constraint from the ε-parameter is that the effective ΛETC $> 10^4$ TeV. Such huge ETC mass scales imply tiny quark and lepton masses and ETC contributions to $M\pi$T of at most a few GeV, in conflict with LEP searches for πT at the Z^0.

Extended technicolor is a very ambitious proposal, requiring that quark and lepton masses and mixing angles arise from experimentally accessible interactions. *If* there exists a successful model, it would not only predict the masses and mixings of quarks and leptons (and technipions), it would explain why there are three families of each: they are the ones that fit into the ETC representations of q, ℓ and T. It should not be surprising that the construction of a successful model has proven to be very difficult.

3.9.4   Walking technicolor

Since quark and lepton masses are proportional to the bilinear technifermion condensate divided by the ETC mass
scale squared, their tiny values can be avoided if the condensate is enhanced above the weak-αTC estimate in Eq.
(2). $\langle \bar{T}T \rangle_{ETC} \cong \langle \bar{T}T \rangle_{TC} \cong 4\pi F^3_{EW}$.

During the 1980s, several dynamical mechanisms were advanced to do this. In 1981 Holdom suggested that, if the
αTC(μ) evolves to a nontrivial fixed point in the ultraviolet, with a large positive anomalous dimension γ_m for TT
, realistic quark and lepton masses could arise with ΛETC
large enough to suppress ETC-induced K--$\bar{K}$ mixing.[14]
However, no example of a nontrivial ultraviolet fixed point
in a four-dimensional gauge theory has been constructed.
In 1985 Holdom analyzed a technicolor theory in which a
"slowly varying" αTC(μ) was envisioned.[15] His focus was
to separate the chiral breaking and confinement scales, but
he also noted that such a theory could enhance $\langle \bar{T}T \rangle_{ETC}$
and thus allow the ETC scale to be raised. In 1986 Ak-
iba and Yanagida also considered enhancing quark and lepton masses, by simply assuming that αTC is constant and
strong all the way up to the ETC scale.[16] In the same year
Yamawaki, Bando and Matumoto again imagined an ultraviolet fixed point in a non-asymptotically free theory to enhance the technifermion condensate.[17]

In 1986 Appelquist, Karabali and Wijewardhana discussed
the enhancement of fermion masses in an asymptotically
free technicolor theory with a slowly running, or "walking",
gauge coupling.[18] The slowness arose from the screening
effect of a large number of technifermions, with the analysis carried out through two-loop perturbation theory. In
1987 Appelquist and Wijewardhana explored this walking
scenario further.[19] They took the analysis to three loops,
noted that the walking can lead to a power law enhancement
of the technifermion condensate, and estimated the resultant quark, lepton, and technipion masses. The condensate
enhancement arises because the associated technifermion
mass decreases slowly, roughly linearly, as a function of its
renormalization scale. This corresponds to the condensate
anomalous dimension γ_m in Eq. (3) approaching unity (see
below).[20]

In the 1990s, the idea emerged more clearly that walking
is naturally described by asymptotically free gauge theories
dominated in the infrared by an approximate fixed point.
Unlike the speculative proposal of ultraviolet fixed points,
fixed points in the infrared are known to exist in asymptotically free theories, arising at two loops in the beta function providing that the fermion count N_f is large enough.
This has been known since the first two-loop computation in
1974 by Caswell.[21] If N_f is close to the value $\hat{N}_f$ at which
asymptotic freedom is lost, the resultant infrared fixed point
is weak, of parametric order $\hat{N}_f - N_f$, and reliably acces-

sible in perturbation theory. This weak-coupling limit was
explored by Banks and Zaks in 1982.[22]

The fixed-point coupling αIR becomes stronger as N_f is reduced from $\hat{N}_f$. Below some critical value N_{fc} the coupling
becomes strong enough ($> \alpha_\chi SB$) to break spontaneously the
massless technifermions' chiral symmetry. Since the analysis must typically go beyond two-loop perturbation theory, the definition of the running coupling αTC(μ), its fixed
point value αIR, and the strength $\alpha_\chi SB$ necessary for chiral
symmetry breaking depend on the particular renormaliza-
tion scheme adopted. For $0 < (\alpha_{IR} - \alpha_\chi SB)/\alpha_{IR} \ll 1$;
i.e., for N_f just below N_{fc}, the evolution of αTC(μ) is governed by the infrared fixed point and it will evolve slowly
(walk) for a range of momenta above the breaking scale
ΛTC. To overcome the M^2_{ETC} -suppression of the masses
of first and second generation quarks involved in K--$\bar{K}$
mixing, this range must extend almost to their ETC scale, of
$\mathcal{O}(10^3$ TeV$)$. Cohen and Georgi argued that $\gamma_m = 1$ is the
signal of spontaneous chiral symmetry breaking, i.e., that
$\gamma_m(\alpha_\chi SB) = 1$.[20] Therefore, in the walking-αTC region,
$\gamma_m \cong 1$ and, from Eqs. (2) and (3), the light quark masses
are enhanced approximately by METC/ΛTC.

The idea that αTC(μ) walks for a large range of momenta
when αIR lies just above $\alpha_\chi SB$ was suggested by Lane and
Ramana.[23] They made an explicit model, discussed the
walking that ensued, and used it in their discussion of walking technicolor phenomenology at hadron colliders. This
idea was developed in some detail by Appelquist, Terning
and Wijewardhana.[24] Combining a perturbative computation of the infrared fixed point with an approximation of α_χ
SB based on the Schwinger-Dyson equation, they estimated
the critical value N_{fc} and explored the resultant electroweak
physics. Since the 1990s, most discussions of walking technicolor are in the framework of theories assumed to be
dominated in the infrared by an approximate fixed point.
Various models have been explored, some with the technifermions in the fundamental representation of the gauge
group and some employing higher representations.[25][26][27]

The possibility that the technicolor condensate can be enhanced beyond that discussed in the walking literature,
has also been considered recently by Luty and Okui under
the name "conformal technicolor".[28] They envision an infrared stable fixed point, but with a very large anomalous dimension for the operator TT . It remains to be seen whether
this can be realized, for example, in the class of theories
currently being examined using lattice techniques.

Top quark mass

The walking enhancement described above may be insufficient to generate the measured top quark mass, even for
an ETC scale as low as a few TeV. However, this problem

could be addressed if the effective four-technifermion coupling resulting from ETC gauge boson exchange is strong and tuned just above a critical value.[29] The analysis of this strong-ETC possibility is that of a Nambu–Jona–Lasinio model with an additional (technicolor) gauge interaction. The technifermion masses are small compared to the ETC scale (the cutoff on the effective theory), but nearly constant out to this scale, leading to a large top quark mass. No fully realistic ETC theory for all quark masses has yet been developed incorporating these ideas. A related study was carried out by Miransky and Yamawaki.[30] A problem with this approach is that it involves some degree of parameter fine-tuning, in conflict with technicolor's guiding principle of naturalness.

Finally, it should be noted that there is a large body of closely related work in which ETC does not generate m_t. These are the top quark condensate,[31] topcolor and top-color-assisted technicolor models,[32] in which new strong interactions are ascribed to the top quark and other third-generation fermions. As with the strong-ETC scenario described above, all these proposals involve a considerable degree of fine-tuning of gauge couplings.

3.9.5 Technicolor on the lattice

Lattice gauge theory is a non-perturbative method applicable to strongly interacting technicolor theories, allowing first-principles exploration of walking and conformal dynamics. In 2007, Catterall and Sannino used lattice gauge theory to study $SU(2)$ gauge theories with two flavors of Dirac fermions in the symmetric representation,[33] finding evidence of conformality that has been confirmed by subsequent studies.[34]

As of 2010, the situation for $SU(3)$ gauge theory with fermions in the fundamental representation is not as clearcut. In 2007, Appelquist, Fleming and Neil reported evidence that a non-trivial infrared fixed point develops in such theories when there are twelve flavors, but not when there are eight.[35] While some subsequent studies confirmed these results, others reported different conclusions, depending on the lattice methods used, and there is not yet consensus.[36]

Further lattice studies exploring these issues, as well as considering the consequences of these theories for precision electroweak measurements, are underway by several research groups.[37]

3.9.6 Technicolor phenomenology

Any framework for physics beyond the Standard Model must conform with precision measurements of the elec-

troweak parameters. Its consequences for physics at existing and future high-energy hadron colliders, and for the dark matter of the universe must also be explored.

Precision electroweak tests

In 1990, the phenomenological parameters S, T, and U were introduced by Peskin and Takeuchi to quantify contributions to electroweak radiative corrections from physics beyond the Standard Model.[38] They have a simple relation to the parameters of the electroweak chiral Lagrangian.[39][40] The Peskin-Takeuchi analysis was based on the general formalism for weak radiative corrections developed by Kennedy, Lynn, Peskin and Stuart,[41] and alternate formulations also exist.[42]

The S, T, and U-parameters describe corrections to the electroweak gauge boson propagators from physics Beyond the Standard Model. They can be written in terms of polarization functions of electroweak currents and their spectral representation as follows:

$$(5) \qquad S = 16\pi \frac{d}{dq^2} \left[\Pi_{33}^{\mathbf{new}}(q^2) - \Pi_{3Q}^{\mathbf{new}}(q^2) \right]_{q^2=0}$$

$$= 4\pi \int \frac{dm^2}{m^4} \left[\sigma_V^3(m^2) - \sigma_A^3(m^2) \right]^{\mathbf{new}};$$

$$(6) \qquad T = \frac{16\pi}{M_Z^2 \sin^2 2\theta_W} \left[\Pi_{11}^{\mathbf{new}}(0) - \Pi_{33}^{\mathbf{new}}(0) \right]$$

$$= \frac{4\pi}{M_Z^2 \sin^2 2\theta_W} \int_0^\infty \frac{dm^2}{m^2}$$

$$\left[\sigma_V^1(m^2) + \sigma_A^1(m^2) - \sigma_V^3(m^2) - \sigma_A^3(m^2) \right]^{\mathbf{new}},$$

where only new, beyond-standard-model physics is included. The quantities are calculated relative to a minimal Standard Model with some chosen reference mass of the Higgs boson, taken to range from the experimental lower bound of 117 GeV to 1000 GeV where its width becomes very large.[43] For these parameters to describe the dominant corrections to the Standard Model, the mass scale of the new physics must be much greater than MW and MZ, and the coupling of quarks and leptons to the new particles must be suppressed relative to their coupling to the gauge bosons. This is the case with technicolor, so long as the lightest technivector mesons, ρT and aT, are heavier than 200–300 GeV. The S-parameter is sensitive to all new physics at the TeV scale, while T is a measure of weak-isospin breaking effects. The U-parameter is generally not useful; most new-physics theories, including technicolor theories, give negligible contributions to it.

The S and T-parameters are determined by global fit to experimental data including Z-pole data from LEP at CERN, top quark and W-mass measurements at Fermilab, and measured levels of atomic parity violation. The resultant bounds on these parameters are given in the Review of Particle Properties.[43] Assuming $U = 0$, the S and T parameters are small and, in fact, consistent with zero:

$$(7) \quad \begin{aligned} S &= -0.04 \pm 0.09 \, (-0.07), \\ T &= 0.02 \pm 0.09 \, (+0.09), \end{aligned}$$

where the central value corresponds to a Higgs mass of 117 GeV and the correction to the central value when the Higgs mass is increased to 300 GeV is given in parentheses. These values place tight restrictions on beyond-standard-model theories—when the relevant corrections can be reliably computed.

The S parameter estimated in QCD-like technicolor theories is significantly greater than the experimentally allowed value.[38][42] The computation was done assuming that the spectral integral for S is dominated by the lightest ρT and aT resonances, or by scaling effective Lagrangian parameters from QCD. In walking technicolor, however, the physics at the TeV scale and beyond must be quite different from that of QCD-like theories. In particular, the vector and axial-vector spectral functions cannot be dominated by just the lowest-lying resonances.[44] It is unknown whether higher energy contributions to $\sigma_{V,A}^3$ are a tower of identifiable ρT and aT states or a smooth continuum. It has been conjectured that ρT and aT partners could be more nearly degenerate in walking theories (approximate parity doubling), reducing their contribution to S.[45] Lattice calculations are underway or planned to test these ideas and obtain reliable estimates of S in walking theories.[2][46]

The restriction on the T-parameter poses a problem for the generation of the top-quark mass in the ETC framework. The enhancement from walking can allow the associated ETC scale to be as large as a few TeV,[24] but—since the ETC interactions must be strongly weak-isospin breaking to allow for the large top-bottom mass splitting—the contribution to the T parameter,[47] as well as the rate for the decay $Z^0 \to \bar{b}b$,[48] could be too large.

Hadron collider phenomenology

Early studies generally assumed the existence of just one electroweak doublet of technifermions, or of one techni-family including one doublet each of color-triplet techni-quarks and color-singlet technileptons (four electroweak doublets in total).[49] The number ND of electroweak doublets determines the decay constant F needed to produce the correct electroweak scale, as $F = F_{EW}/\sqrt{ND} = 246$ GeV$/\sqrt{ND}$. In the minimal, one-doublet model, three Goldstone bosons (technipions, πT) have decay constant $F = F_{EW} = 246$ GeV and are eaten by the electroweak gauge bosons. The most accessible collider signal is the production through $\bar{q}q$ annihilation in a hadron collider of spin-one $\rho_T^{\pm,0}$, and their subsequent decay into a pair of longitudinally polarized weak bosons, $W_L^{\pm} Z_L^0$ and $W_L^+ W_L^-$. At an expected mass of 1.5–2.0 TeV and width of 300–400

GeV, such ρT's would be difficult to discover at the LHC. A one-family model has a large number of physical techni-pions, with $F = F_{EW}/\sqrt{4} = 123$ GeV.[50] There is a collection of correspondingly lower-mass color-singlet and octet technivectors decaying into technipion pairs. The πT's are expected to decay to the heaviest possible quark and lepton pairs. Despite their lower masses, the ρT's are wider than in the minimal model and the backgrounds to the πT decays are likely to be insurmountable at a hadron collider.

This picture changed with the advent of walking technicolor. A walking gauge coupling occurs if $\alpha_\chi SB$ lies just below the IR fixed point value αIR, which requires either a large number of electroweak doublets in the fundamental representation of the gauge group, e.g., or a few doublets in higher-dimensional TC representations.[25][51] In the latter case, the constraints on ETC representations generally imply other technifermions in the fundamental representation as well.[12][23] In either case, there are technipions πT with decay constant $F \ll F_{EW}$. This implies $\Lambda_{TC} \ll F_{EW}$ so that the lightest technivectors accessible at the LHC—ρT, ωT, aT (with $I^G\, J^{PC} = 1^+\, 1^{--}$, $0^-\, 1^{--}$, $1^-\, 1^{++}$)—have masses well below a TeV. The class of theories with many technifermions and thus $F \ll F_{EW}$ is called low-scale technicolor.[52]

A second consequence of walking technicolor concerns the decays of the spin-one technihadrons. Since technipion masses $M_{\pi_T}^2 \propto \langle \bar{T}T\bar{T}T \rangle_{M_{ETC}}$ (see Eq. (4)), walking enhances them much more than it does other technihadron masses. Thus, it is very likely that the lightest M_ρT $< 2M\pi$T and that the two and three-πT decay channels of the light technivectors are closed.[25] This further implies that these technivectors are very narrow. Their most probable two-body channels are $W_L^{\pm,0}\pi_T$, WL WL, $\gamma\,\pi$T and γ WL. The coupling of the lightest technivectors to WL is proportional to F/F_{EW}.[53] Thus, all their decay rates are suppressed by powers of $(F/F_{EW})^2 \ll 1$ or the fine-structure constant, giving total widths of a few GeV (for ρT) to a few tenths of a GeV (for ωT and T).

A more speculative consequence of walking technicolor is motivated by consideration of its contribution to the S-parameter. As noted above, the usual assumptions made to estimate STC are invalid in a walking theory. In particular, the spectral integrals used to evaluate STC cannot be dominated by just the lowest-lying ρT and aT and, if STC is to be small, the masses and weak-current couplings of the ρT and aT could be more nearly equal than they are in QCD.

Low-scale technicolor phenomenology, including the possibility of a more parity-doubled spectrum, has been developed into a set of rules and decay amplitudes.[53] An April 2011 announcement of an excess in jet pairs produced in association with a W boson measured at the Tevatron[54] has been interpreted by Eichten, Lane and Martin as a possible signal of the technipion of low-scale technicolor.[55]

The general scheme of low-scale technicolor makes little sense if the limit on M_{ρ_T} is pushed past about 700 GeV. The LHC should be able to discover it or rule it out. Searches there involving decays to technipions and thence to heavy quark jets are hampered by backgrounds from $\bar{t}t$ production; its rate is 100 times larger than that at the Tevatron. Consequently, the discovery of low-scale technicolor at the LHC relies on all-leptonic final-state channels with favorable signal-to-background ratios: $\rho_T^\pm \to W_L^\pm Z_L^0$, $a_T^\pm \to \gamma W_L^\pm$ and $\omega_T \to \gamma Z_L^0$.[56]

Dark matter

Technicolor theories naturally contain dark matter candidates. Almost certainly, models can be built in which the lowest-lying technibaryon, a technicolor-singlet bound state of technifermions, is stable enough to survive the evolution of the universe.[43][57] If the technicolor theory is low-scale ($F \ll F_{EW}$), the baryon's mass should be no more than 1–2 TeV. If not, it could be much heavier. The technibaryon must be electrically neutral and satisfy constraints on its abundance. Given the limits on spin-independent dark-matter-nucleon cross sections from dark-matter search experiments ($\lesssim 10^{-42}$ cm^2 for the masses of interest[58]), it may have to be electroweak neutral (weak isospin $I = 0$) as well. These considerations suggest that the "old" technicolor dark matter candidates may be difficult to produce at the LHC.

A different class of technicolor dark matter candidates light enough to be accessible at the LHC was introduced by Francesco Sannino and his collaborators.[59] These states are pseudo Goldstone bosons possessing a global charge that makes them stable against decay.

3.9.7   See also

- Higgsless model
- Topcolor
- Top quark condensate

3.9.8   References

[1] For a recent introductions to and reviews of technicolor, see:
Christopher T. Hill & Elizabeth H. Simmons (2003). "Strong Dynamics and Electroweak Symmetry Breaking". *Physics Reports* **381** (4-6): 235–402. arXiv:hep-ph/0203079. Bibcode:2003PhR...381..235H. doi:10.1016/S0370-1573(03)00140-6.
Kenneth Lane (2002). *Two Lectures on Technicolor*. l'Ecole

de GIF at LAPP, Annecy-le-Vieux, France. arXiv:hep-ph/0202255.
Robert Shrock (2007). "Some Recent Results on Models of Dynamical Electroweak Symmetry Breaking". In M. Tanabashi; M. Harada; K. Yamawaki. *Nagoya 2006: The Origin of Mass and Strong Coupling Gauge Theories*. International Workshop on Strongly Coupled Gauge Theories. pp. 227–241. arXiv:hep-ph/0703050.
Adam Martin (2008). *Technicolor Signals at the LHC*. The 46th Course at the International School of Subnuclear Physics: Predicted and Totally Unexpected in the Energy Frontier Opened by LHC. arXiv:0812.1841.
Francesco Sannino (2009). "Conformal Dynamics for TeV Physics and Cosmology". *Acta Physica Polonica* **B40**: 3533–3745. arXiv:0911.0931. Bibcode:2009arXiv0911.0931S.

[2] George Fleming (2008). "Strong Interactions for the LHC". *Proceedings of Science*. LATTICE 2008: 21. arXiv:0812.2035. Bibcode:2008arXiv0812.2035F.

[3] "CERN experiments observe particle consistent with long-sought Higgs boson". CERN press release. 4 July 2012. Retrieved 4 July 2012.

[4] Taylor, Lucas (4 July 2012). "Observation of a New Particle with a Mass of 125 GeV". *CMS Public Web site*. CERN.

[5] "Latest Results from ATLAS Higgs Search". ATLAS. 4 July 2012. Retrieved 4 July 2012.

[6] R. Jackiw & K. Johnson (1973). "Dynamical Model of Spontaneously Broken Gauge Symmetries". *Physical Review* **D8** (8): 2386–2398. Bibcode:1973PhRvD...8.2386J. doi:10.1103/PhysRevD.8.2386.

[7] John M. Cornwall & Richard E. Norton (1973). "Spontaneous Symmetry Breaking Without Scalar Mesons". *Physical Review* **D8** (10): 3338–3346. Bibcode:1973PhRvD...8.3338C. doi:10.1103/PhysRevD.8.3338.

[8] Marvin Weinstein (1973). "Conserved Currents, Their Commutators, and the Symmetry Structure of Renormalizable Theories of Electromagnetic, Weak, and Strong Interactions". *Physical Review* **D8** (8): 2511–2524. Bibcode:1973PhRvD...8.2511W. doi:10.1103/PhysRevD.8.2511.

[9] Steven Weinberg (1976). "Implications of dynamical symmetry breaking". *Physical Review* **D13** (4): 974–996. Bibcode:1976PhRvD..13..974W. doi:10.1103/PhysRevD.13.974.
S. Weinberg (1979). "Implications of dynamical symmetry breaking: An addendum". *Physical Review* **D19** (4): 1277–1280. Bibcode:1979PhRvD..19.1277W. doi:10.1103/PhysRevD.19.1277.

[10] Leonard Susskind (1979). "Dynamics of spontaneous symmetry breaking in the Weinberg-Salam theory". *Physical Review* **D20** (10):

2619–2625.   Bibcode:1979PhRvD..20.2619S.
doi:10.1103/PhysRevD.20.2619.

[11] Savas Dimopoulos & Leonard Susskind (1979). "Mass without scalars". *Nuclear Physics* **B155** (1): 237–252. Bibcode:1979NuPhB.155..237D. doi:10.1016/0550-3213(79)90364-X.

[12] Estia Eichten & Kenneth Lane (1980). "Dynamical breaking of weak interaction symmetries". *Physics Letters* **B90** (1-2): 125–130. Bibcode:1980PhLB...90..125E. doi:10.1016/0370-2693(80)90065-9.

[13] Roger Dashen (1969). "Chiral SU(3)⊗SU(3) as a Symmetry of the Strong Interactions". *Physical Review* **183** (5): 1245–1260. Bibcode:1969PhRv..183.1245D. doi:10.1103/PhysRev.183.1245.
Roger Dashen (1971). "Some Features of Chiral Symmetry Breaking". *Physical Review* **D3** (8): 1879–1889. Bibcode:1971PhRvD...3.1879D. doi:10.1103/PhysRevD.3.1879.

[14] Bob Holdom (1981). "Raising the sideways scale". *Physical Review* **D24** (5): 1441–1444. Bibcode:1981PhRvD..24.1441H. doi:10.1103/PhysRevD.24.1441.

[15] Bob Holdom (1985). "Techniodor". *Physics Letters* **B150** (4): 301–305. Bibcode:1985PhLB..150..301H. doi:10.1016/0370-2693(85)91015-9.

[16] T. Akiba & T. Yanagida (1986). "Hierarchic chiral condensate". *Physics Letters* **B169** (4): 432–435. Bibcode:1986PhLB..169..432A. doi:10.1016/0370-2693(86)90385-0.

[17] Koichi Yamawaki; Masako Bando & Ken-iti Matumoto (1986). "Scale-Invariant Hypercolor Model and a Dilaton". *Physical Review Letters* **56** (13): 1335–1338. Bibcode:1986PhRvL..56.1335Y. doi:10.1103/PhysRevLett.56.1335. PMID 10032641.

[18] Thomas Appelquist; Dimitra Karabali & L. C. R. Wijewardhana (1986). "Chiral Hierarchies and Flavor-Changing Neutral Currents in Hypercolor". *Physical Review Letters* **57** (8): 957–960. Bibcode:1986PhRvL..57..957A. doi:10.1103/PhysRevLett.57.957. PMID 10034209.

[19] Thomas Appelquist & L. C. R. Wijewardhana (1987). "Chiral hierarchies from slowly running couplings in technicolor theories". *Physical Review* **D36** (2): 568–580. Bibcode:1987PhRvD..36..568A. doi:10.1103/PhysRevD.36.568.

[20] Andrew Cohen & Howard Georgi (1989). "Walking beyond the rainbow". *Nuclear Physics* **B314** (1): 7–24. Bibcode:1989NuPhB.314....7C. doi:10.1016/0550-3213(89)90109-0.

[21] William E. Caswell (1974). "Asymptotic Behavior of Non-Abelian Gauge Theories to Two-Loop Order". *Physical Review Letters* **33** (4): 244–246. Bibcode:1974PhRvL..33..244C. doi:10.1103/PhysRevLett.33.244.

[22] T. Banks & A. Zaks (1982). "On the phase structure of vector-like gauge theories with massless fermions". *Nuclear Physics* **B196** (2): 189–204. Bibcode:1982NuPhB.196..189B. doi:10.1016/0550-3213(82)90035-9.

[23] Kenneth Lane & M. V. Ramana (1991). "Walking technicolor signatures at hadron colliders". *Physical Review* **D44** (9): 2678–2700. Bibcode:1991PhRvD..44.2678L. doi:10.1103/PhysRevD.44.2678.

[24] Thomas Appelquist; John Terning & L. C. R. Wijewardhana (1997). "Postmodern Technicolor". *Physical Review Letters* **79** (15): 2767–2770. arXiv:hep-ph/9706238. Bibcode:1997PhRvL..79.2767A. doi:10.1103/PhysRevLett.79.2767.

[25] Kenneth Lane & Estia Eichten (1989). "Two-scale technicolor". *Physics Letters* **B222** (2): 274–280. Bibcode:1989PhLB..222..274L. doi:10.1016/0370-2693(89)91265-3.

[26] Francesco Sannino & Kimmo Tuominen (2005). "Orientifold theory dynamics and symmetry breaking". *Physical Review* **D71** (5): 051901. arXiv:hep-ph/0405209. Bibcode:2005PhRvD..71e1901S. doi:10.1103/PhysRevD.71.051901.

[27] Dennis D. Dietrich; Francesco Sannino & Kimmo Tuominen (2005). "Light composite Higgs boson from higher representations versus electroweak precision measurements: Predictions for CERN LHC". *Physical Review* **D72** (5): 055001. arXiv:hep-ph/0505059. Bibcode:2005PhRvD..72e5001D. doi:10.1103/PhysRevD.72.055001.
Dennis D. Dietrich; Francesco Sannino & Kimmo Tuominen (2006). "Light composite Higgs and precision electroweak measurements on the Z resonance: An update". *Physical Review* **D73** (3): 037701. arXiv:hep-ph/0510217. Bibcode:2006PhRvD..73c7701D. doi:10.1103/PhysRevD.73.037701.
Dennis D. Dietrich & Francesco Sannino (2007). "Conformal window of SU(N) gauge theories with fermions in higher dimensional representations". *Physical Review* **D75** (8): 085018. arXiv:hep-ph/0611341. Bibcode:2007PhRvD..75h5018D. doi:10.1103/PhysRevD.75.085018.
Thomas A. Ryttov & Francesco Sannino (2007). "Conformal windows of SU(N) gauge theories, higher dimensional representations, and the size of the unparticle world". *Physical Review* **D76** (10): 105004. arXiv:0707.3166. Bibcode:2007PhRvD..76j5004R. doi:10.1103/PhysRevD.76.105004.
Thomas A. Ryttov & Francesco Sannino (2008). "Supersymmetry inspired QCD beta function". *Physical Review* **D78** (6): 065001. arXiv:0711.3745. Bibcode:2008PhRvD..78f5001R. doi:10.1103/PhysRevD.78.065001.

[28] Markus A. Luty & Takemichi Okui (2006). "Conformal technicolor". *Journal of High Energy Physics* **0609** (09): 070. arXiv:hep-ph/0409274. Bibcode:2006JHEP...09..070L. doi:10.1088/1126-6708/2006/09/070.

Markus A. Luty (2009). "Strong conformal dynamics at the LHC and on the lattice". *Journal of High Energy Physics* **0904** (04): 050. arXiv:0806.1235. Bibcode:2009JHEP...04..050L. doi:10.1088/1126-6708/2009/04/050.

Jared A. Evans; Jamison Galloway; Markus A. Luty & Ruggero Altair Tacchi (2010). "Minimal conformal technicolor and precision electroweak tests". *Journal of High Energy Physics* **1010** (10): 086. arXiv:1001.1361. Bibcode:2010JHEP...10..086E. doi:10.1007/JHEP10(2010)086.

[29] Thomas Appelquist; T. Takeuchi; Martin Einhorn & L. C. R. Wijewardhana (1989). "Higher mass scales and mass hierarchies". *Physics Letters* **B220** (1-2): 223–228. Bibcode:1989PhLB..220..223A. doi:10.1016/0370-2693(89)90041-5.

[30] V. A. Miransky & K. Yamawaki (1989). "On Gauge Theories with Additional Four Fermion Interaction". *Modern Physics Letters* **A4** (2): 129–135. Bibcode:1989MPLA....4..129M. doi:10.1142/S0217732389000186.

[31] Y. Nambu (1989). "BCS mechanism, quasi supersymmetry, and fermion masses". In Z. Ajduk; S. Pokorski; A. Trautman. *Proceedings of the Kazimierz 1988 Conference on New Theories in Physics*. XI International Symposium on Elementary Particle Physics. pp. 406–415.

V. A. Miransky; Masaharu Tanabashi & Koichi Yamawaki (1989). "Is the t Quark Responsible for the Mass of W and Z Bosons?". *Modern Physics Letters* **A4** (11): 1043–1053. Bibcode:1989MPLA....4.1043M. doi:10.1142/S0217732389001210.

V. A. Miransky; Masaharu Tanabashi & Koichi Yamawaki (1989). "Dynamical electroweak symmetry breaking with large anomalous dimension and t quark condensate". *Physics Letters* **B221** (2): 177–183. Bibcode:1989PhLB..221..177M. doi:10.1016/0370-2693(89)91494-9.

William A. Bardeen; Christopher T. Hill & Manfred Lindner (1990). "Minimal dynamical symmetry breaking of the standard model". *Physical Review* **D41** (5): 1647–1660. Bibcode:1990PhRvD..41.1647B. doi:10.1103/PhysRevD.41.1647.

[32] Christopher T. Hill (1991). "Topcolor: top quark condensation in a gauge extension of the standard model". *Physics Letters* **B266** (3-4): 419–424. Bibcode:1991PhLB..266..419H. doi:10.1016/0370-2693(91)91061-Y.

Christopher T. Hill (1995). "Topcolor assisted technicolor". *Physics Letters* **B345** (4): 483–489. arXiv:hep-ph/9411426. Bibcode:1995PhLB..345..483H. doi:10.1016/0370-2693(94)01660-5.

[33] Simon Catterall & Francesco Sannino (2007). "Minimal Walking on the Lattice". *Physical Review* **D76** (3): 034504. arXiv:0705.1664. Bibcode:2007PhRvD..76c4504C. doi:10.1103/PhysRevD.76.034504.

[34] Simon Catterall; Joel Giedt; Francesco Sannino & Joe Schneible (2008). "Phase diagram of SU(2) with 2 flavors of dynamical adjoint quarks". *Journal of High Energy Physics* **0811** (11): 009. arXiv:0807.0792. Bibcode:2008JHEP...11..009C. doi:10.1088/1126-6708/2008/11/009.

Ari J. Hietanen; Kari Rummukainen & Kimmo Tuominen (2009). "Evolution of the coupling constant in SU(2) lattice gauge theory with two adjoint fermions". *Physical Review* **D80** (9): 094504. arXiv:0904.0864. Bibcode:2009PhRvD..80i4504H. doi:10.1103/PhysRevD.80.094504.

[35] Thomas Appelquist; George T. Fleming & Ethan T. Neil (2008). "Lattice Study of the Conformal Window in QCD-like Theories". *Physical Review Letters* **100** (17): 171607. arXiv:0712.0609. Bibcode:2008PhRvL.100q1607A. doi:10.1103/PhysRevLett.100.171607. PMID 18518277.

[36] Albert Deuzeman; Maria Paola Lombardo & Elisabetta Pallante (2008). "The physics of eight flavours". *Physics Letters* **B670** (1): 41–48. arXiv:0804.2905. Bibcode:2008PhLB..670...41D. doi:10.1016/j.physletb.2008.10.039.

Thomas Appelquist; George T. Fleming & Ethan T. Neil (2009). "Lattice study of conformal behavior in SU(3) Yang-Mills theories". *Physical Review* **D79** (7): 076010. arXiv:0901.3766. Bibcode:2009PhRvD..79g6010A. doi:10.1103/PhysRevD.79.076010.

Erek Bilgici; et al. (2009). "New scheme for the running coupling constant in gauge theories using Wilson loops". *Physical Review* **D80** (3): 034507. arXiv:0902.3768. Bibcode:2009PhRvD..80c4507B. doi:10.1103/PhysRevD.80.034507.

Xiao-Yong Jin & Robert D. Mawhinney (2009). "Lattice QCD with 8 and 12 degenerate quark flavors" (PDF). *Proceedings of Science*. LAT2009: 049.

Zoltan Fodor; Kieran Holland; Julius Kuti; Daniel Nogradi; et al. (2009). "Chiral symmetry breaking in nearly conformal gauge theories" (PDF). *Proceedings of Science*. LAT2009: 058. arXiv:0911.2463. Bibcode:2009arXiv0911.2463F.

Anna Hasenfratz (2010). "Conformal or Walking? Monte Carlo renormalization group studies of SU(3) gauge models with fundamental fermions". *Physical Review* **D82** (1): 014506. arXiv:1004.1004. Bibcode:2010PhRvD..82a4506H. doi:10.1103/PhysRevD.82.014506.

[37] Thomas DeGrand; Yigal Shamir & Benjamin Svetitsky (2009). "Phase structure of SU(3) gauge theory with two flavors of symmetric-representation fermions". *Physical Review* **D79** (3): 034501. arXiv:0812.1427. Bibcode:2009PhRvD..79c4501D. doi:10.1103/PhysRevD.79.034501.

Thomas Appelquist; et al. (2009). "Toward TeV Conformality". *Physical Review Letters* **104** (7): 071601. arXiv:0910.2224. Bibcode:2010PhRvL.104g1601A. doi:10.1103/PhysRevLett.104.071601. PMID 20366870.

[38] Michael E. Peskin & Tatsu Takeuchi (1990). "New constraint on a strongly interacting Higgs sector". *Physical Review Letters* **65** (8): 964–967. Bibcode:1990PhRvL..65..964P. doi:10.1103/PhysRevLett.65.964. PMID 10043071.
Michael E. Peskin & Tatsu Takeuchi (1992). "Estimation of oblique electroweak corrections". *Physical Review* **D46** (1): 381–409. Bibcode:1992PhRvD..46..381P. doi:10.1103/PhysRevD.46.381.

[39] Thomas Appelquist & Claude Bernard (1980). "Strongly interacting Higgs bosons". *Physical Review* **D22** (1): 200–213. Bibcode:1980PhRvD..22..200A. doi:10.1103/PhysRevD.22.200.

[40] Anthony C. Longhitano (1980). "Heavy Higgs bosons in the Weinberg-Salam model". *Physical Review* **D22** (5): 1166–1175. Bibcode:1980PhRvD..22.1166L. doi:10.1103/PhysRevD.22.1166.
Anthony C. Longhitano (1981). "Low-energy impact of a heavy Higgs boson sector". *Nuclear Physics* **B188** (1): 118–154. Bibcode:1981NuPhB.188..118L. doi:10.1016/0550-3213(81)90109-7.

[41] B. W. Lynn; Michael Edward Peskin & R. G. Stuart (1985). "Radiative Corrections in SU(2) x U(1): LEP / SLC". In Bryan W. Lynn & Claudio Verzegnassi. *Tests of electroweak theories: polarized processes and other phenomena*. Second Conference on Tests of Electroweak Theories, Trieste, Italy. 10–12 June 1985. p. 213.
D. C. Kennedy & B. W. Lynn (1989). "Electroweak radiative corrections with an effective lagrangian: Four-fermions processes". *Nuclear Physics* **B322** (1): 1–54. Bibcode:1989NuPhB.322....1K. doi:10.1016/0550-3213(89)90483-5.

[42] Mitchell Golden & Lisa Randall (1991). "Radiative corrections to electroweak parameters in technicolor theories". *Nuclear Physics* **B361** (1): 3–23. Bibcode:1991NuPhB.361....3G. doi:10.1016/0550-3213(91)90614-4.
B. Holdom & J. Terning (1990). "Large corrections to electroweak parameters in technicolor theories". *Physics Letters* **B247** (1): 88–92. Bibcode:1990PhLB..247...88H. doi:10.1016/0370-2693(90)91054-F.
G. Altarelli; R. Barbieri & S. Jadach (1992). "Toward a model-independent analysis of electroweak data". *Nuclear Physics* **B369** (1-2): 3–32. Bibcode:1992NuPhB.369....3A. doi:10.1016/0550-3213(92)90376-M.

[43] Particle Data Group (C. Amsler *et al.*) (2008). "Review of Particle Physics". *Physics Letters* **B667** (1-5): 1. Bibcode:2008PhLB..667....1P. doi:10.1016/j.physletb.2008.07.018.

[44] Kenneth Lane (1994). "An introduction to technicolor". In K. T. Mahantappa. *Boulder 1993 Proceedings: The building blocks of creation*. Theoretical Advanced Study Institute (TASI 93) in Elementary Particle Physics: The Building Blocks of Creation - From Microfermis to Megaparsecs. Boulder, Colorado, 6 June - 2 July 1993. pp. 381–408. arXiv:hep-ph/9401324.
Kenneth Lane (1995). "Technicolor and precision tests of the electroweak interactions". In P. J. Bussey; I. G. Knowles. *High energy physics: Proceedings*. 27th International Conference on High Energy Physics (ICHEP), Glasgow, Scotland, 20–27 July 1994. p. 543. arXiv:hep-ph/9409304.

[45] Thomas Appelquist & Francesco Sannino (1999). "Physical spectrum of conformal SU(N) gauge theories". *Physical Review* **D59** (6): 067702. arXiv:hep-ph/9806409. Bibcode:1999PhRvD..59f7702A. doi:10.1103/PhysRevD.59.067702.
Johannes Hirn & Verónica Sanz (2006). "Negative S Parameter from Holographic Technicolor". *Physical Review Letters* **97** (12): 121803. arXiv:hep-ph/0606086. Bibcode:2006PhRvL..97l1803H. doi:10.1103/PhysRevLett.97.121803. PMID 17025952.
R. Casalbuoni; D. Dominici; A. Deandrea; R. Gatto; et al. (1996). "Low energy strong electroweak sector with decoupling". *Physical Review* **D53** (9): 5201–5221. arXiv:hep-ph/9510431. Bibcode:1996PhRvD..53.5201C. doi:10.1103/PhysRevD.53.5201.

[46] Lattice Strong Dynamics Collaboration.

[47] Thomas Appelquist; Mark J. Bowick; Eugene Cohler & Avi I. Hauser (1985). "Breaking of isospin symmetry in theories with a dynamical Higgs mechanism". *Physical Review* **D31** (7): 1676–1684. Bibcode:1985PhRvD..31.1676A. doi:10.1103/PhysRevD.31.1676.
R. S. Chivukula; B. A. Dobrescu & J. Terning (1995). "Isospin breaking and fine-tuning in top-color assisted technicolor". *Physics Letters* **B353** (2-3): 289–284. arXiv:hep-ph/9503203. Bibcode:1995PhLB..353..289C. doi:10.1016/0370-2693(95)00569-7.

[48] R. Sekhar Chivukula; Stephen B. Selipsky & Elizabeth H. Simmons (1992). "Nonoblique effects in the Zbb̄ vertex from extended technicolor dynamics". *Physical Review Letters* **69** (4): 575–577. arXiv:hep-ph/9204214. Bibcode:1992PhRvL..69..575C. doi:10.1103/PhysRevLett.69.575. PMID 10046976.
Elizabeth H. Simmons; R.S. Chivukula & J. Terning (1996). "Testing extended technicolor with R(b)". *Progress of Theoretical Physics Supplement* **123**: 87–96. arXiv:hep-ph/9509392. Bibcode:1996PThPS.123...87S. doi:10.1143/PTPS.123.87.

[49] E. Eichten; I. Hinchliffe; K. Lane & C. Quigg (1984). "Supercollider physics". *Reviews of Modern Physics* **56** (4): 579–707. Bibcode:1984RvMP...56..579E. doi:10.1103/RevModPhys.56.579.
E. Eichten; I. Hinchliffe; K. Lane & C. Quigg (1986). "Erratum: Supercollider physics". *Reviews of Modern Physics*

58 (4): 1065–1073. Bibcode:1986RvMP...58.1065E. doi:10.1103/RevModPhys.58.1065.

[50] E. Farhi & L. Susskind (1979). "Grand unified theory with heavy color". *Physical Review* **D20** (12): 3404–3411. Bibcode:1979PhRvD..20.3404F. doi:10.1103/PhysRevD.20.3404.

[51] Dennis D. Dietrich; Francesco Sannino & Kimmo Tuominen (2005). "Light composite Higgs boson from higher representations versus electroweak precision measurements: Predictions for CERN LHC". *Physical Review* **D72** (5): 055001. arXiv:hep-ph/0505059. Bibcode:2005PhRvD..72e5001D. doi:10.1103/PhysRevD.72.055001.

[52] Kenneth Lane & Estia Eichten (1995). "Natural topcolor-assisted technicolor". *Physics Letters* **B352** (3-4): 382–387. arXiv:hep-ph/9503433. Bibcode:1995PhLB..352..382L. doi:10.1016/0370-2693(95)00482-Z.
Estia Eichten & Kenneth Lane (1996). "Low-scale technicolor at the Tevatron". *Physics Letters* **B388** (4): 803–807. arXiv:hep-ph/9607213. Bibcode:1996PhLB..388..803E. doi:10.1016/S0370-2693(96)01211-7.
Estia Eichten; Kenneth Lane & John Womersley (1997). "Finding low-scale technicolor at hadron colliders". *Physics Letters* **B405** (3-4): 305–311. arXiv:hep-ph/9704455. Bibcode:1997PhLB..405..305E. doi:10.1016/S0370-2693(97)00637-0.

[53] Kenneth Lane (1999). "Technihadron production and decay in low-scale technicolor". *Physical Review* **D60** (7): 075007. arXiv:hep-ph/9903369. Bibcode:1999PhRvD..60g5007L. doi:10.1103/PhysRevD.60.075007.
Estia Eichten & Kenneth Lane (2008). "Low-scale technicolor at the Tevatron and LHC". *Physics Letters* **B669** (3-4): 235–238. arXiv:0706.2339. Bibcode:2008PhLB..669..235E. doi:10.1016/j.physletb.2008.09.047.

[54] CDF Collaboration (T. Aaltonen *et al.*) (2011). "Invariant Mass Distribution of Jet Pairs Produced in Association with a W boson in ppbar Collisions at sqrt(s) = 1.96 TeV". arXiv:1104.0699.

[55] Estia J. Eichten; Kenneth Lane & Adam Martin (2011). "Technicolor at the Tevatron". arXiv:1104.0976.

[56] Gustaaf H. Brooijmans; New Physics Working Group (2008). "New Physics at the LHC: A Les Houches Report". *Les Houches 2007: Physics at TeV Colliders*. 5th Les Houches Workshop on Physics at TeV Colliders 11–29 June 2007, Les Houches, France. pp. 363–489. arXiv:0802.3715.

[57] S. Nussinov (1985). "Technocosmology — could a technibaryon excess provide a "natural" missing mass candidate?". *Physics Letters* **B165** (1-3): 55–58. Bibcode:1985PhLB..165...55N. doi:10.1016/0370-2693(85)90689-6.
R. S. Chivukula & Terry P. Walker (1990). "Technicolor cosmology". *Nuclear Physics* **B329** (2): 445–463. Bibcode:1990NuPhB.329..445C. doi:10.1016/0550-3213(90)90151-3.
John Bagnasco; Michael Dine & Scott Thomas (1994). "Detecting technibaryon dark matter". *Physics Letters* **B320** (1-2): 99–104. arXiv:hep-ph/9310290. Bibcode:1994PhLB..320...99B. doi:10.1016/0370-2693(94)90830-3.
Sven Bjarke Gudnason; Chris Kouvaris & Francesco Sannino (2006). "Dark matter from new technicolor theories". *Physical Review* **D74** (9): 095008. arXiv:hep-ph/0608055. Bibcode:2006PhRvD..74i5008G. doi:10.1103/PhysRevD.74.095008.

[58] D. McKinsey. "Direct Dark Matter Detection Using Noble Liquids", 2009 Institute for Advanced Study Workshop on Current Trends in Dark Matter.

[59] Sven Bjarke Gudnason; Chris Kouvaris & Francesco Sannino (2006). "Towards working technicolor: Effective theories and dark matter". *Physical Review* **D73** (11): 115003. arXiv:hep-ph/0603014. Bibcode:2006PhRvD..73k5003G. doi:10.1103/PhysRevD.73.115003.
Sven Bjarke Gudnason; Chris Kouvaris & Francesco Sannino (2006). "Dark matter from new technicolor theories". *Physical Review* **D74** (9): 095008. arXiv:hep-ph/0608055. Bibcode:2006PhRvD..74i5008G. doi:10.1103/PhysRevD.74.095008.
Thomas A. Ryttov & Francesco Sannino (2008). "Ultraminimal technicolor and its dark matter technicolor interacting massive particles". *Physical Review* **D78** (11): 115010. arXiv:0809.0713. Bibcode:2008PhRvD..78k5010R. doi:10.1103/PhysRevD.78.115010.
Enrico Nardi; Francesco Sannino & Alessandro Strumia (2009). "Decaying Dark Matter can explain the e± excesses". *Journal of Cosmology and Astroparticle Physics* **0901** (01): 043. arXiv:0811.4153. Bibcode:2009JCAP...01..043N. doi:10.1088/1475-7516/2009/01/043.
Roshan Foadi; Mads T. Frandsen & Francesco Sannino (2009). "Technicolor dark matter". *Physical Review* **D80** (3): 037702. arXiv:0812.3406. Bibcode:2009PhRvD..80c7702F. doi:10.1103/PhysRevD.80.037702.
Mads T. Frandsen & Francesco Sannino (2010). "Isotriplet technicolor interacting massive particle as dark matter". *Physical Review* **D81** (9): 097704. arXiv:0911.1570. Bibcode:2010PhRvD..81i7704F. doi:10.1103/PhysRevD.81.097704.

3.10 Kaluza–Klein theory

This article is about gravitation and electromagnetism. For the mathematical generalization of K theory, see KK-theory.

In physics, **Kaluza–Klein theory** (**KK theory**) is a unified field theory of gravitation and electromagnetism built around the idea of a fifth dimension beyond the usual four of space and time. It is considered to be an important precursor to string theory.

The five-dimensional theory was developed in three steps. The original hypothesis came from Theodor Kaluza, who sent his results to Einstein in 1919,[1] and published them in 1921.[2] Kaluza's theory was a purely classical extension of general relativity to five dimensions. The five-dimensional metric has 15 components. Ten components are identified with the four-dimensional spacetime metric, four components with the electromagnetic vector potential, and one component with an unidentified scalar field sometimes called the "radion" or the "dilaton". Correspondingly, the five-dimensional Einstein equations yield the four-dimensional Einstein field equations, the Maxwell equations for the electromagnetic field, and an equation for the scalar field. Kaluza also introduced the hypothesis known as the "cylinder condition", that no component of the five-dimensional metric depends on the fifth dimension. Without this assumption, the field equations of five-dimensional relativity are enormously more complex. Standard four-dimensional physics seems to manifest the cylinder condition. Kaluza also set the scalar field equal to a constant, in which case standard general relativity and electrodynamics are recovered identically.

In 1926, Oskar Klein gave Kaluza's classical five-dimensional theory a quantum interpretation,[3][4] to accord with the then-recent discoveries of Heisenberg and Schrödinger. Klein introduced the hypothesis that the fifth dimension was curled up and microscopic, to explain the cylinder condition. Klein also calculated a scale for the fifth dimension based on the quantum of charge.

It wasn't until the 1940s that the classical theory was completed, and the full field equations including the scalar field were obtained by three independent research groups:[5] Thiry,[6][7][8] working in France on his dissertation under Lichnerowicz; Jordan, Ludwig, and Müller in Germany,[9][10][11][12][13] with critical input from Pauli and Fierz; and Scherrer [14][15][16] working alone in Switzerland. Jordan's work led to the scalar-tensor theory of Brans & Dicke;[17] Brans and Dicke were apparently unaware of Thiry or Scherrer. The full Kaluza equations under the cylinder condition are quite complex, and most English-language reviews as well as the English translations of Thiry contain some errors. The complete Kaluza equations were evaluated using tensor algebra software in 2015.[18]

3.10.1   Kaluza hypothesis

In his 1921 paper,[2] Kaluza established all the elements of the classical five-dimensional theory: the metric, the field equations, the equations of motion, the stress-energy tensor,

and the cylinder condition. The theory has no free parameters; it merely extends general relativity to five dimensions. One starts by hypothesizing a form of the five-dimensional metric $\tilde{g}_{ab}$, where Roman indices span five dimensions. Let one also introduce the four-dimensional spacetime metric $g_{\mu\nu}$, where Greek indices span the usual four dimensions of space and time; a 4-vector A^μ which will be identified with the electromagnetic vector potential; and a scalar field ϕ. Then decompose the 5D metric so that the 4D metric is framed by the electromagnetic vector potential, with the scalar field at the fifth diagonal. This can be visualized as:

$$\tilde{g}_{ab} \equiv \begin{bmatrix} g_{\mu\nu} + \phi^2 A_\mu A_\nu & \phi^2 A_\mu \\ \phi^2 A_\nu & \phi^2 \end{bmatrix}.$$

More precisely, one can write

$$\tilde{g}_{\mu\nu} \equiv g_{\mu\nu} + \phi^2 A_\mu A_\nu, \quad \tilde{g}_{5\nu} \equiv \tilde{g}_{\nu 5} \equiv \phi^2 A_\nu, \quad \tilde{g}_{55} \equiv \phi^2$$

where the index 5 indicates the fifth coordinate by convention even though the first four coordinates are indexed with 0, 1, 2, and 3. The associated inverse metric is

$$\tilde{g}^{ab} \equiv \begin{bmatrix} g^{\mu\nu} & -A^\mu \\ -A^\nu & g_{\alpha\beta}A^\alpha A^\beta + \frac{1}{\phi^2} \end{bmatrix}.$$

So far, this decomposition is quite general and all terms are dimensionless. Kaluza then applies the machinery of standard general relativity to this metric. The field equations are obtained from five-dimensional Einstein equations, and the equations of motion are obtained from the five-dimensional geodesic hypothesis. The resulting field equations provide both the equations of general relativity and of electrodynamics; the equations of motion provide the four-dimensional geodesic equation and the Lorentz force law, and one finds that electric charge is identified with motion in the fifth dimension.

The hypothesis for the metric implies an invariant five-dimensional length element ds:

$$ds^2 \equiv \tilde{g}_{ab}dx^a dx^b = g_{\mu\nu}dx^\mu dx^\nu + \phi^2(A_\nu dx^\nu + dx^5)^2$$

3.10.2   Field equations from the Kaluza hypothesis

The field equations of the 5-dimensional theory were never adequately provided by Kaluza or Klein, mainly regarding the scalar field. The full Kaluza field equations are generally attributed to Thiry,[7] who most famously obtained vacuum field equations, although Kaluza [2] originally provided a stress-energy tensor for his theory and Thiry included a stress-energy tensor in his thesis. But as described by Gonner,[5] several independent groups worked on the field equations in the 1940s and earlier. Thiry is perhaps best known only because an English translation was provided by Applequist, Chodos, & Freund in their review book.[19] Applequist et al. also provided an English translation of Kaluza's paper. There are no English translations of the Jordan papers.[9][10][12]

To obtain the 5D field equations, the 5D connections $\tilde{\Gamma}^a_{bc}$ are calculated from the 5D metric $\tilde{g}_{ab}$, and the 5D Ricci tensor $\tilde{R}_{ab}$ is calculated from the 5D connections.

The classic results of Thiry and other authors presume the cylinder condition:

$$\frac{\partial \tilde{g}_{ab}}{\partial x^5} = 0$$

Without this assumption, the field equations become much more complex, providing many more degrees of freedom that can be identified with various new fields. Paul Wesson and colleagues have pursued relaxation of the cylinder condition to gain extra terms that can be identified with the matter fields,[20] for which Kaluza [2] otherwise inserted a stress-energy tensor by hand.

It has been an objection to the original Kaluza hypothesis to invoke the fifth dimension only to negate its dynamics. But Thiry argued [5] that the interpretation of the Lorentz force law in terms of a 5-dimensional geodesic mitigates strongly for a fifth dimension irrespective of the cylinder condition. Most authors have therefore employed the cylinder condition in deriving the field equations. Furthermore, vacuum equations are typically assumed for which

$$\tilde{R}_{ab} = 0$$

where

$$\tilde{R}_{ab} \equiv \partial_c \tilde{\Gamma}^c_{ab} - \partial_b \tilde{\Gamma}^c_{ca} + \tilde{\Gamma}^c_{cd}\tilde{\Gamma}^d_{ab} - \tilde{\Gamma}^c_{bd}\tilde{\Gamma}^d_{ac}$$

and

$$\tilde{\Gamma}^a_{bc} \equiv \frac{1}{2}\tilde{g}^{ad}(\partial_b \tilde{g}_{dc} + \partial_c \tilde{g}_{db} - \partial_d \tilde{g}_{bc})$$

The vacuum field equations obtained in this way by Thiry [7] and Jordan's group [9][10][12] are as follows.

The field equation for ϕ is obtained from

$$\tilde{R}_{55} = 0 \Rightarrow \Box\phi = \frac{1}{4}\phi^3 F^{\alpha\beta}F_{\alpha\beta}$$

where $F_{\alpha\beta} \equiv \partial_\alpha A_\beta - \partial_\beta A_\alpha$, where $\Box \equiv g^{\mu\nu}\nabla_\mu\nabla_\nu$, and where ∇_μ is a standard, 4D covariant derivative. It shows that the electromagnetic field is a source for the scalar field. Note that the scalar field cannot be set to a constant without constraining the electromagnetic field. The earlier treatments by Kaluza and Klein did not have an adequate description of the scalar field, and did not realize the implied constraint on the electromagnetic field by assuming the scalar field to be constant.

The field equation for A^ν is obtained from

$$\tilde{R}_{5\alpha} = 0 = \frac{1}{2}g^{\beta\mu}\nabla_\mu(\phi^3 F_{\alpha\beta})$$

It has the form of the vacuum Maxwell equations if the scalar field is constant.

The field equation for the 4D Ricci tensor $R_{\mu\nu}$ is obtained from

$$\tilde{R}_{\mu\nu} - \frac{1}{2}\tilde{g}_{\mu\nu}\tilde{R} = 0 \Rightarrow R_{\mu\nu} - \frac{1}{2}g_{\mu\nu}R = \frac{1}{2}\phi^2 \quad \times$$

$$\left(g^{\alpha\beta}F_{\mu\alpha}F_{\nu\beta} - \frac{1}{4}g_{\mu\nu}F_{\alpha\beta}F^{\alpha\beta}\right) + \frac{1}{\phi}(\nabla_\mu\nabla_\nu\phi - g_{\mu\nu}\Box\phi)$$

where R is the standard 4D Ricci scalar.

This equation shows the remarkable result, called the "Kaluza miracle", that the precise form for the electromagnetic stress-energy tensor emerges from the 5D vacuum equations as a source in the 4D equations: field from the vacuum. This relation allows the definitive identification of A^μ with the electromagnetic vector potential. Therefore, the field needs to be rescaled with a conversion constant k such that $A^\mu \to kA^\mu$.

The relation above shows that we must have

$$\frac{k^2}{2} = \frac{8\pi G}{c^4}\frac{1}{\mu_0} = \frac{2G}{c^2}4\pi\epsilon_0$$

where G is the gravitational constant and μ_0 is the permeability of free space. In the Kaluza theory, the gravitational constant can be understood as an electromagnetic coupling constant in the metric. There is also a stress-energy tensor for the scalar field. The scalar field behaves like a variable gravitational constant, in terms of modulating the coupling of electromagnetic stress energy to spacetime curvature. The sign of ϕ^2 in the metric is fixed by correspondence with 4D theory so that electromagnetic energy densities are positive. This turns out to imply that the 5th coordinate is spacelike in its signature in the metric.

In the presence of matter, the 5D vacuum condition can not be assumed. Indeed, Kaluza did not assume it. The full field equations require evaluation of the 5D Einstein tensor

$$\tilde{G}_{ab} \equiv \tilde{R}_{ab} - \frac{1}{2}\tilde{g}_{ab}\tilde{R}$$

as seen in the recovery of the electromagnetic stress-energy tensor above. The 5D curvature tensors are complex, and most English-language reviews contain errors in either $\tilde{G}_{ab}$ or $\tilde{R}_{ab}$, as does the English translation of.[7] See [18] for a complete set of 5D curvature tensors under the cylinder condition, evaluated using tensor algebra software.

3.10.3 Equations of motion from the Kaluza hypothesis

The equations of motion are obtained from the five-dimensional geodesic hypothesis [2] in terms of a 5-velocity $\tilde{U}^a \equiv dx^a/ds$:

$$\tilde{U}^b\tilde{\nabla}_b\tilde{U}^a = \frac{d\tilde{U}^a}{ds} + \tilde{\Gamma}^a_{bc}\tilde{U}^b\tilde{U}^c = 0$$

This equation can be recast in several ways, and it has been studied in various forms by authors including Kaluza,[2] Pauli,[21] Gross & Perry,[22] Gegenberg & Kunstatter,[23] and Wesson & Ponce de Leon,[24] but it is instructive to convert it back to the usual 4-dimensional length element $c^2d\tau^2 \equiv g_{\mu\nu}dx^\mu dx^\nu$, which is related to the 5-dimensional length element ds as given above:

$$ds^2 = c^2d\tau^2 + \phi^2(kA_\nu dx^\nu + dx^5)^2$$

Then the 5D geodesic equation can be written [25] for the spacetime components of the 4velocity, $U^\nu \equiv dx^\nu/d\tau$: $\frac{dU^\nu}{d\tau} + \tilde{\Gamma}^\mu_{\alpha\beta}U^\alpha U^\beta + 2\tilde{\Gamma}^\mu_{5\alpha}U^\alpha U^5 + \tilde{\Gamma}^\mu_{55}(U^5)^2 + U^\mu\frac{d}{d\tau}\ln\left(\frac{cd\tau}{ds}\right) = 0$

The term quadratic in U^ν provides the 4D geodesic equation plus some electromagnetic terms:

$$\tilde{\Gamma}^\mu_{\alpha\beta} = \Gamma^\mu_{\alpha\beta} + \frac{1}{2}g^{\mu\nu}k^2\phi^2(A_\alpha F_{\beta\nu} + A_\beta F_{\alpha\nu} + A_\alpha A_\beta\partial_\nu\ln\phi^2)$$

The term linear in U^ν provides the Lorentz force law:

$$\tilde{\Gamma}^\mu_{5\alpha} = \frac{1}{2}g^{\mu\nu}k\phi^2(F_{\alpha\nu} - A_\alpha\partial_\nu\ln\phi^2)$$

This is another expression of the "Kaluza miracle". The same hypothesis for the 5D metric that provides electromagnetic stress-energy in the Einstein equations, also provides the Lorentz force law in the equation of motions along

with the 4D geodesic equation. Yet correspondence with the Lorentz force law requires that we identify the component of 5-velocity along the 5th dimension with electric charge:

$$kU^5 = k\frac{dx^5}{d\tau} \rightarrow \frac{q}{mc}$$

where m is particle mass and q is particle electric charge. Thus, electric charge is understood as motion along the 5th dimension. The fact that the Lorentz force law could be understood as a geodesic in 5 dimensions was to Kaluza a primary motivation for considering the 5-dimensional hypothesis, even in the presence of the aesthetically-unpleasing cylinder condition.

Yet there is a problem: the term quadratic in U^5.

$$\tilde{\Gamma}^\mu_{55} = -\frac{1}{2}g^{\mu\alpha}\partial_\alpha\phi^2$$

If there is no gradient in the scalar field, the term quadratic in U^5 vanishes. But otherwise the expression above implies

$$U^5 \sim c\frac{q/m}{G^{1/2}}$$

For elementary particles, $U^5 > 10^{20}c$. The term quadratic in U^5 should dominate the equation, perhaps in contradiction to experience. This was the main shortfall of the 5-dimensional theory as Kaluza saw it,[2] and he gives it some discussion in his original article.

The equation of motion for U^5 is particularly simple under the cylinder condition. Start with the alternate form of the geodesic equation, written for the covariant 5-velocity:

$$\frac{d\tilde{U}_a}{ds} = \frac{1}{2}\tilde{U}^b\tilde{U}^c\frac{\partial\tilde{g}_{bc}}{\partial x^a}$$

This means that under the cylinder condition, $\tilde{U}_5$ is a constant of the 5-dimensional motion:

$$\tilde{U}_5 = \tilde{g}_{5a}\tilde{U}^a = \phi^2\frac{cd\tau}{ds}(kA_\nu U^\nu + U^5) = \text{constant}$$

3.10.4 Kaluza's hypothesis for the matter stress-energy tensor

Kaluza [2] proposed a 5D matter stress tensor $\tilde{T}^{ab}_M$ of the form

$$\tilde{T}_M^{ab} = \rho \frac{dx^a}{ds} \frac{dx^b}{ds}$$

where ρ is a density and the length element ds is as defined above.

Then, the spacetime component gives a typical "dust" stress energy tensor:

$$\tilde{T}_M^{\mu\nu} = \rho \frac{dx^\mu}{ds} \frac{dx^\nu}{ds}$$

The mixed component provides a 4-current source for the Maxwell equations:

$$\tilde{T}_M^{5\mu} = \rho \frac{dx^\mu}{ds} \frac{dx^5}{ds} = \rho U^\mu \frac{q}{kmc}$$

Just as the five-dimensional metric comprises the 4-D metric framed by the electromagnetic vector potential, the 5-dimensional stress-energy tensor comprises the 4-D stress-energy tensor framed by the vector 4-current.

3.10.5   Quantum interpretation of Klein

Kaluza's original hypothesis was purely classical and extended discoveries of general relativity. By the time of Klein's contribution, the discoveries of Heisenberg, Schroedinger, and de Broglie were receiving a lot of attention. Klein's *Nature* paper [4] suggested that the fifth dimension is closed and periodic, and that the identification of electric charge with motion in the fifth dimension be interpreted as standing waves of wavelength λ^5 , much like the electrons around a nucleus in the Bohr model of the atom. The quantization of electric charge could then be nicely understood in terms of integer multiples of fifth-dimensional momentum. Combining the previous Kaluza result for U^5 in terms of electric charge, and a de Broglie relation for momentum $p^5 = h/\lambda^5$, Klein [4] obtained an expression for the 0th mode of such waves:

$$mU^5 = \frac{cq}{G^{1/2}} = \frac{h}{\lambda^5} \rightarrow \lambda^5 \sim \frac{hG^{1/2}}{cq}$$

where h is the Planck constant. Klein found $\lambda^5 \sim 10^{-30}$ cm, and thereby an explanation for the cylinder condition in this small value.

Klein's *Zeitschrift für Physik* paper of the same year,[3] gave a more-detailed treatment that explicitly invoked the techniques of Schroedinger and de Broglie. It recapitulated much of the classical theory of Kaluza described above, and

then departed into Klein's quantum interpretation. Klein solved a Schroedinger-like wave equation using an expansion in terms of fifth-dimensional waves resonating in the closed, compact fifth dimension.

3.10.6   Quantum field theory interpretation

3.10.7   Group theory interpretation

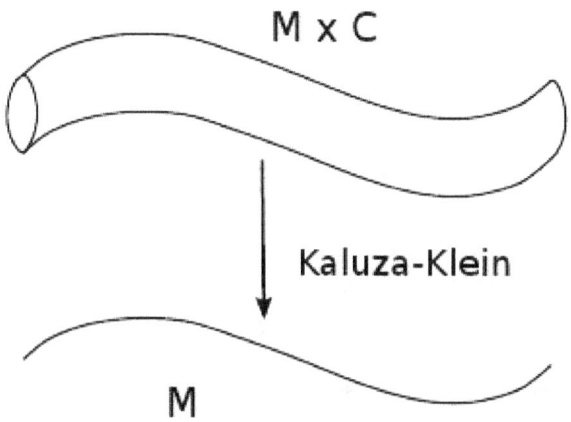

The space M × C *is compactified over the compact set* C, *and after Kaluza–Klein decomposition one has an effective field theory over* M.

A splitting of five-dimensional spacetime into the Einstein equations and Maxwell equations in four dimensions was first discovered by Gunnar Nordström in 1914, in the context of his theory of gravity, but subsequently forgotten. Kaluza published his derivation in 1921 as an attempt to unify electromagnetism with Einstein's general relativity.

In 1926, Oskar Klein proposed that the fourth spatial dimension is curled up in a circle of a very small radius, so that a particle moving a short distance along that axis would return to where it began. The distance a particle can travel before reaching its initial position is said to be the size of the dimension. This extra dimension is a compact set, and the phenomenon of having a space-time with compact dimensions is referred to as compactification.

In modern geometry, the extra fifth dimension can be understood to be the circle group U(1), as electromagnetism can essentially be formulated as a gauge theory on a fiber bundle, the circle bundle, with gauge group U(1). In Kaluza–Klein theory this group suggests that gauge symmetry is the symmetry of circular compact dimensions. Once this geometrical interpretation is understood, it is relatively straightforward to replace $U(1)$ by a general Lie group. Such generalizations are often called Yang–Mills theories. If a distinction is drawn, then it is that Yang–Mills theories occur on a flat space-time, whereas Kaluza–Klein treats the more general case of curved spacetime. The base space of Kaluza–Klein theory need not be four-dimensional space-time; it can be any (pseudo-)Riemannian manifold, or even a supersymmetric manifold or orbifold or even a noncommutative space.

The construction can be outlined, roughly, as follows.[26] One starts by considering a principle fiber bundle P with gauge group G over a manifold M. Given a connection on the bundle, and a metric on the base manifold, and a gauge invariant metric on the tangent of each fiber, one can construct a bundle metric defined on the entire bundle. Computing the scalar curvature of this bundle metric, one finds that it is constant on each fiber: this is the "Kaluza miracle". One did not have to explicitly impose a cylinder condition, or to compactify: by assumption, the gauge group is already compact. Next, one takes this scalar curvature as the Lagrangian density, and, from this, constructs the Einstein–Hilbert action for the bundle, as a whole. The equations of motion, the Euler–Lagrange equations, can be then obtained by considering where the action is stationary with respect to variations of either the metric on the base manifold, or of the gauge connection. Variations with respect to the base metric gives the Einstein field equations on the base manifold, with the energy-momentum tensor given by the curvature (field strength) of the gauge connection. On the flip side, the action is stationary against variations of the gauge connection precisely when the gauge connection solves the Yang-Mills equations. Thus, by applying a single idea: the principle of least action, to a single quantity: the scalar curvature on the bundle (as a whole), one obtains simultaneously all of the needed field equations, for both the space-time and the gauge field.

As an approach to the unification of the forces, it is straightforward to apply the Kaluza–Klein theory in an attempt to unify gravity with the strong and electroweak forces by using the symmetry group of the Standard Model, SU(3) × SU(2) × U(1). However, an attempt to convert this interesting geometrical construction into a bona-fide model of reality founders on a number of issues, including the fact that the fermions must be introduced in an artificial way (in nonsupersymmetric models). Nonetheless, KK remains an important touchstone in theoretical physics and is often embedded in more sophisticated theories. It is studied in its own right as an object of geometric interest in K-theory.

Even in the absence of a completely satisfying theoretical physics framework, the idea of exploring extra, compactified, dimensions is of considerable interest in the experimental physics and astrophysics communities. A variety of predictions, with real experimental consequences, can be made (in the case of large extra dimensions and warped models). For example, on the simplest of principles, one might expect to have standing waves in the extra compactified dimension(s). If a spatial extra dimension is of radius R, the invariant mass of such standing waves would be $Mn = nh/Rc$ with n an integer, h being Planck's constant and c the speed of light. This set of possible mass values is often called the **Kaluza–Klein tower**. Similarly, in Thermal quantum field theory a compactification of the euclidean time dimension leads to the Matsubara frequencies and thus to a discretized thermal energy spectrum.

However, Klein's approach to a quantum theory is flawed and, for example, leads to a calculated electron mass of $3 \times (10^{30})$ MeV instead of the measured value 0.511 MeV.

Examples of experimental pursuits include work by the CDF collaboration, which has re-analyzed particle collider data for the signature of effects associated with large extra dimensions/warped models.

Brandenberger and Vafa have speculated that in the early universe, cosmic inflation causes three of the space dimensions to expand to cosmological size while the remaining dimensions of space remained microscopic.

3.10.8   Space–time–matter theory

One particular variant of Kaluza–Klein theory is **space-time–matter theory** or **induced matter theory**, chiefly promulgated by Paul Wesson and other members of the so-called Space–Time–Matter Consortium.[27] In this version of the theory, it is noted that solutions to the equation

$$\tilde{R}_{ab} = 0$$

may be re-expressed so that in four dimensions, these solutions satisfy Einstein's equations

$$G_{\mu\nu} = 8\pi T_{\mu\nu}$$

with the precise form of the $T\mu\nu$ following from the Ricci-flat condition on the five-dimensional space. In other words, the cylinder condition of the previous development is dropped, and the stress–energy now comes from the derivatives of the 5D metric with respect to the fifth coordinate. Because the energy–momentum tensor is normally understood to be due to concentrations of matter in four-dimensional space, the above result is interpreted as saying that four-dimensional matter is induced from geometry in five-dimensional space.

In particular, the soliton solutions of $\tilde{R}_{ab} = 0$ can be shown to contain the Friedmann–Lemaître–Robertson–Walker metric in both radiation-dominated (early universe) and matter-dominated (later universe) forms. The general equations can be shown to be sufficiently consistent with classical tests of general relativity to be acceptable on physical principles, while still leaving considerable freedom to also provide interesting cosmological models.

3.10.9 Geometric interpretation

The Kaluza–Klein theory has a particularly elegant presentation in terms of geometry. In a certain sense, it looks just like ordinary gravity in free space, except that it is phrased in five dimensions instead of four.

Einstein equations

The equations governing ordinary gravity in free space can be obtained from an action, by applying the variational principle to a certain action. Let M be a (pseudo-)Riemannian manifold, which may be taken as the spacetime of general relativity. If g is the metric on this manifold, one defines the action $S(g)$ as

$$S(g) = \int_M R(g)\mathrm{vol}(g)$$

where $R(g)$ is the scalar curvature and $\mathrm{vol}(g)$ is the volume element. By applying the variational principle to the action

$$\frac{\delta S(g)}{\delta g} = 0$$

one obtains precisely the Einstein equations for free space:

$$R_{ij} - \frac{1}{2}g_{ij}R = 0$$

Here, R_{ij} is the Ricci tensor.

Maxwell equations

By contrast, the Maxwell equations describing electromagnetism can be understood to be the Hodge equations of a principal U(1)-bundle or circle bundle π: $P \to M$ with fiber U(1). That is, the electromagnetic field F is a harmonic 2-form in the space $\Omega^2(M)$ of differentiable 2-forms on the manifold M. In the absence of charges and currents, the free-field Maxwell equations are

$$\mathrm{d}F = 0 \text{ and } \mathrm{d}^*F = 0.$$

where * is the Hodge star.

Kaluza–Klein geometry

To build the Kaluza–Klein theory, one picks an invariant metric on the circle $\mathbf{S}^1$ that is the fiber of the U(1)-bundle of electromagnetism. In this discussion, an *invariant metric* is simply one that is invariant under rotations of the circle. Suppose this metric gives the circle a total length of Λ. One then considers metrics $\widehat{g}$ on the bundle P that are consistent with both the fiber metric, and the metric on the underlying manifold M. The consistency conditions are:

- The projection of $\widehat{g}$ to the vertical subspace $\mathrm{Vert}_p P \subset T_p P$ needs to agree with metric on the fiber over a point in the manifold M.

- The projection of $\widehat{g}$ to the horizontal subspace $\mathrm{Hor}_p P \subset T_p P$ of the tangent space at point $p \in P$ must be isomorphic to the metric g on M at $\pi(p)$.

The Kaluza–Klein action for such a metric is given by

$$S(\widehat{g}) = \int_P R(\widehat{g})\,\mathrm{vol}(\widehat{g})$$

The scalar curvature, written in components, then expands to

$$R(\widehat{g}) = \pi^* \left(R(g) - \frac{\Lambda^2}{2}|F|^2 \right)$$

where π^* is the pullback of the fiber bundle projection π: $P \to M$. The connection A on the fiber bundle is related to the electromagnetic field strength as

$$\pi^* F = \mathrm{d}A$$

That there always exists such a connection, even for fiber bundles of arbitrarily complex topology, is a result from homology and specifically, K-theory. Applying Fubini's theorem and integrating on the fiber, one gets

$$S(\widehat{g}) = \Lambda \int_M \left(R(g) - \frac{1}{\Lambda^2}|F|^2 \right)\,\mathrm{vol}(g)$$

Varying the action with respect to the component A, one regains the Maxwell equations. Applying the variational principle to the base metric g, one gets the Einstein equations

$$R_{ij} - \frac{1}{2}g_{ij}R = \frac{1}{\Lambda^2}T_{ij}$$

with the stress–energy tensor being given by

$$T^{ij} = F^{ik}F^{jl}g_{kl} - \frac{1}{4}g^{ij}|F|^2.$$

sometimes called the **Maxwell stress tensor**.

The original theory identifies Λ with the fiber metric g_{55}, and allows Λ to vary from fiber to fiber. In this case, the coupling between gravity and the electromagnetic field is not constant, but has its own dynamical field, the radion.

Generalizations

In the above, the size of the loop Λ acts as a coupling constant between the gravitational field and the electromagnetic field. If the base manifold is four-dimensional, the Kaluza–Klein manifold P is five-dimensional. The fifth dimension is a compact space, and is called the **compact dimension**. The technique of introducing compact dimensions to obtain a higher-dimensional manifold is referred to as compactification. Compactification does not produce group actions on chiral fermions except in very specific cases: the dimension of the total space must be 2 mod 8 and the G-index of the Dirac operator of the compact space must be nonzero.[28]

The above development generalizes in a more-or-less straightforward fashion to general principal G-bundles for some arbitrary Lie group G taking the place of U(1). In such a case, the theory is often referred to as a Yang–Mills theory, and is sometimes taken to be synonymous. If the underlying manifold is supersymmetric, the resulting theory is a super-symmetric Yang–Mills theory.

3.10.10   Empirical tests

Up to now, no experimental or observational signs of extra dimensions have been officially reported. Many theoretical search techniques for detecting Kaluza–Klein resonances have been proposed using the mass couplings of such resonances with the top quark, however until the Large Hadron Collider (LHC) reaches full operational power observation of such resonances are unlikely. An analysis of results from the LHC in December 2010 severely constrains theories with large extra dimensions.[29]

The observation of a Higgs-like boson at the LHC puts a brand new empirical test in the search for Kaluza–Klein resonances and supersymmetric particles. The loop Feynman diagrams that exist in the Higgs interactions allow any particle with electric charge and mass to run in such a loop. Standard Model particles besides the top quark and W boson do not make big contributions to the cross-section observed in

the H $\to \gamma\gamma$ decay, but if there are new particles beyond the Standard Model, they could potentially change the ratio of the predicted Standard Model H $\to \gamma\gamma$ cross-section to the experimentally observed cross-section. Hence a measurement of any dramatic change to the H $\to \gamma\gamma$ cross section predicted by the Standard Model is crucial in probing the physics beyond it.

3.10.11   See also

- Classical theories of gravitation

- DGP model

- Quantum gravity

- Randall–Sundrum model

- String theory

- Supergravity

- Superstring theory

3.10.12   Notes

[1] Pais, Abraham (1982). *Subtle is the Lord ...: The Science and the Life of Albert Einstein*. Oxford: Oxford University Press. pp. 329–330.

[2] Kaluza, Theodor (1921). "Zum Unitätsproblem in der Physik". *Sitzungsber. Preuss. Akad. Wiss. Berlin. (Math. Phys.)*: 966–972.

[3] Klein, Oskar (1926). "Quantentheorie und fünfdimensionale Relativitätstheorie". *Zeitschrift für Physik A* **37** (12): 895–906. Bibcode:1926ZPhy...37..895K. doi:10.1007/BF01397481.

[4] Klein, Oskar (1926). "The Atomicity of Electricity as a Quantum Theory Law". *Nature* **118**: 516. Bibcode:1926Natur.118..516K. doi:10.1038/118516a0.

[5] Goenner, H. (2012). "Some remarks on the genesis of scalar-tensor theories". *General Relativity and Gravitation* **44**: 2077–2097. arXiv:1204.3455. Bibcode:2012GReGr..44.2077G. doi:10.1007/s10714-012-1378-8.

[6] Lichnerowicz, A.; Thiry, M.Y. (1947). *Compt. Rend. Acad. Sci. Paris* **224**: 529–531. Missing or empty |title= (help)

[7] Thiry, M.Y. (1948). *Compt. Rend. Acad. Sci. Paris* **226**: 216–218. Missing or empty |title= (help)

[8] Thiry, M.Y. (1948). *Compt. Rend. Acad. Sci. Paris* **226**: 1881–1882. Missing or empty |title= (help)

[9] Jordan, P. (1946). *Naturwiss.* **11**: 250–251. Missing or empty |title= (help)

[10] Jordan, P.; Müller, C. (1947). *Z. Naturforsch.* **2a**: 1–2. Bibcode:1947ZNatA...2....1J. doi:10.1515/zna-1947-0102. Missing or empty |title= (help)

[11] Ludwig, G. (1947). *Z. Naturforsch.* **2a**: 3–5. Bibcode:1947ZNatA...2....3L. doi:10.1515/zna-1947-0103. Missing or empty |title= (help)

[12] Jordan, P. (1948). *Astron. Nachr.* **276**: 193–208. Bibcode:1948AN....276..193J. doi:10.1002/asna.19482760502. Missing or empty |title= (help)

[13] Ludwig, G.; Müller, C. (1948). *Annalen der Physik* **2** (6): 76–84. Missing or empty |title= (help)

[14] Scherrer, W. (1941). *Helv. Phys. Acta* **14** (2): 130. Missing or empty |title= (help)

[15] Scherrer, W. (1949). *Helv. Phys. Acta* **22**: 537–551. Missing or empty |title= (help)

[16] Scherrer, W. (1949). *Helv. Phys. Acta* **23**: 547–555. Missing or empty |title= (help)

[17] Brans, C. H.; Dicke, R. H. (November 1, 1961). "Mach's Principle and a Relativistic Theory of Gravitation". *Physical Review* **124** (3): 925–935. Bibcode:1961PhRv..124..925B. doi:10.1103/PhysRev.124.925.

[18] Williams, L.L. (2015). "Field Equations and Lagrangian for the Kaluza Metric Evaluated with Tensor Algebra Software". *Journal of Gravitation* **2015**: 901870. doi:10.1155/2015/901870.

[19] Appelquist, Thomas; Chodos, Alan; Freund, Peter G. O. (1987). *Modern Kaluza–Klein Theories*. Menlo Park, Cal.: Addison–Wesley. ISBN 0-201-09829-6.

[20] Wesson, Paul S. (1999). *Space-Time-Matter, Modern Kaluza-Klein Theory*. Singapore: World Scientific. ISBN 981-02-3588-7.

[21] Pauli, Wolfgang (1958). *Theory of Relativity* (translated by George Field ed.). New York: Pergamon Press. pp. Supplement 23.

[22] Gross, D.J.; Perry, M.J. (1983). "Magnetic monopoles in Kaluza-Klein theories". *Nucl. Phys.* **B 226**: 29–48. Bibcode:1983NuPhB.226...29G. doi:10.1016/0550-3213(83)90462-5.

[23] Gegenberg, J.; Kunstatter, G. (1984). *Phys. Lett.* **106A**: 410. Missing or empty |title= (help)

[24] Wesson, P.S.; Ponce de Leon, J. (1995). *Astronomy and Astrophysics* **294**: 1. Bibcode:1995A&A...294....1W. Missing or empty |title= (help)

[25] Williams, L.L. (2012). "Physics of the Electromagnetic Control of Spacetime and Gravity". *Proceedings of 48th AIAA Joint Propulsion Conference*. AIAA 2012-3916. doi:10.2514/6.2012-3916.

[26] David Bleecker, "Gauge Theory and Variational Principles" (1982) D. Reidel Publishing *(See chapter 9)*

[27] 5Dstm.org

[28] L. Castellani et al., Supergravity and superstrings, Vol 2, chapter V.11

[29] CMS Collaboration, "Search for Microscopic Black Hole Signatures at the Large Hadron Collider", http://arxiv.org/abs/1012.3375

3.10.13 References

● Nordström, Gunnar (1914). "Über die Möglichkeit, das elektromagnetische Feld und das Gravitationsfeld zu vereinigen". *Physikalische Zeitschrift* **15**: 504–506. OCLC 1762351.

● Kaluza, Theodor (1921). "Zum Unitätsproblem in der Physik". *Sitzungsber. Preuss. Akad. Wiss. Berlin. (Math. Phys.)*: 966–972. https://archive.org/details/sitzungsberichte1921preussi

● Klein, Oskar (1926). "Quantentheorie und fünfdimensionale Relativitätstheorie". *Zeitschrift für Physik A* **37** (12): 895–906. Bibcode:1926ZPhy...37..895K. doi:10.1007/BF01397481.

● Witten, Edward (1981). "Search for a realistic Kaluza–Klein theory". *Nuclear Physics B* **186** (3): 412–428. Bibcode:1981NuPhB.186..412W. doi:10.1016/0550-3213(81)90021-3.

● Appelquist, Thomas; Chodos, Alan; Freund, Peter G. O. (1987). *Modern Kaluza–Klein Theories*. Menlo Park, Cal.: Addison–Wesley. ISBN 0-201-09829-6. *(Includes reprints of the above articles as well as those of other important papers relating to Kaluza–Klein theory.)*

● Brandenberger, Robert; Vafa, Cumrun (1989). "Superstrings in the early universe". *Nuclear Physics B* **316** (2): 391–410. Bibcode:1989NuPhB.316..391B. doi:10.1016/0550-3213(89)90037-0.

● Duff, M. J. (1994). "Kaluza–Klein Theory in Perspective". In Lindström, Ulf (ed.). *Proceedings of the Symposium 'The Oskar Klein Centenary'*. Singapore: World Scientific. pp. 22–35. ISBN 981-02-2332-3.

● Overduin, J. M.; Wesson, P. S. (1997). "Kaluza–Klein Gravity". *Physics Reports* **283** (5): 303–378. arXiv:gr-qc/9805018. Bibcode:1997PhR...283..303O. doi:10.1016/S0370-1573(96)00046-4.

● Wesson, Paul S. (1999). *Space-Time-Matter, Modern Kaluza-Klein Theory*. Singapore: World Scientific. ISBN 981-02-3588-7.

- Wesson, Paul S. (2006). *Five-Dimensional Physics: Classical and Quantum Consequences of Kaluza-Klein Cosmology*. Singapore: World Scientific. ISBN 981-256-661-9.

- Coquereaux, R.; Esposito-Farese, G. (1990). "The Theory of Kaluza-Klein-Jordan-Thiry revisited". *Annales de l'I.H.P., Section A* **52**: 113–150.

3.10.14   Further reading

- Grøn, Øyvind; Hervik, Sigbjørn (2007). *Einstein's General Theory of Relativity*. New York: Springer. ISBN 978-0-387-69199-2.

- Kaku, Michio and Robert O'Keefe. *Hyperspace: A Scientific Odyssey Through Parallel Universes, Time Warps, and the Tenth Dimension*. New York: Oxford University Press, 1994. ISBN 0-19-286189-1

- The CDF Collaboration, *Search for Extra Dimensions using Missing Energy at CDF*. (2004) *(A simplified presentation of the search made for extra dimensions at the Collider Detector at Fermilab (CDF) particle physics facility.)*

- John M. Pierre, *SUPERSTRINGS! Extra Dimensions*, (2003).

- TeV scale gravity, mirror universe, and ... dinosaurs Article from Acta Physica Polonica B by Z.K. Silagadze.

- Chris Pope, *Lectures on Kaluza–Klein Theory*.

- Edward Witten (2014). "A Note On Einstein, Bergmann, and the Fifth Dimension". arXiv:1401.8048: pdf

3.11   Grand Unified Theory

For the album, see Grand Unification (album).

A **Grand Unified Theory** (**GUT**) is a model in particle physics in which at high energy, the three gauge interactions of the Standard Model which define the electromagnetic, weak, and strong interactions or forces, are merged into one single force. This unified interaction is characterized by one larger gauge symmetry and thus several force carriers, but one unified coupling constant. If Grand Unification is realized in nature, there is the possibility of a grand unification epoch in the early universe in which the fundamental forces are not yet distinct.

Models that do not unify all interactions using one simple group as the gauge symmetry, but do so using semisimple groups, can exhibit similar properties and are sometimes referred to as Grand Unified Theories as well.

Unifying gravity with the other three interactions would provide a theory of everything (TOE), rather than a GUT. Nevertheless, GUTs are often seen as an intermediate step towards a TOE.

The novel particles predicted by GUT models are expected to have masses around the GUT scale—just a few orders of magnitude below the Planck scale—and so will be well beyond the reach of any foreseen particle collider experiments. Therefore, the particles predicted by GUT models will be unable to be observed directly and instead the effects of grand unification might be detected through indirect observations such as proton decay, electric dipole moments of elementary particles, or the properties of neutrinos.[1] Some GUTs such as the Pati-Salam model, predict the existence of magnetic monopoles.

As of 2012, all GUT models which aim to be completely realistic are quite complicated, even compared to the Standard Model, because they need to introduce additional fields and interactions, or even additional dimensions of space. The main reason for this complexity lies in the difficulty of reproducing the observed fermion masses and mixing angles. Due to this difficulty, and due to the lack of any observed effect of grand unification so far, there is no generally accepted GUT model.

3.11.1   History

Historically, the first true GUT which was based on the simple Lie group SU(5), was proposed by Howard Georgi and Sheldon Glashow in 1974.[2] The Georgi–Glashow model was preceded by the semisimple Lie algebra Pati–Salam model by Abdus Salam and Jogesh Pati,[3] who pioneered the idea to unify gauge interactions.

The acronym GUT was first coined in 1978 by CERN researchers John Ellis, Andrzej Buras, Mary K. Gaillard, and Dimitri Nanopoulos, however in the final version of their paper[4] they opted for the less anatomical *GUM* (Grand Unification Mass). Nanopoulos later that year was the first to use[5] the acronym in a paper.[6]

3.11.2   Motivation

The fact that the electric charges of electrons and protons seem to cancel each other exactly to extreme precision is essential for the existence of the macroscopic world as we know it, but this important property of elementary particles is not explained in the Standard Model of particle physics. While the description of strong and weak interactions within the Standard Model is based on gauge symmetries governed by the simple symmetry groups SU(3) and SU(2) which allow only discrete charges, the remaining component, the weak hypercharge interaction is described by an abelian symmetry U(1) which in principle allows for arbitrary charge assignments.[note 1] The observed charge quantization, namely the fact that all known elementary particles carry electric charges which appear to be exact multiples of ⅓ of the "elementary" charge, has led to the idea that hypercharge interactions and possibly the strong and weak interactions might be embedded in one Grand Unified interaction described by a single, larger simple symmetry group containing the Standard Model. This would automatically predict the quantized nature and values of all elementary particle charges. Since this also results in a prediction for the relative strengths of the fundamental interactions which we observe, in particular the weak mixing angle, Grand Unification ideally reduces the number of independent input parameters, but is also constrained by observations.

Grand Unification is reminiscent of the unification of electric and magnetic forces by Maxwell's theory of electromagnetism in the 19th century, but its physical implications and mathematical structure are qualitatively different.

3.11.3   Unification of matter particles

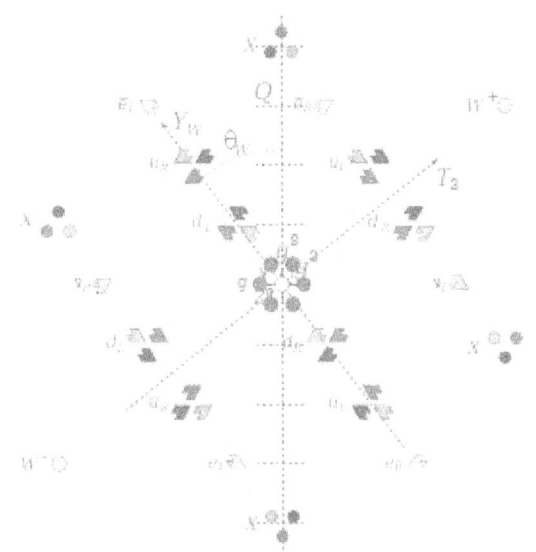

Schematic representation of fermions and bosons in SU(5) GUT showing **5 + 10** split in the multiplets. Neutral bosons (photon, Z-boson, and neutral gluons) are not shown but occupy the diagonal entries of the matrix in complex superpositions

For an elementary introduction to how Lie algebras are related to particle physics, see the article Particle physics and representation theory.

SU(5)

Main article: Georgi–Glashow model

SU(5) is the simplest GUT. The smallest simple Lie group

The pattern of weak isospins, weak hypercharges, and strong charges for particles in the SU(5) model, rotated by the predicted weak mixing angle, showing electric charge roughly along the vertical. In addition to Standard Model particles, the theory includes twelve colored X bosons, responsible for proton decay.

which contains the standard model, and upon which the first Grand Unified Theory was based, is

$$SU(5) \supset SU(3) \times SU(2) \times U(1)$$

Such group symmetries allow the reinterpretation of several known particles as different states of a single particle field. However, it is not obvious that the simplest possible choices for the extended "Grand Unified" symmetry should yield the correct inventory of elementary particles. The fact that all currently known (2009) matter particles fit nicely into three copies of the smallest group representations of SU(5) and immediately carry the correct observed charges, is one of the first and most important reasons why people believe that a Grand Unified Theory might actually be realized in nature.

The two smallest irreducible representations of SU(5) are **5** and **10**. In the standard assignment, the **5** contains the charge conjugates of the right-handed down-type quark color triplet and a left-handed lepton isospin doublet, while the **10** contains the six up-type quark components, the left-handed down-type quark color triplet, and the right-handed electron. This scheme has to be replicated for each of the three known generations of matter. It is notable that the theory is anomaly free with this matter content.

The hypothetical right-handed neutrinos are a singlet of SU(5), which makes that its mass is not forbidden by any symmetry so it doesn't need a spontaneous symmetry breaking which explains why its mass would be heavy. (see seesaw mechanism).

SO(10)

Main article: SO(10) (physics)

The next simple Lie group which contains the standard

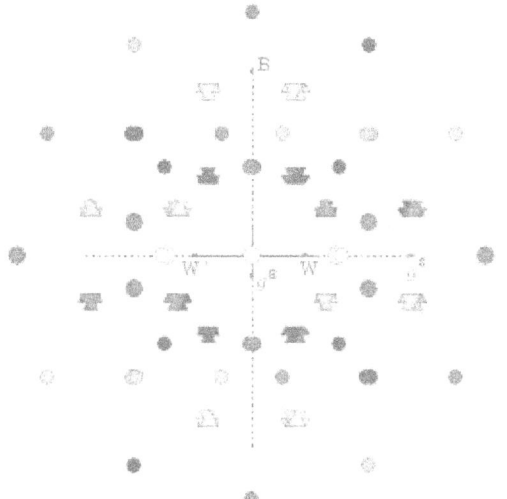

The pattern of weak isospin, W, weaker isospin, W', strong g3 and g8, and baryon minus lepton, B, charges for particles in the SO(10) Grand Unified Theory, rotated to show the embedding in E_6.

model is

$$SO(10) \supset SU(5) \supset SU(3) \times SU(2) \times U(1)$$

Here, the unification of matter is even more complete, since the irreducible spinor representation **16** contains both the **5** and **10** of SU(5) and a right-handed neutrino, and thus the complete particle content of one generation of the extended standard model with neutrino masses. This is already the largest simple group which achieves the unification of matter in a scheme involving only the already known matter particles (apart from the Higgs sector).

Since different standard model fermions are grouped together in larger representations, GUTs specifically predict relations among the fermion masses, such as between the electron and the down quark, the muon and the strange

quark, and the tau lepton and the bottom quark for SU(5) and SO(10). Some of these mass relations hold approximately, but most don't (see Georgi-Jarlskog mass relation).

The boson matrix for SO(10) is found by taking the 15 × 15 matrix from the **10 + 5** representation of SU(5) and adding an extra row and column for the right handed neutrino. The bosons are found by adding a partner to each of the 20 charged bosons (2 right-handed W bosons, 6 massive charged gluons and 12 X/Y type bosons) and adding an extra heavy neutral Z-boson to make 5 neutral bosons in total. The boson matrix will have a boson or its new partner in each row and column. These pairs combine to create the familiar 16D Dirac spinor matrices of SO(10).

SU(8)

Assuming 4 generations of fermions instead of 3 makes a total of **64** types of particles. These can be put into **64 = 8 + 56** representations of SU(8). This can be divided into SU(5) × SU(3)F × U(1) which is the SU(5) theory together with some heavy bosons which act on the generation number.

O(16)

Again assuming 4 generations of fermions, the **128** particles and anti-particles can be put into a single spinor representation of O(16).

Symplectic groups and quaternion representations

Symplectic gauge groups could also be considered. For example, Sp(8) (which is called Sp(4) in the article symplectic group) has a representation in terms of 4 × 4 quaternion unitary matrices which has a **16** dimensional real representation and so might be considered as a candidate for a gauge group. Sp(8) has 32 charged bosons and 4 neutral bosons. Its subgroups include SU(4) so can at least contain the gluons and photon of SU(3) × U(1). Although it's probably not possible to have weak bosons acting on chiral fermions in this representation. A quaternion representation of the fermions might be:

$$\begin{bmatrix} e + i\bar{e} + jv + k\overline{v} \\ u_r + i\overline{u_r} + jd_r + k\overline{d_r} \\ u_g + i\overline{u_g} + jd_g + k\overline{d_g} \\ u_b + i\overline{u_b} + jd_b + k\overline{d_b} \end{bmatrix}_L$$

A further complication with quaternion representations of fermions is that there are two types of multiplication: left multiplication and right multiplication which must be taken into account. It turns out that including left and right-handed 4 × 4 quaternion matrices is equivalent to including a single right-multiplication by a unit quaternion which adds an extra SU(2) and so has an extra neutral boson and two more charged bosons. Thus the group of left and right handed 4 × 4 quaternion matrcies is Sp(8) × SU(2) which does include the standard model bosons:

SU(4, H)L×HR = Sp(8)×SU(2) ⊃ SU(4)×SU(2)

⊃ SU(3)×SU(2)×U(1)

If ψ is a quaternion valued spinor, A_μ^{ab} is quaternion hermitian 4 × 4 matrix coming from Sp(8) and B_μ is a pure imaginary quaternion (both of which are 4-vector bosons) then the interaction term is:

$$\overline{\psi^a}\gamma_\mu \left(A_\mu^{ab}\psi^b + \psi^a B_\mu \right)$$

Octonion representations

It can be noted that a generation of 16 fermions can be put into the form of an octonion with each element of the octonion being an 8-vector. If the 3 generations are then put in a 3x3 hermitian matrix with certain additions for the diagonal elements then these matrices form an exceptional (grassman-) Jordan algebra, which has the symmetry group of one of the exceptional Lie groups (F_4, E_6, E_7 or E_8) depending on the details.

$$\psi = \begin{bmatrix} a & e & \mu \\ \overline{e} & b & \tau \\ \overline{\mu} & \overline{\tau} & c \end{bmatrix}$$

$[\psi_A, \psi_B] \subset J_3(O)$

Because they are fermions the anti-commutators of the Jordan algebra become commutators. It is known that E_6 has subgroup O(10) and so is big enough to include the Standard Model. An E_8 gauge group, for example, would have 8 neutral bosons, 120 charged bosons and 120 charged anti-bosons. To account for the 248 fermions in the lowest multiplet of E_8, these would either have to include anti-particles (and so have baryogenesis), have new undiscovered particles, or have gravity-like (spin connection) bosons affecting elements of the particles spin direction. Each of these poses theoretical problems.

Beyond Lie groups

Other structures have been suggested including Lie 3-algebras and Lie superalgebras. Neither of these fit with Yang–Mills theory. In particular Lie superalgebras would introduce bosons with the wrong statistics. Supersymmetry however does fit with Yang–Mills. For example, N=4 Super Yang Mills Theory requires an SU(N) gauge group.

3.11.4 Unification of forces and the role of supersymmetry

The unification of forces is possible due to the energy scale dependence of force coupling parameters in quantum field theory called renormalization group running, which allows parameters with vastly different values at usual energies to converge to a single value at a much higher energy scale.[7]

The renormalization group running of the three gauge couplings in the Standard Model has been found to nearly, but not quite, meet at the same point if the hypercharge is normalized so that it is consistent with SU(5) or SO(10) GUTs, which are precisely the GUT groups which lead to a simple fermion unification. This is a significant result, as other Lie groups lead to different normalizations. However, if the supersymmetric extension MSSM is used instead of the Standard Model, the match becomes much more accurate. In this case, the coupling constants of the strong and electroweak interactions meet at the grand unification energy, also known as the GUT scale:

$$\Lambda_{GUT} \approx 10^{16}\,\text{GeV}$$

It is commonly believed that this matching is unlikely to be a coincidence, and is often quoted as one of the main motivations to further investigate supersymmetric theories despite the fact that no supersymmetric partner particles have been experimentally observed (May 2015). Also, most model builders simply assume supersymmetry because it solves the hierarchy problem — i.e., it stabilizes the electroweak Higgs mass against radiative corrections.

3.11.5 Neutrino masses

Since Majorana masses of the right-handed neutrino are forbidden by SO(10) symmetry, SO(10) GUTs predict the Majorana masses of right-handed neutrinos to be close to the GUT scale where the symmetry is spontaneously broken in those models. In supersymmetric GUTs, this scale tends to be larger than would be desirable to obtain realistic masses of the light, mostly left-handed neutrinos (see neutrino oscillation) via the seesaw mechanism.

3.11.6   Proposed theories

Several such theories have been proposed, but none is currently universally accepted. An even more ambitious theory that includes *all* fundamental forces, including gravitation, is termed a theory of everything. Some common mainstream GUT models are:

Not quite GUTs:

Note: These models refer to Lie algebras not to Lie groups. The Lie group could be $[SU(4) \times SU(2) \times SU(2)]/Z_2$, just to take a random example.

The most promising candidate is SO(10). (Minimal) SO(10) does not contain any exotic fermions (i.e. additional fermions besides the Standard Model fermions and the right-handed neutrino), and it unifies each generation into a single irreducible representation. A number of other GUT models are based upon subgroups of SO(10). They are the minimal left-right model, SU(5), flipped SU(5) and the Pati–Salam model. The GUT group E_6 contains SO(10), but models based upon it are significantly more complicated. The primary reason for studying E_6 models comes from $E_8 \times E_8$ heterotic string theory.

GUT models generically predict the existence of topological defects such as monopoles, cosmic strings, domain walls, and others. But none have been observed. Their absence is known as the monopole problem in cosmology. Most GUT models also predict proton decay, although not the Pati–Salam model; current experiments still haven't detected proton decay. This experimental limit on the proton's lifetime pretty much rules out minimal SU(5).

- Proton Decay. These graphics refer to the X bosons and Higgs bosons.

- Dimension 6 proton decay mediated by the X boson in SU(5) GUT

- Dimension 6 proton decay mediated by the X boson in flipped SU(5) GUT

- Dimension 6 proton decay mediated by the triplet Higgs and the anti-triplet Higgs in SU(5) GUT

Some GUT theories like SU(5) and SO(10) suffer from what is called the doublet-triplet problem. These theories predict that for each electroweak Higgs doublet, there is a corresponding colored Higgs triplet field with a very small mass (many orders of magnitude smaller than the GUT scale here). In theory, unifying quarks with leptons, the Higgs doublet would also be unified with a Higgs triplet. Such triplets have not been observed. They would also

cause extremely rapid proton decay (far below current experimental limits) and prevent the gauge coupling strengths from running together in the renormalization group.

Most GUT models require a threefold replication of the matter fields. As such, they do not explain why there are three generations of fermions. Most GUT models also fail to explain the little hierarchy between the fermion masses for different generations.

3.11.7   Ingredients

A GUT model basically consists of a gauge group which is a compact Lie group, a connection form for that Lie group, a Yang–Mills action for that connection given by an invariant symmetric bilinear form over its Lie algebra (which is specified by a coupling constant for each factor), a Higgs sector consisting of a number of scalar fields taking on values within real/complex representations of the Lie group and chiral Weyl fermions taking on values within a complex rep of the Lie group. The Lie group contains the Standard Model group and the Higgs fields acquire VEVs leading to a spontaneous symmetry breaking to the Standard Model. The Weyl fermions represent matter.

3.11.8   Current status

As of 2012, there is still no hard evidence that nature is described by a Grand Unified Theory. The discovery of neutrino oscillations indicates that the Standard Model is incomplete and has led to renewed interest toward certain GUT such as SO(10). One of the few possible experimental tests of certain GUT is proton decay and also fermion masses. There are a few more special tests for supersymmetric GUT.

The gauge coupling strengths of QCD, the weak interaction and hypercharge seem to meet at a common length scale called the GUT scale and equal approximately to 10^{16} GeV, which is slightly suggestive. This interesting numerical observation is called the **gauge coupling unification**, and it works particularly well if one assumes the existence of superpartners of the Standard Model particles. Still it is possible to achieve the same by postulating, for instance, that ordinary (non supersymmetric) SO(10) models break with an intermediate gauge scale, such as the one of Pati–Salam group

3.11.9   See also

- Paradigm shift

- Classical unified field theories

- X and Y bosons

- B – L quantum number

3.11.10 Notes

[1] There are however certain constraints on the choice of particle charges from theoretical consistency, in particular anomaly cancellation.

3.11.11 References

[1] Ross, G. (1984). *Grand Unified Theories*. Westview Press. ISBN 978-0-8053-6968-7.

[2] Georgi, H.; Glashow, S.L. (1974). "Unity of All Elementary Particle Forces". *Physical Review Letters* **32**: 438–441. Bibcode:1974PhRvL..32..438G. doi:10.1103/PhysRevLett.32.438.

[3] Pati, J.; Salam, A. (1974). "Lepton Number as the Fourth Color". *Physical Review D* **10**: 275–289. Bibcode:1974PhRvD..10..275P. doi:10.1103/PhysRevD.10.275.

[4] Buras, A.J.; Ellis, J.; Gaillard, M.K.; Nanopoulos, D.V. (1978). "Aspects of the grand unification of strong, weak and electromagnetic interactions" (PDF). *Nuclear Physics B* **135** (1): 66–92. Bibcode:1978NuPhB.135...66B. doi:10.1016/0550-3213(78)90214-6. Retrieved 2011-03-21.

[5] Nanopoulos, D.V. (1979). "Protons Are Not Forever". *Orbis Scientiae* **1**: 91. Harvard Preprint HUTP-78/A062.

[6] Ellis, J. (2002). "Physics gets physical". *Nature* **415** (6875): 957. Bibcode:2002Natur.415..957E. doi:10.1038/415957b.

[7] Ross, G. (1984). *Grand Unified Theories*. Westview Press. ISBN 978-0-8053-6968-7.

3.11.12 Further reading

- Stephen Hawking, A Brief History of Time, includes a brief popular overview.

- Grand unification Paul Langacker Scholarpedia 7(10):11419. doi:10.4249/scholarpedia.11419

3.11.13 External links

- The Algebra of Grand Unified Theories

3.12   Theory of everything

This article is about the physical concept. For other uses, see Theory of everything (disambiguation).

A **theory of everything** (**ToE**) or **final theory**, **ultimate theory**, or **master theory** is a hypothetical single, all-encompassing, coherent theoretical framework of physics that fully explains and links together all physical aspects of the universe.[1]:6 Finding a ToE is one of the major unsolved problems in physics. Over the past few centuries, two theoretical frameworks have been developed that, as a whole, most closely resemble a ToE. These two theories upon which all modern physics rests are general relativity (GR) and quantum field theory (QFT). GR is a theoretical framework that only focuses on gravity for understanding the universe in regions of both large-scale and high-mass: stars, galaxies, clusters of galaxies, etc. On the other hand, QFT is a theoretical framework that only focuses on three non-gravitational forces for understanding the universe in regions of both small scale and low mass: sub-atomic particles, atoms, molecules, etc. QFT successfully implemented the Standard Model and unified the interactions (so-called Grand Unified Theory) between the three non-gravitational forces: weak, strong, and electromagnetic force.[2]:122

Through years of research, physicists have experimentally confirmed with tremendous accuracy virtually every prediction made by these two theories when in their appropriate domains of applicability. In accordance with their findings, scientists also learned that GR and QFT, as they are currently formulated, are mutually incompatible – they cannot both be right. Since the usual domains of applicability of GR and QFT are so different, most situations require that only one of the two theories be used.[3][4]:842–844 As it turns out, this incompatibility between GR and QFT is only an apparent issue in regions of extremely small-scale and high-mass, such as those that exist within a black hole or during the beginning stages of the universe (i.e., the moment immediately following the Big Bang). To resolve this conflict, a theoretical framework revealing a deeper underlying reality, unifying gravity with the other three interactions, must be discovered to harmoniously integrate the realms of GR and QFT into a seamless whole: a single theory that, in principle, is capable of describing all phenomena. In pursuit of this goal, quantum gravity has become an area of active research.

Over the past few decades, a single explanatory framework, called "string theory", has emerged that intends to be the ultimate theory of the universe. Some physicists believe that, at the beginning of the universe (up to 10^{-43} seconds after the Big Bang), the four fundamental forces were once a single fundamental force. According to string theory, every particle in the universe, at its most microscopic level (Planck length), consists of varying combinations of vibrating strings (or strands) with preferred patterns of vibration. String theory claims that it is through these specific oscillatory patterns of strings that a particle of unique mass and force charge is created (that is to say, the electron is a type of string that vibrates one way, while the up-quark is a type of string vibrating another way, and so forth).

Initially, the term *theory of everything* was used with an ironic connotation to refer to various overgeneralized theories. For example, a grandfather of Ijon Tichy — a character from a cycle of Stanisław Lem's science fiction stories of the 1960s — was known to work on the "General Theory of Everything". Physicist John Ellis[5] claims to have introduced the term into the technical literature in an article in *Nature* in 1986.[6] Over time, the term stuck in popularizations of theoretical physics research.

3.12.1   Historical antecedents

From ancient Greece to Einstein

Archimedes was possibly the first scientist known to have described nature with axioms (or principles) and then deduce new results from them.[7] He thus tried to describe "everything" starting from a few axioms. Any "theory of everything" is similarly expected to be based on axioms and to deduce all observable phenomena from them.[8]:340

The concept of 'atom', introduced by Democritus, unified all phenomena observed in nature as the motion of atoms. In ancient Greek times philosophers speculated that the apparent diversity of observed phenomena was due to a single type of interaction, namely the collisions of atoms. Following atomism, the mechanical philosophy of the 17th century posited that all forces could be ultimately reduced to contact forces between the atoms, then imagined as tiny solid particles.[9]:184[10]

In the late 17th century, Isaac Newton's description of the long-distance force of gravity implied that not all forces in nature result from things coming into contact. Newton's work in his *Mathematical Principles of Natural Philosophy* dealt with this in a further example of unification, in this case unifying Galileo's work on terrestrial gravity, Kepler's laws of planetary motion and the phenomenon of tides by explaining these apparent actions at a distance under one single law: the law of universal gravitation.[11]

In 1814, building on these results, Laplace famously suggested that a sufficiently powerful intellect could, if it knew the position and velocity of every particle at a given time, along with the laws of nature, calculate the position of any particle at any other time:[12]:ch 7

An intellect which at a certain moment would know all forces that set nature in motion, and all positions of all items of which nature is composed, if this intellect were also vast enough to submit these data to analysis, it would embrace in a single formula the movements of the greatest bodies of the universe and those of the tiniest atom; for such an intellect nothing would be uncertain and the future just like the past would be present before its eyes.
— *Essai philosophique sur les probabilités*, Introduction, 1814

Laplace thus envisaged a combination of gravitation and mechanics as a theory of everything. Modern quantum mechanics implies that uncertainty is inescapable, and thus that Laplace's vision has to be amended: a theory of everything must include gravitation and quantum mechanics.

In 1820, Hans Christian Ørsted discovered a connection between electricity and magnetism, triggering decades of work that culminated in 1865, in James Clerk Maxwell's theory of electromagnetism. During the 19th and early 20th centuries, it gradually became apparent that many common examples of forces – contact forces, elasticity, viscosity, friction, and pressure – result from electrical interactions between the smallest particles of matter.

In his experiments of 1849–50, Michael Faraday was the first to search for a unification of gravity with electricity and magnetism.[13] However, he found no connection.

In 1900, David Hilbert published a famous list of mathematical problems. In Hilbert's sixth problem, he challenged researchers to find an axiomatic basis to all of physics. In this problem he thus asked for what today would be called a theory of everything.[14]

In the late 1920s, the new quantum mechanics showed that the chemical bonds between atoms were examples of (quantum) electrical forces, justifying Dirac's boast that "the underlying physical laws necessary for the mathematical theory of a large part of physics and the whole of chemistry are thus completely known".[15]

After 1915, when Albert Einstein published the theory of gravity (general relativity), the search for a unified field theory combining gravity with electromagnetism began with a renewed interest. In Einstein's day, the strong and the weak forces had not yet been discovered, yet, he found the potential existence of two other distinct forces -gravity and electromagnetism- far more alluring. This launched his thirty-year voyage in search of the so-called "unified field theory" that he hoped would show that these two forces are really manifestations of one grand underlying principle. During these last few decades of his life, this quixotic quest isolated Einstein from the mainstream of physics. Understandably, the mainstream was instead far more excited about the newly emerging framework of quantum mechanics. Einstein wrote to a friend in the early 1940s, "I have become a lonely old chap who is mainly known because he doesn't wear socks and who is exhibited as a curiosity on special occasions." Prominent contributors were Gunnar Nordström, Hermann Weyl, Arthur Eddington, Theodor Kaluza, Oskar Klein, and most notably, Albert Einstein and his collaborators. Einstein intensely searched for, but ultimately failed to find, a unifying theory.[16]:ch 17 (But see:Einstein–Maxwell–Dirac equations.) More than a half a century later, Einstein's dream of discovering a unified theory has become the Holy Grail of modern physics.

Twentieth century and the nuclear interactions

In the twentieth century, the search for a unifying theory was interrupted by the discovery of the strong and weak nuclear forces (or interactions), which differ both from gravity and from electromagnetism. A further hurdle was the acceptance that in a ToE, quantum mechanics had to be incorporated from the start, rather than emerging as a consequence of a deterministic unified theory, as Einstein had hoped.

Gravity and electromagnetism could always peacefully coexist as entries in a list of classical forces, but for many years it seemed that gravity could not even be incorporated into the quantum framework, let alone unified with the other fundamental forces. For this reason, work on unification, for much of the twentieth century, focused on understanding the three "quantum" forces: electromagnetism and the weak and strong forces. The first two were combined in 1967–68 by Sheldon Glashow, Steven Weinberg, and Abdus Salam into the "electroweak" force.[17] Electroweak unification is a broken symmetry: the electromagnetic and weak forces appear distinct at low energies because the particles carrying the weak force, the W and Z bosons, have non-zero masses of 80.4 GeV/c^2 and 91.2 GeV/c^2, whereas the photon, which carries the electromagnetic force, is massless. At higher energies Ws and Zs can be created easily and the unified nature of the force becomes apparent.

While the strong and electroweak forces peacefully coexist in the Standard Model of particle physics, they remain distinct. So far, the quest for a theory of everything is thus unsuccessful on two points: neither a unification of the strong and electroweak forces – which Laplace would have called 'contact forces' – has been achieved, nor has a unification of these forces with gravitation been achieved.

3.12.2 Modern physics

Conventional sequence of theories

A Theory of Everything would unify all the fundamental interactions of nature: gravitation, strong interaction, weak interaction, and electromagnetism. Because the weak interaction can transform elementary particles from one kind into another, the ToE should also yield a deep understanding of the various different kinds of possible particles. The usual assumed path of theories is given in the following graph, where each unification step leads one level up:

In this graph, electroweak unification occurs at around 100 GeV, grand unification is predicted to occur at 10^{16} GeV, and unification of the GUT force with gravity is expected at the Planck energy, roughly 10^{19} GeV.

Several Grand Unified Theories (GUTs) have been proposed to unify electromagnetism and the weak and strong forces. Grand unification would imply the existence of an electronuclear force; it is expected to set in at energies of the order of 10^{16} GeV, far greater than could be reached by any possible Earth-based particle accelerator. Although the simplest GUTs have been experimentally ruled out, the general idea, especially when linked with supersymmetry, remains a favorite candidate in the theoretical physics community. Supersymmetric GUTs seem plausible not only for their theoretical "beauty", but because they naturally produce large quantities of dark matter, and because the inflationary force may be related to GUT physics (although it does not seem to form an inevitable part of the theory). Yet GUTs are clearly not the final answer; both the current standard model and all proposed GUTs are quantum field theories which require the problematic technique of renormalization to yield sensible answers. This is usually regarded as a sign that these are only effective field theories, omitting crucial phenomena relevant only at very high energies.[3]

The final step in the graph requires resolving the separation between quantum mechanics and gravitation, often equated with general relativity. Numerous researchers concentrate their efforts on this specific step; nevertheless, no accepted theory of quantum gravity – and thus no accepted theory of everything – has emerged yet. It is usually assumed that the ToE will also solve the remaining problems of GUTs.

In addition to explaining the forces listed in the graph, a ToE may also explain the status of at least two candidate forces suggested by modern cosmology: an inflationary force and dark energy. Furthermore, cosmological experiments also suggest the existence of dark matter, supposedly composed of fundamental particles outside the scheme of the standard model. However, the existence of these forces and particles has not been proven yet.

String theory and M-theory

Since the 1990s, some physicists believe that 11-dimensional M-theory, which is described in some limits by one of the five perturbative superstring theories, and in another by the maximally-supersymmetric 11-dimensional supergravity, is the theory of everything. However, there is no widespread consensus on this issue.

A surprising property of string/M-theory is that extra dimensions are required for the theory's consistency. In this regard, string theory can be seen as building on the insights of the Kaluza–Klein theory, in which it was realized that applying general relativity to a five-dimensional universe (with one of them small and curled up) looks from the four-dimensional perspective like the usual general relativity together with Maxwell's electrodynamics. This lent credence to the idea of unifying gauge and gravity interactions, and to extra dimensions, but did not address the detailed experimental requirements. Another important property of string theory is its supersymmetry, which together with extra dimensions are the two main proposals for resolving the hierarchy problem of the standard model, which is (roughly) the question of why gravity is so much weaker than any other force. The extra-dimensional solution involves allowing gravity to propagate into the other dimensions while keeping other forces confined to a four-dimensional spacetime, an idea that has been realized with explicit stringy mechanisms.[18]

Research into string theory has been encouraged by a variety of theoretical and experimental factors. On the experimental side, the particle content of the standard model supplemented with neutrino masses fits into a spinor representation of SO(10), a subgroup of E8 that routinely emerges in string theory, such as in heterotic string theory[19] or (sometimes equivalently) in F-theory.[20][21] String theory has mechanisms that may explain why fermions come in three hierarchical generations, and explain the mixing rates between quark generations.[22] On the theoretical side, it has begun to address some of the key questions in quantum gravity, such as resolving the black hole information paradox, counting the correct entropy of black holes[23][24] and allowing for topology-changing processes.[25][26][27] It has also led to many insights in pure mathematics and in ordinary, strongly-coupled gauge theory due to the Gauge/String duality.

In the late 1990s, it was noted that one major hurdle in this endeavor is that the number of possible four-dimensional universes is incredibly large. The small, "curled up" extra dimensions can be compactified in an enormous number of different ways (one estimate is 10^{500}) each of which leads to different properties for the low-energy particles and forces. This array of models is known as the string theory landscape.[8]:347

One proposed solution is that many or all of these possibilities are realised in one or another of a huge number of universes, but that only a small number of them are habitable. Hence what we normally conceive as the fundamental constants of the universe are ultimately the result of the anthropic principle rather than dictated by theory. This has led to criticism of string theory,[28] arguing that it cannot make useful (i.e., original, falsifiable, and verifiable) predictions and regarding it as a pseudoscience. Others disagree,[29] and string theory remains an extremely active topic of investigation in theoretical physics.

Loop quantum gravity

Current research on loop quantum gravity may eventually play a fundamental role in a ToE, but that is not its primary aim.[30] Also loop quantum gravity introduces a lower bound on the possible length scales.

There have been recent claims that loop quantum gravity may be able to reproduce features resembling the Standard Model. So far only the first generation of fermions (leptons and quarks) with correct parity properties have been modelled by Sundance Bilson-Thompson using preons constituted of braids of spacetime as the building blocks.[31] However, there is no derivation of the Lagrangian that would describe the interactions of such particles, nor is it possible to show that such particles are fermions, nor that the gauge groups or interactions of the Standard Model are realised. Utilization of quantum computing concepts made it possible to demonstrate that the particles are able to survive quantum fluctuations.[32]

This model leads to an interpretation of electric and colour charge as topological quantities (electric as number and chirality of twists carried on the individual ribbons and colour as variants of such twisting for fixed electric charge).

Bilson-Thompson's original paper suggested that the higher-generation fermions could be represented by more complicated braidings, although explicit constructions of these structures were not given. The electric charge, colour, and parity properties of such fermions would arise in the same way as for the first generation. The model was expressly generalized for an infinite number of generations and for the weak force bosons (but not for photons or gluons) in a 2008 paper by Bilson-Thompson, Hackett, Kauffman and Smolin.[33]

Other attempts

A recent development is the theory of causal fermion systems,[34] giving all three current physical theories (quantum mechanics, general relativity and quantum field theory) as limiting cases.

A recent and very prolific attempt is called Causal Sets. As some of the approaches mentioned above, its direct goal isn't necessarily to achieve a ToE but primarily a working theory of quantum gravity, which might eventually include the standard model and become a candidate for a ToE. Its founding principle is that spacetime is fundamentally discrete and that the spacetime events are related by a partial order. This partial order has the physical meaning of the causality relations between relative past and future distinguishing spacetime events.

Outside the previously mentioned attempts there is Garrett Lisi's E8 proposal. This theory provides an attempt of identifying general relativity and the standard model within the Lie group E8. The theory doesn't provide a novel quantization procedure and the author suggests its quantization might follow the Loop Quantum Gravity approach above mentioned.[35]

Christoph Schiller's Strand Model attempts to account for the gauge symmetry of the Standard Model of particle physics, U(1)×SU(2)×SU(3), with the three Reidemeister moves of knot theory by equating each elementary particle to a different tangle of one, two, or three strands (selectively a long prime knot or unknotted curve, a rational tangle, or a braided tangle respectively).

Present status

At present, there is no candidate theory of everything that includes the standard model of particle physics and general relativity. For example, no candidate theory is able to calculate the fine structure constant or the mass of the electron. Most particle physicists expect that the outcome of the ongoing experiments – the search for new particles at the large particle accelerators and for dark matter – are needed in order to provide further input for a ToE.

3.12.3 Theory of everything and philosophy

Main article: Theory of everything (philosophy)

The philosophical implications of a physical ToE are frequently debated. For example, if philosophical physicalism is true, a physical ToE will coincide with a philosophical theory of everything.

The "system building" style of metaphysics attempts to answer *all* the important questions in a coherent way, providing a complete picture of the world. Plato and Aristotle could be said to have created early examples of comprehensive systems. In the early modern period (17th and 18th centuries), the system-building *scope* of philosophy is often linked to the rationalist *method* of philosophy, which

is the technique of deducing the nature of the world by pure *a priori* reason. Examples from the early modern period include the Leibniz's Monadology, Descarte's Dualism, and Spinoza's Monism. Hegel's Absolute idealism and Whitehead's Process philosophy were later systems.

3.12.4 Arguments against a theory of everything

In parallel to the intense search for a ToE, various scholars have seriously debated the possibility of its discovery.

Gödel's incompleteness theorem

A number of scholars claim that Gödel's incompleteness theorem suggests that any attempt to construct a ToE is bound to fail. Gödel's theorem, informally stated, asserts that any formal theory expressive enough for elementary arithmetical facts to be expressed and strong enough for them to be proved is either inconsistent (both a statement and its denial can be derived from its axioms) or incomplete, in the sense that there is a true statement that can't be derived in the formal theory.

Stanley Jaki, in his 1966 book *The Relevance of Physics*, pointed out that, because any "theory of everything" will certainly be a consistent non-trivial mathematical theory, it must be incomplete. He claims that this dooms searches for a deterministic theory of everything.[36] In a later reflection, Jaki states that it is wrong to say that a final theory is impossible, but rather that "when it is on hand one cannot know rigorously that it is a final theory."[37]

Freeman Dyson has stated that "Gödel's theorem implies that pure mathematics is inexhaustible. No matter how many problems we solve, there will always be other problems that cannot be solved within the existing rules. [...] Because of Gödel's theorem, physics is inexhaustible too. The laws of physics are a finite set of rules, and include the rules for doing mathematics, so that Gödel's theorem applies to them."[38]

Stephen Hawking was originally a believer in the Theory of Everything but, after considering Gödel's Theorem, concluded that one was not obtainable: "Some people will be very disappointed if there is not an ultimate theory, that can be formulated as a finite number of principles. I used to belong to that camp, but I have changed my mind."[39]

Jürgen Schmidhuber (1997) has argued against this view; he points out that Gödel's theorems are irrelevant for computable physics.[40] In 2000, Schmidhuber explicitly constructed limit-computable, deterministic universes whose pseudo-randomness based on undecidable, Gödel-like halting problems is extremely hard to detect but does not at all prevent formal ToEs describable by very few bits of information.[41]

Related critique was offered by Solomon Feferman,[42] among others. Douglas S. Robertson offers Conway's game of life as an example:[43] The underlying rules are simple and complete, but there are formally undecidable questions about the game's behaviors. Analogously, it may (or may not) be possible to completely state the underlying rules of physics with a finite number of well-defined laws, but there is little doubt that there are questions about the behavior of physical systems which are formally undecidable on the basis of those underlying laws.

Since most physicists would consider the statement of the underlying rules to suffice as the definition of a "theory of everything", most physicists argue that Gödel's Theorem does *not* mean that a ToE cannot exist. On the other hand, the scholars invoking Gödel's Theorem appear, at least in some cases, to be referring not to the underlying rules, but to the understandability of the behavior of all physical systems, as when Hawking mentions arranging blocks into rectangles, turning the computation of prime numbers into a physical question.[44] This definitional discrepancy may explain some of the disagreement among researchers.

Fundamental limits in accuracy

No physical theory to date is believed to be precisely accurate. Instead, physics has proceeded by a series of "successive approximations" allowing more and more accurate predictions over a wider and wider range of phenomena. Some physicists believe that it is therefore a mistake to confuse theoretical models with the true nature of reality, and hold that the series of approximations will never terminate in the "truth". Einstein himself expressed this view on occasions.[45] Following this view, we may reasonably hope for *a* theory of everything which self-consistently incorporates all currently known forces, but we should not expect it to be the final answer.

On the other hand, it is often claimed that, despite the apparently ever-increasing complexity of the mathematics of each new theory, in a deep sense associated with their underlying gauge symmetry and the number of fundamental physical constants, the theories are becoming simpler. If this is the case, the process of simplification cannot continue indefinitely.

Lack of fundamental laws

There is a philosophical debate within the physics community as to whether a theory of everything deserves to be called *the* fundamental law of the universe.[46] One view

is the hard reductionist position that the ToE is the fundamental law and that all other theories that apply within the universe are a consequence of the ToE. Another view is that emergent laws, which govern the behavior of complex systems, should be seen as equally fundamental. Examples of emergent laws are the second law of thermodynamics and the theory of natural selection. The advocates of emergence argue that emergent laws, especially those describing complex or living systems are independent of the low-level, microscopic laws. In this view, emergent laws are as fundamental as a ToE.

The debates do not make the point at issue clear. Possibly the only issue at stake is the right to apply the high-status term "fundamental" to the respective subjects of research. A well-known one took place between Steven Weinberg and Philip Anderson

Impossibility of being "of everything"

Although the name "theory of everything" suggests the determinism of Laplace's quotation, this gives a very misleading impression. Determinism is frustrated by the probabilistic nature of quantum mechanical predictions, by the extreme sensitivity to initial conditions that leads to mathematical chaos, by the limitations due to event horizons, and by the extreme mathematical difficulty of applying the theory. Thus, although the current standard model of particle physics "in principle" predicts almost all known non-gravitational phenomena, in practice only a few quantitative results have been derived from the full theory (e.g., the masses of some of the simplest hadrons), and these results (especially the particle masses which are most relevant for low-energy physics) are less accurate than existing experimental measurements. The ToE would almost certainly be even harder to apply for the prediction of experimental results, and thus might be of limited use.

A motive for seeking a ToE, apart from the pure intellectual satisfaction of completing a centuries-long quest, is that prior examples of unification have predicted new phenomena, some of which (e.g., electrical generators) have proved of great practical importance. And like in these prior examples of unification, the ToE would probably allow us to confidently define the domain of validity and residual error of low-energy approximations to the full theory.

Infinite number of onion layers

Lee Smolin regularly argues that the layers of nature may be like the layers of an onion, and that the number of layers might be infinite. This would imply an infinite sequence of physical theories.

The argument is not universally accepted, because it is not

obvious that infinity is a concept that applies to the foundations of nature.

Impossibility of calculation

Weinberg[47] points out that calculating the precise motion of an actual projectile in the Earth's atmosphere is impossible. So how can we know we have an adequate theory for describing the motion of projectiles? Weinberg suggests that we know *principles* (Newton's laws of motion and gravitation) that work "well enough" for simple examples, like the motion of planets in empty space. These principles have worked so well on simple examples that we can be reasonably confident they will work for more complex examples. For example, although general relativity includes equations that do not have exact solutions, it is widely accepted as a valid theory because all of its equations with exact solutions have been experimentally verified. Likewise, a ToE must work for a wide range of simple examples in such a way that we can be reasonably confident it will work for every situation in physics.

3.12.5 See also

- Absolute (philosophy)
- An Exceptionally Simple Theory of Everything
- Argument from beauty
- Attractor
- Beyond black holes
- Beyond the standard model
- Big Bang
- Brownian motion
- Chaos theory
- Chronology of the universe
- Electroweak interaction
- Holographic principle
- Mathematical beauty
- Mathematical universe hypothesis
- Multiverse
- Standard Model (mathematical formulation)
- Superfluid vacuum theory (SVT)

- *The Theory of Everything (2014 film)* – a feature film about Prof. Stephen Hawking and his first wife Jane Hawking
- Timeline of the Big Bang
- Zero-energy universe

3.12.6 References

Footnotes

[1] Steven Weinberg. *Dreams of a Final Theory: The Scientist's Search for the Ultimate Laws of Nature*. Knopf Doubleday Publishing Group. ISBN 978-0-307-78786-6.

[2] Stephen W. Hawking (28 February 2006). *The Theory of Everything: The Origin and Fate of the Universe*. Phoenix Books; Special Anniv. ISBN 978-1-59777-508-3.

[3] Carlip, Steven (2001). "Quantum Gravity: a Progress Report". *Reports on Progress in Physics* **64** (8): 885. arXiv:gr-qc/0108040. Bibcode:2001RPPh...64..885C. doi:10.1088/0034-4885/64/8/301.

[4] Susanna Hornig Priest (14 July 2010). *Encyclopedia of Science and Technology Communication*. SAGE Publications. ISBN 978-1-4522-6578-0.

[5] Ellis, John (2002). "Physics gets physical (correspondence)". *Nature* **415** (6875): 957. Bibcode:2002Natur.415..957E. doi:10.1038/415957b.

[6] Ellis, John (1986). "The Superstring: Theory of Everything, or of Nothing?". *Nature* **323** (6089): 595–598. Bibcode:1986Natur.323..595E. doi:10.1038/323595a0.

[7] Rorres, Chris (2009). "ARCHIMEDES AND THE QUEST FOR THE THEORY OF EVERYTHING".

[8] Chris Impey (26 March 2012). *How It Began: A Time-Traveler's Guide to the Universe*. W. W. Norton. ISBN 978-0-393-08002-5.

[9] William E. Burns (1 January 2001). *The Scientific Revolution: An Encyclopedia*. ABC-CLIO. ISBN 978-0-87436-875-8.

[10] Shapin, Steven (1996). *The Scientific Revolution*. University of Chicago Press. ISBN 0-226-75021-3.

[11] Newton, Sir Isaac (1729). *The Mathematical Principles of Natural Philosophy* **II**. p. 255.

[12] Sean Carroll (7 January 2010). *From Eternity to Here: The Quest for the Ultimate Theory of Time*. Penguin Group US. ISBN 978-1-101-15215-7.

[13] Faraday, M. (1850). "Experimental Researches in Electricity. Twenty-Fourth Series. On the Possible Relation of Gravity to Electricity". *Abstracts of the Papers Communicated to the Royal Society of London* **5**: 994–995. doi:10.1098/rspl.1843.0267.

[14] Gorban, Alexander N.; Karlin, Ilya (2013). "Hilbert's 6th Problem: Exact and approximate hydrodynamic manifolds for kinetic equations". *Bulletin of the American Mathematical Society* **51** (2): 187. doi:10.1090/S0273-0979-2013-01439-3.

[15] Dirac, P.A.M. (1929). "Quantum mechanics of many-electron systems". *Proceedings of the Royal Society of London A* **123** (792): 714. Bibcode:1929RSPSA.123..714D. doi:10.1098/rspa.1929.0094.

[16] Abraham Pais (23 September 1982). *Subtle is the Lord : The Science and the Life of Albert Einstein: The Science and the Life of Albert Einstein*. Oxford University Press. ISBN 978-0-19-152402-8.

[17] Weinberg (1993), Ch. 5

[18] Holloway, M (2005). "The Beauty of Branes" (PDF). *Scientific American* (Scientific American) **293** (4): 38. Bibcode:2005SciAm.293d..38H. doi:10.1038/scientificamerican1005-38. PMID 16196251. Retrieved August 13, 2012.

[19] Nilles, Hans Peter; Ramos-Sánchez, Saúl; Ratz, Michael; Vaudrevange, Patrick K. S. (2008). "From strings to the MSSM". *The European Physical Journal C* **59** (2): 249. arXiv:0806.3905. Bibcode:2009EPJC...59..249N. doi:10.1140/epjc/s10052-008-0740-1.

[20] Beasley, Chris; Heckman, Jonathan J; Vafa, Cumrun (2009). "GUTs and exceptional branes in F-theory — I". *Journal of High Energy Physics* **2009**: 058. arXiv:0802.3391. Bibcode:2009JHEP...01..058B. doi:10.1088/1126-6708/2009/01/058.

[21] Donagi, Ron; Wijnholt, Martijn (2008). "Model Building with F-Theory". arXiv:0802.2969v3 [hep-th].

[22] Heckman, Jonathan J.; Vafa, Cumrun (2008). "Flavor Hierarchy from F-theory". *Nuclear Physics B* **837**: 137–151. arXiv:0811.2417v3. Bibcode:2010NuPhB.837..137H. doi:10.1016/j.nuclphysb.2010.05.009.

[23] Strominger, Andrew; Vafa, Cumrun (1996). "Microscopic origin of the Bekenstein-Hawking entropy". *Physics Letters B* **379**: 99. arXiv:hep-th/9601029. Bibcode:1996PhLB..379...99S. doi:10.1016/0370-2693(96)00345-0.

[24] Horowitz, Gary (1996). "Gravitational Wave Astronomy". *The Origin of Black Hole Entropy in String Theory*. Astrophysics and Space Science Library **211**. p. 95. arXiv:gr-qc/9604051. doi:10.1007/978-94-011-5812-1_7. ISBN 978-94-010-6455-2.

[25] Greene, Brian R.; Morrison, David R.; Strominger, Andrew (1995). "Black hole condensation and the unification of string vacua". *Nuclear Physics B* **451**: 109. arXiv:hep-th/9504145. Bibcode:1995NuPhB.451..109G. doi:10.1016/0550-3213(95)00371-X.

[26] Aspinwall, Paul S.; Greene, Brian R.; Morrison, David R. (1994). "Calabi-Yau moduli space, mirror manifolds and spacetime topology change in string theory". *Nuclear Physics B* **416** (2): 414. arXiv:hep-th/9309097. Bibcode:1994NuPhB.416..414A. doi:10.1016/0550-3213(94)90321-2.

[27] Adams, Allan; Liu, Xiao; McGreevy, John; Saltman, Alex; Silverstein, Eva (2005). "Things fall apart: Topology change from winding tachyons". *Journal of High Energy Physics* **2005** (10): 033. arXiv:hep-th/0502021. Bibcode:2005JHEP...10..033A. doi:10.1088/1126-6708/2005/10/033.

[28] Smolin, Lee (2006). *The Trouble With Physics: The Rise of String Theory, the Fall of a Science, and What Comes Next*. Houghton Mifflin. ISBN 978-0-618-55105-7.

[29] Duff, M. J. (2011). "String and M-Theory: Answering the Critics". *Foundations of Physics* **43**: 182. arXiv:1112.0788. Bibcode:2013FoPh...43..182D. doi:10.1007/s10701-011-9618-4.

[30] Potter, Franklin (15 February 2005). "Leptons And Quarks In A Discrete Spacetime" (PDF). *Frank Potter's Science Gems*. Retrieved 2009-12-01.

[31] Bilson-Thompson, Sundance O.; Markopoulou, Fotini; Smolin, Lee (2007). "Quantum gravity and the standard model". *Classical and Quantum Gravity* **24** (16): 3975–3994. arXiv:hep-th/0603022. Bibcode:2007CQGra..24.3975B. doi:10.1088/0264-9381/24/16/002.

[32] Castelvecchi, Davide; Valerie Jamieson (August 12, 2006). "You are made of space-time". *New Scientist* (2564).

[33] Sundance Bilson-Thompson; Jonathan Hackett; Lou Kauffman; Lee Smolin (2008). "Particle Identifications from Symmetries of Braided Ribbon Network Invariants". arXiv:0804.0037 [hep-th].

[34] F. Finster; J. Kleiner (2015). "Causal fermion systems as a candidate for a unified physical theory". *Journal of Physics: Conference Series* **626** (2015): 012020. arXiv:1502.03587. Bibcode:2015JPhCS.626a2020F. doi:10.1088/1742-6596/626/1/012020.

[35] A. G. Lisi (2007). "An Exceptionally Simple Theory of Everything". arXiv:0711.0770 [hep-th].

[36] Jaki, S.L. (1966). *The Relevance of Physics*. Chicago Press, pp. 127–130.

[37] Stanley L. Jaki (2004) "A Late Awakening to Gödel in Physics", pp. 8–9.

[38] Freeman Dyson, NYRB, May 13, 2004

[39] Stephen Hawking, Gödel and the end of physics, July 20, 2002

[40] Schmidhuber, Jürgen (1997). *A Computer Scientist's View of Life, the Universe, and Everything*. *Lecture Notes in Computer Science*. Springer. pp. 201–208. doi:10.1007/BFb0052071. ISBN 978-3-540-63746-2.

[41] Schmidhuber, Jürgen (2002). "Hierarchies of generalized Kolmogorov complexities and nonenumerable universal measures computable in the limit". *Sections in: Hierarchies of generalized Kolmogorov complexities and nonenumerable universal measures computable in the limit. International Journal of Foundations of Computer Science* ():587-612 (2002). *Section 6 in: the Speed Prior: A New Simplicity Measure Yielding Near-Optimal Computable Predictions. in J. Kivinen and R. H. Sloan, editors, Proceedings of the 15th Annual Conference on Computational Learning Theory (COLT 2002), Sydney, Australia, Lecture Notes in Artificial Intelligence, pages 216--228. Springer, 2002* **13** (4): 1–5. arXiv:quant-ph/0011122. Bibcode:2000quant.ph.11122S.

[42] Feferman, Solomon (17 November 2006). "The nature and significance of Gödel's incompleteness theorems" (PDF). Institute for Advanced Study. Retrieved 2009-01-12.

[43] Robertson, Douglas S. (2007). "Goedel's Theorem, the Theory of Everything, and the Future of Science and Mathematics". *Complexity* **5** (5): 22–27. doi:10.1002/1099-0526(200005/06)5:5<22::AID-CPLX4>3.0.CO;2-0.

[44] Hawking, Stephen (20 July 2002). "Gödel and the end of physics". Retrieved 2009-12-01.

[45] Einstein, letter to Felix Klein, 1917. (On determinism and approximations.) Quoted in Pais (1982), Ch. 17.

[46] Weinberg (1993), Ch 2.

[47] Weinberg (1993) p. 5

Bibliography

- Pais, Abraham (1982) *Subtle is the Lord...: The Science and the Life of Albert Einstein* (Oxford University Press, Oxford, . Ch. 17, ISBN 0-19-853907-X

- Weinberg, Steven (1993) *Dreams of a Final Theory: The Search for the Fundamental Laws of Nature*, Hutchinson Radius, London, ISBN 0-09-177395-4

3.12.7 External links

- The Elegant Universe, *Nova* episode about the search for the theory of everything and string theory.

- Theory of Everything, freeview video by the Vega Science Trust, BBC and Open University.

- The Theory of Everything: Are we getting closer, or is a final theory of matter and the universe impossible? Debate between John Ellis (physicist), Frank Close and Nicholas Maxwell.

- Why The World Exists, a discussion between physicist Laura Mersini-Houghton, cosmologist George Francis Rayner Ellis and philosopher David Wallace about dark matter, parallel universes and explaining why these and the present Universe exist.

Chapter 4

Text and image sources, contributors, and licenses

4.1   Text

- **Physics beyond the Standard Model** *Source:* https://en.wikipedia.org/wiki/Physics_beyond_the_Standard_Model?oldid=723758668 *Contributors:* David spector, Ewen, Michael Hardy, Andrewman327, Sanxiyn, IceKarma, Donarreiskoffer, Nurg, Rursus, David Gerard, Alison, David Schaich, RJHall, El C, Kwamikagami, I9Q79oL78KiL0QTFHgyc, Jeodesic, 4v4l0n42, Alinor, Count Iblis, Rjwilmsi, Koavf, Strait, Eyu100, HappyCamper, Lmatt, BradBeattie, Ohwilleke, Arado, Bhny, SCZenz, CecilWard, Karl Andrews, Ntu, Dna-webmaster, Pawyilee, 2over0, Caco de vidro, Jaysbro, SmackBot, Mdj, Nickst, Chris the speller, Bluebot, Scwlong, QFT, Pepsidrinka, Jgwacker, Yevgeny Kats, Doug Bell, John, Dspitzle, RandomCritic, JarahE, Kurtan~enwiki, Headbomb, Peter Gulutzan, N shaji, Lenny Kaufman, VoABot II, Email4mobile, Maliz, R'n'B, HEL, Natsirtguy, Rod57, Tarotcards, DadaNeem, Goop Goop, Fences and windows, Michael H 34, Venny85, Lamro, Wing gundam, Beast of traal, Randy Kryn, Bhuna71, Mild Bill Hiccup, Djr32, Excirial, RCalabraro, Brews ohare, Mastertek, TimothyRias, XLinkBot, Truthnlove, Addbot, Luckas-bot, Zhitelew, Yobot, AnomieBOT, Citation bot, LilHelpa, Smk65536, Stevebow, Omnipaedista, Seeleschneider, A. di M., Kenneth Dawson, Steve Quinn, Citation bot 2, Aturen, Jonesey95, Tom.Reding, ErgSlider, Physics therapist, Gistmass, Bj norge, Vstarsky, Dcirovic, Serketan, ZéroBot, Galaktiker, Arbnos, Suslindisambiguator, Wiggles007, Maschen, Smtchahal, ClaudeDes, Braincricket, Widr, Helpful Pixie Bot, Mike9110, DryRun, Bibcode Bot, BG19bot, Brainssturm, Qtom.masters, ThePeriodicTable123, M0532062613, Kryomaxim, Andyhowlett, Cinaro, I am One of Many, Kowrje, CtrlAltBackspace, 22merlin, Monkbot, Delbert7, Tetra quark, TQuentin, Christos Theopoulos, MauiPhoenix, MongoMikeLambert, Bigben037, Mark Goldes and Anonymous: 78

- **Mass generation** *Source:* https://en.wikipedia.org/wiki/Mass_generation?oldid=676957128 *Contributors:* Rjwilmsi, Bhny, JorisvS, Niceguyedc, Ktr101, Tom.Reding, ClueBot NG, Accedie, Bibcode Bot, Itchmean, Evensteven, Prestigiouzman, Ambermae2002 and Anonymous: 10

- **Strong CP problem** *Source:* https://en.wikipedia.org/wiki/Strong_CP_problem?oldid=699491628 *Contributors:* Phys, Bearcat, Aetheling, Christopherlin, Lumidek, Hidaspal, TheParanoidOne, Linas, Rjwilmsi, YurikBot, Bambaiah, Conscious, Malcolma, JorisvS, David Cherney, Andre.holzner, Muro Bot, Yobot, Redirect fixer, AnomieBOT, Omnipaedista, Ofercomay, Suslindisambiguator, Maschen, BG19bot and Anonymous: 13

- **Neutrino oscillation** *Source:* https://en.wikipedia.org/wiki/Neutrino_oscillation?oldid=725000873 *Contributors:* Edward, Michael Hardy, SebastianHelm, Taxman, BenRG, Robbot, Nurg, Giftlite, Jmnbpt, Xerxes314, ConradPino, Tubedogg, B.d.mills, Lazarus666, Mike Rosoft, Rich Farmbrough, ESkog, Cedders, Worldtraveller, Wiki-uk, Keenan Pepper, Bsadowski1, Falcorian, Flying fish, Linas, JFG, Jugger90, Pirk, Yurik, Rjwilmsi, Strait, Mike Peel, Dudegalea, Erkcan, Itinerant1, Goudzovski, Chobot, Hairy Dude, Kordas, Aaronwinborn, Salsb, Jamesg, Thiseye, Santaduck, Gadget850, Banus, Tosus, Teply, GrinBot~enwiki, Eatcacti, MacsBug, SmackBot, Haymaker, Arbe, Stepa, Saros136, Bluebot, Timothy Clemans, QFT, DMacks, Q9a, Mets501, DI2000, IRevLinas, Kurtan~enwiki, Harold f, MrFizyx, Rotiro, Michael C Price, Hugozam, Headbomb, D.H, Credema, Yellowdesk, Spartaz, DanPMK, Leyo, Patar knight, Maurice Carbonaro, Choihei, Higgsino, Lseixas, Cuzkatzimhut, Pleasantville, Improve~enwiki, EverGreg, Jasondet, Paulfharrison, Aardvarkleg, Al Leween, Von Crayola, Mild Bill Hiccup, Hyh1048576, Alexbot, Jwfvalle, SchreiberBike, Asf107, SkyLined, Luismarques83, Addbot, Zorrobot, Luckas-bot, Yobot, WikiDan61, Wireader, AnomieBOT, Citation bot, Blennow, Omnipaedista, Mnmngb, Paine Ellsworth, Ysyoon, Jonesey95, Pmokeefe, Trappist the monk, LilyKitty, RjwilmsiBot, CaptRik, Dcirovic, Hhhippo, Arbnos, Cymru.lass, Timetraveler3.14, Brandmeister, Billingd, Xronon, ClueBot NG, Gareth Griffith-Jones, Ben morphett, Mightyname, Bibcode Bot, BG19bot, Dwightboone, MskKrieger, Toni 001, Arcandam, FoCuSandLeArN, Zanpan, Klingerdinger, XXX8906, Drscientific, Sircier, Trackteur, Soham92, Nøkkenbuer, Esmith227, Kcher12, Die bad guy and Anonymous: 113

- **Baryon asymmetry** *Source:* https://en.wikipedia.org/wiki/Baryon_asymmetry?oldid=715525674 *Contributors:* Timwi, Lupin, Mike Rosoft, Laurascudder, Shenme, PaulHanson, Stigocki, Woohookitty, Aristotle Pagaltzis, Strait, Bubba73, Klortho, Ohwilleke, MacsBug, SmackBot, Skizzik, Bluebot, RDBrown, Basalisk, Colonies Chris, SQGibbon, Banedon, Thijs!bot, Headbomb, Escarbot, Grant Gussie, Magioladitis, HEL, TallNapoleon, Trombone boy89, Anonymous Dissident, Someguy1221, Paradoctor, Likebox, DragonBot, Djr32, Addbot, Proxima Centauri, Traitor, Luckas-bot, Yobot, Amirobot, AnomieBOT, Kwiki, Tom.Reding, RjwilmsiBot, Slightsmile, ZéroBot, Cogiati, Cobaltcigs, Suslindisambiguator, Widr, Bibcode Bot, BG19bot, WikiWanderer3, MLearry, M0532062613, Kryomaxim, SomeFreakOnTheInternet, MatthewJ00, AloisKabelschacht, Mfb, 22merlin, Monkbot and Anonymous: 39

266

- **Dark matter** *Source:* https://en.wikipedia.org/wiki/Dark_matter?oldid=725279916 *Contributors:* AxelBoldt, Chenyu, Derek Ross, CYD, BF, Bryan Derksen, The Anome, Tarquin, Taw, XJaM, Arvindn, William Avery, Roadrunner, Mintguy, Bth, Stevertigo, Edward, Nealmcb, Boud, FrankH, Cprompt, DopefishJustin, Bobby D. Bryant, Ixfd64, SebastianHelm, Alfio, CesarB, Looxix~enwiki, Mkweise, William M. Connolley, JWSchmidt, Glenn, Mxn, Charles Matthews, Timwi, Fuzheado, Rednblu, Haukurth, DW40, Dragons flight, Furrykef, Saltine, Dogface, Populus, Jusjih, Finlay McWalter, Bearcat, Robbot, Zandpert, Korath, Nurg, Naddy, Arkuat, Gandalf61, Pingveno, Rursus, Rfhsher, Wereon, Diberri, Adam78, Aasim75, Marc Venot, Ancheta Wis, Giftlite, Graeme Bartlett, Laudaka, Barbara Shack, Herbee, Fropuff, Xerxes314, Dratman, Curps, Joconnor, Jdavidb, Unconcerned, Eequor, Bobblewik, Andycjp, Alexf, Geni, Antandrus, HorsePunchKid, Melikamp, PDH, Rdsmith4, Anythingyouwant, Bosmon, Bbbl67, Icairns, Sam Hocevar, Cynical, Lumidek, Iantresman, Burschik, Joyous!, Adashiel, Urvabara, Discospinster, Rich Farmbrough, Oliver Lineham, Vsmith, Jpk, ArnoldReinhold, Murtasa, D-Notice, JPX7, KaiSeun, SpookyMulder, Bender235, Kjoonlee, Kaisershatner, Pk2000, PsychoDave, RJHall, Mr. Billion, El C, Bletch, PhilHibbs, Shanes, Frankenschulz, Art LaPella, RoyBoy, Themusicgod1, Bobo192, Smalljim, Shenme, Cmdrjameson, Reuben, Kmaguire, I9Q79oL78KiL0QTFHgyc, Zelda~enwiki, Mr. Brownstone, E is for Ian, Jumbuck, Storm Rider, Alansohn, Gary, Anthony Appleyard, Guy Harris, Eric Kvaalen, Arthena, Keenan Pepper, Kocio, Bart133, RPellessier, Benna, ClockworkSoul, Cal 1234, Count Iblis, Guthrie, H2g2bob, Bsadowski1, GabrielF, Pauli133, Leondz, DV8 2XL, Gene Nygaard, Feline1, Oleg Alexandrov, Brookie, Natalya, Flying fish, WilliamKF, Yeastbeast, Mindmatrix, RHaworth, Piek, BillC, JPFlip, Benbest, JFG, ^demon, WadeSimMiser, Gxojo, MONGO, Jwanders, Torqueing, 游戏人间, Joke137, Wisq, Christopher Thomas, Palica, Mandarax, RedBLACKand-BURN, Aarghdvaark, RichardWeiss, Ashmoo, Graham87, Malangthon, Mamling, Jclemens, Drbogdan, Loris Bennett, Rjwilmsi, Lars T., Strait, Patrick Gill, Tangotango, Tawker, Smithfarm, Stevenscollege, Mike Peel, HappyCamper, SeanMack, ScottJ, Bubba73, Krash, Dermeister, Rangek, Madcat87, FlaBot, Ian Pitchford, PlatypeanArchcow, A scientist, Margosbot~enwiki, Gark, Nivix, Gparker, Pathoschild, Gurch, Stevenfruitsmaak, Goudzovski, Tomer Ish Shalom, Smithbrenon, Chobot, Moocha, DVdm, Bgwhite, Gwernol, The Rambling Man, YurikBot, Wavelength, RobotE, Koveras, Hairy Dude, Huw Powell, Phmer, Hillman, RussBot, Michael Slone, Ohwilleke, Bhny, JabberWok, GLaDOS, DanMS, Zelmerszoetrop, Stephenb, Eleassar, Merick, Big Brother 1984, NawlinWiki, Alpertron, Dlugosz, Schlafly, FFLaguna, BlackAndy, Dbmag9, SCZenz, Haoie, Raven4x4x, Ospalh, Durval, Bota47, Supspirit, Pegship, Noosfractal, Charlie Wiederhold, WAS 4.250, Smoggyrob, Reyk, Tvaughan, Joedixon, Eric TF Bat, Emc2, Ilmari Karonen, Allens, Bernd in Japan, InsayneWrapper, Bclayabt, Attilios, MacsBug, SmackBot, Cubs Fan, Ashill, IddoGenuth, Tomer yaffe, Stellea, InverseHypercube, KnowledgeOfSelf, Allixpeeke, Clpo13, Nickst, RedSpruce, Nightbat, Doc Strange, Herbm, Edgar181, HalfShadow, Flux.books, Dheerajkakar, Yamaguchi先生, Richmeister, Gilliam, Folajimi, The Gnome, Oscarthecat, Skizzik, Kmarinas86, Chris the speller, SuperBuuBuu, Quinsareth, Persian Poet Gal, Sirex98, MalafayaBot, Silly rabbit, Sangrolu, Villarinho, DHN-bot~enwiki, Sbharris, Hongooi, Jdthood, CheerLeone, Gtkysor, Can't sleep, clown will eat me, Nick Levine, Tamfang, Kelvin Case, Vladislav, Vanished User 0001, Rrburke, Jgoulden, Auvii, Krich, Wen D House, Radagast83, Engwar, Nakon, VegaDark, John D. Croft, Alexander110, KimO, Adrigon, SpiderJon, Ultraexactzz, Zadignose, Tesseran, Byelf2007, L337p4wn, K7lim, SashatoBot, Mchavez, Swatjester, Leftydan6, Minaker, Attys, Brillow, John, Ashoat, Scientizzle, Acitrano, Linnell, JoshuaZ, James.S, JorisvS, Coredesat, Goodnightmush, ICBB, Plunge, JHunterJ, Hypnosifl, Silverthorn, Descubes, Freederick, Dr.K., Vanished user, Iridescent, Darkerprojects, Astrobayes, Newone, MOBle, Igoldste, CapitalR, AGK, Courcelles, Tawkerbot2, Dlohcierekim, Chetvorno, Hammer Raccoon, Owen214, Eastlaw, Peledre, Pukkie, Anakata, Banedon, Runningonbrains, DKOH, NickW557, Gregbard, MikeWren, Vttoth, Necessary Evil, Ryan, Viciouspiggy, Gogo Dodo, Anonymi, Xxanthippe, A Softer Answer, Odie5533, Tawkerbot4, DumbBOT, Robertinventor, Kozuch, Mtpaley, Philza85, Starship Trooper, UberScienceNerd, Crum375, Thijs!bot, Epbr123, Astroceltica, Passaggio, Barbarina, Mbell, Eugenespeed, N5iln, Mojo Hand, Carlif, Headbomb, Tonyle, Marek69, RickinBaltimore, Lars Lindberg Christensen, OtterSmith, SusanLesch, Dawnseeker2000, Mmortal03, Hmrox, Hires an editor, AntiVandalBot, Seaphoto, Orionus, Opelio, Shirt58, Rehnn83, Joehodge, AaronY, Jj137, TTN, Dylan Lake, Chill doubt, Spencer, Yellowdesk, Sniktaw, Lfstevens, CPitt76, Gökhan, Jcarter1, Res2216firestar, JAnDbot, Leuko, Husond, MER-C, CosineKitty, Plantsurfer, Mcorazao, Therealintellectual, Folkform, Balbers, 100110100, Autotheist, Wasell, Magioladitis, Bongwarrior, VoABot II, Timothy McVeigh, Charlesrkiss, AuburnPilot, Krkaiser, Mbarbier, Kaivosukeltaja, Foroa, Swpb, Stigmj, T a y l o s, Ekantik, Brusegadi, Bubba hotep, Fabrictramp, Catgut, Lilian.Kaufmann, Zhanghia, Acornwithwings, BatteryIncluded, Vssun, LtHija, Whisky5, DerHexer, Prisca6023, PeteSF, KenyaSong, Rickard Vogelberg, NatureA16, DancingPenguin, MartinBot, Schmloof, STBot, Pagw, Fs644, Nikpapag, Anaxial, CommonsDelinker, Jean-Pierre Petit~enwiki, PrestonH, WelshMatt, Chrishy man, Tgeairn, J.delanoy, Pharaoh of the Wizards, Trusilver, Adavidb, Kpvats, Kudpung, Rod57, Arion 3x3, PedEye1, McSly, Tarotcards, Davy p, HiLo48, NewEnglandYankee, Ohms law, Jorfer, Bickavnger, Potatoswatter, Kyl-eTastic, Joshua Issac, Infiniteglitch, Remember the dot, Pitpif, Vanished user 39948282, Neekap, Natl1, Ldebain, BernardZ, SoCalSuperEagle, Squids and Chips, Borat fan, CardinalDan, Idioma-bot, Sheliak, Funandtrvl, Lights, Hammersoft, VolkovBot, Craigheinke, Itsfullofstars, ColdCase, Jeff G., JohnBlackburne, Mocirne, AlnoktaBOT, Scikid, Grammarmonger, Leojohns, Larry R. Holmgren, Philip Trueman, TXiK-iBoT, Oshwah, Docanton, Authorized User, Theophilus reed, Drestros power, Strichek, MarekMahut, Monkey Bounce, Lradrama, Sintaku, Carillonatreides, Martin451, Broadbot, Wiae, Mazarin07, Inductiveload, Knightshield, Telecineguy, Spiral5800, Kurowoofwoof111, Greswik, RobertFritzius, SwordSmurf, Falcon8765, Hellothere17, Enviroboy, Littlehollah, Wanchung Hu, Illumini85, SonOfMog Worf, Jazzman123, PGWG, 19merlin69, FlyingLeopard2014, Neparis, Bfpage, S-n-ushakov, SieBot, Calliopejen1, Tresiden, Wibubba48, Tachyonics, Pallab1234, Paradoctor, KGyST, Bentogoa, Jimlester51, Battlepuse, Oda Mari, Aaarnooo, Suomichris, Crowstar, PromX1, Lightmouse, Tombomp, Cyberplasm, Diego Grez-Cañete, Spartan-James, Thinghy, Mygerardromance, Hamiltondaniel, Superbeecat, Denisarona, JL-Bot, Escape Orbit, Starcluster, Troy 07, Atif.t2, ArepoEn, Ak47gforce, Ratemonth, Sfan00 IMG, ClueBot, Phoenix-wiki, GorillaWarfare, The Thing That Should Not Be, ArdClose, Rodhullandemu, Cptmurdok, Drmies, Frmorrison, Uncle Milty, Iuhkjhk87y678, Niceguyedc, MrBosnia, Bhaskarns, Andwor, Ktr101, Excirial, Dombom12, Cromescythe, Barbarinaz, FOARP, Brews ohare, Jotterbot, Iohannes Animosus, R.Andrae, Kentgen1, Ordovico, Mastertek, Rgoogin, Thehelpfulone, 1ForTheMoney, Versus22, Palmer666palmer, PCHS-NJROTC, Burner0718, Pillar of Babel, SoxBot III, Erodium, Vanished user uih38riiw4hjlsd, 1ofhissheep, TimothyRias, Arianewiki1, XLinkBot, DCCougar, Oldnoah, Rror, Gwark, Feinoha, Ost316, Avoided, Webmaster369, Gthomson, Tugrul irmak, Noctibus, Ploversegg, ZooFari, Parejkoj, Tayste, Addbot, Xp54321, Grayfell, Experimental Hobo Infiltration Droid, Willking1979, Some jerk on the Internet, Uruk2008, 04aeverington, DOI bot, Tcncv, Nohomers48, CharlesChandler, Gmeyerowitz, Haasfelix, Download, Proxima Centauri, Ashirgo, RTG, Redheylin, Luke Maurer, Glane23, Darkmatter654, SamatBot, Nanzilla, Lzkelley, Clone 209, Tassedethe, Numbo3-bot, Peridon, Chinchinthehun, Evildeathmath, Tide rolls, Lightbot, OlEnglish, Qemist, Gail, North Polaris, Legobot, Artichoke-Boy, Luckas-bot, Yobot, WikiDan61, Cosoce, Dov Henis, Aldebaran66, KillYourLove, Czech-Falcon, Amble, Mmxx, CinchBug, Perusnarpk, IW.HG, Einstein vs Dark energys, Eric-Wester, Tempodivalse, Synchronism, AnomieBOT, Letuño, Girl Scout cookie, IRP, JackieBot, RBM 72, AdjustShift, Nicolaas Vroom, Henrykandrup, Iluziat, Materialscientist, Dendlai, Im-peratorExercitus, The High Fin Sperm Whale, Citation bot, Ternity0127, Maxis ftw, Frankenpuppy, Quebec99, LilHelpa, Aksel89, Xqbot, Stiwebs, Random astronomer, Sionus, Cureden, Jradis1337, Capricorn42, Wperdue, Julianhyde, Deleance, Raspw, Tomwsulcer, Magicxcian, Gap9551, AbigailAbernathy, Srich32977, NOrbeck, Artemis6234, Almabot, Abell 1367, Feldhaus, False vacuum, RibotBOT, Waleswatcher,

Mikedr, Kongkokhaw, Rvnieuwe, Shadowjams, MeDrewNotYou, A. di M., Peter470, Sageman7, 神話, Luminique, Captain-n00dle, Imyfujita, FrescoBot, Andyradke0, Ag allstar, Paine Ellsworth, Originalwana, Styxpaint, Mark Renier, VS6507, PhysicsExplorer, Dbirkhofer, Steve Quinn, Nestlefolife, Adrian Akau, 1414rwbt, SF88, Citation bot 1, Redrose64, DUUJEEGWEEM, Tyler6298, Pinethicket, I dream of horses, Grammarspellchecker, Danlot, 10metreh, Jonesey95, Tom Reding, Pmokeele, A8UDI, For.a.limited.time.only, Elentirno, TedderBot, Aknochel, Sky-Machine, IVAN3MAN, Kgrad, Nieuwenh, Trappist the monk, Puzl bustr, Fama Clamosa, Domeinthebumhole, Michael9422, UrukHaiLoR, Allen4names, JLincoln, Jeffrd10, Lovemybluetooth, Diannaa, Fastilysock, Innotata, DrCrisp, Whisky drinker, Onel5969, RjwilmsiBot, 5mgoblue5, Blakelewis122, Þorri, Mathewsyriac, Leandro.lelas, Mserard313, Mdznr, Sbugnon, Ultima821, EmausBot, Francophile124, Grrow, Super48paul, GoingBatty, RA0808, Gimmetoo, Solarra, Jmencisom, Slightsmile, Tommy2010, Winner 42, SusanaMultidark, Gocows2, Wikipelli, Dcirovic, Serketan, Kriiferjel, Zurich Astro, Hhhippo, Mz7, Mhatthei, Svolin, Micahqgecko, JSquish, Josve05a, Trojanmice, MithrandirAgain, Edwinkaren, Devilaza, Arbnos, Oraclan, Suslindisambiguator, SporkBot, AlbertusmagnusOP, Mcmatter, Tolly4bolly, L1A1 FAL, Ancient Anomaly, L Kensington, Maj den, Corabilek, Donner60, Aldnonymous, Ihardlythinkso, RockMagnetist, Terra Novus, TYelliot, DASHBotAV, Kroupap, D Phoesheezey, Mannix Chan, Travies10, Jxraynor, TheTimesAreAChanging, ClueBot NG, Rich Smith, Gareth Griffith-Jones, Afjvanraan, Lyla1205, Crystal7878, Catinthehat93, Bped1985, Inhnifold, Wiggit002, Jj1236, PapaMike, MonEyshOt42069210, Muon, Esdacosta, Asukite, Masssly, Ph.d Carl edenburgh, Widr, Gavin.perch, Helpful Pixie Bot, Curb Chain, Calabe1992, Bibcode Bot, BG19bot, Dualus, Kishanparekh, Stevenwilkins, NacowY, Cheeseray1, Cyberguy5, Darkmatter adam, Yomomma8102, Hza a 9, Rarelight, Cyberpower678, Cosmologist77, MusikAnimal, தமிழ்க்குமரன் சரவணகுமார், Dahliamtl, Dodshe, Mark Arsten, Darkmatterotheruniverses, Samcstewart, Cadiomals, Trevayne08, Zedshort, Achowat, Rolandwilliamson, A2Die, Clint55555, Mgka79, NotEither, BattyBot, Millennium bug, Ronin712, Babymushrooms, Davidmexican, Drphilmarshall, Dilaton, Quin71901, U-95, ChrisGualtieri, Npmay, Kvark92, Lukasz.astrus, Ducknish, JYBot, Davidwinkler, Astrohap, Hunterf12, Dexbot, Caroline1981, Gravityking100, Junavia, Fredrikdn, CuriousMind01, Lugia2453, Wjs64, Andwor42, Frosty, Honneydewp243, Junjunone, DrHowzer73, JustAMuggle, Me, Myself, and I are Here, WadiElNatrun, Reatlas, Rfassbind, Acetotyce, I am One of Many, DirkXcal, Melonkelon, Ybidzian, Gig9876, M ashratnia, Trolololman12, Ilikedeletingstufffromhere, DavidLeighEllis, Onecreation, LahmacunKebab, The Herald, Zenibus, Jernahthern, Hipposaregrey, Frinthruit, Derekdoth, Stamptrader, Cyberalchemyst, Aaronknowsitall, FelixRosch, Darkmer, Doubleknockout, Monkbot, Yikkayaya, Wardinstrument, Leegrc, Vikas Rauniyar, Apipia, Upsalla, Jkvaternik, Lol kaptyn troll, HMSLavender, Mohammedshukoor, Callum92, Stefania.deluca, Ashweigh, Mrpiecjohnson, Oldstone James, Astezar, 39Debangshu, YoYoDude012, Anunaki truth, Pyrotle, Tetra quark, Carazmatic, God of matterrr, Silversparkcontributions, Isambard Kingdom, VexorAbVikipædia, Rizi0909, Absolutelypuremilk, Anand2202, Kbap2002, Kb2002, DN-boards1, Yohoona, Denniscabrams, KasparBot, I love trains sooo much, Sir Cumference, Ricardo A. Olea, Id6040, MHolland85, TychosElk, Boowiebear, Stephane Le Corre, Mustachman71, Abdelrahmam shawky, Huritisho, Jeffman257, Outedexits, Reg7d88, Maka Tree, Eslam nsr, Incendiary Iconoclasm, GSS-1987, Boobety boop boop boop, Atharv4321, Zhakhan9er, XABHISHEK23x, Elaysungur, Not a creative person, The Voidwalker, Shane ducharme, Tyler kwejew, Igojamon, Iamcoolswag, Riki71144, Kdkskdfj, Soopdish, Mr ST 142, Portal da Ciência, WikiEditor79, Wiki62838351936, MoneyMonkey112, Aidos the great, Trefoily+ and Anonymous: 1310

- **Dark energy** *Source:* https://en.wikipedia.org/wiki/Dark_energy?oldid=724683658 *Contributors:* The Anome, Dachshund, Roadrunner, Schewek, Stevertigo, Thesteve, Nealmcb, Michael Hardy, Tim Starling, FrankH, Bobby D. Bryant, SebastianHelm, Ahoerstemeier, Glenn, Tristanb, Reddi, Wik, DW40, Dragons flight, Anupamsr, Pierre Boreal, BenRG, Jellq, Donarreiskoffer, Robbot, Zandperl, Korath, Scott McNay, Vespristiano, Peak, Gandalf61, Rursus, Mlaine, UtherSRG, SC, Mattflaschen, Acm, Ancheta Wis, Giftlite, Graeme Bartlett, Awolf002, Jyril, Art Carlson, Herbee, Perl, Curps, Henry Flower, Gzornenplatz, Manuel Anastácio, Andycjp, BruceR, LucasVB, Antandrus, Beland, Karol Langner, Kevin B12, Bbbl67, Urvabara, JimJast, Discospinster, Rich Farmbrough, Pjacobi, Vsmith, D-Notice, Dbachmann, Bender235, Eric Forste, RJHall, JustinWick, Omnibus, El C, Lycurgus, Jomel, Kwamikagami, Frankenschulz, RoyBoy, Stesmo, Reuben, Russ3Z, I9Q79oL78KiL0QTFHgyc, Diego Moya, Keenan Pepper, Slugmaster, Axl, Benna, Wtmitchell, RainbowOfLight, Mikeo, Vuo, Freyr, DV8 2XL, Kazvorpal, Falcorian, Velho, Batintherain, Hottscubbard, OwenX, Mindmatrix, FeanorStar7, Velvetsmog, Uncle G, Netdragon, Jeff3000, GregorB, Isnow, SDC, 嘉嘉嘉嘉嘉, Joke137, Abd, Christopher Thomas, Sneakums, Dysepsion, BD2412, Doc Savage, Malangthon, RadioActive~enwiki, Drbogdan, Loris Bennett, Rjwilmsi, Strait, TheRingess, Salleman, HappyCamper, Sohmc, Ems57fcva, DonJuan~enwiki, BitterMan, Tomer Ish Shalom, Srleffler, Smithbrenon, CJLL Wright, Chobot, DVdm, Wavelength, RobotE, SamuelR, Diliff, Bhny, Stephenb, Cambridge-BayWeather, Merick, NawlinWiki, Msikma, FFLaguna, LiamE, SCZenz, FoolsWar, Bota47, Rwxrwxrwx, Daniel C, Enormousdude, 2over0, Helge Rosé, Pb30, Dr.alf, Joedixon, Rlove, Geoffrey.landis, Ilmari Karonen, Moonsleeper7, Kungfuadam, Bernd in Japan, GrinBot~enwiki, Treesmill, SmackBot, Ashill, Saravask, Bayardo, Tom Lougheed, InverseHypercube, KnowledgeOfSelf, Melchoir, J.Sarfatti, Nickst, Silverhand, Edgar181, Vixus, Gilliam, Skizzik, Jlsilva, Andy M. Wang, Tyciol, Sirex98, Oli Filth, DHN-bot~enwiki, Sbharris, Colonies Chris, Jdthood, Can't sleep, clown will eat me, ThePromenader, PoiZaN, Chlewbot, Joema, Cybercobra, Lpgeffen, Rpf, Kendrick7, Ccchambers, Byelf2007, Rory096, Boradis, Titus III, Richard L. Peterson, Xerxesx18, Writtenonsand, JorisvS, Mgiganteus1, Ckatz, Hypnosifl, Megane~enwiki, Ryulong, Quaeler, Dan Gluck, Spebudmak, Paul venter, Cxat, UncleDouggie, Courcelles, Tawkerbot2, JRSpriggs, Atomobot, Trevor.tombe, JForget, CRGreathouse, Lavateraguy, Nadyes, Mlsmith10, Arnavion, Logical2u, Rob Maguire, Cydebot, Stebbins, Gmusser, 879(CoDe), Rracecarr, Soetermans, Michael C Price, Chrislk02, Kozuch, Landroo, Thijs!bot, Headbomb, Marek69, Electron9, Second Quantization, Chris goulet, Davidhorman, Turelli, Dawnseeker2000, AntiVandalBot, Orionus, Gnixon, Fayenatic london, Tim Shuba, Empyrius, Archmagusrm, AstroPaul, Bagster, JAnDbot, Carl1011, Davewho2, MER-C, CosineKitty, Rkomatsu, Michael Wood-Vasey, Felix116, Acroterion, Bongwarrior, VoABot II, Tripbeetle, LordCémOnur, Seleucus, Kevinwiatrowski, Ours18, DerHexer, Nevit, Robin S, Simplizissimus, NatureA16, Johann1870, Jimmilu, ARCG, Nikpapag, TechnoFaye, Christian424, Tgeairn, J.delanoy, Trusilver, Maurice Carbonaro, Natty4bumpo, Komowkwa, OttoMäkelä, Jtechem, Tsuite, SJP, Videokunst~enwiki, Malerin, Jorfer, Potatoswatter, Cmichael, DorganBot, Jcmargeson, Ja 62, JHussein, Jjabellar, Sheliak, Johnassassin, Caribbean H.Q., VolkovBot, ColdCase, JohnBlackburne, D A Patriarche, AlnoktaBOT, Fences and windows, Philip Trueman, Darren22, HowardFrampton, TXiKiBoT, Oshwah, Dwight666, Zanardm, Someguy1221, Oxfordwang, Jackfork, UnitedStatesian, Mazarin07, Venny85, Goaliemaster121, SwordSmurf, Lamro, RayNorris, Fourthark, Wanchung Hu, Obsidianmile, Radical Robert, Noncompliant one, Donauland~enwiki, PlanetStar, TrulyBlue, Murad.Shibli, Likebox, Flyer22 Reborn, Hotdiggity, Avidalired, Faradayplank, Poindexter Propellerhead, OKBot, Aquijex, Loren.wilton, Martarius, BillWilliam, ClueBot, Dead10ck, The Thing That Should Not Be, Rodhullandemu, SuperHamster, Andwor, Tms9, Jusdafax, Da rulz07, Barbarinaz, Kentgen1, Razorflame, Stevecrye, AC+79 3888, Pillar of Babel, TimothyRias, Gwark, Ost316, PL290, MikeSmith10, Parejkoj, Andreaprins, Dgirl1723, HexaChord, D.M. from Ukraine, Addbot, Gravitophoton, Uruk2008, DOI bot, Nernom, LaaknorBot, Adfellin, Glane23, Delaszk, ChenzwBot, Sophia8891, Combatman~enwiki, Craigsjones, Arbitrarily0, Gurusof12, Cosmos72, Luckas-bot, Yobot, Cosoce, Systemizer, Aldebaran66, Fulcanelli, Amble, AnomieBOT, Iluziat, Materialscientist, Citation bot, Icosmology, ArthurBot, Xqbot, S h i v a (Visnu), Sionus, Drilnoth, Wperdue, Tomwsulcer, BLP-outrageous move logs, ProtectionTaggingBot, Mathonius, Shadowjams, Finncarey, PrimeMatter, FrescoBot, Paine Ellsworth, Tobby72, Sławomir Biały, Zero Thrust, Kvgyarmati, Woodingdean, Alpha plus (a+), Citation bot 1, Redrose64, Pinethicket, I dream of horses, Jonesey95, Three887, Tom.Reding, Shahidur Rahman Sikder, Efficiency1101e, Casimir9999, Aknochel, IVAN3MAN, Meier99, BradTheBadWiki, Trappist the monk, TADEET, Jordgette, Heurisko, Michael9422, Adi4094, Earthandmoon, Wellsmax, RjwilmsiBot, Alph Bot, EmausBot, Mmpcq, Grrow, Quantanew, RA0808, Slightsmile, Dcirovic, Italia2006, NicatronTg, H3llBot, Suslindisambiguator, Paulstarpaulstar, Frigotoni, Colin.campbell.27, Iiar, HCPotter, Tunborough, RockMagnetist, Herk1955, Deathglass, DASHBotAV, Fire Vortex, Mjbmrbot, Yceren Loq, ClueBot NG, Ccalen, Chester Markel, Matias Pocobi, Jj1236, Frietjes, Helvitica Bold, Curb Chain, Bibcode Bot, BG19bot, Gordonben, Cheeseray1, FiveColourMap, Hippokrateszholdacskai, Yizlpku, Snow Blizzard, Gerhardtschmerhardt, Migrainus, Mcspaans, Szczureq, Unclejoe0306, Guanghuilin, Akshay Lattimardi, Dexbot, City-OfUr, CuriousMind01, Wjs64, JustAMuggle, WorldWideJuan, Epicgenius, Yheyma, MiceEater, LindaYeah, DavidLeighEllis, Federicoturner, Babitaarora, Isateach, Onecreation, Prokaryotes, Christophe1946, BerdanII, Anrrusna, Stamptrader, Suelru, Monkbot, Mlsmith55, Haxxorz596, THemanRES%S23, Jnojha007, Richard.drapeau, UrDreamViola, MF22, ChamithN, Larsyxa, EpicLX, Tibenas, Mediavalia, ScrapIronIV, 39Debangshu, Anunaki truth, Tetra quark, Isambard Kingdom, Anand2202, GeneralizationsAreBad, Jman135, Grammarian3.14159265359, KasparBot, ShankZeTank, Tgorewic, Sir Cumference, ShiningSword, Esadri21, Phseek, Srisri19962003, TychosElk, Buckbill10, Alopresti777, Themalina, SireWonton, Adithya2804, Khrpr, Kigaei, Gerald wish, Dr. Hung M. Choi, Soopdish and Anonymous: 535

- **General relativity** *Source:* https://en.wikipedia.org/wiki/General_relativity?oldid=725436400 *Contributors:* AxelBoldt, Mav, Bryan Derksen, The Anome, AstroNomer, Ap, RK, Andre Engels, XJaM, Christintott, JeLuF, Christian List, William Avery, Roadrunner, Ktsquare, B4hand, Stevertigo, Frecklefoot, Edward, Patrick, Boud, Michael Hardy, Menchi, Ixfd64, Bcrowell, Nimrod~enwiki, TakuyaMurata, Mcarling, Minesweeper, Alfio, Looxix~enwiki, ArnoLagrange, Ellywa, Ahoerstemeier, Stevenj, William M. Connolley, Snoyes, Angela, Mark Foskey, Julesd, Salsa Shark, AugPi, Andres, Evercat, Hectorthebat, Hick ninja, A.Tigges~enwiki, Gingekerr, Jitse Niesen, Gutza, Rednblu, Doradus, Wik, Dragons flight, Tero~enwiki, Phys, Shizhao, Elwoz, Jerzy, BenRG, Banno, Northgrove, Phil Boswell, Robbot, Craig Stuntz, Sdedeo, Bvc2000, Goethean, Altenmann, Romanm, Lowellian, Mayooranathan, Gandalf61, Blainster, Diderot, DHN, Hadal, Alba, Johnstone, Fuelbottle, Isopropyl, Xanzzibar, Carnildo, Tea2min, Enochlau, Ancheta Wis, Tosha, Giftlite, JamesMLane, Graeme Bartlett, Mikez, BenFrantzDale, Lethe, Tom harrison, Fropuff, Everyking, Physman, Curps, Michael Devore, Jason Quinn, Alvestrand, SWAdair, Glengarry, Bobblewik, Edcolins, DefLog~enwiki, Pgan002, Knutux, GeneralPatton, HorsePunchKid, Robert Brockway, Kaldari, MadIce, Karol Langner, Rjpetti, Rdsmith4, JimWae, Anythingyouwant, Martin Wisse, Thincat, Euphoria, Icairns, Zfr, AmarChandra, Zondor, Econrad, JimJast, Discospinster, Rich Farmbrough, Guanabot, Pak21, ThomasK, Masudr, Pjacobi, Vsmith, Cdyson37, Jowr, Paul August, SpookyMulder, Dmr2, Bender235, Dcabrilo, Ground, Ben Standeven, Nabla, Livajo, El C, Worldtraveller, Shanes, Etimbo, Causa sui, Bobo192, Robotje, Smalljim, Rbj, JW1805, ParticleMan, I9Q79oL78KiL0QTFHgyc, Mr2001, Matt McIrvin, PWilkinson, Haham hanuka, Schnolle, Varuna, Jumbuck, Jérôme, Alansohn, Hackwrench, Cctoide, Crebbin, Wikidea, SlimVirgin, Benefros, Alexwg, Wtmitchell, Orionix, CloudNine, Bsadowski1, DV8 2XL, LordLoki, HenryLi, Oleg Alexandrov, Kelly Martin, Linas, FeanorStar7, Sabejias, Moneky, Kzollman, Cleonis, Mpatel, Jok2000, Schzmo, Pdn~enwiki, GregorB, Plrk, Wayward, Joke137, Christopher Thomas, Mandarax, Colodia, Canderson7, Rjwilmsi, WCFrancis, MarSch, Eyu100, JoshuacUK, JHMM13, Mike Peel, SanitysEdge, R.e.b., Ems57fcva, Bubba73, Gringo300, Ian Pitchford, RobertG, Mishuletz, Arnero, Mathbot, Nihiltres, Vsion, Perfect Tommy~enwiki, Itinerant1, Alfred Centauri, Gparker, Slant, Carrionluggage, Srleffler, Chobot, DVdm, Bgwhite, Dresdnhope, Manscher, PointedEars, Roboto de Ajvol, YurikBot, Wavelength, Bcarm1185, Splintercellguy, Hillman, EDG, MattWright, RussBot, Loom91, AVM, KSmrq, DanMS, SpuriousQ, Shawn81, Eleassar, Shanel, Syth, Madcoverboy, Tailpig, Schlafly, Dputig07, Beanyk, Tony1, Dna-webmaster, Enormousdude, 2over0, KGasso, Petri Krohn, GraemeL, Rlove, Sambc, LeonardoRob0t, Geoffrey.landis, HereToHelp, Willtron, Caballero1967, Meegs, Bsod2, Finell, Luk, Sardanaphalus, SmackBot, Kurochka, Hydrogen Iodide, Pavlović, Gnangarra, Unyoyega, Nickst, Delldot, Motorneuron, Cessator, Harald88, Edgar181, Shai-kun, Sectryan, Gilliam, Skizzik, Dauto, Saros136, Silly rabbit, Complexica, Colonies Chris, Zven, Abyssal, RProgrammer, Hve, RedHillian, BentSm, Phaedriel, Khoikhoi, Cybercobra, Downwards, Coolbho3000, Nakon, Peterwhy, SkyWriter, DMacks, Nairebis, Henning Makholm, UncleFester, Bidabadi~enwiki, Byelf2007, SashatoBot, Lambiam, Lapaz, Cronholm144, Gizzakk, CPMcE, JorisvS, Goodnightmush, Ckatz, Frokor, Garthbarber, SirFozzie, SandyGeorgia, Midnightblueowl, RichardF, Novangelis, Peter Horn, MTSbot~enwiki, Kvng, JarahE, Licorne, Quaeler, Fan-1967, Editor.singapore, MFago, JoeBot, ShyK, MOBle, RekishiEJ, CapitalR, MD:astronomer, Courcelles, Tawkerbot2, JRSpriggs, Kurtan~enwiki, Harold f, JForget, Sakurambo, Thermochap, Avanu, NickW557, MarsRover, Harrigan, Ian Beynon, Cydebot, Jasperdoomen, WillowW, Fl, MC10, Mato, Pascal.Tesson, Michael C Price, Christian75, DumbBOT, Biblbroks, Omicronpersei8, Crum375, N. Macchiavelli, Epbr123, Fisherjs, Markus Pössel, Martin Hogbin, MrXow, Oliver202, Headbomb, Pjvpjv, Tom Barlow, Davidhorman, D.H, AntiVandalBot, Abu-Fool Danyal ibn Amir al-Makhiri, Tkirkman, Gnixon, VectorPosse, TimVickers, Scepia, Dawz, Billevans~enwiki, Tim Shuba, Rico402, Archmagusrm, Jaredroberts, JAnDbot, Vorpal blade, Hut 8.5, YK Times, Acroterion, Pervect, Magioladitis, Connormah, RogierBrussee, WolfmanSF, JamesBWatson, Swpb, Ling.Nut, Soulbot, Pixel :-), KConWiki, WhatamIdoing, BatteryIncluded, Eldumpo, Allstarecho, User A1, Mollwollfumble, Chris G, Archen~enwiki, Thompson.matthew, STBot, Mermaid from the Baltic Sea, Shentino, Mschel, CommonsDelinker, Pbroks13, J.delanoy, DrKay, R. Baley, Numbo3, Leafsfan85, Aveh8, Lantonov, M C Y 1008, Mathlabster, Zedmelon, Aboutmovies, C quest000, Tcisco, Marrilpet, Nwbeeson, Aatomic1, Potatoswatter, Kolja21, Lseixas, Rémih, Caracalocelot, DemonicInfluence, Sheliak, Deor, Part Deux, JohnBlackburne, Philip Trueman, TXiKiBoT, Oshwah, Coder Dan, GimmeBot, Gombo, Hqb, Rei-bot, IPSOS, Qxz, T dohing, Molinogi, Fizzackerly, JhsBot, Leafyplant, Geometry guy, Ilyushka88, Thebigbendizzle, SwordSmurf, Andy Dingley, Gabrielsleitao, Lamro, Antixt, Vector Potential, James-Chin, Arcfrk, Ccheese4, StevenJohnston, Katzmik, YohanN7, Dnarby, SieBot, Tiddly Tom, Work permit, Yintan, RadicalOne, Wizzard2k, SteakNShake, Arbor to SJ, Babareddeer, JSpung, Phil Bridger, Wmpearl, Oxymoron83, Henry Delforn (old), Csloomis, Thehotelambush, Lightmouse, BrightRoundCircle, OpTioNiGhT, The-G-Unit-Boss, Emgg, AWeishaupt, Divinestuff, Coldcreation, Adam Cuerden, Duae Quartunciae, Heptarchy of teh Anglo-Saxons, baby, Randomblue, TFCforever, Danthewhale, Martarius, Sfan00 IMG, ClueBot, The Thing That Should Not Be, Rjd0060, Metaprimer, Wwheaton, Der Golem, JTBX, TheAmigo42, CounterVandalismBot, Viran, Blanchardb, Rotational, Agge1000, Itzguru, Tanketz, CohesionBot, Eeekster, Stealth500, Brews ohare, NuclearWarfare, PhySusie, SockPuppetForTomruen, SchreiberBike, Another Believer, RubenGarciaHernandez, AC+79 3888, MasterOfHisOwnDomain, He6kd, TimothyRias, Lazyrussian, PseudoOne, Skarebo, NellieBly, JinJian, Truthnlove, Everydayidiot, Tayste, Balungfrancis, Addbot, Mortense, Some jerk on the Internet, Fizzycyst, DOI bot, Mistyocean3, Metagraph, Stariki, Fluffernutter, Schmoolik, MrOllie, Download, EconoPhysicist, Delaszk, Favonian, LinkFA-Bot, Tuition, Tassedethe, Nnedass, Tide rolls, Lightbot, Knutls, Luckas-bot, Ptbotgourou, Legobot II, Julia W, Anypodetos, Trickyboarder93, Superamoeba, AnomieBOT, Kristen Eriksen, Giordano.ferdinandi, Jim1138, Jo3sampl, Materialscientist, Wandering Courier, The High Fin Sperm Whale, Citation bot, Xqbot, Stlwebs, Sionus, Amareto2, Unigfjkl, Nickkid5, Stsang, Coretheapple, GrouchoBot, Collin21594, RibotBOT, Rucko123, GhalyBot, Acannas, LucienBOT, Paine Ellsworth, Lagelspeil, Steve Quinn, Knowandgive, Pokyrek, Citation bot 1, Citation bot 4, Electrozity8, Pinethicket, LittleWink, Jonesey95, A412, Tom.Reding, Yougeeaw, Barras, Jauhienij, Meier99, Citator, Comet Tuttle, Hughston, Defender of torch, Duoduoduo, Aribashka, Iibbmm, Diannaa, Earthandmoon, Tbhotch, Brambleclawx, Marie Poise, RjwilmsiBot, Aznhero3793, Ripchip Bot, EmausBot,

WikitanvirBot, Immunize, Zhaskey, Fly by Night, DuKu, GoingBatty, Jmencisom, Slightsmile, Dcirovic, Hhhippo, JSquish, ZéroBot, Cogiati, Stanford96, Empty Buffer, Sanford123456, H3llBot, Quondum, REkaxkjdsc, Monterey Bay, Mr little irish, TonyMath, Brandmeister, Maschen, Puffin, Carmichael, Newstv11, RockMagnetist, Sona11235, WizardofCalculus, Milk Coffee, Whoop whoop pull up, Mjbmrbot, Helpsome, ClueBot NG, Manubot, Hagenfeldt, This lousy T-shirt, SusikMkr, Ggonzalm, Jj1236, Mgvongoeden, Snotbot, Widr, Jamester234, Pluma, Ginger.spice14, Bibcode Bot, Jeraphine Gryphon, Lowercase sigmabot, BG19bot, Quarkgluonsoup, Bolatbek, Marsambe, Amp71, Mark Arsten, Lovepool1220, Marsambe1, Benzband, ENG.F.Younis, 123matt123, DeviantFrog, IrishDevil2, F=q(E+v^B), Egbertus2, Harizotoh9, Doctor Lipschitz, Snow Blizzard, Physicsch, Zoldyick, Roozitaa, BattyBot, Reed07, Vanobamo, JoshuSasori, Stigmatella aurantiaca, Cyberbot II, Abhay ravi, ChrisGualtieri, Maestro814, Deathlasersonline, Ptokijnu, Billyshiverstick, Read Blooded, Theeditor6079, Flyer1997, Dexbot, Suffian Akhtar, Irondome, Kryomaxim, Twhitguy14, CuriousMind01, J0437-4715, Jamesx12345, Among Men, Leprof 7272, WorldWideJuan, Devinray1991, 1888software, EvergreenFir, Enchantedscience, Mohamed F. El-Hewie, Vai ra'a toa Taina, NeapleBerlina, Jwratner1, Gigantmozg, Ginsulott, SirKesuma, Anrnusna, JaconaFrere, Osamabin7, Juenni32, Filedelinkerbot, SantiLak, Aryabhatt 21, Willbh15, S11027158, Cjsmith.us, ChamithN, Cris Cyborg, PeterShawhan, Evgeniy E., Sweeeeeeeed, Tetra quark, Absolutelypuremilk, Praveece, JLT2045, LL221W, Jf2839, GeneralizationsAreBad, KasparBot, Jmc76, Sir Cumference, Lemonberry622, Pizzaman62, Dgray101, Amrespi2007, Narasimha Kanduri, דוקימרון, J1738, Soopdish and Anonymous: 735

- **Big Bang** *Source*: https://en.wikipedia.org/wiki/Big_Bang?oldid=725333250 *Contributors*: AxelBoldt, Chenyu, Trelvis, The Epopt, CYD, Mav, Bryan Derksen, Zundark, Timo Honkasalo, The Anome, AstroNomer, Manning Bartlett, Malcolm Farmer, Tim Chambers, Ed Poor, Andre Engels, Josh Grosse, Danny, XJaM, William Avery, Roadrunner, SimonP, Aoineko, FvdP, David spector, Heron, GrahamN, Bth, Montrealais, Youandme, Branko, Modemac, Bignose, Hephaestos, Stevertigo, Edward, Nealmcb, Patrick, D. Michael Hardy, EddEdmondson, Modster, Bewildebeast, MartinHarper, Collabi, Tannin, Bobby D. Bryant, Ixfd64, Bcrowell, Sannse, Huboluv, Kosebamse, CesarB, Egil, NuclearWinner, Looxix~enwiki, Mkweise, Ahoerstemeier, KAMiKAZOW, Dgaubin, Stevenj, William M. Connolley, Muriel Gottrop~enwiki, Angela, Jebba, Kingturtle, Bueller 007, Alvaro, Aarchiba, Александър, Glenn, Chimpa, Susurrus, Andres, Evercat, Rob Hooft, Edaelon, Smack, Schneelocke, Hike395, Hashar, Feedmecereal, Disdero, Ec5618, Charles Matthews, Timwi, Reddi, Viajero, Jitse Niesen, Wayne~enwiki, The Anomebot, Doradus, Bjh21, DJ Clayworth, CBDunkerson, Saltine, Paul-L~enwiki, Phys, Omegatron, Ionized, Bevo, Topbanana, Raul654, BenRG, Pollinator, Rossnixon, MrJones, Astronautics~enwiki, Fredrik, Korath, Schutz, Goethean, Peak, Yelyos, Romanm, Lowellian, Gandalf61, Academic Challenger, Rursus, Blainster, DHN, Sunray, Hadal, Fuelbottle, Racky, Isopropyl, JerryFriedman, Jooler, Ancheta Wis, Apol0gies, Centrx, Giftlite, Graeme Bartlett, Christopher Parham, Jacoplane, Andy, Qartis, Harp, Lee J Haywood, Lupin, Timpo, Fastfission, Obli, Peruvianllama, Noone~enwiki, Wwoods, Everyking, Plautus satire, Jacob1207, Anville, Curps, Joe Kress, Jtdwolff, Guanaco, Jason Quinn, Sundar, Gracefool, Eequor, Bobblewik, Deus Ex, Golbez, Wmahan, Stevietheman, Masterhomer, Gadfium, Utcursch, Andycjp, Sohailstyle, Mendel, Sonjaaa, Antandrus, HorsePunchKid, Dialog, Eroica, MisfitToys, Piotrus, Kaldari, Jossi, Karol Langner, OwenBlacker, Latitude0116, RetiredUser2, DanielDemaret, Icairns, JDoolin, Iantresman, Neutrality, Urhixidur, Jcw69, JohnArmagh, Djyang~enwiki, Klemen Kocjancic, Sonett72, Adashiel, Lacrimosus, Mike Rosoft, Ta bu shi da yu, HedgeHog, Jayjg, Rll, Archer3, Ultratomio, Deadlock, Carl Henderson, Arensb, KeyStroke, JimJast, Discospinster, ElTyrant, Rich Farmbrough, Rhobite, Guanabot, FT2, Kdammers, Oliver Lineham, Pjacobi, Vsmith, Jpk, Silence, Arthur Holland, Mani1, SpookyMulder, Bender235, ESkog, Kbh3rd, Kaisershatner, Ben Standeven, Plugwash, Violetriga, Brian0918, RJHall, Carlon, Lycurgus, Rgdboer, Jomel, Worldtraveller, Shanes, Art LaPella, RoyBoy, JeremyLydellHaugen, Causa sui, Bobo192, Army1987, Viriditas, Cohesion, Adrian~enwiki, L33tminion, DaveGorman, I9Q79oL78KiL0QTFHgyc, Giraffedata, Joe Jarvis, Man vyi, Nk, TheProject, Ajdlinux, Obradovic Goran, Sam Korn, Jonathunder, Mdd, Orangemartin, Ranveig, Jumbuck, Schissel, Danski14, Gary, JYolkowski, Raintaster, Comrade Tassadar, Diego Moya, Mr Adequate, Andrewpmk, Paleorthid, Plumbago, Sp82, Punarbhava, Riana, Thorns among our leaves, Lectonar, Lightdarkness, Sligocki, JHG, Snowolf, Wtmitchell, Dschwen, Schapel, Orionix, Fourthords, Rick Sidwell, Rafti Institute, Knowledge Seeker, BrandonYusufToropov, Staeiou, Стамм, Aaron Bruce, Deathphoenix, Itsmine, DV8 2XL, Gene Nygaard, HenryLi, Kazvorpal, Ott, Matevzk, Dmitry Brant, Hq3473, WilliamKF, JarlaxleArtemis, Anilocra, Twobitsprite, Braxeus, Bwallum, Jacobolus, Nameneko, Ruud Koot, JeremyA, Chochopk, K Lepo, Jok2000, Pi@k~enwiki, Tabletop, Schzmo, Pdn~enwiki, Al E., I64s, Optichan, Cigsandalcohol, GregorB, Dataphiliac, Preisler, CharlesC, Wayward, Joke137, Prashanthns, Gimboid13, Cedrus-Libani, Christopher Thomas, Srkpriv, Bebenko, Paxsimius, GSlicer, Larman, RichardWeiss, Rn120, Graham87, Magister Mathematicae, BD2412, Chun-hian, Fleisher, DianaS, Nairb~enwiki, David Levy, FreplySpang, Khronos21, RxS, Icey, BorgHunter, GBoehm, Tlroche, Ketiltrout, Sjö, SteveW, Drbogdan, Rjwilmsi, Wikibofh, Vary, Xosé, Tangotango, Aximilli, MZMcBride, Mike s, ErikHaugen, Mike Peel, Vegaswikian, Oblivious, Justin Hirsh, HappyCamper, Ligulem, ElKevbo, Bubba73, Brighterorange, Krash, Dar-Ape, Sango123, Raprat0, Yamamoto Ichiro, FayssalF, Drrngrvy, FlaBot, RobertG, Old Moonraker, Nihiltres, Dantecubed, HiddenWolf, RexNL, Gurch, ElfQrin, Intgr, TeaDrinker, Diza, Consumed Crustacean, Phoenix2~enwiki, Mongreilf, Chobot, Moocha, HKT, DVdm, Mhking, JesseGarrett, UkPaolo, Satanael, YurikBot, Wavelength, Spacepotato, RobotE, Sceptre, Jimp, Mahahahaneapneap, JustSomeKid, RussBot, BNL52577, Icarus3, Bhny, JabberWok, Chris Mid, CanadianCaesar, Stephenb, Madyasiwi, The1physicist, Thryllkill, Gaius Cornelius, Eleassar, Oni Lukos, Wimt, Anomalocaris, Fnorp, NawlinWiki, SEWilcoBot, Wiki alf, Bachrach44, ChadThomson, Grafen, Jaxl, Dlugosz, Willbown, Dureo, JocK, Apokryltaros, Irishguy, Sir48, Kdbuffalo, Anetode, Chrisbri88, Cholmes75, PhilipO, Raven4x4x, Grafikm fr, Syrthiss, Aaron Schulz, Samir, BOT-Superzerocool, Gadget850, Alex cole, Kortoso, DeadEyeArrow, Bota47, T-rex, Mistercow, Petergaltco, Smaines, Wknight94, Boivie, FF2010, Enormousdude, 2over0, Ageekgal, Thnidu, Endomion, Dr.alf, Varith, CharlesHBennett, Sean Whitton, GraemeL, JoanneB, CWenger, Carabinieri, Ilmari Karonen, Caco de vidro, DisambigBot, Katieh5584, Kungfuadam, Maxamegalon2000, Infinity0, Amberrock, DVD R W, Finell, Andrewwang90, Luk, Robertd, Anton n, Amalthea, Crystallina, KnightRider~enwiki, SmackBot, YellowMonkey, Scorpiona, Mehranwahid, Joeljostin, Dav2008, KnowledgeOfSelf, Melchoir, Kimon, Pgk, Blue520, WilyD, KocjoBot~enwiki, Tchernobog, Jagged 85, Davewild, Bodhis005a, Dsouza, Hew~enwiki, Delldot, Hardyplants, Hbackman, Harald88, Edgar181, Alsandro, Trystan, Markeer, Moralis, Onsly, Gilliam, Portillo, Skizzik, Chaojoker, Ppntori, Raghav t, Cabe6403, Qtoktok, Jeffro77, Fetofs, Pope523, Aryeztur, Persian Poet Gal, MalafayaBot, Papa November, MidgleyDJ, Elerner, Dustimagic, Ikiroid, Ted87, Kungming2, Mohamed Al-Dabbagh, DHN-bot~enwiki, Terraguy, Rlevse, Mikker, Scwlong, Can't sleep, clown will eat me, Jefffire, Writtenright, Ioscius, Hve, Vanished User 0001, Nixeagle, Paales, Dannylim, Yidisheryid, TheKMan, EvelinaB, Xiner, Andy120290, Mr.Z-man, ClairSamoht, King Vegita, Jmlk17, Gwaka Lumpa, Daqu, Nakon, Miketzhu, TedE, VegaDark, MHoerich, MichaelBillington, Alexandra lb, Weregerbil, Jklin, Duke nemmerle, Metamagician3000, Twir, Kalathalan, Shushruth, Marcus Brute, Captainbeefart, Ck lostsword, Samuel Sol, Pilotguy, Kukini, Rossp, Byelf2007, SashatoBot, Lambiam, Nishkid64, Rory096, Harryboyles, Rklawton, Dasune, Sophia, T-dot, Kuru, John, AmiDaniel, Scientizzle, J 1982, Rijkbenik, Benesch, JoshuaZ, Robert Stevens, Reuvenk, NathanLee, Bjankuloski06en~enwiki, Gnevin, IronGargoyle, Bilby, Heliogabulus, Kakadinho2210, 041744, Ckatz, RandomCritic, Ezrarez, JHunterJ, Digger3000, Kyphe, Muadd, Martinp23, George The Dragon, Mr Stephen, FredrickS, Hypnosifl, Waggers, Geologyguy, AdultSwim, Potable potables, Condem, Novangelis, Sasata, Hu12, Keith-264, Vanished user, Dekaels~enwiki, Astrobayes, Paul venter, Cxat, Lottamiata, Shoeofdeath, Newone,

Twas Now, RekishiEJ, CapitalR, Az1568, Wikidude1, Tawkerbot2, Dlohcierekim, Chetvorno, Tommysun, WikiMarshall, Kurtan~enwiki, OAP boba, Idols of Mud, VinnieCool, JForget, CmdrObot, Geremia, Irwangatot, Insanephantom, Memetics, Tom33, Agathman, Mohitkhullar, Olaf Davis, BeenAroundAWhile, Leopoldhausen, CWY2190, GHe, Aquirata, Dgw, NickW557, WeggeBot, SelfStudyBuddy, Casper2k3, Neelix, Nnp, Cnj, Myasuda, Awptics, Gregbard, Steel, Iceman14n, CovenantD, Gogo Dodo, Zarcom, Zgystardst, Travelbird, Bridgecross, Wa2ise, Frosty0814snowman, Ricuda, 879(CoDe), Roketjack, Aeiownusir, Michael C Price, Tawkerbot4, Christian75, DumbBOT, Chrislk02, Dinnerbone, Wazzz, Kozuch, Septagram, Helvetica, Scarpy, Omicronpersei8, Voldemortuet, PaladinWriter, Landroo, PamD, LilDice, Malleus Fatuorum, Ulnevets, Qwyrxian, KimDabelsteinPetersen, Skyfire.michael, Kablammo, Sry85, UniverseToday, Keraunos, Gamer007, Anupam, ANIMAL~enwiki, Sopranosmob781, Publicola, Headbomb, Simeon H, West Brom 4ever, Warrior m4, Tapir Terrific, James086, Second Quantization, Peter Gulutzan, X201, Davidhorman, Rosencrantz1, Rhrad, Aristox, Greg L, BlytheG, Will Bradshaw, Dawnseeker2000, Navigatr85, Escarbot, KrakatoaKatie, AntiVandalBot, Martyn Smith, Luna Santin, CodeWeasel, Bigtimepeace, Benjaburns, Gnixon, Doc Tropics, The Hut, Rabbi-m, Ste4k, Dane 1981, Dr. Submillimeter, PhJ, John Gibbons2, Danger, Farosdaughter, Skynet1216, Tim Shuba, Chill doubt, DarthShrine, Lonestar662p3, Rico402, Byrgenwulf, DrMacrophage, Kaini, AstroPaul, Nate Slayer0, MER-C, Matthew Fennell, Db099221, JamesAgain, Andonic, Tstrobaugh, Viriathus, LittleOldMe, .anacondabot, Pervect, Propaniac, Magioladitis, YishaiMagelMoganim, Murgh, Bongwarrior, VoABot II, Jvasu 2000, P64, Bakken, Bobby McGehee, Feeeshboy, Marhadiasa, Godwillwin, OoTV, TheMusicalGenius, Out0ftunevi0lin, Outoftuneviolin 5, Avicennasis, Bubba hotep, Fabricebaro, Theroadislong, Animum, Truthiness34, Cgingold, Outoftuneviolin Returns, Outoftuneviolin VS Wikipedia, Charliet, Kevinwiatrowski, Another sockpuppet of Outoftuneviolin, BatteryIncluded, JJ Harrison, MiPe, Joe hill, Allstarecho, Mike Payne, Lowmax2, Sontearemoreequal, Shijualex, Glen, GAH GAH, DerHexer, Teardrop onthefire, Benra, Kheider, Rickard Vogelberg, Joshua Davis, Sambop, Misarxist, Hdt83, MartinBot, Bhenderson, Verkle, Yasinmiyar, LordPhobos, Liam159, Anaxial, R'n'B, Soccerpro, CommonsDelinker, Nwhitehair, Crisnumbertwo, Fellwalker57, Down.with.conformity, HEL, Dromoreboy, DrKay, C Ronald, Svetovid, C T, Paulamicela, Leon kennedy8, Maurice Carbonaro, Kemiv, Extransit, WarthogDemon, TomS TDotO, 1tephania, It Is Me Here, BrokenSphere, Bot-Schafter, Katalaveno, Sonalchagi, Hiathitler, Bejerbel, Ben robbins, Grosscha, AMERICA SUPREME, Skier Dude, Tarotcards, Hammiolo, NewEnglandYankee, Wesino, SJP, Student7, Umair82, Pinea, KylieTastic, Triangulator, Cometstyles, STBotD, Kenneth M Burke, Que-Can, Johnston213, Jtankers, Sarregouset, DMCer, Ollie 9045, Cazlo0, Furrypig, Squids and Chips, Muchclag, Azndragon2131, Idiomabot, Funandtrvl, Spellcast, Jamesaf123, Fainites, Wikieditor06, Lights, Urrg.... X!, UnicornTapestry, VolkovBot, Ghustug, Johan1298~enwiki, Gwuen Galeus1978, AlnoktaBOT, Kyle the bot, FergusM1970, George Adam Horváth, Philip Trueman, Alvevind~enwiki, Childhoodsend, TXiKiBoT, Joopercoopers, NDUTU~enwiki, 99DBSIMLR, Tricky Victoria, Kww, Udufruduhu, Flarblesarefun, Smalls1652, Dchall1, Zybez, AlysTarr, Qxz, Someguy1221, Lradrama, Meisaran, Fizzackerly, PDFbot, UnitedStatesian, Henryodell, Reevesastronomy, Mazarin07, Telecineguy, Billinghurst, SwordSmurf, Lamro, Tbtkorg, Petej010, Synthebot, Falcon8765, Moose-32, Havs84, WatermelonPotion, Insanity Incarnate, Thealltruth, AlleborgoBot, Michael Frind, FlyingLeopard2014, EmxBot, PaddyLeahy, Drschawrz, KyZan, SieBot, Kmasters0, Soccermonkey 77, Duguti, Lalala98, Restre419, Tiddly Tom, Nihil novi, Invmog, Bubuntu, Phe-bot, Parhamr, Caltas, Gravitan, Radcliffe777677, Luke2thab, Keilana, Atl braves, Abhishikt, Flyer22 Reborn, Tiptoety, Oda Mari, Mirkoruckels, Skaggamoo, Jc-SOCO, L.P, Cheshunt, Gilmiciak, Scooby1257, Oxymoron83, Likemk687, Yeenar, Nuttycoconut, Zharradan.angelfire, Steamboatdude, Lightmouse, Jruderman, Sunrise, Pediainsight, Dsmith7707, Jay Turner, Coldcreation, Vanished user ewfisn2348tui2f8n2fio2utjfeoi210r39jf, Duae Quartunciae, Anchor Link Bot, Phral, Hamiltondaniel, Firefly322, Superbeecat, Denisarona, Escape Orbit, Into The Fray, C0nanPayne, Myrvin, Asher196, Invertzoo, SallyForth123, Nondistinguished, Faithlessthewonderboy, ArepoEn, CrunchyAvocado000, Ixodeth, ClueBot, Wildie, RedGav, GorillaWarfare, Ryanborgz, Double kz77, Fyyer, Kotniski, Sammmtttt, Petersburg, Taffboyz, Surfer9986, Vacio, EMC125, Dean Wormer, Jaygeisler, Unbuttered Parsnip, Tanglewood4, UserDoe, Slater bob, Drmies, Russ143, Joshua Gonsalves, Altenhofen, Polyamorph, SuperHamster, Nursebhayes, CounterVandalismBot, NovaDog, VandalCruncher, Agge1000, Neverquick, ChandlerMapBot, Ninjorturtle456, BlueAmethyst, Markarkrkk, Chimesmonster, DragonBot, Darian Tang, Welsh-girl-Lowri, Heaney21, MissLadyZara, Peacetoyomama, Santokh01, Thebeast373, CidVSReno, Utopial, Millionsandbillions, Cenarium, Jotterbot, PhySusie, Biochem67, M.N.Qunson, Kentgen1, Scog, Revotfel, Truth is relative, understanding is limited, Chaosdruid, Panos84, ModestMouse2, Inlovewithaboyscout, Aitias, Spinoff, Dana boomer, Byeahman, Johnuniq, Oore, DumZiBoT, TimothyRias, JWhitt433, DCCougar, Stickee, MilnerJames, Gwark, Spinner198, Purnajitphukon, Ummhihello, Jprw, Jdude3, Ilikepie2221, SilvonenBot, Shailaja87, Cow457, Stragler, Fairdeal08, Padfoot79, Kaiwhakahaere, Aunt Entropy, Nishbond, Leondoneit2, Zinger0, Good Olfactory, Parejkoj, Chris-constant, Tre2, Tayste, KirbyManiac, Maldek, Addbot, Basilicofresco, Ocrasaroon, Uruk2008, DOI bot, Yoenit, Jazza18, Fosftots, KitchM, WFPM, NjardarBot, LaaknorBot, BepBot, Blueman 7, The C of E, TStein, Knightofbaghdad, LinkFA-Bot, Patton123, Tassedethe, VASANTH S.N., Astro-norte, TundraGreen, John.St, Krukouski, आश्रीष मशताना, Yinweichen, Luckas-bot, Yobot, Ht686rg90, Aldebaran66, Whileactor~enwiki, KamikazeBot, Azcolvin429, Farsight001, Armchair info guy, AnomieBOT, 1exec1, Six words, Jim1138, RBM 72, LlywelynII, Dr. Günter Bechly, HowDumbAreYou, Csigabi, Mann jess, Ameki, Materialscientist, Citation bot, Stronach, MauritsBot, Xqbot, TinucherianBot II, A455bcd9, Hanberke, Gilo1969, Gap9551, GrouchoBot, Sirrontail, Silence-is-infinite, Ashershow1, RibotBOT, Waleswatcher, DASDBILL2, Championpork, GhalyBot, Canned Soul, A. di M., Jbananal, 新, Alexmaxbir, FreeKnowledgeCreator, CES1596, Nagualdesign, Lionelt, FrescoBot, LucienBOT, Paine Ellsworth, Ilikesealife, Flygongengar, Rotideypoc41352, D'ohBot, EmilTyf, SF88, Kwiki, Airborne84, Citation bot 1, Careful With That Axe, Eugene, Redrose64, Theory2reality, Dogaru Florin, Pinethicket, Edderso, Tom.Reding, AmphBot, Shahidur Rahman Sikder, Longview32, Efficiency1101e, RedBot, Btilm, MastiBot, IVAN3MAN, Gamewizard71, TobeBot, Trappist the monk, Puzl bustr, Fama Clamosa, Lotje, Callanecc, Extra999, JLincoln, Tbhotch, DARTH SIDIOUS 2, RjwilmsiBot, Bento00, Ripchip Bot, Androstachys, Charlieadam, Techhead7890, Tesseract2, DASHBot, Jpatros, EmausBot, Nathanl1192, WikitanvirBot, Gfoley4, Dominus Vobisdu, Quantanew, Katherine, MikeyTMNT, Joseph507357, Canprog, Bt8257, Tommy2010, Wikipelli, Dcirovic, Italia2006, Grondilu, Ida Shaw, A2soup, JosueM, Kpreet1996, Everard Proudfoot, Barbara.fischerclark, Aeonx, H3llBot, SporkBot, Cymru.lass, Hubbabubba3, Mcmatter, Mop head155, Brandmeister, Vanished user fijtji34toksdcknqrjn54yoimascj, L Kensington, Yobingfrog, Mahendra.sharma83, Brandon82694, Krsaurabhbca, Harmi.banik, FelixG1995, 1800reverse, RockMagnetist, Blarg123456789, Linette18, Czeror, Ebehn, ClueBot NG, Smtchahal, Gilderien, Satellizer, Joefromrandb, Movses-bot, Hiperfelix, Widr, North Atlanticist Usonian, Helpful Pixie Bot, Rsercher, Bibcode Bot, BG19bot, Ymblanter, ArtifexMayhem, Stephfo, Quarkgluonsoup, Knowledge Examiner, Hurricane2u, Dr.Toonhattan, FiveColourMap, Cadiomals, Harizotoh9, MrBill3, Cky2250, Emaha, Russianamerican1, TheGoodBadWorst, Amphibio, BattyBot, Tomh903, Judiakok1985, Cyberbot II, ChrisGualtieri, Soulbust, Arcandam, SD5bot, Khazar2, Rhlozier, Lelouch Di Britannia, JYBot, Wassup234, Dexbot, Inayity, Jamesx12345, Sowlos, Junjunone, Choor monster, Rfassbind, FamAD123, Linuxgal, Everymorning, Hardcoreromancatholic, Praemonitus, Curatiotech, Comp.arch, Crow, Arjunkrishna90, Rajgopal iyer, Mahusha, LOrmston, Monkbot, Yikkayaya, Chuckleheimersy, Owais Khursheed, Pingumeister, Wikipedian 2, Spumuq, Hamnus, Rubbish computer, I'm your Grandma., Li Da Mo II, Tetra quark, Douglask1835, Isambard Kingdom, Anand2202, SocraticOath, Jerodlycett, Supdiop, KasparBot, Greyhatrex, Adam9007, Sir Cumference, Intelligent Mr Toad 2, TychosElk, Youknowwhatimsayin, Milku3459, Reuben Miguel Felix, Wasd2333, Worldandhistory, Xx Cool Guy7202 xX and Anonymous: 1068

- **Black hole** *Source:* https://en.wikipedia.org/wiki/Black_hole?oldid=725487065 *Contributors:* Chenyu, CYD, Ansible, Bryan Derksen, Zundark, Timo Honkasalo, The Anome, Tarquin, AstroNomer, Gareth Owen, Ed Poor, Wayne Hardman, Eclecticology, Graham Chapman, XJaM, Arvindn, Roadrunner, SimonP, Ben-Zin~enwiki, Apollia, Modemac, Chris Q, Dbundy, Stevertigo, Frecklefoot, Edward, Patrick, RTC, Boud, JohnOwens, PhilipMW, Ken Arromdee, Michael Hardy, Tim Starling, EddEdmondson, Kwertii, Isomorphic, Fuzzie, Jketola, Sam Francis, Ixfd64, Bcrowell, Iluvcapra, AquaRichy, Minesweeper, Alfio, Kosebamse, Stw, Looxix~enwiki, Ahoerstemeier, Cyp, Anders Feder, Ronz, William M. Connolley, Theresa knott, Snoyes, Suisui, Angela, Den fjättrade ankan~enwiki, Jebba, Glenn, Susurrus, Evercat, Samuel~enwiki, Mxn, Schneelocke, Hike395, Emperorbma, Frieda, Fry-kun, Vanished user 5zariu3jisj0j4irj, Wikiborg, Paul Stansifer, Jwrosenzweig, The Anomebot, Doradus, Tpbradbury, Marshman, Maximus Rex, Furrykef, Morwen, Saltine, Taxman, Rei, Ed g2s, Rnbc, Thue, Lord Emsworth, Joy, Raul654, BenRG, Banno, Jhobson1, Jeffq, Owen, RadicalBender, Mrdice, Northgrove, SD6-Agent, Phil Boswell, Vt-aoe, AlexPlank, Robbot, Sander123, Craig Stuntz, TomPhil, Alrasheedan, RedWolf, Donreed, Altenmann, Romanm, Lowellian, Merovingian, Sverdrup, Meelar, Auric, JB82, DHN, Davodd, Hadal, Quincy, JesseW, Wikibot, Wereon, Borislav, Reid, Jheise, JerryFriedman, Diberri, Jholman, Dina, Tea2min, Alan Liefting, Enochlau, Giftlite, DocWatson42, Christopher Parham, MPF, Gtrmp, Awolf002, Andy, Barbara Shack, Castaa, Cobaltbluetony, Lethe, Tom harrison, Art Carlson, Lupin, Herbee, SheikYerBooty, Xerxes314, Paul Pogonyshev, Peruvianllama, Everyking, Plautus satire, Anville, NASA~enwiki, Curps, David Johnson, Home Row Keysplurge, Sriehl, Joe Kress, Cantus, Rpyle731, Andris, Guanaco, Avsa, Jorge Stolfi, Sundar, Eequor, Solipsist, Nathan Hamblen, Foobar, Dan Gardner, PlatinumX, SWAdair, D.A.ugosz, Bccomm, Bobblewik, Joseph Dwayne, RcktScientistX, Stevietheman, StuartH, Chowbok, Geni, Gdr, Fpahl, Antandrus, HorsePunchKid, Beland, Onco p53, MadIce, Noirum, Rdsmith4, Maximaximax, Jokestress, Aranoff, Jobrober, Variant, Kevin B12, Bosmon, Satori, GeoGreg, Sam Hocevar, Tzarius, Gscshoyru, Iantresman, Neutrality, Joyous!, Quota, TJSwoboda, Jewbacca, Deglr6328, Temujin9, Jwtidnet, Ayager, Mike Rosoft, Ouro, Simonides, Freakofnurture, Spiffy sperry, Poccil, Bactram, Indosauros, یک, Discospinster, Solitude, Rich Farmbrough, Guanabot, Yuval madar, Igorivanov~enwiki, FT2, Pjacobi, Vsmith, Jpk, Ponder, Antaeus Feldspar, Manil, Olau, Paul August, Xjaymanx, Dmr2, MJSS, Bender235, ESkog, Zaslav, Kjoonlee, Cucumberslumber, Kalel, Nabla, Brian0918, RJHall, Livajo, El C, Lankiveil, Parklandspanaway, Edward Z. Yang, Shanes, Arete~enwiki, Spearhead, Susvolans, Rsmelt, Art LaPella, RoyBoy, Dalf, Jpgordon, Iridia, Causa sui, Shoujun, Bobo192, 23skidoo, Billymac00, Flxmghvgvk, Draco2, Reuben, Jguk 2, JW1805, Redquark, I9Q79oL78KiL0QTFHgyc, Timl, Chbarts, Toh, La goutte de pluie, Nk, BM, NickSchweitzer, Doozer, Hi3221 10, Tuskey, Cherlin, Apostrophe, Haham hanuka, Hagerman, Tms, Wayfarer, Solocommand, MetalMilitia, Papeschr, Knucmo2, Jumbuck, Jérôme, Alansohn, Jamyskis, Tek022, Keenan Pepper, Andrewpmk, Tezeti, Ricky81682, Andrew Gray, Lord Pistachio, Punarbhava, Riana, Wikidea, AzaToth, Keflavich, Lectonar, Axl, R Calvete, Mac Davis, Cdc, Grobertson, Transcend~enwiki, Wtmitchell, BanyanTree, BRW, QuixoticKate, Almafeta, Cecil, Yuckfoo, Jheald, Count Iblis, H2g2bob, ThomasWinwood, Gortu, Computerjoe, Jchillerup, Freyr, DV8 2XL, Mordero, Gene Nygaard, Axeman89, Nick Mks, Kazvorpal, Markaci, Njk, Dmitry Brant, Bobrayner, ChrisJMoor, Richard Arthur Norton (1958-), Rorschach, OwenX, JarlaxleArtemis, Camw, LOL, Pinball22, Merlinme, Prophile, Orchew, BillC, Mazca, HFarmer, JeremyA, Direwolf, Mpatel, Tabletop, Ianweller, Schzmo, Nirmalya, 164s, MFH, PhoenixPinion, GregorB, Macaddct1984, El Suizo, CharlesC, Jon Harald Søby, Joke137, Christopher Thomas, Rufous, Rgbea, Bebenko, GSlicer, Rnt20, Graham87, Magister Mathematicae, GoldRingChip, BD2412, Qwertyus, Chun-hian, Kbdank71, Eteq, RxS, BorgHunter, Drbogdan, Akubhai, Coneslayer, Sjakkalle, Rjwilmsi, Zbxgscqf, Phileas, Arie~enwiki, WCFrancis, Wikibofh, Vary, Strait, Marasama, Strake, Sdornan, Captain Disdain, HandyAndy, Mike s, Nick R. Elkester, Ligulem, Jehochman, SeanMack, Ems57fcva, AndyKali, Bhadani, Jackdriscoll, Maurog, GregAsche, Sango123, Yamamoto Ichiro, KaiMartin, W00d, Lionelbrits, FayssalF, FlaBot, Patrick1982, Ian Pitchford, SchuminWeb, Vegardw, RobertG, Musical Linguist, SiriusB, Nihltres, Harmil, Nivix, Chanting Fox, Itinerant1, RexNL, Gurch, Schumps, Hansamurai, Algri, Gmz1023, Poderis, Fresheneesz, Pete.Hurd, Preslethe, Jesse0986, Alphachimp, Diza, Tedder, Kri, Imnotminkus, Ahsankhan, GringoCroco, Essaregee, Chobot, Fourdee, DVdm, RashBold, GreyedOut, Bgwhite, Ahpook, Cactus.man, Eric B. NSR, Tone, Amaurea, CaseKid, Mike5904, Wiserd911, Siddhant, McGinnis, YurikBot, Wavelength, Extraordinary Machine, Splintercellguy, Sceptre, Hairy Dude, Deeptrivia, Rt66lt, Jimp, Hillman, Brandmeister (old), StuffOfInterest, Tznkai, Phantomsteve, RussBot, Arado, Gunblade~enwiki, TheDoober, Xihr, Splash, Chris Capoccia, SnoopY~enwiki, JabberWok, Jengelh, Anomaly1, SpuriousQ, Stephenb, Argentino, Gaius Cornelius, CambridgeBayWeather, Lavenderbunny, Morphh, Salsb, Tavilis, Anomalocaris, NawlinWiki, Wiki alf, Joshdboz, ErkDemon, ThunderE6, John Newbury, Joelr31, Thiseye, JocK, SCZenz, Irishguy, Nick, Aaron Brenneman, ArmadniGeneral, Jpowell, Ravedave, Eipipuz, Schmock, EverettColdwell, Raven4x4x, Moe Epsilon, Farmanesh, Swen, Lomn, Semperf, Beanyk, Raskolnikov The Penguin, Tony1, Bucketsofg, Linkofazeroth, Gadget850, DeadEyeArrow, Rjrawlings~enwiki, .marc., RyanJones, Mistercow, Wknight94, JECompton, SamuelRiv, Deeday-UK, Richardcavell, BazookaJoe, WAS 4.250, Light current, Albus Dumbledore~enwiki, Enormousdude, TheKoG, Chesnok, Ageekgal, Oysteinp, Chase me ladies, I'm the Cavalry, Theda, Ketsuekigata, Fang Aili, Brz7, Aeon1006, Alias Flood, CWenger, Alain r, LeonardoRob0t, Fram, HereToHelp, Emc2, JLaTondre, ArietGold, PhS, Caco de vidro, Stuhacking, Nsevs, Banus, RG2, Benandorsqueaks, Infinity0, GrinBot~enwiki, Serendipodous, DVD R W, DocendoDiscimus, Sardanaphalus, Snottily, MacsBug, SmackBot, Aim Here, Jo marie, Terrancommander, JoeColiver, Imz, Kurochka, Varunbhalerao, EinsteinIV, Brianyoumans, Tom Lougheed, Herostratus, Stellea, Prodego, Melchoir, MJMyers2~enwiki, Brokenfrog, Unyoyega, CyclePat, Jim62sch, Prototime, WilyD, KocjoBot~enwiki, Davewild, Silpion, Evanhatesspam, Canthusus, Hbackman, Jpvinall, Man with two legs, HalfShadow, Bb1, CorvinZahn, Aksi great, Gilliam, Ohnoitsjamie, Oscarthecat, GwydionM, Cabe6403, Cowman109, Saros136, Izehar, Scaife, Kurykh, Keegan, Basejumper123~enwiki, Temiree, Njerseyguy, Persian Poet Gal, Omghgomg, HubHikari, Jfsamper, Kungming2, DHN-bot~enwiki, Cassivs, Colonies Chris, Firetrap9254, Scwlong, Audriusa, Brainblaster52, Can't sleep, clown will eat me, Rtdlow, Scott3, Jefffire, Oscar Bravo, Skidude9950, D roc16, Sephirothrr, God of War, TheKMan, Starexplorer, Andy120290, LeContexte, Kcordina, Ishanz, Grover cleveland, CamXV, Portcho, Khoikhoi, Jmlk17, Sloverlord, Ftyguy649, Fuhghettaboutit, Iapetus, Mwmoretti, Tiki2099, Nakon, Jiddisch~enwiki, Brithackermack, Mustanglover, John D. Croft, Hoof Hearted, "alyosha", Aidepolcycne, Dream out loud, Pwjb, Richard001, Eran of Arcadia, Invincible Ninja, Uriel-238, Kellyprice, Maximum bobby, DMacks, Doooook, Daniel.Cardenas, Tangsyde, Pilotguy, Yevgeny Kats, MegaHasher, The undertow, SashatoBot, Nishkid64, Rory096, Bcasterline, Robomaeyhem, Richard L. Peterson, John, Thedoj, Swienz, Sfuerst, Philosophus, J 1982, Dog Eat Dog World, Filthish, Cronholm144, Kipala, Alex Arnold, The Infidel, Rijkbenik, Soumyasch, Dhesi, Shadowlynk, AstroChemist, JorisvS, Mancroft, Mgiganteus1, CredoFromStart, Shawdow, Berrick, Ben Moore, Zzzzzzzzzzz, Stratadrake, Slakr, TheHYPO, George.howitt, Optimale, Aeluwas, Childzy, Hypnosifl, InedibleHulk, Waggers, Mets501, Funnybunny, Ryulong, Serlin, Citicat, EEPROM Eagle, MTSbot~enwiki, SmokeyJoe, John F, Amitch, KJS77, Hetar, T boyd, TaggedJC, ILovePlankton, Buntykawale, Michaelbusch, TerryE, Clarityfiend, Abel Cavaşi, D Hill, Dreftymac, Joseph Solis in Australia, Chyko, JoeBot, T.O. Rainy Day, Newone, Icefox2k, Turbokoala, MOBle, UncleDouggie, Solipse, Fsotrain09, Tony Fox, CapitalR, Domitori, Tuttt, Humanperson0, Anger22, Dpeters11, Laplace's Demon, Rwst, Tawkerbot2, JRSpriggs, Filelakeshoe, Chetvorno, Cryptic C62, Flubeca, Hammer Raccoon, IronChris, Orangutan, Hsjawanda, Fvasconcellos, Kotepho, SkyWalker, Firehawk1717, JForget, Danras, CRGreathouse, Calmargulis, Nityann, Lenky, Crescentnebula, Capefeather, Wikifried, D.N.Parrish, ClovisHopman, Syphondu, Lmcelhiney, Benwildeboer, Green caterpillar, Crabnebula, ShelfSkewed, S.Bowen, Some P. Erson, Moreschi, Kjknohw, Rotiro, Terre, Cydebot, Natasha2006,

Kanags, Zima65, Reywas92, Ramitmahajan, BobQQ, Gogo Dodo, Gagueci, JFreeman, Boardhead, Bazzargh, Wikipediarules2221, Difluoroethene, Jlmorgan, Dancter, Codingmasters, Michael C Price, Tdvance, Tawkerbot4, BMG~enwiki, DumbBOT, Chrislk02, Jay32183, Blm22, Narcosa, Omicronpersei8, Sharkbait784, Ephyon, Quophnix, Lo2u, Gimmetrow, Graham21kidd, Blobpic, BetacommandBot, CieloEstrellado, Thergvk, Thijs!bot, Lord Hawk, Crockspot, Mercury~enwiki, Qwyrxian, Waynesun, Markus Pössel, Kablammo, N5iln, Wahlin, Oerjan, Berria, MrXow, Headbomb, Newton2, Simeon H, Bobblehead, Kathovo, Tellyaddict, Cool Blue, Gvbn, Dfrg.msc, Infophile, Dgies, CharlotteWebb, Greg L, Srose, FreeKresge, Sam42, Wikidenizen, Anarchopedia, Dawnseeker2000, Elert, Escarbot, KrakatoaKatie, WikiSlasher, AntiVandalBot, Ais523, Macmanui, User's name, Majorly, Yonatan, Luna Santin, CodeWeasel, StantheGarbageMan, Yomangani, Voortle, Doc Tropics, Edokter, PhilLa, Messiah23, Dr. Submillimeter, Rsocol, LibLord, Danger, Science History, Glennwells, Spartaz, Daniels 9212, Archmagusrm, Byrgenwulf, Elaragirl, OGGVOB, Myanw, Lklundin, Uusitunnus, Kigali1, Bobvila2, Komponisto, MER-C, Never been to spain, Instinct, IanOsgood, Tonyrocks922, Davidpage, Andonic, 100110100, Cameron.walsh, Kirrages, Denimadept, Bigresearcher, LittleOldMe, DavidLaurenson, Acroterion, Yahel Guhan, Pervect, Gtation, Magioladitis, Mikemill, Gekedo, WolfmanSF, VoABot II, Raduberinde, Myopic, Antientropic, Praveenp, Farquaadhnchmn, Xeddy, Michele123, Lonewolf79 04, Loqi, Niele2006, Matt Bartlett74, SparrowsWing, BrianGV, JaKoBay, Crunchy Numbers, Giggy, Tuncrypt, Jeroje, Dyert, Mlsquad, Fluffy snowey, Disney freak!, Ceolwulf~enwiki, Chris G, DerHexer, Irishchieftain, Jomom, Patstuart, Jman73, Olsonist, Robin S, Joshua Davis, NatureA16, Otvaltak, DancingPenguin, Dr. Morbius, MartinBot, Mogus0226, Shentino, Vanessaezekowitz, Arjun01, NAHID, Gigaknight, UnfriendlyFire, John okell, Waynephinney, John Millikin, InnerJustice, Rettetast, Jay Litman, Filksinger, Loof1, Mschel, CommonsDelinker, 4.18GB, AlexiusHoratius, Pbroks13, MapleTree, Popeye Doyle, Siliconov, Sheila Rogers, LedgendGamer, Nucleartusion, Ssolbergj, Natsirtguy, RockMFR, J.delanoy, Trusilver, Ledzep3012, Allbraves08, Svetovid, Rgoodermote, Philcha, JamesR, Bogey97, Wa3frp, Melamed katz, BillWSmithJr, Catmoongirl, Uncle Dick, Mike Winters, Jonpro, Qatter, Jreferee, Tomgibbons, Bumblebee55555, Lantonov, Dargaud, St.daniel, Turtlebean2, Mozzley, LordAnubisBOT, Whilding87, Jimbothechicken, Jerggp, Crakkpot, Zedmelon, Adam Snapp, Territory, ReekRend, AntiSpamBot, Lordaal, Plasticup, Mcaig jt, Anton1234, Glens userspace watcher, Goingstuckey, NewEnglandYankee, Ryan858, Charmander trainer, Cobi, Touch Of Light, Seanskusindinmamma, Fui fui moi moi2, Minesweeper.007, Han Solar de Harmonics, Angular, Brancron, BrettAllen, Reversepolarity, SBKT, Natl1, Gtg204y, TWCarlson, Andy Marchbanks, H1voltage, Lseixas, TKM625, Billebrooks, Idioma-bot, Sheliak, Azuriteking, Signalhead, Scunnane, ACSE, Cactus Guru, Vranak, Ironrooster, VolkovBot, Parker2010, Mocirne, Milenita~enwiki, PureJadeKid, WOSlinker, Wolfnix, Philip Trueman, Fran Rogers, DarkShroom, Canopus27, TXiKiBoT, Rollo44, Jreut, Mrkwtrs, Z.E.R.O., Anonymous Dissident, Ryan shell, Oplek, Italiandevil0505, Ask123, JayC, Vanished user ikijeirw34iuaeolaseriffic, Lradrama, Melsaran, Gekritzl, Corvus cornix, Mzmadmike, Abdullais4u, Driski555, Cremepuff222, PouponOnToast, Maxim, Maksdo, Zvbxrpl, Tfmmushroom, ViresetHonestas, Happycore3, Rex Imperator, Brittadudette, Andy Dingley, MP 12, Jon1992, Lamro, Rouhibeki, Miko3k, James McBride, Ridow, Falcon8765, Spinningspark, WatermelonPotion, Bigevan1, Brianga, Mike4ty4, Gunnville, Bobo The Ninja, Bk2001050, Wisamzaqoot, The Mad Genius, Radical Robert, San Diablo, Iamnotastarwarsfan, Bufrost, LuigiManiac, DarthBotto, Harshil8, Cowlinator, Sfmammamia, Callix, FlyingLeopard2014, Nogood202, Kvncrtr, Steven Weston, D. Recorder, Bigev1, Brattbratt, Kbrose, Sureshonsearch, Tutszilla, Gaelen S., Bob freeman1, Lylefor, SieBot, BalanceRestored, Netgem21, Nabiki87, Timb66, Taftgod, NonChalance, JamesA, Work permit, Euryalus, I Like Cheeseburgers, Clissold07, Paradoctor, Bengal fan13, Joncam, Viskonsas, Caltas, ConfuciusOrnis, Poopstix, Smenge32, SolusX, Wayne317, Andersmusician, Siegel.ord, Yulu, Ujjwol, Iames, Likebox, RadicalOne, Oysterguitarist, Poopypoopypiepie, Emperorfurkan, JetLover, Hello4719, Stilkver, Godfinger, DevOhm, Oxymoron83, Bfesta14, Cmac16, Nuttycoconut, Canadianboyjd, Zharradan.angelfire, JBauer24, John fromer, Gangsterls, Lightmouse, Mydoggcoco, Poindexter Propellerhead, The Great Attractor, The-G-Unit-Boss, BenoniBot~enwiki, A kaldenhoven, DivineBurner, SteakNotShake, Dsmith7707, Coldcreation, Soulofdarkness01, David xie, Forser5, Jeroen888, Cosmo0, Randomblue, Hamiltondaniel, Movieguru2006, Vanished user 8902317830, Dust Filter, Thekingofspain, Payno, Gantuya eng, Phantomkaiser, Monmmom, Colin012, Dstebbins, Mr. Granger, Saltwell1986, Aidan180495, ArepoEn, Martarius, Beeblebrox, ClueBot, Stevekirst7, Robwalsh, Ander549, Suti1000, Andrew Nutter, PipepBot, Snigbrook, Scribble07, Patrickfongfong106, The Thing That Should Not Be, ArdClose, Kapohenry, Supersonicstars, Techdawg667, Vikasatkin, Wwheaton, Docbillnet, CyrilThePig4, Arakunem, Andr0o, Drmies, Jimbo jones9, Control-alt-delete, Russ143, Chewlett, BlackJunebug, Mcnurse, Tyguth123, Boing! said Zebedee, Rotational, Agge1000, Phenylalanine, *blissfully ignorant*BETCHES, Aua, DragonBot, Snaxalotl, Ktr101, CohesionBot, Three-quarter-ten, GoldenGoose100, Carninia, Eujin16, Timsdad, Jemxia, Leonard^Bloom, Gwguffey, Josephmd, Cenarium, Nmoo, Bracton, Jotterbot, PhySusie, Scog, Dleiter, Jwaits12, C628, 3CUTiE--PiE, Thingg, Pisceesumsprecan, Lx 121, AC+79 3888, Trulystand700, Armhouse, DumZiBoT, TimothyRias, BarretB, Baron von HoopleDoople, Oldnoah, Psycholian, WikHead, Holoeconomics, Benjamnjoel2, SilvonenBot, Zetsubo666, Sweetpoet, Padfoot79, JinJian, ZooFari, MaizeAndBlue86, Fiskbil, ElMeBot, Lemmey, Parejkoj, Whtrz, Supermonkey443, Pogozelski123, Addbot, Lkvlamen, Crissyman, TheNightRyder, 11341134a, Uruk2008, DOI bot, JJ606, Snakeboy144, Gnatbuzz, Crazysane, Artie bristles, Jugbug2, Bte99, Groundsquirrel13, WFPM, Haasfelix, Proxima Centauri, Delaszk, Syber Sid, Debresser, AnnaFrance, LinkFA-Bot, Elen of the Roads, Prim Ethics, Harvardstudent, Blmichel, Ryttaren, Tide rolls, Whatintheworldisthat, OlEnglish, Potekhin, Samuel Pepys, ScienceApe, Snookerman, Krukouski, Legobot, Luckas-bot, Yobot, LoneRubberDragon, Bunnyhop11, Tohd8BohaithuGh1, Legobot II, Lolchanges, VZ9, Gum Stuck on Bottom of Shoe, Anypodetos, Pigetrational, KamikazeBot, Rubin16, Ayrton Prost, Szajci, AnomieBOT, AndrooUK, Archon 2488, Grey Fox-9589, Message From Xenu, AdjustShift, Ornamentalone, Asoer, Powerzilla, Materialscientist, RobertEves92, Citation bot, Eumolpo, Palitzsch250, Xqbot, Meewam, Emerydora, DSisyphBot, Hanberke, Tad Lincoln, NASCAR Nathan, Runaway9995, Gap9551, UlmPhysiker, GrouchoBot, Mpe.mpg.de, ProtectionTaggingBot, Nlilovic, Omnipaedista, Kurtdriver, RibotBOT, Seeleschneider, Der Falke, JediMaster362, Moxy, WillMall, Imperators II. A. di M., Interstellar Man, Sesu Prime, FrescoBot, Feneeth of Borg, Akuvar, Originalwana, Goodbye Galaxy, Worrycharm, THENEWMIKON8ER, Europi3n, Mfwitten, Steve Quinn, Citation bot 2, Skull33, Robo37, Fruit.Smoothie, Citation bot 1, Javert, Careful with That Axe, Eugene, Gil987, Jonesey95, Tom.Reding, Achim1999, Concernedresident's butler, SpaceFlight89, Xaviertan, An elite, OldManNick, Savemaxim, MertyWiki, Tempk, SanDiego7, Ashishg1984, TheInforment, Revenge12345678, Seattle Jörg, Nora lives, IVAN3MAN, Lemmiwinks2, Thames Aldwych W. Mines, Rajeev Goutam, Meier99, FoxBot, TobeBot, Trappist the monk, Belchman, Randomlogan, D climacus, Jordgette, Lolcakes1414, Williame3, Jamie s w, Lotje, Extra999, EventHorizon5488, Spikescape, Ugly Ketchup, RjwilmsiBot, Mifield, Mrfencey, Hardikvasa, NameIsRon, Chriss.2, Mchcopl, Burmiester, Newty23125, Salvio giuliano, Billare, EmausBot, John of Reading, WikitanvirBot, JCRules, AlexUT, Grrow, Quantanew, Racerx11, Joseph507357, PoeticVerse, Dangoerman, Jmencisom, Challisrussia, Cpl-pike, Chricho, Italia2006, Hhhippo, Ida Shaw, Stanford96, Socioj, StringTheory11, Xabier Armendaritz, Nicolas Eynaud, Aeonx, H3llBot, Brandmeister, Y-barton, Crux007, ChuispastonBot, RockMagnetist, One.Ouch.Zero, Herk1955, ClueBot NG, Gilderien, Iloveandrea, Nijiravipp, Jj1236, Tabletrack, Garlikguy2, Rezabot, JoetheMoe25, Danim, Pluma, Helpful Pixie Bot, Asdfjkl1235, Bibcode Bot, BG19bot, Pine, Furkhaocean, Badon, Pascal yuiop, Cadiomals, Blaspie55, Zedshort, BattyBot, U-95, ChrisGualtieri, Khazar2, Ducknish, Dexbot, Webclient101, Mogism, Stas1995, Cerabot~enwiki, CuriousMind01, SFK2, Graphium, RobH103, Cserez, Max14182000, Corn cheese, Among Men, Reatlas, Joeinwiki, Anastronomer, Rfassbind, Donfbreed2, Greengreengreenred, MatthewJ00, Smortypi, Light Peak, Ryenocerous, Jakec, Dustin V. S.,

Rolf h nelson, SuicideRider003, Space core192, Blakethecake333, Comp.arch, Kharkiv07, Bacontry, Ritviksaharan, Kogge, Mark Matthew Dalton, Anrnusna, Sudoiusudo, Signoredexter, Elenceq, Monkbot, Zhermes, Paul Masson, Garfield Garfield, SkyFlubbler, ChamithN, DangerousJXD, Chaloagarcia, Freshness For Lettuce, Sb2s3, Tetra quark, JLT2045, DN-boards1, Jerodlycett, Fogbannana, KasparBot, Ceannlann gorm, EternalNomad, CheeseStick1, Brandon Defrise Carter, BowlAndSpoon, Astro4686, Mociaty, Zhakhan9er, Sardeis, Bensinio and Anonymous: 1420

- **Hierarchy problem** *Source:* https://en.wikipedia.org/wiki/Hierarchy_problem?oldid=723045435 *Contributors:* The Anome, WhisperToMe, Phys, AnonMoos, Jni, Giftlite, Xerxes314, Thincat, Lumidek, Rich Farmbrough, FT2, Pt, Jag123, I9Q79oL78KiL0QTFHgyc, Mindmatrix, GregorB, VermillionBird, Coemgenus, Koavf, Mattmartin, Strait, Salix alba, UkPaolo, Ugha, Bhny, Netrapt, Ephraim33, QFT, Jgwacker, Derek R Bullamore, NNemec, Marek69, Ninjakannon, Shambolic Entity, Dr. Morbius, Drgnrave, X!, James Banogon, Megalekaitrane, D.scain.farenzena, Copyeditor42, Alexbot, Lalegria, Addbot, Mixen Dixon, TutterMouse, Aboctok, Debresser, Topquark22, Yobot, AnomieBOT, Yemibedu, Materialscientist, Citation bot, Neurolysis, ArthurBot, Pra1998, Omnipaedista, A. di M., Erik9bot, Banak, FrescoBot, Paine Ellsworth, Puzl bustr, Bj norge, Hauntedpz, RjwilmsiBot, EmausBot, Dcirovic, AsceticRose, Arbnos, Suslindisambiguator, Quondum, Jbackroyd, Bibcode Bot, Ervin Goldfain, Drcooljoe, IluvatarBot, Ownedroad9, MSUGRA, Prokaryotes, Mfb and Anonymous: 49

- **Quantum triviality** *Source:* https://en.wikipedia.org/wiki/Quantum_triviality?oldid=723789051 *Contributors:* Michael Hardy, D6, David Schaich, Guy Harris, Drbreznjev, Rjwilmsi, Dresdnhope, Jmnbatista, EdGl, JorisvS, Ruslik0, Cydebot, Headbomb, Magioladitis, Acalamari, CardinalDan, Cuzkatzimhut, TXiKiBoT, Addbot, Luka666, Yobot, Citation bot, Pra1998, Kenneth Dawson, Citation bot 1, Physics therapist, Taxpaying nonscientist, RjwilmsiBot, George Hanratty, Dcirovic, Maroansika, Aturzillo, Helpful Pixie Bot, Vikram Shastry, Bibcode Bot, Tor Vergata Fisica, Dexbot, Yikkayaya and Anonymous: 7

- **CP violation** *Source:* https://en.wikipedia.org/wiki/CP_violation?oldid=722522991 *Contributors:* Roadrunner, Stevertigo, Michael Hardy, Albertplanck, TakuyaMurata, Angela, Julesd, Netsnipe, Palfrey, Raven in Orbit, Coren, Charles Matthews, The Anomebot, Phys, Donarreiskoffer, Pigsonthewing, COGDEN, Ruakh, Giftlite, Jmnbpt, Harp, Xerxes314, Gracefool, ConradPino, HorsePunchKid, Mako098765, WhiteDragon, Karol Langner, Pmanderson, Lumidek, Rich Farmbrough, Hidaspal, Bobo192, Davidruben, Elipongo, Foobaz, I9Q79oL78KiL0QTFHgyc, M0rph, MPerel, Pearle, Sligocki, Evil Monkey, Dirac1933, W7KyzmJt, Kusma, Kay Dekker, Oleg Alexandrov, Linas, Nopherox, Marudubshinki, Strait, Tawker, Ligulem, Mathbot, Lmatt, Goudzovski, Chobot, Gdrbot, Bhny, Limulus, JabberWok, NawlinWiki, Grafen, Crasshopper, Tony1, Tonywalton, Square87~enwiki, Fram, ArielGold, GrinBot~enwiki, MacsBug, SmackBot, HalfShadow, PeterSymonds, Dauto, Chris the speller, Tigerhawkvok, Can't sleep, clown will eat me, QFT, Voyajer, Wen D House, Pwjb, Ligulembot, Drunken Pirate, GTFleming, Erwin, Ryulong, Dan Gluck, Lottamiata, Qqs83, IRevLinas, CRGreathouse, CmdrObot, Wafulz, Vyznev Xnebara, Friendofthehose, Vanished user vjhsduheuiui4t5hjri, Simon Brady, DumbBOT, Gimmetrow, Raoul NK, Mbell, Cosmi, Applecore91, Headbomb, WilliamH, Insane99, Dingaling, Leevclarke, Txomin, Igodard, Kaonslau~enwiki, Thasaidon, Parsecboy, Kevinmon, Homunq, Tonyfault, DerHexer, Dr. Morbius, MartinBot, Rettetast, Warrickball, Felixbecker2, The dark lord trombonator, Extransit, I310342~enwiki, Larryisgood, Rich Janis, Jackfork, Venny85, Rknase, PaddyLeahy, SieBot, Nintendostar, Wing gundam, Scasa~enwiki, LonelyMarble, ClueBot, Likebreakfe, Rotational, Chimesmonster, Yakrami, NuclearWarfare, DumZiBoT, Saeed.Veradi, MystBot, Airplaneman, Addbot, Cxz111, T.c.w7468, Landon1980, Leszek Jańczuk, Debresser, Tide rolls, Lightbot, Zorrobot, Micko.hjort~enwiki, LuK3, Legobot, Luckas-bot, Yobot, Amirobot, AnomieBOT, Killiondude, Citation bot, Quebec99, Xqbot, Spidern, False vacuum, FrescoBot, Paine Ellsworth, Citation bot 1, Sunandclouds, Elockid, Mutinus, Thinking of England, Sanomi, Higgshunter, The Perfection, Dizanl, Ofercomay, John of Reading, Bphyswiki, Zerkroz, Dcirovic, Fæ, Arbnos, Ebrambot, AManWithNoPlan, Kweckzilber, AfroScientist, Maschen, QuantumSquirrel, ClueBot NG, Jj1236, Snotbot, Ghartshaw, Fascismsucks555, Keithtacokeithsta, Sndfnsdfsdffdd, Togtto, Honestguy55543, Anonymous5555, Helpful Pixie Bot, Bibcode Bot, Petermahizahn, Solomon7968, MeanMotherJr, ChrisGualtieri, Ranze, Andyhowlett, Mtdevans, DavidLeighEllis, Nigellwh, Spyglasses, Prokaryotes, Justinvasel, 22merlin, Monkbot, Soham92 and Anonymous: 111

- **Cosmological constant** *Source:* https://en.wikipedia.org/wiki/Cosmological_constant?oldid=723876791 *Contributors:* AxelBoldt, Magnus Manske, Vicki Rosenzweig, Bryan Derksen, The Anome, Ed Poor, Enchanter, William Avery, Roadrunner, Schewek, Hephaestos, Boud, Bcrowell, Lquilter, TakuyaMurata, Minesweeper, Stevenj, Kimiko, Samw, Timwi, Reddi, Asar~enwiki, Dogface, Bevo, Anupamsr, Johnleemk, BenRG, Phil Boswell, Robbot, Goethean, Wereon, Giftlite, Bobblewik, Jonel, Rjpetti, Icairns, Rgrg, Burschik, JimJast, 4pq1injbok, Pjacobi, Vsmith, StephanKetz, Dbachmann, Pavel Vozenilek, Dmr2, Bender235, RJHall, Pt, El C, Frankenschulz, RoyBoy, Rbj, I9Q79oL78KiL0QTFHgyc, Knucmo2, Jumbuck, Falcorian, Angr, OwenX, Linas, StradivariusTV, Kzollman, Mpatel, Joke137, Wisq, Christopher Thomas, Rnt20, Ashmoo, Coneslayer, Rjwilmsi, Coemgenus, Nightscream, RE, Itinerant1, Srleffler, Chobot, PointedEars, YurikBot, Hillman, RussBot, Ytrottier, SpuriousQ, Gaius Cornelius, Salsb, Sir48, Muu-karhu, DeadEyeArrow, Helge Rosé, Petri Krohn, KasugaHuang, SmackBot, Incnis Mrsi, WilyD, Nickst, Cush, Colonies Chris, Avb, Cybercobra, Ligulembot, Yevgeny Kats, Lambiam, Matt489, Paladinwannabe2, Ckatz, Onionmon, Basicdesign, Newone, Sirwhiteout, Chetvorno, CmdrObot, Orannis, Hardrada, Mlsmith10, MaxEnt, Phatom87, Forthommel, Frostlion, Dr.enh, Michael C Price, Tawkerbot4, Clovis Sangrail, Christian75, Thijs!bot, Mathmoclaire, Headbomb, Peter Gulutzan, Gnartyocelot, Escarbot, AntiVandalBot, Tim Shuba, JAnDbot, LinkinPark, .anacondabot, WolfmanSF, SHCarter, Ling.Nut, Jlerner, DAGwyn, Nikopopl, MartinBot, Mschel, Morris729, Lantonov, BobEnyart, Jorfer, Blckavnger, Fylwind, Atheuz, TXiKiBoT, Rei-bot, Mathwhiz 29, Thrawn562, Venny85, SieBot, El Wray, Puzhok, Gerakibot, BartekChom, OKBot, ClueBot, The Thing That Should Not Be, Frdayeen, Niceguyedc, Excirial, Bender2k14, Brews ohare, Kentgen1, Scog, Panos84, Louis925, Alphatronic, XLinkBot, DCCougar, Sesquihypercerebral, Torchflame, Addbot, DOI bot, Zahd, Delaszk, Legobot, Luckas-bot, Yobot, Aldebaran66, Amble, Perusnarpk, AnomieBOT, Materialscientist, Citation bot, Louelle, Srich32977, Omnipaedista, Waleswatcher, A. di M., 晰, Paine Ellsworth, Citation bot 1, Newt Scamander, Gil987, Tom.Reding, BlackHades, Jordgette, Michael9422, Earthandmoon, Vekov, UpdateNerd, RjwilmsiBot, Racerx11, Dcirovic, Solomonfromfinland, Italia2006, Hhhippo, ZéroBot, Liquidmetalrob, Arbnos, Quondum, Ewa5050, Iiar, Zueignung, Khestwol, ClueBot NG, Astrocog, Frietjes, Jhmmok, Rezabot, Const.S, Helpful Pixie Bot, Bibcode Bot, Rascal Sage, Jeffloiselle, Hippokrateszholdacskai, RiseUpAgain, Makecat-bot, Kryomaxim, Wjs64, Andyhowlett, Jp4gs, Blackbombchu, Prokaryotes, Inanygivenhole, Kogge, Paspaspas, Christophe1946, RandomAgentNation, Monkbot, Tetra quark, KasparBot, Jmc76, Maha Abdelmoneim, RandomEditor99 and Anonymous: 145

- **Loop quantum gravity** *Source:* https://en.wikipedia.org/wiki/Loop_quantum_gravity?oldid=726193855 *Contributors:* Bryan Derksen, The Anome, AstroNomer, RK, Toby Bartels, Miguel~enwiki, Schewek, Ewen, Michael Hardy, TakuyaMurata, Islandboy99, GTBacchus, Mcarling, Looxix~enwiki, Ahoerstemeier, Cyp, Kimiko, Palfrey, Jordi Burguet Castell, Mxn, Charles Matthews, Sanxiyn, Maximus Rex, Phys, Omegatron, Finlay McWalter, Dmytro, Sdedeo, Astronautics~enwiki, Peak, Chris Roy, Mirv, Sverdrup, Kn1kda, Hadal, Jheise, Clementi, Connelly, Giftlite, Sj, Fastfission, Herbee, Anville, Dratman, Curps, JeffBobFrank, Jason Quinn, Gzornenplatz, C17GMaster, DÀ.ugosz, PhiloVivero,

DefLog~enwiki, Gadfium, HorsePunchKid, Sam Hocevar, Lumidek, Tdent, Joyous!, M1sslontomars2k4, Eep², Poccil, Rich Farmbrough, Avriette, Pjacobi, Vsmith, MuDavid, Pavel Vozenilek, Bender235, ESkog, Clement Cherlin, Peter M Gerdes, Drhex, John Vandenberg, C S, Cmdrjameson, GTubio, Tweet Tweet, Slicky, Ral315, Lysdexia, Arthena, Xaphan9966, Wtmitchell, Greg Kuperberg, Count Iblis, Egg, Lee-Anne, Kazvorpal, Killing Vector, Linas, Merlinme, HFarmer, Sympleko, Hfarmer, Mpatel, GregorB, J M Rice, Ae7flux, Tjbk tjb, Alienus, BD2412, Fleisher, Sjö, Rjwilmsi, Nightscream, Zbxgscqf, Bubba73, FlaBot, John Baez, Don Gosiewski, Smithbrenon, Chobot, Spasemunki, Bgwhite, Roboto de Ajvol, YurikBot, Wavelength, RobotE, Rt66lt, Hillman, DanMS, Chaos, Salsb, Welsh, Schmock, Crasshopper, Beanyk, Akashmitra, Bota47, JonathanD, Endomion, Modify, Petri Krohn, Ilmari Karonen, Caco de vidro, Benandorsqueaks, SmackBot, Bayardo, FlashSheridan, Unyoyega, Vald, JMiall, Chris the speller, IvanAndreevich, DHN-bot~enwiki, Colonies Chris, Chlewbot, Pepsidrinka, Chrylis, MegaHasher, TriTertButoxy, Lambiam, Vincenzo.romano, Loadmaster, Konklone, K. G-W, Kurtan~enwiki, Harold f, Will314159, Friendly Neighbour, Vyznev Xnebara, Ian Beynon, Myasuda, Gmusser, Rjm656s, Fournax, Headbomb, Marek69, Nick Number, MichaelMaggs, Edokter, Byrgenwulf, Knotwork, Arch dude, Igodard, Yill577, WolfmanSF, Tonyfault, Skylights76, Rickard Vogelberg, Gwern, AltiusBimm, Melamed katz, Vanished user 47736712, WJBscribe, Izno, KittyHawker, Sheliak, Maxzimet, AlnoktaBOT, Nxavar, Jackfork, Carlorovelli, Anotherak, SieBot, Keskiyat, AS, Robdunst, Hugh16, Senderista~enwiki, Bnsreenath, Caidh, Oxymoron83, Dcattell, Swiebodzice, Sk8hack, Danthewhale, Martarius, Sfan00 IMG, Shaded0, Djr32, CohesionBot, JavierReynaldo, Arjayay, SchreiberBike, Pqnelson, Mjaniec, DumZiBoT, Ianbay, Neuralwarp, XLinkBot, Fastily, Tenner47, Arthur chos, Avoided, Tenderbuttons, Benplusnumber, Balungifrancis, Addbot, DOI bot, 15lsoucy, Tarosic, Debresser, Favonian, SamatBot, Yobot, Ibayn, 4th-otaku, AnomieBOT, VanishedUser sdu9aya9fasdsopa, Archon 2488, Francois33, Citation bot, Xqbot, Imushhq, MIRROR, Pra1998, Dumontierc, Omnipaedista, Franco3450, Rr2000, FrescoBot, Paine Ellsworth, Nunc aut numquam, Martlet1215, Citation bot 1, Jonesey95, Tom.Reding, Schiefesfragezeichen, ROMVLVS, Casimir9999, RobinK, Meier99, Orenburg1, Trappist the monk, Dinamik-bot, Bj norge, ElPeste, Afteread, EmausBot, Detogain, John of Reading, Racerx11, GoingBatty, XinaNicole, Ensabah6, Uploadvirus, ZéroBot, Arbnos, Zueignung, WaterCrane, Crown Prince, LaurentRDC, Isocliff, Vodkacannon, Raidr, Helpful Pixie Bot, Titodutta, Bibcode Bot, BG19bot, Spaligo, KateWishing, PhnomPencil, Sylvain.maurin, Kecchina, Halfb1t, JMtB03, Brad7777, Fylbecatulous, Jimw338, MyTuppence, Mogism, LTWoods, Andyhowlett, Jawa0, &reasNink, SomeFreakOnTheInternet, Tentinator, EvergreenFir, DimReg, Pedarkwa, Db9199 24, Anrnusna, Notspelly, Ntomlin1996, Monkbot, Isbromberg, Dsprc, Dakroth, YeOldeGentleman, Tetra quark, Gvprtskvnis, Srednuas Lenoroc, Tomuel99, Wulframm, Chemistry1111 and Anonymous: 347

- **Causal dynamical triangulation** *Source:* https://en.wikipedia.org/wiki/Causal_dynamical_triangulation?oldid=699695930 *Contributors:* The Anome, Charles Matthews, Jeffq, Tea2min, Dratman, Pjacobi, Alamino, Anthony Appleyard, Tabletop, Tiroche, MarSch, Itinerant1, Xihr, JocK, JonathanD, SmackBot, HTeutsch, Schmiteye, Colonies Chris, Fuhghettaboutit, Vincenzo.romano, Gmusser, Tonyfault, Gibimi, Jemather, Melamed katz, Lantonov, Sigmundur, Sheliak, Foresyte, AlleborgoBot, Pallab1234, Thehotelambush, Ideal gas equation, Greennature2, Shamanchill, Addbot, Eric Drexler, AnomieBOT, Prari, Paine Ellsworth, Steve Quinn, ZéroBot, Arbnos, Maschen, Raidr, DavidRideout, Fraulein451, &reasNink, Dimension10, Tetra quark, Chemistry1111 and Anonymous: 25

- **Canonical quantum gravity** *Source:* https://en.wikipedia.org/wiki/Canonical_quantum_gravity?oldid=723716388 *Contributors:* Michael Hardy, TakuyaMurata, Jordi Burguet Castell, Chuunen Baka, Gandalf61, Slicky, J Heath, Mpatel, Joke137, Slgrandson, Eyu100, Anonymous editor, Perry Middlemiss, WAS 4.250, Ligulembot, St Cyrill, AlphaNumeric, Headbomb, Marek69, Adavidb, Chrystomath, Sheliak, VolkovBot, Mastertek, Muro Bot, Addbot, DOI bot, Favonian, Ibayn, Citation bot, Xqbot, Paine Ellsworth, Afteread, John of Reading, Arbnos, SemanticMantis, Raidr, Helpful Pixie Bot, Bibcode Bot, A.bt(w) and Anonymous: 19

- **Superfluid vacuum theory** *Source:* https://en.wikipedia.org/wiki/Superfluid_vacuum_theory?oldid=708609465 *Contributors:* Rich Farmbrough, Cedders, Mu301, DeadlyAssassin, Serendipodous, Tom Morris, Colonies Chris, Thijs!bot, R'n'B, Uncle Milty, SchreiberBike, TimothyRias, Addbot, Favonian, Yobot, Edstamos, AnomieBOT, Omnipaedista, FrescoBot, Paine Ellsworth, Tom.Reding, RjwilmsiBot, Arbnos, Staszek Lem, Baseball Watcher, Bibcode Bot, Machoota, Brainssturm, Uiopik, Account12098, Intogain891, Andyhowlett, Rolf h nelson, Aroomwhile, Gravytacky67, Soaring and Anonymous: 14

- **Twistor theory** *Source:* https://en.wikipedia.org/wiki/Twistor_theory?oldid=719970187 *Contributors:* Bryan Derksen, Stevertigo, Jimfbleak, Tracian, BenRG, JorgeGG, Sverdrup, Giftlite, Mporter, Quinwound, Nkocharh, Rdsmith4, Starx, Lumidek, Zro, Mormegil, Francis Davey, Dmr2, Danhash, Linas, Mpatel, GregorB, Frankie1969, BD2412, Rjwilmsi, MarSch, R.e.b., YurikBot, Wavelength, Hillman, Bhny, Crasshopper, Bota47, CLW, Ilmari Karonen, Benly~enwiki, BWDuncan, Artie p, Vina-iwbot~enwiki, George100, Thijs!bot, Headbomb, JAnDbot, Husond, Bdalevin, Logolego, WJBscribe, Telecomtom, Mercurywoodrose, Anonymous Dissident, Sunmoonstars, YohanN7, Robdunst, JL-Bot, ArepoEn, Stevekirst7, Alexbot, CaDyTrOn, AnonyScientist, XLinkBot, Addbot, Uruk2008, Lightbot, Luckas-bot, Citation bot, GrouchoBot, Omnipaedista, FrescoBot, LucienBOT, BenzolBot, Gil987, Casimir9999, Miracle Pen, Le Docteur, Everyplace, ClueBot NG, Raidr, Lurscher, Bibcode Bot, BG19bot, Jimw338, AHusain314, Dimension10, Polytope24, YeOldeGentleman and Anonymous: 52

- **Supersymmetry** *Source:* https://en.wikipedia.org/wiki/Supersymmetry?oldid=721808506 *Contributors:* Bryan Derksen, Taw, Andre Engels, Roadrunner, Maury Markowitz, Ewen, Stevertigo, Edward, Michael Hardy, Arpingstone, Theresa knott, IMSoP, Jeandré du Toit, Samw, Smack, Charles Matthews, Maximus Rex, Phys, Raul654, BenRG, Rursus, Mor~enwiki, Ancheta Wis, Giftlite, Mporter, Ferkelparade, Monedula, Froputf, Xerxes314, Anville, Gus Polly, Moyogo, Unconcerned, DO'Neil, Maarten van Vliet, Pharotic, LiDaobing, Sam Hocevar, Lumidek, Degir6328, Arivero, Rich Farmbrough, Roybb95~enwiki, Bender235, El C, Nornagon~enwiki, Duk, Tweet Tweet, Russ3Z, LostLeviathan, Pearle, Gary, Francescog~enwiki, Wtmitchell, RJFJR, Reaverdrop, Blaxthos, Killing Vector, Jordan14, Ted BJ, MONGO, Mpatel, MFH, SeventyThree, Bodera, VermillionBird, Drbogdan, Rjwilmsi, Josiah Rowe, R.e.b., Bubba73, Maxim Razin, Drrngrvy, FlaBot, Cless Alvein, Nowhither, Itinerant1, Gparker, KFP, Lmatt, Chobot, Bgwhite, Vyroglyph, YurikBot, Wavelength, RussBot, Ohwilleke, Bhny, Epolk, Sasuke Sarutobi, Maxim Leyenson, Chaos, Romanc19s, Bota47, Mgnbar, Closedmouth, Arthur Rubin, RG2, That Guy, From That Show!, A bit iffy, SmackBot, Mira, Kurochka, Wangjiaji, Gilliam, Bluebot, Cadmasteradam, Complexica, Bazonka, Colonies Chris, Can't sleep, clown will eat me, QFT, Ruff ilb, Wen D House, Solarapex, Radagast83, Jgwacker, TheMaster42, Martijn Hoekstra, Ligulembot, Acjohnson55, Yevgeny Kats, Charleswestbrook, TriTertButoxy, Lambiam, Tktktk, Xiaphias, JarahE, Mdanziger, Dan Gluck, Newone, Marysunshine, Tawkerbot2, Banedon, Cydebot, Hydraton31, Bazzargh, David edwards, Michael C Price, Crum375, Koeplinger, Headbomb, J.christianson, Escarbot, Salgueiro~enwiki, Kborland, Jpod2, Cgingold, Maliz, TimidGuy, C9, Kostisl, R'n'B, Zentropa77, Natsirtguy, Maurice Carbonaro, Kevin Hickerson, Shawn in Montreal, Idioma-bot, Sheliak, Cuzkatzimhut, Nxavar, Kawakameha, Cuboidal, Ptrslv72, PhysPhD, Kbrose, SieBot, Nn123645, ClueBot, Jcpilman, Chessmaster7m, Kitsunegami, Rhododendrites, Mastertek, Mishas42, Scrabby~enwiki, TimothyRias, WikHead, MystBot, Addbot, DOI bot, Zahd, Barak Sh, F Notebook, Lightbot, Windward1, Luckas-bot, Yobot, Ibayn, TaBOT-zerem, Amirobot, Nonnormalizable, AnomieBOT, Girl Scout cookie, Materialscientist, Citation bot, ArthurBot, Plumpurple, Tomwsulcer, Omnipaedista, Gsard, CES1596, FrescoBot, HaloStereo1, Paine

Ellsworth, Xmikywayx, Citation bot 1, Gil987, Kikeku, Jonesey95, Eddie Nixon, MondalorBot, Aknochel, Tom1661, Gagoga ju, TobeBot, Puzl bustr, Andraas, EmausBot, Djloststylez, Klbrain, Ddimensões, Arbnos, Susy is it, ChuispastonBot, Isocliff, ClueBot NG, KagakuKyouju, IJVin, Frietjes, Helpful Pixie Bot, Bibcode Bot, BG19bot, Teika kazura, JayBeeEye, Ninmacer20, ChrisGualtieri, Dexbot, Logosun, AHusain314, NA48, Rfassbind, Katherine Pendleton, Lioinnisfree, Laplacemat, Liquidityinsta, TaiSakuma, Stamptrader, Kdmeaney, Qxxxxxq, Almaionescu, Monkbot, Janhaithabu, Asympto, Mammoth2011, Jwill530, Stacie Croquet, Cuttlas1, AHusain3141, Wave system, Archaon593 and Anonymous: 178

- **String theory** *Source:* https://en.wikipedia.org/wiki/String_theory?oldid=724772544 *Contributors:* AxelBoldt, Sodium, Mav, Bryan Derksen, Zundark, The Anome, Tarquin, Taw, Eean, Malcolm Farmer, Hephaestos, Olivier, Drseudo, Stevertigo, Spiff~enwiki, Edward, PhilipMW, Michael Hardy, Bewildebeast, Dante Alighieri, Gabbe, Graue, Tgeorgescu, Mcarling, CesarB, Looxix~enwiki, Ahoerstemeier, Theresa knott, Suisui, Angela, Den fjättrade ankan~enwiki, Jdforrester, Julesd, Salsa Shark, Schneelocke, Charles Matthews, Timwi, Bemoeial, Jitse Niesen, 4lex, Greenrd, ErikStewart, Furrykef, Saltine, Phys, Omegatron, Bevo, Topbanana, Trent, Nufy8, Robbot, Craig Stuntz, Fredrik, Chris 73, R3m0t, COGDEN, Mirv, Wjhonson, Sverdrup, Academic Challenger, DHN, Hadal, Khlo, ElBenevolente, HaeB, Xanzzibar, Tea2min, Giftlite, DocWatson42, Christopher Parham, Awolf002, Mporter, Amorim Parga, Mikez, Harp, Kim Bruning, Tom harrison, Ferkelparade, Leflyman, Froputf, No Guru, Anville, Moyogo, Curps, Pashute, Nomad~enwiki, Mboverload, Solipsist, SWAdair, DemonThing, Wmahan, Btphelps, MSTCrow, Decoy, Chowbok, Gadfium, Steuard, Pgan002, Quadell, Carandol~enwiki, Antandrus, Beland, JoJan, Khaosworks, Tothebarricades.tk, Thincat, Tomruen, Shidobu, Icairns, Lumidek, NoPetrol, Avihu, Fanghong~enwiki, Trevor MacInnis, Lacrimosus, Zro, Mike Rosoft, D6, Urvabara, Felix Wan, Jkl, Discospinster, ElTyrant, Rich Farmbrough, Rhobite, Pjacobi, Alien life form, Vapour, Silence, Kzzl, LindsayH, Mani1, Pavel Vozenilek, Paul August, Bender235, Kjoonlee, Mashford, Kelvinc, Perlman10s, Panu~enwiki, Brian0918, Dpotter, Livajo, El C, Laurascudder, Shanes, Zegona beach, RoyBoy, Causa sui, Bobo192, Directorstratton, Janna Isabot, Smalljim, John Vandenberg, Flxmghvgvk, I9Q79oL78KiL0QTFHgyc, Physicistjedi, Bongoo, 4v4l0n42, Merope, Geschichte, Linuxlad, Phils, Merenta, Alansohn, Gary, JYolkowski, Enirac Sum, Ryanmcdaniel, Arthena, Borisblue, Rd232, Plumbago, Axl, R Calvete, Lightdarkness, Kocio, Bart133, Wtmitchell, Isaac, Tycho, Cal 1234, Fadereu, CloudNine, Sciurinæ, Computerjoe, Kusma, DV8 2XL, Pwqn, Gene Nygaard, Ringbang, Ceyockey, Falcorian, Bobrayner, Joriki, Mel Etitis, Linas, BillC, Jacobolus, HFarmer, Before My Ken, Netdragon, MONGO, GeorgeOrr, Mpatel, Bbatsell, GregorB, , Joke137, Christopher Thomas, Dysepsion, GSlicer, Jan.bannister, Graham87, Magister Mathematicae, Hillbrand, BD2412, Elvey, Galwhaa, Raymond Hill, JIP, RxS, Athelwulf, Edison, Sjakkalle, Rjwilmsi, Xgamer4, Jake Wartenberg, Arabani, MarSch, TheRingess, Jmcc150, Aero66, Crazynas, Juan Marquez, R.e.b., Bubba73, DoubleBlue, Zelos, AlisonW, Asafavi, Lionelbrits, Conorihe, Zunz, Mathbot, Crazycomputers, RexNL, Gurch, Algri, TeaDrinker, Zifnabxar, XAXISx, Erik4, Phoenix 2~enwiki, Antimatter15, Ggb667, Chobot, Visor, DVdm, Mhking, VolatileChemical, Bgwhite, Algebraist, Ben Tibbetts, YurikBot, Ugha, Wavelength, Borgx, NuclearFusion ~enwiki, Angus Lepper, Hairy Dude, Jimp, Hillman, Cyferx, Wolfmankurd, Pip2andahalf, RussBot, Moronoman, Crazytales, Pippo2001, Bhny, Pigman, SpuriousQ, Branman515, Stephenb, Gaius Cornelius, Eleassar, Rsrikanth05, Bovineone, Cheesus, Shanel, NawlinWiki, Tong~enwiki, Mike18xx, SCZenz, Cleared as filed, Bdiah, Pym98, SColombo, Haemo, FF2010, Closedmouth, Reyk, Brina700, Chris Brennan, Vicarious, Brianlucas, Geoffrey.landis, Hitch- hiker89, Spliffy, Pred, ArielGold, Roy Fulton, Ilmari Karonen, Katieh5584, Pentasyllabic, Lunch, DVD R W, WikiFew, That Guy, From That Show !, Street Scholar, AndrewWTaylor, QSquared, Sardanaphalus, Vanka5, MacsBug, Hvitlys, SmackBot, Kurochka, Zazaban, Tom Lougheed, Prodego, KnowledgeOfSelf, Hydrogen Iodide, Melchoir, Vald, Skrewtape, Atomota, Canthusus, GaeusOctavius, Cool3, Andyvn22, Gilliam, Skizzik, RobertM525, Dauto, Bluebot, SSJ5, Keegan, Aidan Croft, Thumperward, Oli Filth, Silly rabbit, Timneu22, SchfiftyThree, Moshe Constantine Hassan Al-Silverburg, Complexica, Rediahs, RayAYang, Aero77, Adamstevenson, Ikiroid, Epastore, Baronnet, Ned Scott, Sbharris, Colonies Chris, Konstable, Sct72, Scwlong, Can't sleep, clown will eat me, Timothy Clemans, Onorem, Neilanderson, EvelinaB, TKD, KerathFreeman, Addshore, UU, The tooth, Pepsidrinka, Somebody2292, --=The Doctor=--, Fuhghettaboutit, Cybercobra, Irish Souffle, Nakon, Jdlambert, James McNally, MichaelBillington, Lostart, Insineratehymn, Drphilharmonic, SpiderJon, DMacks, Ihatetoregister, Where, Michael IFA, Yevgeny Kats, Vasiliy Faronov, Byelf2007, Angela26, Visium, Rory096, Zymurgy, Harryboyles, Mdl 53711, T-dot, Titus III, Ergative rlt, MagnaMopus, UberCryxic, Vgy7ujm, Lazylaces, Linnell, Mgiganteus 1, Nonsuch, IronGargoyle, Ckatz, DoItAgain, AstroGod, Kirbytime, Jimbo Mahoney, FredrickS, Invisifan, Ryulong, Ryanjunk, MathStuf, Mike Doughney, Norm mit, Hindol, Dan Gluck, Huntscorpio, Irides-cent, K.Sunoco, You? Me? Us?, CzarB, Rabinzkaman, JoeBot, Lottamiata, Tony Fox, Vrkaul, Torrazzo, Gil Gamesh, Areldyb, Courcelles, Tawkerbot2, Gebrah, Shamvil, Fdssdf, DKqwerty, Lbr123, Harold f, Heqs, Devourer09, Duduong, Sarvagnya, Dewayne76, JForget, Cg-realms, InvisibleK, CRGreathouse, CmdrObot, Earthlyreason, Van helsing, Olaf Davis, CBM, Rawling, Jibal, Witten Is God, Nunquam Dormio, Giko, KnightLago, Thubsch, Leujohn, SlashDot, TheTito, Karenjc, Myasuda, Emarv, Cydebot, Gmusser, Gogo Dodo, Jkokavec, Kahananite, Qua-jafrie, Michael C Price, Doug Weller, DumbBOT, Narayanese, AlphaNumeric, SRoughsedge, Vanished User jdksfajtlasd, Woland37, Zalgo, Daniel Olsen, UberScienceNerd, Bkazaz, DJBullfish, Thijs!bot, Epbr123, Rwmnau, Babemachine, Pimpin101, Mbell.O, Faigl.ladislav, Ucan-lookitup, Andyjsmith, Headbomb, Tcturner2002, Marek69, Brahmajnani, Arthureprado~enwiki, Y.t., D3gtrd, Babemonkey, Dark dude, Duncan McB, EdJohnston, MichaelMaggs, Ancientanubis, Natalie Erin, Hemptel, Jomoal99, Mmortal03, Mentifisto, Geekdom 04, AntiVandalBot, Luna Santin, Seaphoto, Ed270791, Opelio, Doc Tropics, David 136a, NithinBekal, Dotdotdotdash, Helicoptor, Poshzombie, MontanNito, Dylan Lake, Maximilian 77, Shlomi Hillel, Db63376, SamIAmNot, Knotwork, Res2216firestar, Superior IQ Genius, MER-C, Andonic, Sitethief, 100110100, TallulahBelle, Nestamachine, Savant13, Daynightrader, Goldenglove, Charibdis, Acroterion, Ophion, Aigisthos, Editmyhandman, Aruben537, Magioladitis, WolfmanSF, Bongwarrior, VoABot II, Yandman, JamesBWatson, رامبله, Qutt, Jespinos, Kevinmon, Aka042, Froid, DAGwyn, Catgut, Panser Born, Ensign beedrill, Perspectival, JJ Harrison, Dirac66, Justanother, Aziz1005, Cpl Syx, ChazBeckett, Teardrop onthefire, WLU, Stephen shenker, Robin S, SkepticVK, Joshua Davis, Mkroh, B9hummingbird hovering, S3000, Hdt 83, MartinBot, FlieGerFaUstMe 262, Ytomem, Shimwell, Arjun01, KrishSundaresan, Anaxial, Jay Litman, Alexcalamaro, Andrej.westermann, Smokizzy, LedgendGamer, Cyrus Andiron, Peteryoung144, Tgeairn, Artaxiad, HEL, AlphaEta, J.delanoy, AstroHurricane001, Maurice Carbonaro, Yonidebot, Morris729, M C Y 1008, 69gangsta420, It Is Me Here, Shawn in Montreal, Janus Shadowsong, Bailo26, Fredsie, Madagaskar 07, Duchesserin, AntiSpam-Bot, CHIAGEHYANG, Chiswick Chap, Watsup1313, Belovedfreak, HaloInverse, NewEnglandYankee, Scott1329m, Thesis4Eva, Policron, Jrcla2, KylieTastic, WJBscribe, Rnricklefs, Jamesofur, Eyelidlessness, Jonnyk aus, Kvdveer, JavierMC, Izno, Xiahou, CardinalDan, Sheliak, HamatoKameko, Malik Shabazz, Concertmusic, JohnBlackburne, JustinHagstrom, Fences and windows, Wooba doob, Philip Trueman, Door-sAjar, HowardFrampton, TXiKiBoT, Oshwah, Zidonuke, Red Act, Kriak, Catwiki, Technopat, Hqb, Andrius.v, Anonymous Dissident, Crohnie, AlysTarr, Qxz, Vanished user ikijeirw34iuaeolaseriffic, Impunv, Seraphim, Martin451, Don4of4, ABigGreenHippo, Huperphu ff, LeaveSleaves, Kaenneth, StringyGuy, Maxim, Erth64net, Meters, Lamro, Rickstauduhar, Enviroboy, Turgan, Anna512, PhysPhD, Northfox, NPguy, Matthew Sanders, Luke Watkerson, Newbyguesses, MissMJ, SieBot, Escher 26, J.A.Ireland, BAt IHPST }, 4wajzkd02, Robdunst, Dreamafter, Pallab1234, Dbelange, MTHarden, Lemonflash, Kylemew, Yintan, GlassCobra, Discrete, Bentogoa, Likebox, Flyer22Reborn, Exert, ProGeek314, Arbor to SJ, Babawhitemoose, Caidh, Dhatfield, Audree, Oxymoron83, Pretty Green, Weaselstomp, Manway, Alex.muller, Taco Manipulator, Tschach.

Manheat84, Anchor Link Bot, Mikebernstein, ImperialismGo, Nergaal, Ionheld, Ayleuss, Sh4wz0r, Naturespace, ImageRemovalBot, Martarius, Phyte, ClueBot, The Thing That Should Not Be, String4d, Illusion96, Polyamorph, Mpd1989, Alexdeburca18, Wiggl3sLimited, Excirial, Kjramesh, Jusdafax, Resoru, WikiZorro, Eeekster, Verum~enwiki, Tamaratrouts, Gtstricky, Humanino, Brews ohare, NuclearWarfare, Cenarium, Arjayay, Razorflame, Scoobey, BOTarate, Sideswiper, Thingg, Capudo, BVBede, Versus22, Introductory adverb clause, MelonBot, SoxBot III, Egmontaz, Notpayingthepsychiatrist, DumZiBoT, BahTab, TimothyRias, Aj00200, Reaperfromhell, Dunkaroo207, XLinkBot, AlexGWU, Impshum, Saeed.Veradi, Little Mountain 5, Guy392, David424, Truthnlove, Qweeveen, Tayste, Addbot, Steven66s, Denali134, Elemented9, Varrey280303, Eric Drexler, Some jerk on the Internet, Fizzycyst, Uruk2008, DOI bot, Jojhutton, AngryBacon, Non-dropframe, Captaintucker, Auspex1729, Kongr43gpen, Fgnievinski, Rhetoric Of A Sophist, Ronhjones, CanadianLinuxUser, Cst17, Download, Glane23, Bassbonerocks, Chzz, Favonian, Kronix35, LinkFA-Bot, Udugunit, Aktsu, Tassedethe, Numbo3-bot, Anpecota, Tide rolls, HerpesVirus, SDJ, OlEnglish, Scourge of God, Davidmedlar, Couldbenoway66, Yobot, Maxdamantus, Terrisknickers, Kartano, TaBOT-zerem, Julia W, Unique and proud of it, FireMouseHQ, Terrifictriffid, ArchonMagnus, CinchBug, Synchronism, AnomieBOT, Cleeseheb, 1exec1, Charlesvi, Bigdaddy4x4, Gitman4, Jim1138, IRP, Mintrick, Drweetmola, Ornamentalone, M00npirate, Gautam10, Csigabi, Poti-Psy, Materialscientist, 90 Auto, Citation bot, Teleprinter Sleuth, Vuerqex, Twri, Frankenpuppy, Fuzzy Bob Saget, DirlBot, Georgepowell2008, Heidisql, Cureden, Ekwos, Capricorn42, Gensanders, NFD9001, Anna Frodesiak, Tomwsulcer, A23649, Pra1998, Coretheapple, RadiX, Jagbag2, Vandalism destroyer, Ab1, Omnipaedista, Bandit5005, Shirik, RibotBOT, Wateswatcher, Saalstin, Amaury, Aaron35510, Caz34, Doulos Christos, Sewbton, Born Gay, Capricorn24, SchnitzelMannGreek, A. di M., SpacePyjamas, Kierkkadon, A.amitkumar, Dougofborg, StringLove, Nobelprizewinner, Astiburg, FrescoBot, Fortdj33, Paine Ellsworth, Goodbye Galaxy, HJ Mitchell, Steve Quinn, Vhann, Kwiki, Xhaoz, Citation bot 1, Batong, Gil987, Pinethicket, I dream of horses, Tallboyhoops1991, Three887, Steveo27five, RedBot, Sardinita, Serols, Vhsatheeshkumar, Swisstingle, DeletionUK, Reconsider the static, IVAN3MAN, Remingtonhill1, Orenburg1, Coltonhs, Angus Guilherme, Smamaret, Bethovenn, Dinamik-bot, Dc987, Oswaldo Zapata, Egemont, Syebo, Alaithiran, Reaper Eternal, Seahorseruler, Ybungalobill, Quaker phil, Specs112, Dr. Aakash Patel, Bj norge, Tbhotch, StormbringerUK, Minimac, Mathgenius3141592, Keegscee, Omgwaffels, Mick le pick, Solancel, Aznhero3793, Dwielark, Afteread, Enauspeaker, EmausBot, MaooaM, Immunize, Az29, Milkocookie, Faolin42, Fotoni, RA0808, RenamedUser01302013, 8digits, Yukiseaside, Slightsmile, Tommy2010, Winner 42, Wikipelli, Dcirovic, JonezyKiDx, Joe Gazz84, ZéroBot, Timeitsways, John Cline, Cogiati, Quaqa, Chrispaps2413, Nasulikid, Vollrath2323, Benjamin1414141414141414, Arbnos, Green Lane, A930913, Bamyers99, Azeraphale, H3llBot, Encyclopadia, Danga1988, Ollainen, PoisonGM, Wayne Slam, OnePt618, Knome335, L Kensington, Lulzprotuns, Kranix, Rpcappello, Maschen, Vastly~enwiki, Donner60, CatFiggy, CountMacula, Orange Suede Sofa, Etov, M1k3 101, Bill william compton, Wakabaloola, TERBAFAN, Nickslspride34, NeuralLotus, Isocliff, Brechbill123, Xanchester, ClueBot NG, Martti Muukkonen, KagakuKyouju, Jeff Song, This lousy T-shirt, Satellizer, Name Omitted, Marcdean123, Wiki incorp, Frietjes, O.Koslowski, Alexdamaino9, Dream of Nyx, Blackhall616, Widr, Sashhere, WikiPuppies, Stu181, T00g00d96, Pluma, Storm.sarup, Helpful Pixie Bot, Manzeet, Waffleboy36, HMSSolent, Mikeshelton1, Bibcode Bot, 2001:db8, Phillip.phillipson, Hoaxinator, Lowercase sigmabot, Thor cherubim, BG19bot, Mrshaban, Nishch, Flowerhat15, AvocatoBot, Housegeek224, MahRanch, Benzband, Altaïr, Benhenchdickthomas, Shreyakstring, Sweaty maori sphincter, DaFalk, Dsabo74, Ratanmaitra, MM4EVAH, Steven.w.kowalski, Minsbot, JGallardo2600, Dylanlatham, Myfriendganesha, OCCullens, Likeaboss189, Sean271293, LinusE8, BattyBot, Several Pending, Aldrich2122, CommanderMoka, Cyberbot II, The Illusive Man, ChrisGualtieri, KoalamaN2, Trevorkid45, Catsloveit07, Alex Modzz, Rustyjamsen, Goh ryangoh, Dexbot, Exolius, Hilander316, Alman1234321, SuperCalzer, LightandDark2000, MeekMelange, BQND, Cdarrai1, Kephir, TheMonkeyboy524, Michael Anon, TwoTwoHello, Mattfat8, Lugia2453, Anruy, Rachel weld, Jamesx12345, AHusain314, BossEditors, Hillbillyholiday, Joeinwiki, Mattninja, Theshadow444, Asua82, Jakemarz197, Kzhang1025, Epicgenius, Spongbob456789, , TestMaster, Ianreisterariola, GrapperJ, Makeitnasty, Moemajdi, I am One of Many, Nualalvy, BAZINGASS, St3fanPC, Eyesnore, Isaac grozd, Jordanissexyaf1999, Baruch6525, Mosbruckerej, Ihatedirac2k13, Jonamithy121314, 123physicsquantum, Jt198, RaphaelQS, HeyJude70, AParker628, DimReg, A.k.blaze1, Joshuk, Zenibus, Nianoobasik, Ihelpapplen, Gamo To Apoel, SacredLabyrinth, Ginsuloft, Vampre1122, Dimension10, Howard Wolowitz, AddWittyNameHere, Polytope24, Elysion, Tutun12S, Longerboats5, SimonWombat8, Konveyor Belt, Vtank54, Micheal545, Hck24, Caliae19, Hexafish, Simpick, TheRealTheKoi, Bballbro62, Monkbot, ArmyPath, Gabero.88, TheQ Editor, Jtsmith098, Joshmiller1, Hanseer360, XXvPIEvXx, Dbennett 24, Ghikpenos, Nick65633, Saundra03, Thehippothatknows, Sewwgers, Teelaskeletor, Cirksena, Balockaye1234, PloppyDoo, Yesufu29, Lumpy2k14, Podayeruma, Abstract92, Sbenfiel, Monkman2k4, Swegwegdgfyetkfoffkkfkfkv, John95541234, Poopman224, ScrapIronIV, Tetra quark, GeneralizationsAreBad, Shivansh2014n, KasparBot, SHUCKYLUCKY, Fabiotheoto, FartGoblin, Joca potato, Joshcool246, Theoretical Physisist4444, JanetTom55, Reg7d88, CHANDLER MERRILL, Baking Soda, FklfjDKFd bfl, Rajputclann, Entranced98, Jahziahk, Mjhog, Strong81, WikiTikiDude007, ILoveShukii, Qwerty2345B, Irene000, A1D1A2D2, Ricshaw021 and Anonymous: 1601

- **Superstring theory** Source: https://en.wikipedia.org/wiki/Superstring_theory?oldid=722090009 Contributors: Mav, Bryan Derksen, Stevertigo, Michael Hardy, Erik Zachte, Minesweeper, Looixx~enwiki, Ahoerstemeier, JWSchmidt, Cyan, Palfrey, Evercat, Schneelocke, Hashar, Charles Matthews, Tpbradbury, Motor, David Shay, Omegatron, Bevo, Bcorr, Robbot, Fredrik, Hadal, Vuara, Tea2min, Giftlite, Barbara Shack, Herbee, Fropuff, Anville, Maarten van Vliet, WalkinDownThirtyThree, Christopherlin, Steuard, Karol Langner, Lumidek, Prestonmarkstone, Rich Farmbrough, Igorivanov~enwiki, Autiger, Pavel Vozenilek, El C, Rgdboer, Shadow demon, Causa sui, Billymac00, BM, Gary, Pion, Tycho, Cal 1234, Redvers, Postrach, Supercool Dude, Mindmatrix, Mpatel, Joke137, Mandarax, Bill37212, Yamamoto Ichiro, Bubbleboys, Chobot, Ben Tibbetts, Wavelength, RussBot, Chris Capoccia, Chensiyuan, Cate, Chaos, NawlinWiki, Astral, Voidxor, TheMadBaron, Zerodamage, Allens, SmackBot, Android 93, Kurochka, McGeddon, Kintetsubuffalo, Cesoid, Silly rabbit, Stevage, Baronnet, Colonies Chris, Scwlong, Mesons, Kurrupt3d, Bjankuloski06en~enwiki, Makyen, MathStuf, Hu12, Iridescent, Kahalachan, Gatortpk, Mattbr, Neelix, Gregbard, Nauticashades, Cydebot, Davidzaxldua, ChKa, Headbomb, Escarbot, Jj137, Shlomi Hillel, Dougher, JAnDbot, 100110100, 28421u2232nfenfcenc, Aziz1005, Connor Behan, Rickard Vogelberg, Jean-Pierre Petit~enwiki, Hans Dunkelberg, Maurice Carbonaro, Bot-Schafter, SmilesALot, Student7, Cmichael, Sheliak, JayCo777, Catwiki, Andrius.v, Molinogi, Billinghurst, Lamro, Enviroboy, Seraphita~enwiki, Drschawrz, Henry Delforn (old), Lightmouse, Altzinn, Gratedparmesan, ClueBot, Arakunem, Vergil 577, Drmies, Frdayeen, Vizzini101, Niceguyedc, Neverquick, Resoru, Mastertek, Kakofonous, Princess Janay, Alex123irish123, Madeinmexico567, Oldnoah, Madeinmexico566, Truthnlove, YeAaMsLtA, Addbot, Physicman123, CWatchman, Cuaxdon, Semdino, AnnaFrance, LinkFA-Bot, TaBOT-zerem, Evans1982, Gerixau, Eric-Wester, Magog the Ogre, AnomieBOT, DemocraticLuntz, ^musaz, Josh Guffin, Jim1138, Materialscientist, Citation bot, Renaissancee, BLP-outrageous move logs, Omnipaedista, RibotBOT, Paine Ellsworth, Steve Quinn, Tom.Reding, Klavesin, Tkachyk, Dinamik-bot, Bj norge, Orphan Wiki, Idh0854, Arbnos, Vramasub, L Kensington, Particle hep, Isocliff, ClueBot NG, Widr, Adminium, Delivernews, Bibcode Bot, Khanduras, Quarkgluonsoup, Flowerhat15, MythosMagic, OCCullens, Aldrich2122, Graphium, AHusain314, Jochen Burghardt, WorldWideJuan, Jakec, Liquidityinsta, E8xE8, Polytope24, FlaviusCorcoata, Cirksena, BakedLikaBiscuit, Crito10, KasparBot, Rantonels, Baking Soda, MisterRandomized and Anonymous: 185

- **Supergravity** *Source:* https://en.wikipedia.org/wiki/Supergravity?oldid=723039665 *Contributors:* AxelBoldt, Michael Hardy, TakuyaMurata, Angela, Charles Matthews, Phys, Bevo, Robbot, Gandalf61, Giftlite, Herbee, LeYaYa, Fropuff, Moyogo, Jeremy Henty, Leonard G., Urvabara, Arivero, Masudr, Dmr2, Srbauer, Markryherd, Physicistjedi, Axl, Wtmitchell, Japanese Searobin, Linas, Kzollman, Mpatel, GregorB, Canderson7, Marasama, Gurch, LeCire~enwiki, Chobot, Roboto de Ajvol, Hillman, Conscious, E. Menay, Hellbus, Wimt, Smoggyrob, QmunkE, Ilmari Karonen, Caco de vidro, SmackBot, Melchoir, FlashSheridan, Vald, Chris the speller, Colonies Chris, QFT, BWDuncan, TheST, Kuru, Jim.belk, JarahE, Michaelbusch, Zero sharp, CapitalR, Jorbesch, Crichigno, CmdrObot, Myasuda, Equendil, Phatom87, Pyro95819, Mbell, Marek69, WVhybrid, West Brom 4ever, Icep, Shlomi Hillel, Yill577, David Eppstein, N.Nahber, Andre.holzner, Mschel, EdBever, Freeboson, Wesino, WJBscribe, Fuenfundachtzig, Signalhead, Cuzkatzimhut, Jickle, Robdunst, WereSpielChequers, Caltas, Wing gundam, Paolo.dL, Oxymoron83, Lightmouse, JL-Bot, EmanWilm, RS1900, ClueBot, ArdClose, Mild Bill Hiccup, JavierReynaldo, Vivio Testarossa, Mastertek, Pqnelson, AnonyScientist, Goulu, Truthnlove, Addbot, Some jerk on the Internet, Wentuq, Luckas-bot, Yobot, Bility, AnomieBOT, ArthurBot, Omnipaedista, Gsard, Hep thinker, FrescoBot, Paine Ellsworth, Pxpt, Tom.Reding, Casimir9999, Wornsear, EmausBot, Slightsmile, Wikipelli, HiW-Bot, ZéroBot, Cogiati, Arbnos, Quantumor, Terraflorin, Bbeehvh, ClueBot NG, Joefromrandb, Helpful Pixie Bot, Bibcode Bot, BG19bot, Altair, BattyBot, Jeremy112233, M0532062613, Jamesx12345, Mamzypig99, Bitprior, Monkbot, AHusain3141, KasparBot, Profusionex, John-Starling and Anonymous: 76

- **M-theory** *Source:* https://en.wikipedia.org/wiki/M-theory?oldid=725848351 *Contributors:* AxelBoldt, CYD, Eloquence, BF, Bryan Derksen, Zundark, The Anome, Ap, Tim Chambers, Hari, Maury Markowitz, Stevertigo, Michael Hardy, Tim Starling, Gabbe, Tompagenet, Ixfd64, CesarB, Looxix~enwiki, JWSchmidt, Darkwind, Marco Krohn, Jeandré du Toit, Evercat, Schneelocke, Charles Matthews, Timwi, Reddi, Malcohol, Bevo, Jusjih, Slawojarek, Sander123, Fredrik, R3m0t, RedWolf, Blainster, DHN, Hadal, HaeB, Tea2min, David Gerard, Giftlite, DocWatson42, Jmnbpt, Barbara Shack, Fropuff, Moyogo, Sigfpe, Daen, Antandrus, Lumidek, ChrisCostello, Mike Rosoft, Spiffy sperry, Urvabara, Noisy, Discospinster, H0riz0n, Vsmith, Loren36, El C, Momotaro, Shanes, RoyBoy, Triona, Constantine, Smalljim, I9Q79oL78KiL0QTFHgyc, Giraffedata, Wolfrider~enwiki, Physicistjedi, MPerel, Gsklee, ShardPhoenix, Axl, Mac Davis, Kocio, Burn, Hu, Wtmitchell, SidP, DV8 2XL, Ringbang, Kazvorpal, Omnist, Sharkie, Joelpt, Angr, Firsfron, FeanorStar7, Pol098, WadeSimMiser, Mpatel, GregorB, Jugger90, Paxsimius, Mandarax, Chun-hian, Grammarbot, Rjwilmsi, Nightscream, Koavf, Zbxgscqf, Oblivious, Yug, Lionelbrits, Ruidlopes, The ARK, Latka, Mathbot, Diza, Phoenix2~enwiki, DVdm, Eric B, Bomb319, Loom91, Zaftroblue05, Bhny, Stephenb, KSchutte, Bovineone, Salsb, Eriethonan, Bobak, Asarelah, Dna-webmaster, Sandstein, Superdude99, Zzuuzz, Imaninjapirate, Arthur Rubin, Ilmari Karonen, Caco de vidro, DVD R W, Hide&Reason, Jmeden2000, Teo64x, Sardanaphalus, MartinGugino, RupertMillard, SmackBot, Kurochka, K-UNIT, Rwp, Rlbates99, Ajt, Ian Rose, Gilliam, Wlmg, DividedByNegativeZero, Mirokado, Bluebot, Cush, SMP, Ben.c.roberts, MalafayaBot, Nbarth, DHN-bot~enwiki, Colonies Chris, Joemah, N.MacInnes, Xiner, Nunocordeiro, Mbertsch, Addshore, EPM, Nakon, Kiplantt, Bigmantonyd, Martijn Hoekstra, Kabain52, Brdforallseasons, Sayden, Doug Bell, Jaganath, Shadowlynk, IronGargoyle, Jochietoch, Hu12, EMQYKqMKzpdS, Tawkerbot2, Valoem, Gebrah, Albertod4, Kurtan~enwiki, Harold f, Devourer09, Cyrusc, CRGreathouse, Olaf Davis, Lambertian, Friendlystar, Rowellcf, Bmk, Myasuda, DepartedUser2, Ekajati, Cydebot, Fluence, Mike Christie, Meno25, Gagueci, Kahananite, Michael C Price, Alexnye, IComputerSaysNo, Lord Satorious, Krowe, Mrockman, Thijs!bot, Epbr123, Daniel, Headbomb, NeilHalfway, James086, KrakatoaKatie, AntiVandalBot, Blue Tie, Alphachimpbot, J rowley, Shambolic Entity, SuperLuigi31, Buchhemi, Fetchcomms, 100110100, WolfmanSF, VoABot II, Madevin314, SHCarter, Rami R, Jqshenker, Just H, War wizard90, Rickard Vogelberg, Stephen Shenker, Theoretic, MartinBot, Kostisl, R'n'B, Euku, Numbo3, Maurice Carbonaro, Nly8nchz, Thucydides411, LordAnubisBOT, Janus Shadowsong, Peskydan, Isoko, Belovedfreak, Antony-22, Wesino, WJBscribe, Thomas795135, Blood Oath Bot, Idioma-bot, Sheliak, Gogobera, Jeff G., Rei-bot, Ask123, Pennstatephil, JhsBot, Mazarin07, Peace keeper II, Lamro, Antixt, Why Not A Duck, PhysPhD, Rknasc, Guystout, Drschawrz, YohanN7, SieBot, Robdunst, Paradoctor, Wing gundam, Holt27, Astroboyretro, Caidh, OKBot, Divinestuff, Wpac5, Ayleuss, Beofluff, Loren.wilton, ClueBot, Master Shake 9, The Thing That Should Not Be, Haemorrhage, Arakunem, Drmies, IMNTU, Yupjohnny, Huntthetroll, Patrik Andersson, Dank, Gardv, DumZiBoT, Jhosc, Maky, Truthnlove, Autocoast~enwiki, Albambot, Addbot, Uruk2008, Cuaxdon, CanadianLinuxUser, WikiUserPedia, Barak Sh, Tassedethe, Carapheonix, Togekiss101, Tide rolls, OlEnglish, Snaily, Legobot, Luckas-bot, Yobot, Fraggle81, Pcap, Foolo~enwiki, CinchBug, Tempodivalse, AnomieBOT, KDS4444, Götz, Charlesvi, Dalton h, Marcka, Alexzabbey, Jim1138, IRP, AdjustShift, Materialscientist, Citation bot, Quebec99, Ruike, TinucherianBot II, Ekwos, Techwiz2000, Omnipaedista, Peanuts4life, Pinethicket, Vicenarian, Tom.Reding, EDG161, Jusses2, Serols, ActivExpression, SkyMachine, Gerda Arendt, Tkachyk, 122589423KM, தமிழ் பேசுவோம், Reaper Eternal, Apb91781, 786 zikhar, LcawteHuggle, Adam1217, EmausBot, GoingBatty, Pyschobbens, StringTheory11, Smiwi, Suslindisambiguator, SporkBot, PoisonGM, Besneatte, Maschen, SBaker43, Denholm Reynholm, RockMagnetist, ClueBot NG, Blueshift333, Rgwkenyon, Helpful Pixie Bot, Bibcode Bot, BG19bot, SharkinthePool, Msaunier, MusikAnimal, Copernicus01, Elginfball10, Qed3, Blaspie55, Zujua, Kooky2, Mediran, Chris5631, FEYKATD, Ecila3, Ubed junejo, Lugia2453, Frosty, AHusain314, Armanschwarz, Among Men, Faizan, Epicgenius, Diekildie, EddieHugh, Dustin V. S., RaphaelQS, Beakr, DavidLeighEllis, Tedsanders, TFA Protector Bot, Vampire1122, Polytope24, Evandas, Oneidiotsavant, Pretickle, TheRealTheKoi, Shantsforeverandalways, QuantumMatt101, AKS.9955, FACBot, Kh3368, Sizeohnt, Jyhtgqwqsdfghjydwq, Mberkson12, KasparBot, Cmealo, BD2412bot, Yadav.aakash.500, Baking Soda, Chemistry1111 and Anonymous: 457

- **Minimal Supersymmetric Standard Model** *Source:* https://en.wikipedia.org/wiki/Minimal_Supersymmetric_Standard_Model?oldid= 715985338 *Contributors:* Phys, Dmytro, Gandalf61, Rursus, Connelly, Marcika, Waltpohl, Pharotic, Carandol~enwiki, HorsePunchKid, Grunt, Pjacobi, Jensbn, El C, Jag123, JohnyDog, RJFJR, DV8 2XL, Woohookitty, Mpatel, VermillionBird, Rjwilmsi, Goudzovski, Bhny, JabberWok, Shawn81, SCZenz, Closedmouth, Caco de vidro, Tom Lougheed, Stepa, Dauto, Chris the speller, Bluebot, Colonies Chris, SI1982, QFT, MBlume, Jgwacker, Pulu, CenozoicEra, NNemec, Waggers, Dan Gluck, Iridescent, Antonio Prates, Lottamiata, CmdrObot, Myasuda, Michael C Price, Dchristie, RoadMap, Headbomb, CannedhamX, Knotwork, Yill577, Paulnilsson, Matiz, Dr. Morbius, Andre.holzner, Freeboson, Wilsonge, Red Act, Pjoef, StewartMH, PipepBot, ArdClose, Mastertek, Chaosdruid, Rreagan007, SkyLined, Addbot, DOI bot, Mjamja, Tokikake, Luckas-bot, Yobot, Wireader, AnomieBOT, Archon 2488, Citation bot, GenQuest, GrouchoBot, Omnipaedista, Ernsts, Paine Ellsworth, Identitaamore, Citation bot 1, PigFlu Oink, Puzl bustr, RjwilmsiBot, Akrose, EmausBot, WCEngineer, Arbnos, Suslindisambiguator, AManWithNoPlan, Isocliff, Zukertort, Bibcode Bot, BG19bot, ElphiBot, Physlad, ChrisGualtieri, Cinaro, Stamptrader and Anonymous: 50

- **Next-to-Minimal Supersymmetric Standard Model** *Source:* https://en.wikipedia.org/wiki/Next-to-Minimal_Supersymmetric_Standard_ Model?oldid=724017911 *Contributors:* HarryHenryGebel, Lmatt, Malcolma, Maniatis, Nberger, QFT, Jgwacker, Leo C Stein, Dan Gluck, Headbomb, Appraiser, Originalname37, Adavidb, Baba O RLY, TimothyRias, Addbot, Yobot, Wireader, Citation bot, Omnipaedista, Jonesey95, ZéroBot, Bibcode Bot and Anonymous: 9

- **Extra dimensions** *Source:* https://en.wikipedia.org/wiki/Extra_dimensions?oldid=705556627 *Contributors:* Michael Hardy, Lumidek, Jon Awbrey, A876, Mojo Hand, R'n'B, Mild Bill Hiccup, Addbot, Davdde, 陳, Dadonene89, ZéroBot, D.Lazard, Invadibot, SoledadKabocha, CAP-

TAIN RAJU and Anonymous: 2

- **Technicolor (physics)** *Source:* https://en.wikipedia.org/wiki/Technicolor_(physics)?oldid=715556559 *Contributors:* Maury Markowitz, Michael Hardy, IMSoP, Timwi, Grendelkhan, Phys, Xerxes314, CryptoDerk, Pjacobi, David Schaich, Dbachmann, Bender235, RJHall, Jag123, Boredzo, Guy Harris, Rjwilmsi, Brendan Moody, Chobot, Bgwhite, Conscious, SCZenz, Pyrotec, SmackBot, Melchoir, Chris the speller, Njerseyguy, Vichka, Valoem, JRSpriggs, Michael C Price, Headbomb, OrenBochman, Maliz, Laager, Lseixas, Miztli, BotKung, Snideology, ClueBot, Mild Bill Hiccup, Wikisannino, WikiMSM, Addbot, WikiMSSM, Luckas-bot, Wireader, AnomieBOT, Rjanag, Citation bot, Patlatus, Paine Ellsworth, Citation bot 2, Citation bot 1, Tom.Reding, RjwilmsiBot, DacodaNelson, Arbnos, Suslindisambiguator, Bibcode Bot, ChrisGualtieri, Жаворонок, Drscientific, Doctor Dashiki and Anonymous: 23

- **Kaluza–Klein theory** *Source:* https://en.wikipedia.org/wiki/Kaluza%E2%80%93Klein_theory?oldid=725763166 *Contributors:* Sodium, The Anome, XJaM, Roadrunner, Rlee0001, Stevertigo, JohnOwens, Michael Hardy, Looxix~enwiki, Ahoerstemeier, Susurrus, Smack, Charles Matthews, Timwi, Reddi, Wik, Phys, Carbuncle, Ancheta Wis, Kim Bruning, Fropuff, Mdob, Iantresman, Rauyran, Brianhe, Rich Farmbrough, Roo72, Ponder, Paul August, Bender235, Szquirrel, John Vandenberg, I9Q79oL78KiLOQTFHgyc, Geschichte, RJFJR, Notjim, Linas, Lgallindo, Trapolator, Mpatel, Joke137, Rjwilmsi, MarSch, GünniX, YurikBot, Rt66lt, Hillman, Geologician, Buster79, Gillis, Kewp, Petri Krohn, Pred, Caco de vidro, KasugaHuang, Jodarom, Kurochka, Nihonjoe, Zazaban, Unyoyega, Hmains, Chris the speller, MalafayaBot, Colonies Chris, QFT, Legaleagle86, Jon Awbrey, JorisvS, Beetstra, DabMachine, Rschwieb, FrEd 00, Tawkerbot2, CmdrObot, Mattbr, Wfdavis, Moyerjax, MaxEnt, Epbr123, Mojo Hand, Headbomb, JustAGal, Isilanes, Golf Bravo, Magioladitis, JoseAntonioOrtegaRuiz, Jpod2, Mbc362, Maliz, Cpiral, TomyDuby, Quantling, Lseixas, Sheliak, Cuzkatzimhut, Red Act, Impunv, Thomas.schick, AlleborgoBot, YohanN7, ArdClose, EoGuy, Mild Bill Hiccup, Masterpiece2000, Canis Lupus, EverettYou, AnonyScientist, Albambot, Addbot, Gravitophoton, Protonk, Lightbot, Matěj Grabovský, Yobot, Turul2, Jo3sampl, Citation bot, Omnipaedista, RibotBOT, Paine Ellsworth, Quiden711, Tom.Reding, RockSolidCosmo, Crabhiggins, Bj norge, David.c.stone, Arbnos, Quondum, TonyMath, Helpful Pixie Bot, Bibcode Bot, BG19bot, Zerothat, Ownedroad9, Metsfreak2121, MSUGRA, Mogism, Lianatajo, MuonRay, Orderofmagnitudeapproximation, Frinthruit, Monkbot, ManitouLance, Claudio Orzalesi and Anonymous: 88

- **Grand Unified Theory** *Source:* https://en.wikipedia.org/wiki/Grand_Unified_Theory?oldid=726169981 *Contributors:* AxelBoldt, Lee Daniel Crocker, Mav, AstroNomer, XJaM, Heron, Michael Hardy, Zocky, CesarB, Looxix~enwiki, Emperorbma, Dysprosia, Phys, Omegatron, Northgrove, Robbot, Securiger, Lowellian, Meelar, Caknuck, Giftlite, Jmnbpt, Herbee, Fropuff, Xerxes314, Golbez, Ary29, Sam Hocevar, Lumidek, IcycleMort, M1ss1ontomars2k4, Mike Rosoft, Discospinster, 4pq1injbok, Pjacobi, Silence, JustinWick, Bobo192, Smalljim, John Vandenberg, Apyule, Foobaz, I9Q79oL78KiLOQTFHgyc, Jeodesic, Physicistjedi, Jérôme, Alansohn, Krischik, Sligocki, Mac Davis, GeorgeStepanek, RJFJR, Lee-Anne, DV8 2XL, Falcorian, Simetrical, Linas, Mindmatrix, FeanorStar7, GregorB, Ruziklan, Mekong Bluesman, Ashmoo, Rachel1, Rjwilmsi, Strait, Drrngrvy, FlaBot, Margosbot~enwiki, DannyWilde, Rune.welsh, BradBeattie, Snailwalker, Phoenix2~enwiki, Guanxi, DVdm, YurikBot, Ugha, Hairy Dude, Michael Slone, Gaius Cornelius, CambridgeBayWeather, NawlinWiki, Wiki alf, Joel7687, JocK, Zwobot, IceCreamAntisocial, Ms2ger, Noclip, Moogsi, CWenger, Smurrayinchester, Curpsbot-unicodify, Caco de vidro, Jaysbro, Sbyrnes321, SmackBot, Tom Lougheed, Eskimbot, Dauto, Silly rabbit, DHN-bot~enwiki, Colonies Chris, Scwlong, QFT, Addshore, Wen D House, Dreadstar, Pwjb, Gbinal, Thorsen, Vina-iwbot~enwiki, ArglebargleIV, Rory096, Ben Jos, Mr. Lefty, Ckatz, SirFozzie, Quaeler, Baderyp, Richwhite10, Cydebot, Peripitus, Michael C Price, Tawkerbot4, Thijs!bot, Headbomb, Marek69, J.christianson, Luna Santin, Alphachimpbot, JAnDbot, Nyq, Satarsa~enwiki, Homy, Mbarbier, Durianking, Danmctaggart, Maliz, JCarlos, AstroHurricane001, Adavidb, Bogey97, Qatter, Jeepday, Econofire, Lseixas, Jaffar33, Eismc2, Alphanon, Praveen pillay, KabbalistPhysicist, PaddyLeahy, Hemadh, Will Scot 55, Datpol, Moffitma, ClueBot, DFRussia, James edmiston, Ordinaterr, Djr32, PixelBot, Weysheehai, Sun Creator, Subdolous, Dekisugi, Louis925, AnonyScientist, TimothyRias, Forbes72, SilvonenBot, Bywater100, Truthnlove, Balungifrancis, Addbot, Micromaster, Favonian, Mohitsridhar, 84user, OlEnglish, WikiDan61, Amirobot, AnomieBOT, Girl Scout cookie, Theunify, Karanmohan, Materialscientist, Citation bot, Obersachsebot, Blennow, Under22Entreprenuer, Dale Ritter, Senouf, Ernsts, FrescoBot, Paine Ellsworth, Steven Avraham Rosten, Ironboy11, Thamntamil, Slawomir Biały, GreenRoot, Ysyoon, John85, Gil987, Stupidsimple, Tom.Reding, Casimir9999, RobinK, Aknochel, Grandunifier, RjwilmsiBot, Afteread, EmausBot, Slawekb, Arbnos, L Kensington, ClueBot NG, ClaudeDes, Widr, Helpful Pixie Bot, Bibcode Bot, BG19bot, Bernard Rementilla, Kkumer, Wer900, Dilaton, Hilander316, Ryanr666, CuriousMind01, Davidyevgeny, Cjean42, Franzl aus tirol, Sagnac, GabeIglesia, Lmboyer04, Ovidiu cupsa, Jwratner1, Gilitejman1, Soumilm, Tetra quark, BuzzBloom, JD Wilcox, KasparBot, 8SEAL9 and Anonymous: 138

- **Theory of everything** *Source:* https://en.wikipedia.org/wiki/Theory_of_everything?oldid=725933808 *Contributors:* AxelBoldt, Paul Drye, CYD, The Anome, Eclecticology, Toby Bartels, Roadrunner, Zippy, Stevertigo, Lorenzarius, Michael Hardy, Rojclague, Nixdorf, Takuya-Murata, Karada, Skysmith, Kosebamse, CesarB, Anders Feder, Angela, Julesd, Salsa Shark, Ugen64, Poor Yorick, Evercat, Schneelocke, Feedmecereal, Timwi, Dcoetzee, Dysprosia, Jitse Niesen, Wik, Jakenelson, Omegatron, Raul654, Nnh, Kevin M C Harkess, UninvitedCompany, Fredrik, Altenmann, Nurg, Naddy, Gandalf61, Mirv, Academic Challenger, Rursus, Blainster, Caknuck, Wereon, Diberri, Pengo, Tea2min, Hooloovoo, Ancheta Wis, Dbenbenn, Mporter, Jabra, Ferkelparade, Btinn, Xerxes314, Curps, Alison, FeloniousMonk, McGravin, Behnam, Gzornenplatz, JRR Trollkien, Steuard, Andycjp, Sonjaaa, Antandrus, Kim54, Tomruen, Lumidek, Gscshoyru, WpZurp, TJSwoboda, Zondor, Mike Rosoft, JimJast, Discospinster, Rich Farmbrough, H0riz0n, Pjacobi, Vsmith, Pluke, Autiger, Mal~enwiki, Pavel Vozenilek, Floorsheim, El C, Lycurgus, Sourcecode, Oldsoul, PhilHibbs, Sietse Snel, Jpgordon, Atraxani, Smalljim, Slicky, LostLeviathan, Matpitka, Juesch, Danski14, Alansohn, Gary, DariuszT, ShardPhoenix, Kocio, Pion, Hdeasy, Bart133, Schaefer, BanyanTree, ClockworkSoul, Tycho, Suruena, Count Iblis, DV8 2XL, Gene Nygaard, Euphrosyne, Squidwina, Ott, Siafu, Roylee, Woohookitty, Mindmatrix, RHaworth, TigerShark, Savantnavas, MrDarcy, Mpatel, GregorB, Athletec64, Christopher Thomas, Aarghdvaark, Ashmoo, BD2412, Drbogdan, Rjwilmsi, Kinu, Strait, Lordsatri, Dennis Estenson II, HappyCamper, LjL, Bubba73, The wub, Yamamoto Ichiro, JohnDBuell, FayssalF, ColinJF, Wragge, Windchaser, Musical Linguist, Mindloss, RexNL, Gurch, Pete.Hurd, Lmatt, Diza, Zayani, Spencerk, Chobot, Sharkface217, DVdm, Hmonroe, Bgwhite, Ptah~enwiki, Ugha, Wavelength, Hillman, StuffOfInterest, Arado, John Smith's, Zigamorph, SpuriousQ, Jobe457, Stephenb, CambridgeBayWeather, Rsrikanth05, Vibritannia, Neilbeach, Salsb, Big Brother 1984, Anomalocaris, NawlinWiki, Joncolvin, ErkDemon, Trovatore, ETTan, Schrei, THB, Syrthiss, Wknight94, Richardcavell, FF2010, CWenger, Kevin, Caco de vidro, Katieh5584, Banus, Sbyrnes321, Narkstraws, SmackBot, R.E. Freak, Kurochka, DuoDeathscyther 02, Bayardo, McGeddon, Delldot, Kintetsubuffalo, Portillo, Rmosler2100, Bluebot, Jjalexand, 7777777s, Silly rabbit, George Church, Colonies Chris, A. B., Calc rulz, Nicknitro71, Zsinj, TallyJoe, John Hyams, Jamse, Scott3, Jeffhre, Serenity-Fr, Bilgrau, Avb, Rrburke, Addshore, Justin Stafford, DrL, Mr.LMNOP, Rassisi, Spanyard, Byelf2007, Nishkid64, Giovanni33, Soap, Cronholm144, Loadmaster, Stupid Corn, Benjaminlobato, FredrickS, SirFozzie, Waggers, Alexander Gieg, Gcavep, JMK, Abel Cavaşi, Newone, Courcelles, Tubezone, Esn, Dave Runger, Valoem, JRSpriggs, Kurtan~enwiki, 0-8, Duduong, Friendly Neighbour, CRGreathouse,

4.2 Images

- **File:Collage_of_six_cluster_collisions_with_dark_matter_maps.jpg** *Source:* https://upload.wikimedia.org/wikipedia/commons/0/03/Collage_of_six_cluster_collisions_with_dark_matter_maps.jpg *License:* CC BY 3.0 *Contributors:* http://www.spacetelescope.org/images/heic1506a/ *Original artist:* NASA, ESA, D. Harvey (École Polytechnique Fédérale de Lausanne, Switzerland), R. Massey (Durham University, UK), the Hubble SM4 ERO Team, ST-ECF, ESO, D. Coe (STScI), J. Merten (Heidelberg/Bologna), HST Frontier Fields, Harald Ebeling(University of Hawaii at Manoa), Jean-Paul Kneib (LAM)and Johan Richard (Caltech, USA)

- **File:Commons-logo.svg** *Source:* https://upload.wikimedia.org/wikipedia/en/4/4a/Commons-logo.svg *License:* CC-BY-SA-3.0 *Contributors:* ? *Original artist:* ?

- **File:Compactification_example.svg** *Source:* https://upload.wikimedia.org/wikipedia/commons/1/15/Compactification_example.svg *License:* CC BY-SA 4.0 *Contributors:* Brian Greene (2004). The Elegant Universe (DVD). Part II (String's the thing): WGBH Boston Video. Event occurs at 43:55. OCLC 54019786 *Original artist:* Alex Dunkel (Maky)

- **File:Constellation_Fornax,_EXtreme_Deep_Field.jpg** *Source:* https://upload.wikimedia.org/wikipedia/commons/e/ed/Constellation_Fornax%2C_EXtreme_Deep_Field.jpg *License:* Public domain *Contributors:* http://hubblesite.org/newscenter/archive/releases/2012/37/image/a/warn/, http://www.nasa.gov/images/content/690958main_p1237a1.jpg *Original artist:* NASA, ESA; G. Illingworth, D. Magee, and P. Oesch, University of California, Santa Cruz; R. Bouwens, Leiden University; and the HUDF09 Team

- **File:Cosmological_Composition_-_Pie_Chart.svg** *Source:* https://upload.wikimedia.org/wikipedia/commons/7/76/Cosmological_Composition_%E2%80%93_Pie_Chart.svg *License:* CC BY 3.0 *Contributors:* Own work *Original artist:* Ben Finney

- **File:Coupled.svg** *Source:* https://upload.wikimedia.org/wikipedia/commons/d/d1/Coupled.svg *License:* Public domain *Contributors:* Own work *Original artist:* Daniel Schaal (Farbing)

- **File:Crab_Nebula.jpg** *Source:* https://upload.wikimedia.org/wikipedia/commons/0/00/Crab_Nebula.jpg *License:* Public domain *Contributors:* HubbleSite: gallery, release. *Original artist:* NASA, ESA, J. Hester and A. Loll (Arizona State University)

- **File:D3-brane_et_D2-brane.PNG** *Source:* https://upload.wikimedia.org/wikipedia/commons/8/88/D3-brane_et_D2-brane.PNG *License:* Public domain *Contributors:* Image:D-brane.PNG, oeuvre personnelle. *Original artist:* <a href='//commons.wikimedia.org/w/index.php?title=User:Rogilbert..&..action=edit..&..redlink=1' class='new' title='User:Rogilbert (page does not exist)'>Rogilbert</a>

- **File:DMPie_2013.svg** *Source:* https://upload.wikimedia.org/wikipedia/commons/1/11/DMPie_2013.svg *License:* CC BY-SA 3.0 *Contributors:* Own work *Original artist:* Szczureq

- **File:Dark_Energy.jpg** *Source:* https://upload.wikimedia.org/wikipedia/commons/c/ce/Dark_Energy.jpg *License:* Public domain *Contributors:* http://hubblesite.org/newscenter/archive/releases/2001/09/image/g/ OR http://science.nasa.gov/astrophysics/focus-areas/what-is-dark-energy/ *Original artist:* Ann Feild (STScI)

- **File:E6GUT.svg** *Source:* https://upload.wikimedia.org/wikipedia/commons/9/9c/E6GUT.svg *License:* CC BY-SA 3.0 *Contributors:* Own work. Created from Garret Lisi's Elementary Particle Explorer *Original artist:* Cjean42

- **File:Edit-clear.svg** *Source:* https://upload.wikimedia.org/wikipedia/en/f/f2/Edit-clear.svg *License:* Public domain *Contributors:* The Tango! Desktop Project. *Original artist:*
  The people from the Tango! project. And according to the meta-data in the file, specifically: "Andreas Nilsson, and Jakub Steiner (although minimally)."

- **File:Edward_Witten.jpg** *Source:* https://upload.wikimedia.org/wikipedia/commons/9/97/Edward_Witten.jpg *License:* Public domain *Contributors:* Own work *Original artist:* Ojan

- **File:Einstein_cross.jpg** *Source:* https://upload.wikimedia.org/wikipedia/commons/c/c8/Einstein_cross.jpg *License:* Public domain *Contributors:* http://hubblesite.org/newscenter/archive/releases/1990/20/image/a/ *Original artist:* NASA, ESA, and STScI

- **File:Elevator_gravity.svg** *Source:* https://upload.wikimedia.org/wikipedia/commons/1/11/Elevator_gravity.svg *License:* CC BY-SA 3.0 *Contributors:*

- Elevator_gravity2.png *Original artist:*

- derivative work: Pbroks13 (talk)

- **File:En-BigBang.ogg** *Source:* https://upload.wikimedia.org/wikipedia/commons/7/78/En-BigBang.ogg *License:* CC BY-SA 3.0 *Contributors:*

- Derivative of Big Bang *Original artist:* **Speaker:** Dmitry Brant
  Authors of the article

- **File:End_of_universe.jpg** *Source:* https://upload.wikimedia.org/wikipedia/commons/9/98/End_of_universe.jpg *License:* Public domain *Contributors:* ? *Original artist:* ?

- **File:Ergosphere.svg** *Source:* https://upload.wikimedia.org/wikipedia/commons/0/0c/Ergosphere.svg *License:* CC-BY-SA-3.0 *Contributors:* own work based on the graphic uploaded by IMeowbot *Original artist:* MesserWoland

- **File:Fermi_Observations_of_Dwarf_Galaxies_Provide_New_Insights_on_Dark_Matter.ogv** *Source:* https://upload.wikimedia.org/wikipedia/commons/a/a9/Fermi_Observations_of_Dwarf_Galaxies_Provide_New_Insights_on_Dark_Matter.ogv *License:* Public domain *Contributors:* Goddard Multimedia *Original artist:* NASA/Goddard Space Flight Center

- **File:Folder_Hexagonal_Icon.svg** *Source:* https://upload.wikimedia.org/wikipedia/en/4/48/Folder_Hexagonal_Icon.svg *License:* Cc-by-sa-3.0 *Contributors:* ? *Original artist:* ?

- **File:GabrieleVeneziano.jpg** *Source:* https://upload.wikimedia.org/wikipedia/commons/9/95/GabrieleVeneziano.jpg *License:* CC BY-SA 2.5 *Contributors:* Taken by Betsythedevine *Original artist:* The original uploader was Betsythedevine at English Wikipedia

- **File:GalacticRotation2.svg** *Source:* https://upload.wikimedia.org/wikipedia/commons/b/b9/GalacticRotation2.svg *License:* CC-BY-SA-3.0 *Contributors:* Own work in Inkscape 0.42 *Original artist:* PhilHibbs

- **File:Labeled_Triangle_Reflections.svg** *Source:* https://upload.wikimedia.org/wikipedia/commons/3/38/Labeled_Triangle_Reflections.svg *License:* Public domain *Contributors:* Own work *Original artist:* Jim.belk

- **File:Lensshoe_hubble.jpg** *Source:* https://upload.wikimedia.org/wikipedia/commons/a/a9/Lensshoe_hubble.jpg *License:* Public domain *Contributors:* http://apod.nasa.gov/apod/image/1112/lensshoe_hubble_3235.jpg *Original artist:* ESA/Hubble & NASA

- **File:LeonardSusskindStanford2009_cropped.jpg** *Source:* https://upload.wikimedia.org/wikipedia/commons/f/f8/LeonardSusskindStanford2009_cropped.jpg *License:* CC BY-SA 3.0 *Contributors:* File:LeonardSusskindStanford2009.jpg *Original artist:* Jonathan Maltz

- **File:Light_cone.svg** *Source:* https://upload.wikimedia.org/wikipedia/commons/2/27/Light_cone.svg *License:* Public domain *Contributors:* Own work *Original artist:* Sakurambo

- **File:Light_deflection.png** *Source:* https://upload.wikimedia.org/wikipedia/commons/c/c2/Light_deflection.png *License:* CC BY-SA 3.0 *Contributors:* self-made, using numerical integration methods to solve the geodetic equation for light near a spherical massive object (Schwarzschild metric) *Original artist:* Markus Poessel (Mapos)

- **File:Limits_of_M-theory.svg** *Source:* https://upload.wikimedia.org/wikipedia/commons/b/b8/Limits_of_M-theory.svg *License:* CC BY-SA 3.0 *Contributors:*

  Limits of M-theory.png

  Original artist:
- derivative work: Alex Dunkel (Maky)

- **File:MSSM_Flavor_Changing.svg** *Source:* https://upload.wikimedia.org/wikipedia/commons/c/cb/MSSM_Flavor_Changing.svg *License:* CC BY-SA 3.0 *Contributors:* Transferred from en.wikipedia to Commons. *Original artist:* JabberWok at English Wikipedia

- **File:Meissner_effect_p1390048.jpg** *Source:* https://upload.wikimedia.org/wikipedia/commons/5/55/Meissner_effect_p1390048.jpg *License:* CC-BY-SA-3.0 *Contributors:* self photo *Original artist:* Mai-Linh Doan

- **File:MontreGousset001.jpg** *Source:* https://upload.wikimedia.org/wikipedia/commons/4/45/MontreGousset001.jpg *License:* CC-BY-SA-3.0 *Contributors:* Self-published work by ZA *Original artist:* Isabelle Grosjean ZA

- **File:Nuvola_apps_edu_mathematics_blue-p.svg** *Source:* https://upload.wikimedia.org/wikipedia/commons/3/3e/Nuvola_apps_edu_mathematics_blue-p.svg *License:* GPL *Contributors:* Derivative work from Image:Nuvola apps edu mathematics.png and Image:Nuvola apps edu mathematics-p.svg *Original artist:* David Vignoni (original icon); Flamurai (SVG convertion); bayo (color)

- **File:Nuvola_apps_kalzium.svg** *Source:* https://upload.wikimedia.org/wikipedia/commons/8/8b/Nuvola_apps_kalzium.svg *License:* LGPL *Contributors:* Own work *Original artist:* David Vignoni, SVG version by Bobarino

- **File:Open_and_closed_strings.svg** *Source:* https://upload.wikimedia.org/wikipedia/commons/5/56/Open_and_closed_strings.svg *License:* Public domain *Contributors:* Own work *Original artist:* Xoneca

- **File:Oscillations_electron_long.svg** *Source:* https://upload.wikimedia.org/wikipedia/en/7/73/Oscillations_electron_long.svg *License:* CC0 *Contributors:* ? *Original artist:* ?

- **File:Oscillations_electron_short.svg** *Source:* https://upload.wikimedia.org/wikipedia/en/3/3e/Oscillations_electron_short.svg *License:* CC0 *Contributors:* ? *Original artist:* ?

- **File:Oscillations_muon_long.svg** *Source:* https://upload.wikimedia.org/wikipedia/en/1/12/Oscillations_muon_long.svg *License:* CC0 *Contributors:* ? *Original artist:* ?

- **File:Oscillations_muon_short.svg** *Source:* https://upload.wikimedia.org/wikipedia/en/1/1f/Oscillations_muon_short.svg *License:* CC0 *Contributors:* ? *Original artist:* ?

- **File:Oscillations_tau_long.svg** *Source:* https://upload.wikimedia.org/wikipedia/en/2/28/Oscillations_tau_long.svg *License:* CC0 *Contributors:* ? *Original artist:* ?

- **File:Oscillations_tau_short.svg** *Source:* https://upload.wikimedia.org/wikipedia/en/4/4b/Oscillations_tau_short.svg *License:* CC0 *Contributors:* ? *Original artist:* ?

- **File:Oscillations_two_neutrino.svg** *Source:* https://upload.wikimedia.org/wikipedia/en/8/86/Oscillations_two_neutrino.svg *License:* CC0 *Contributors:* ? *Original artist:* ?

- **File:PIA17993-DetectorsForInfantUniverseStudies-20140317.jpg** *Source:* https://upload.wikimedia.org/wikipedia/commons/1/1a/PIA17993-DetectorsForInfantUniverseStudies-20140317.jpg *License:* Public domain *Contributors:* http://photojournal.jpl.nasa.gov/jpeg/PIA17993.jpg *Original artist:* NASA/JPL-Caltech

- **File:PIA18467-NuSTAR-Plot-BlackHole-BlursLight-20140812.png** *Source:* https://upload.wikimedia.org/wikipedia/commons/d/d9/PIA18467-NuSTAR-Plot-BlackHole-BlursLight-20140812.png *License:* Public domain *Contributors:* http://photojournal.jpl.nasa.gov/jpeg/PIA18467.jpg *Original artist:* NASA/JPL-Caltech/Institute for Astronomy, Cambridge

- **File:PIA19822-MagneticBlackHoleWaves-AlfvenS-waves-20150709.jpg** *Source:* https://upload.wikimedia.org/wikipedia/commons/f/fe/PIA19822-MagneticBlackHoleWaves-AlfvenS-waves-20150709.jpg *License:* Public domain *Contributors:* http://photojournal.jpl.nasa.gov/jpeg/PIA19822.jpg *Original artist:* NASA/JPL-Caltech

- **File:Penrose.svg** *Source:* https://upload.wikimedia.org/wikipedia/commons/a/a8/Penrose.svg *License:* Public domain *Contributors:* Transferred from en.wikipedia to Commons by Andrei Stroe using CommonsHelper. *Original artist:* Cronholm144 at English Wikipedia

- **File:Portal-puzzle.svg** *Source:* https://upload.wikimedia.org/wikipedia/en/f/fd/Portal-puzzle.svg *License:* Public domain *Contributors:* ? *Original artist:* ?

4.3 Content license